Land-Atmosphere Interactions and Effects on the Climate of the Tibetan Plateau and Surrounding Regions II

Land-Atmosphere Interactions and Effects on the Climate of the Tibetan Plateau and Surrounding Regions II

Editors

Yaoming Ma

Li Jia

Massimo Menenti

Lei Zhong

MDPI

Basel • Beijing • Wuhan • Barcelona • Belgrade • Novi Sad • Cluj • Manchester

Editors
Yaoming Ma
Institute of Tibetan Plateau Research, Chinese Academy of Sciences
Beijing, China

Li Jia
Aerospace Information Research Institute, Chinese Academy of Sciences
Beijing, China

Massimo Menenti
Department of Geosciennce and Remote Sensing, Technische Universiteit Delft
Delft, The Netherlands

Lei Zhong
School of Earth and Space Sciences, University of Science and Technology of China
Hefei, China

Editorial Office
MDPI
St. Alban-Anlage 66
4052 Basel, Switzerland

This is a reprint of articles from the Special Issue published online in the open access journal *Remote Sensing* (ISSN 2072-4292) (available at: https://www.mdpi.com/journal/remotesensing/special_issues/687823FQ9Q).

For citation purposes, cite each article independently as indicated on the article page online and as indicated below:

Lastname, A.A.; Lastname, B.B. Article Title. *Journal Name* **Year**, *Volume Number*, Page Range.

ISBN 978-3-0365-9136-0 (Hbk)
ISBN 978-3-0365-9137-7 (PDF)
doi.org/10.3390/books978-3-0365-9137-7

Cover image courtesy of Yaoming Ma

Contents

About the Editors

Yaoming Ma

Prof. Dr. Yaoming Ma is a professor at Institute of Tibetan Plateau Research, Chinese Academy of Sciences. He obtained a doctor's degree in atmospheric physics from Okayama University in Japan and a doctor's degree in environmental science from Wageningen University in the Netherlands. He is a distinguished research fellow of the Chinese Academy of Sciences and post professor of the University of the Chinese Academy of Sciences. He is the academic director of the National Observation and Research Station for Qomolongma Special Atmospheric Processes and Environmental Changes. His research interests are atmospheric boundary layer observation and satellite remote sensing application. He was responsible for nearly 20 national major scientific research programs, including key programs of the National Natural Science Foundation of China, the National Science Fund for Distinguished Young Scholars, major international cooperation programs, etc. At the same time, he is one of the two chief coordinators of the major international cooperation research programs 'GAME/Tibet' and 'CAMP/Tibet'.

Li Jia

Prof. Dr. Li Jia is a full professor and director of the Laboratory for Earth Observation for Terrestrial Water Cycle and Climate Change in the Aerospace Information Research Institute (AIR), Chinese Academy of Sciences (CAS). She received her PhD degree in Environmental Science from Wageningen University of the Netherlands in 2004. Her main research disciplines are quantitative remote sensing and remote sensing applications to hydrometeorology, eco-hydrology, water resources, carbon assimilation, drought process and monitoring, and climate change. She has more than 300 publications, with more than 90 peer-reviewed ISI journal papers. She is a member of the WCRP/GEWEX Hydroclimatology Panel (GHP) (since 2019) and an IEEE member. She is the co-chair of a working group on drought monitoring and evaluation in the AOGEO (Asia–Oceania Group on Earth Observations) (since 2017). She acts as the Editor/Associate Editor of several professional journals.

Massimo Menenti

Prof. Dr. Massimo Menenti is an internationally renowned scientist in the fields of earth observation and global terrestrial water cycles. He held senior research positions in the Netherlands, France, USA, China, and Italy; received research awards in the Netherlands, France, USA, and China; and led large projects with participants from Europe, Asia, America, and Africa. His best-known achievements have been attained in the topics of surface parameter retrievals from remote sensing, remote sensing-based evapotranspiration (ET) estimation, time series analysis of remote sensing data, and the application of remote sensing technology in hydrology and climate models. Prof. Menenti initiated the use of RS to assess and monitor crop water requirements and irrigation performance in the late 1980s. He was one of the earliest researchers to use laser technology to measure surface aerodynamic roughness. He initiated the use of time series analysis of satellite data. He presented the surface energy balance index (SEBI) theory for ET estimation, which is the prototype of the S-SEBI, SEBS, and SEBAL models.

Lei Zhong

Prof. Dr. Lei Zhong is a professor at the School of Earth and Space Sciences, University of Science and Technology of China. He received his PhD degree in Natural Geography in 2008 at the Institute of Tibetan Plateau Research, Chinese Academy of Sciences. His main research interests mainly focus on land–atmosphere interaction, energy and water cycles, the application of remote sensing, and monsoon climate. He was recognized by theNational Natural Science Foundation for Distinguished Young Scholars in 2015. He got the first prize of the Tibet Autonomous Region Science and Technology Award in 2022. He has been actively involved in major international cooperation programs, such as 'CAMP/Tibet' and ESA-MOST Dragon programme. He is a director of the China Association for Scientific Exploration. He is also a member of the IAMAS Youth Working Group.

Preface

This book focused on recent advances in land-atmosphere interactions and their effects on the climate change over the Tibetan Plateau and surrounding regions using multisource remote sensing data and in situ measurements.

Retrieval of land surface variables and surface heat fluxes, as well as change monitoring in snow, glaciers, lakes, and other land-surface covers are of particular interest. Special attention is given to quantitative assessments of land surface temperature, sensible heat flux, soil moisture, vegetation and drought indices, groundwater storage, runoff, condensation, and desublimation, as well as the distinct surface processes over lakes and glaciers coupling with climate warming.

This book was funded by the the Second Tibetan Plateau Scientific Expedition and Research (STEP) Program, Ministry of Science and Technology of the People's Republic of China (Grant No. 2019QZKK0103); the National Natural Science Foundation of China (Grant Nos. 42375071, 42230610 and 41875031) and CLIMATE-Pan-TPE (ID 58516) in the framework of the ESA-MOST Dragon 5 program.

Yaoming Ma, Li Jia, Massimo Menenti, and Lei Zhong
Editors

remote sensing

MDPI

Editorial

Land–Atmosphere Interactions and Effects on the Climate of the Tibetan Plateau and Surrounding Regions II

Yaoming Ma [1,2,3,4,5,6], Lei Zhong [7,8,9,10,*], Li Jia [11] and Massimo Menenti [11,12]

1 State Key Laboratory of Tibetan Plateau Earth System, Environment and Resources (TPESER), Institute of Tibetan Plateau Research, Chinese Academy of Sciences, Beijing 100101, China; ymma@itpcas.ac.cn
2 College of Earth and Planetary Sciences, University of Chinese Academy of Sciences, Beijing 100049, China
3 College of Atmospheric Science, Lanzhou University, Lanzhou 730000, China
4 National Observation and Research Station for Qomolongma Special Atmospheric Processes and Environmental Changes, Dingri 858200, China
5 Kathmandu Center of Research and Education, Chinese Academy of Sciences, Beijing 100101, China
6 China-Pakistan Joint Research Center on Earth Sciences, Chinese Academy of Sciences, Islamabad 45320, Pakistan
7 School of Earth and Space Sciences, University of Science and Technology of China, Hefei 230026, China
8 CAS Center for Excellence in Comparative Planetology, University of Science and Technology of China, Hefei 230026, China
9 Frontiers Science Center for Planetary Exploration and Emerging Technologies, University of Science and Technology of China, Hefei 230026, China
10 Jiangsu Collaborative Innovation Center for Climate Change, Nanjing 210023, China
11 Aerospace Information Research Institute, Chinese Academy of Sciences, Beijing 100101, China; jiali@aircas.ac.cn (L.J.); m.menenti@tudelft.nl (M.M.)
12 Department of Geoscience and Remote Sensing, Technische Universiteit Delft, 2600 GA Delft, The Netherlands
* Correspondence: zhonglei@ustc.edu.cn

Citation: Ma, Y.; Zhong, L.; Jia, L.; Menenti, M. Land–Atmosphere Interactions and Effects on the Climate of the Tibetan Plateau and Surrounding Regions II. *Remote Sens.* **2023**, *15*, 4540. https://doi.org/10.3390/rs15184540

Received: 8 September 2023
Accepted: 11 September 2023
Published: 15 September 2023

1. Introduction

As the world's highest and largest plateau, the Tibetan Plateau (TP) is referred to as 'the Asian Water Tower' and 'the Third Pole of the World' [1]. A better understanding of the water and energy cycles in the TP is not only critical for revealing the mechanisms of regional land–atmosphere interactions, but also essential for assessing the causes of changes in the cryosphere and hydrosphere in relation to changes in the plateau atmosphere in the Asian monsoon system [2]. Since the TP is an ecologically fragile region that is sensitive to climate change [3], the systematic evaluation of land–atmosphere interactions in this region also contributes to the quantitative understanding of climate change.

To this end, the aim of this Special Issue was to present recent advances in quantifying (1) processes in the atmospheric boundary layer, (2) soil properties, drought, and freezing–thawing processes, (3) lake and glacier monitoring, (4) hydrological processes, and (5) data assimilation and validation by applying in situ measurements, remote sensing or numerical modelling approaches to the TP region.

Eighteen papers (sixteen articles, one communication, and one technical note) are published in this Special Issue, covering the quantitative assessments of land surface temperature, sensible heat flux, soil moisture, vegetation and drought indices, groundwater storage, runoff, condensation, and desublimation, as well as the distinct surface processes over lakes and glaciers driven by climate warming. The MODIS and GIMMS datasets are validated, and a new high-resolution assimilated dataset is released. In addition, the application of coherent Doppler wind LiDARs is analysed. Additionally, the formation and climatic–environmental significance of the yardangs surrounding the Suoyang city ruins are also discussed.

2. Overview

The surface air temperature difference and sensible heat flux are critical variables in the atmospheric boundary layer, both of which have undergone significant changes due to climate change. Wang et al. [4] reported that although the entire TP was found to be dominated by a positive surface air temperature difference both annually and seasonally, from 1950 to 2021 the TP experienced a significant decreasing trend in the annual surface air temperature difference at a rate of −0.07 K/decade. Additionally, a decreasing trend in sensible heat flux from the mid-1980s to the beginning of the 21st century was also reported by Wang et al. [5] via Noah-MP simulations. The decrease in sensible heat flux was found to be linked to the decrease in both the surface air temperature difference and wind speed [5].

Using the Community Land Model version 4.5, Fu et al. [6] demonstrated that the TP has become warmer and wetter from 1981 to 2016, with increases in both regional average temperature and precipitation. In addition, soil temperature and moisture in most areas of the TP were affected by air temperature and precipitation in turn, and both showed an upwards trend. Consequently, the duration of the freeze–thaw process over the TP has shortened. Despite the increase in the northwest TP, the freeze–thaw duration decreased in the rest of the whole plateau [6]. Additionally, Fang et al. [7] also found that an area of 0.60×106 km^2 of permafrost in the TP degraded to seasonally frozen ground in the 1960s–2000s, and the primary shrinkage period occurred in the 2000s. However, the seasonal and diurnal variation characteristics of soil moisture are diverse in different in situ stations. According to Li et al. [8], the soil moisture at depths of 5 and 10 cm for the Lhari, Biru, Nyainrong, Amdo, Nagqu, Baingoin, and Seng-ge Kambab stations was measured to be 0.55, 0.4, 0.34, 0.3, 0.25, 0.14, and 0.1 cm^3/cm^3, respectively. These large differences also indicated that it was unreasonable to use only the soil moisture of several stations to represent the overall soil moisture of one region [8].

Drought is a major disaster across the TP, and drought indices that can describe drought evolution at a fine temporal scale are still scarce. Cheng et al. [9] constructed daily drought indices based on multisource remote sensing and reanalysis data using four machine learning methods, and a new daily drought index, the standardized integrated drought index (SIDI), was developed via the extreme gradient boosting regression model, which showed the best performance for monitoring agricultural drought. In addition, to analyse the mechanism of drought in Southwest China, Ye et al. [10] constructed a binary linear regression forecast model, which successfully relates precipitation to the anomaly of 500 hPa relative vorticity and relative divergence. Additionally, the spatiotemporal variation characteristics of drought on the TP from 2016 to 2099 were predicted by Liu et al. [11], showing that the overall future climate of the TP will still develop towards warm and humid conditions. However, as the concentration of carbon dioxide emissions increases in the future, the proportion of extremely significant aridification and humidification areas in the TP will significantly increase, and the possibility of extreme disasters will also increase [11].

Changes in lake water volume and glacier mass are sensitive indicators of regional environmental change. Ma et al. [12] systematically analysed the interannual changes from 1970 to 2021 in three typical inland lake basins using multisource remote sensing and water level observations. The results showed that lakes in the monsoon-dominated region showed a significant trend of expansion from 2000 to 2014, but the trend slowed down and stabilized after 2014; lakes in the westerlies-dominated region showed a small expansion trend, while lakes in the region affected by both westerlies and the monsoon showed an overall shrinking trend [12]. On the other hand, Yao et al. [13] reported that the Yala Glacier and the Qiyi Glacier were shrinking, with change rates of −736 mm w.e./a and −567 mm w.e./a, respectively, both undergoing a state of intensive and accelerating mass loss [13].

Condensation, desublimation, runoff, and groundwater storage are all important processes in land–atmosphere interactions. Li et al. [14] evaluated the spatiotemporal variations in condensation and desublimation from 1950 to 2020 on the TP using hourly ERA5-Land and ERA5 reanalysis datasets. The annual mean condensation was estimated to be 8.45 mm, with an increasing trend of 0.24 mm/10a, and the annual mean desublimation was 11.45 mm, with a decreasing trend of −0.26 mm/10a, with the total annual mean condensation and desublimation reaching 19.89 mm, with a weak decreasing trend on the TP overall. Wei et al. [15] investigated the relationships between fourteen landscape patterns and four hydrological indices for ten watersheds in the TP and found that runoff increases when a watershed is dominated by a small patch of landscape. Regarding the variability in groundwater storage, Ren et al. [16] reported that the groundwater storage of the TP decreased at an average rate of −0.89 mm/a from January 2003 to December 2021. However, since January 2016, it has gradually recovered at a rate of 1.47 mm/a. It was found that the rising temperature may result in an increase in groundwater storage in regions where glaciers are distributed [16].

Several articles have evaluated the current products or released new datasets via data assimilation. Lazhu et al. [17] validated the MODIS lake surface water temperature dataset over the TP region. The MODIS LSWT agrees well with the in situ measurements, with a root mean square error < 1 K at nighttime and <2 K in the daytime, indicating a high accuracy of the MODIS LSWT data. However, the MODIS lake surface water temperature data were questionable in the monsoon-controlled region [17]. In addition, Wang et al. [18] applied the MODIS and GIMMS datasets to analyse the annual and seasonal trends in vegetation responses and feedback to temperature on the TP. The results showed that both MODIS and GIMMS data showed a common increase in the normalized difference vegetation index on the TP for all timescales, while the former has had a larger greening area since 2000 [18]. In addition, Wen et al. [19] utilized the weather research and forecasting (WRF) model and a three-dimensional variational assimilation method to create a high-resolution assimilated dataset (HRAD) with a spatial resolution of $0.05° \times 0.05°$ and a temporal resolution of 1 h. The correlation coefficients of the 2 m temperature, surface temperature and surface pressure were relatively high, all above 0.9, and those of relative humidity and wind speed were approximately 0.7 and 0.5, respectively [19].

Moreover, Song et al. [20] evaluated the observations of the wind field and boundary layer height from coherent Doppler wind LiDARs located at the northern edge of the TP from 1 May to 30 August 2021, showing that coherent Doppler wind LiDAR has good applicability in reproducing wind fields in dust, precipitation, and clear-sky conditions. Fan et al. [21] analysed the formation of yardangs surrounding the Suoyang City ruins in the Hexi Corridor of northwestern China. According to ^{14}C dating and historical records of local human activities, the formation of yardangs in the Suoyang City oasis has been suggested to start in the mid-Yuan Dynasty of China, around AD 1291 [21].

3. Summary

In summary, this Special Issue mainly presents up-to-date advances in the quantitative assessments of land surface temperature, sensible heat flux, soil moisture, vegetation and drought indices, groundwater storage, runoff, condensation, and desublimation, as well as the distinct surface processes over lakes and glaciers on the TP. These selected papers are novel and timely in informing the knowledge on land–atmosphere interactions driven by climate warming.

We trust that the collation of these papers will provide quantitative references for the better assessment and prediction of the land–atmosphere interactions in the "Third Pole".

Funding: This research was jointly funded by the Second Tibetan Plateau Scientific Expedition and Research (STEP) Program (Grant No. 2019QZKK0103), the National Natural Science Foundation of China (Grant Nos. 42375071, 41875031 and 42230610) and CLIMATE-Pan-TPE (ID 58516) in the framework of the ESA-MOST Dragon 5 program.

Conflicts of Interest: The author declares no conflict of interest.

References

1. Ma, Y.; Hu, Z.; Xie, Z.; Ma, W.; Wang, B.; Chen, X.; Li, M.; Zhong, L.; Sun, F.; Gu, L.; et al. A long-term (2005–2016) dataset of hourly integrated land-atmosphere interaction observations on the Tibetan Plateau. *Earth Syst. Sci. Data* **2020**, *12*, 2937–2957. [CrossRef]
2. Su, Z.; Ma, Y.; Chen, X.; Dong, X.; Du, J.; Han, C.; He, Y.; Hofste, J.; Li, M.; Li, M.; et al. Monitoring water and energy cycles at climate scale in the Third Pole Environment (CLIMATE-TPE). *Remote Sens.* **2021**, *13*, 3661. [CrossRef]
3. Zhong, L.; Ma, Y.; Xue, Y.; Piao, S. Climate change trends and impacts on vegetation greening over the Tibetan Plateau. *J. Geophys. Res. Atmos.* **2019**, *124*, 7540–7552. [CrossRef]
4. Wang, X.; Chen, R. Evaluation of Spatial and Temporal Variations in the Difference between Soil and Air Temperatures on the Qinghai–Tibetan Plateau Using Reanalysis Data Products. *Remote Sens.* **2023**, *15*, 1894. [CrossRef]
5. Wang, S.; Ma, Y.; Liu, Y. Simulated Trends in Land Surface Sensible Heat Flux on the Tibetan Plateau in Recent Decades. *Remote Sens.* **2023**, *15*, 714. [CrossRef]
6. Fu, C.; Hu, Z.; Yang, Y.; Deng, M.; Yu, H.; Lu, S.; Wu, D.; Fan, W. Responses of Soil Freeze–Thaw Processes to Climate on the Tibetan Plateau from 1980 to 2016. *Remote Sens.* **2022**, *14*, 5907. [CrossRef]
7. Fang, X.; Wang, A.; Lyu, S.; Fraedrich, K. Dynamics of Freezing/Thawing Indices and Frozen Ground from 1961 to 2010 on the Qinghai-Tibet Plateau. *Remote Sens.* **2023**, *15*, 3478. [CrossRef]
8. Li, H.; Xiao, Z.; Wei, J.; Wang, G. The Seasonal and Diurnal Variation Characteristics of Soil Moisture at Different Depths from Observational Sites over the Tibetan Plateau. *Remote Sens.* **2022**, *14*, 5010. [CrossRef]
9. Cheng, M.; Zhong, L.; Ma, Y.; Wang, X.; Li, P.; Wang, Z.; Qi, Y. A New Drought Monitoring Index on the Tibetan Plateau Based on Multisource Data and Machine Learning Methods. *Remote Sens.* **2023**, *15*, 512. [CrossRef]
10. Ye, Y.; Tian, R.; Jin, Z. Dynamic Effects of Atmosphere over and around the Tibetan Plateau on the Sustained Drought in Southwest China from 2009 to 2014. *Remote Sens.* **2023**, *15*, 2198. [CrossRef]
11. Liu, Y.; Jia, Z.; Ma, X.; Wang, Y.; Guan, R.; Guan, Z.; Gu, Y.; Zhao, W. Analysis of Drought Characteristics Projections for the Tibetan Plateau Based on the GFDL-ESM2M Climate Model. *Remote Sens.* **2022**, *14*, 5084. [CrossRef]
12. Ma, W.; Bai, L.; Ma, W.; Hu, W.; Xie, Z.; Su, R.; Wang, B.; Ma, Y. Interannual and Monthly Variability of Typical Inland Lakes on the Tibetan Plateau Located in Three Different Climatic Zones. *Remote Sens.* **2022**, *14*, 5015. [CrossRef]
13. Yao, R.; Shi, J. Glacier Mass Loss Simulation Based on Remote Sensing Data: A Case Study of the Yala Glacier and the Qiyi Glacier in the Third Pole. *Remote Sens.* **2022**, *14*, 5190. [CrossRef]
14. Li, H.; Chen, R.; Han, C.; Yang, Y. Evaluation of the Spatial and Temporal Variations of Condensation and Desublimation over the Qinghai–Tibet Plateau Based on Penman Model Using Hourly ERA5-Land and ERA5 Reanalysis Datasets. *Remote Sens.* **2022**, *14*, 5815. [CrossRef]
15. Wei, C.; Dong, X.; Ma, Y.; Leng, M.; Zhao, W.; Zhang, C.; Yu, D.; Su, B. Relationships between Landscape Patterns and Hydrological Processes in the Subtropical Monsoon Climate Zone of Southeastern China. *Remote Sens.* **2023**, *15*, 2290. [CrossRef]
16. Ren, W.; Gao, Y.; Qian, H.; Ma, Y.; Su, Z.; Ma, W.; Liu, Y.; Xu, P. Spatiotemporal Variation Characteristics of Groundwater Storage and Its Driving Factors and Ecological Effects in Tibetan Plateau. *Remote Sens.* **2023**, *15*, 2418. [CrossRef]
17. Yang, K.; Qin, J.; Hou, J.; Lei, Y.; Wang, J.; Huang, A.; Chen, Y.; Ding, B.; Li, X. A Strict Validation of MODIS Lake Surface Water Temperature on the Tibetan Plateau. *Remote Sens.* **2022**, *14*, 5454.
18. Wang, F.; Ma, Y.; Darvishzadeh, R.; Han, C. Annual and Seasonal Trends of Vegetation Responses and Feedback to Temperature on the Tibetan Plateau since the 1980s. *Remote Sens.* **2023**, *15*, 2475. [CrossRef]
19. Wen, X.; Zhu, X.; Li, M.; Chen, M.; Zhang, S.; Yang, X.; Zheng, Z.; Qin, Y.; Zhang, Y.; Lv, S. Creation and Verification of a High-Resolution Multi-Parameter Surface Meteorological Assimilation Dataset for the Tibetan Plateau for 2010–2020 Available Online. *Remote Sens.* **2023**, *15*, 2906. [CrossRef]
20. Song, M.; Wang, Y.; Mamtimin, A.; Gao, J.; Aihaiti, A.; Zhou, C.; Yang, F.; Huo, W.; Wen, C.; Wang, B. Applicability Assessment of Coherent Doppler Wind LiDAR for Monitoring during Dusty Weather at the Northern Edge of the Tibetan Plateau. *Remote Sens.* **2022**, *14*, 5264. [CrossRef]
21. Fan, Q.; Liao, J.; Li, Y.; Ye, W.; Wang, T.; Yu, L.; Feng, X.; Han, W. The Formation of Yardangs Surrounding the Suoyang City Ruins in the Hexi Corridor of Northwestern China and Its Climatic–Environmental Significance. *Remote Sens.* **2022**, *14*, 5628. [CrossRef]

remote sensing

MDPI

Article

Dynamics of Freezing/Thawing Indices and Frozen Ground from 1961 to 2010 on the Qinghai-Tibet Plateau

Xuewei Fang [1,2,*], Anqi Wang [1], Shihua Lyu [1,3] and Klaus Fraedrich [2]

1 Plateau Atmosphere and Environment Key Laboratory of Sichuan Province, School of Atmospheric Sciences, Chengdu University of Information Technology, Chengdu 610225, China; wanganqi0619@gmail.com (A.W.); slu@cuit.edu.cn (S.L.)
2 Max Planck Institute for Meteorology, 20146 Hamburg, Germany; klaus.fraedrich@mipmet.mpg.de
3 Collaborative Innovation Center on Forecast and Evaluation of Meteorological Disasters, Nanjing University of Information Science & Technology, Nanjing 210044, China
* Correspondence: fangxw@cuit.edu.cn

Abstract: Freezing/thawing indices are important indicators of the dynamics of frozen ground on the Qinghai-Tibet Plateau (QTP), especially in areas with limited observations. Based on the numerical outputs of Community Land Surface Model version 4.5 (CLM4.5) from 1961 to 2010, this study compared the spatial and temporal variations between air freezing/thawing indices (2 m above the ground) and ground surface freezing/thawing indices in permafrost and seasonally frozen ground (SFG) across the QTP after presenting changes in frozen ground distribution in each decade in the context of warming and wetting. The results indicate that an area of 0.60×10^6 km^2 of permafrost in the QTP degraded to SFG in the 1960s–2000s, and the primary shrinkage period occurred in the 2000s. The air freezing index (AFI) and ground freezing index (GFI) decreased dramatically at rates of 71.00 °C·d/decade and 34.33 °C·d/decade from 1961 to 2010, respectively. In contrast, the air thawing index (ATI) and ground thawing index (GTI) increased strikingly, with values of 48.13 °C·d/decade and 40.37 °C·d/decade in the past five decades, respectively. Permafrost showed more pronounced changes in freezing/thawing indices since the 1990s compared to SFG. The changes in thermal regimes in frozen ground showed close relations to air warming until the late 1990s, especially in 1998, when the QTP underwent the most progressive warming. However, a sharp increase in the annual precipitation from 1998 began to play a more controlling role in thermal degradation in frozen ground than the air warming in the 2000s. Meanwhile, the following vegetation expansion hiatus further promotes the thermal instability of frozen ground in this highly wet period.

Keywords: freezing/thawing indices; permafrost dynamics; Community Land Surface Model; Qinghai-Tibet Plateau

Citation: Fang, X.; Wang, A.; Lyu, S.; Fraedrich, K. Dynamics of Freezing/Thawing Indices and Frozen Ground from 1961 to 2010 on the Qinghai-Tibet Plateau. *Remote Sens.* **2023**, *15*, 3478. https://doi.org/10.3390/rs15143478

Academic Editor: Nereida Rodriguez-Alvarez

Received: 1 June 2023
Revised: 29 June 2023
Accepted: 7 July 2023
Published: 10 July 2023

1. Introduction

The Qinghai-Tibet Plateau (QTP) ranks as the highest plateau in low-latitude areas around the world. It has also been called the "third pole" for its high average elevation [1–4]. In terms of the global mid-latitude area, the QTP has the largest distribution of permafrost (~1.06 million km^2), which is defined as the rock or soil that remains at or below 0 °C for two or more consecutive years and is a key component in the cryosphere [5–7]. Due to its lower latitude and higher elevation, the permafrost on the QTP is more sensitive to climate change compared with permafrost in high-altitude areas such as Canada and Russia [8,9]. The dynamic change of the frozen ground distribution under the background of global warming is of great importance for hydrology cycles, ecosystems, engineering infrastructure, and climate change [10–13].

Previous studies have reported that, as a result of global warming, the frozen ground on the QTP had warmed up in recent decades and the warming is more intense than the Arctic and similar mid-latitude regions [14,15]. Wu and Zhang [16] monitored 10 boreholes in permafrost areas along the Qinghai-Tibetan Highway up to 10.7 m depth half-monthly from 1996 to 2006. They reported that the mean annual temperatures at 6.0 m depth have increased from 0.12 °C to 0.67 °C, with an average value of 0.43 °C. Simultaneously, they conducted a further investigation on the active layer thickness and found that it increased sharply (about 39 cm) from 1983 to 2005 [17]. To characterize the freezing and thawing states of the frozen ground, changes in near-surface air freezing/thawing indices (AFI/ATI) and ground-surface freezing/thawing indices (GFI/GTI) on the QTP have also been analyzed using in situ observations [18]. The results show that the QTP has undergone a prominent increasing trend in the thawing index and a decreasing trend in the freezing index since 1998 [19]. It is consistent with the pivotal year of 1998, when the QTP experienced dramatic wetting and warming thereafter [20]. Climate warming also benefits vegetation growth [21–24] and can induce an earlier start date of freezing through feedbacks to regional climates [25]. Furthermore, observational analysis has shown that the greening QTP can amplify the warming impact of spring snow cover on surface seasonally frozen ground (SFG) and can also intensify the warming effect of summer rainfall on top permafrost [26]. These studies have demonstrated the uneven thermal responses of frozen ground to the accelerating climate change and the related vegetation growth on the QTP, serving as a potential connection. Due to the complex terrain and harsh natural conditions, the monitoring sites on the QTP are relatively sparse, and there are uncertainties in the satellite products, especially in the western high-altitude region [27]. Consequently, studies on the changes in the thermal state of the frozen ground on the wetting and warming QTP are still spatially and temporally confined and need further investigations.

Numerical simulation can be an appropriate method for expanding a site study to regional and long-term time scales. Recently, considerable studies regarding the thermal dynamics of frozen ground on the QTP have been conducted employing numerical model simulations [28–31]. Guo and Wang [32] investigated the extent of permafrost degradation on the QTP using the Community Land Surface Model (CLM). The results indicated that the near-surface permafrost area decreased at a rate of 0.09×10^6 km^2/decade, and the average active layer thickness increased by 0.15 m/decade from 1981 to 2010. It was projected that the shrinkage of permafrost area will exceed 58% by the end of the 21st century, and the most sustainable permafrost may only exist in the northwestern QTP [9,33]. Actually, the total area of thermally degraded permafrost was about 1.54×10^6 km^2 in the past, and the key period of degradation was from the 1960s to the 1970s and from the 1990s to the 2000s [14]. A recent numerical experiment further reveals that winter warming has amplified the thermal degradation of permafrost since 2000 [34]. These works have extended our understanding of the thermal responses of frozen ground to climate warming, whereas the temperature gradient transition between near-surface atmosphere and frozen ground in long-term time series and the associated thermal dynamics in the context of climate change still need further investigation. It is of great importance to reasonably project the changes in the frozen ground on the QTP in the foreseeable future. For this purpose, Liu et al. [6] analyzed the spatial and temporal variations of the air freezing/thawing indices and the ground surface freezing/thawing indices in the southwestern QTP from 1900 to 2017. Their results suggest that these indices consistently experienced abrupt changes around the 2000s. It has been recognized that these thermal changes in permafrost were influenced by air warming. Nonetheless, it should also be noted that the QTP has undergone prominent wetting process since the late 1990s [20,35,36], where super-heavy precipitation can cause dramatic rises in soil temperatures by 0.3 to 0.5 °C at shallow depths and advancement thawing of the active layer by half a month in permafrost regions on the northeastern QTP [37]. However, there still exist some uncertainties regarding the thermal responses of permafrost and SFG to intensified wetting and warming conditions, especially with the occurrence of vegetation cover expansion across the QTP in the long run.

This study aims to characterize the thermal dynamics in both permafrost and SFG regions on the QTP from 1961 to 2010 by investigating the differences in freezing/thawing indices between the near-surface atmosphere and the ground surface. In terms of temporal and spatial changes, the analysis is based on a high-resolution (0.05° × 0.05°) simulation conducted by the Community Land Surface Model version 4.5 (CLM4.5). The reminder in this paper is organized as follows: Section 2 describes the data and methods; Section 3 presents the spatial transitions between permafrost and SFG distribution in each decade and the variations of freezing/thawing indices between atmosphere and ground surface in permafrost and SFG regions, respectively; Section 4 provides a discussion of the ground surface freezing/thawing index anomalies and the air freezing/thawing index anomalies and their relation to precipitation and the vegetation cover conditions in the permafrost and SFG areas on the QTP; conclusions and an outlook are presented in Section 5.

2. Materials and Methods

2.1. Data

2.1.1. Land Surface Model Data

An atmospheric forcing dataset at a temporal-spatial resolution of 3 h and 5 km from 1961 to 2010 was used to force the land surface model (http://globalchange.bnu.edu.cn/research/forcing (accessed on 3 January 2020)). It includes seven meteorological factors, including gridded near-surface temperature, relative humidity, wind speed, surface pressure, precipitation, and downward shortwave (longwave) radiation. The forcing fields are generated through a recently proposed approach based on observations collected at approximately 700 stations on mainland China. Before it was released by the Beijing Normal University (China, hereafter BNU), the reasonability of the dataset was validated through comparisons of the corresponding components with the National Centers for Environmental Prediction Climate Forecast System Reanalysis dataset and the Princeton meteorological forcing dataset [38]. Moreover, the reliabilities of the air temperature and precipitation components of the dataset have also been confirmed using the observational records on the QTP prior to the numerical simulation [39]. The precipitation amount across the QTP utilized in this study was also derived from the BNU dataset. Additionally, the in-situ observations across the QTP also confirmed the simulation ability of the CLM4.5 to reproduce the soil temperature values in a long run [36,39], which were used to calculate the ground freezing/thawing indices (GFI/GTI) in this study. The air freezing/thawing indices (AFI/ATI) presented in this study were also calculated using the near-surface (2 m above the ground) air temperatures derived from this data set.

2.1.2. Remote Sensing Data

The Normalized Difference Vegetation Index (NDVI) from July 1981 to December 2015 was used to characterize vegetation cover conditions across the QTP. It is the latest release of the long-sequence product of the NOAA Global Inventory Monitoring and Modeling System (GIMMS), version 3g.v1. The temporal resolution of the NDVI is half a month, and the spatial resolution is 0.08° × 0.08°, respectively. It is the classic dataset used to detect vegetation dynamics, and their influences on the thermal regimes of frozen ground across the QTP [40]. The data set is available from the National Tibetan Plateau Data Center (https://data.tpdc.ac.cn (accessed on 12 June 2022)).

2.2. Methods

2.2.1. Freezing/Thawing Indices

Freezing/thawing indices were calculated using the accumulated monthly air temperature or ground surface temperature of numerical outputs [41,42]. The AFI and GFI can be conceptualized as the accumulated near-surface air temperature (2 m above the ground) and ground surface temperature (4.51 cm under the ground) to avoid the potential impact of random weather processes on the movement from one soil phase to the next when determining the onset of freezing or thawing in the whole freezing period from 1 July to 30 June in the next year. Similarly, the ATI and GTI are the sum of the air and ground surface temperatures with positive monthly mean values in a whole thawing period from

1 January to 31 December within a calendar year [6,19,36]. Specific equations are shown as follows:

$$FI = \sum_{i=7}^{M_F} \left|\bar{T_i}\right| \cdot D_i\left(\bar{T_i} < 0\right) \quad (1)$$

$$TI = \sum_{i=1}^{M_T} \left|\bar{T_i}\right| \cdot D_i\left(\bar{T_i} > 0\right) \quad (2)$$

where FI and TI (°C·d) are freezing and thawing indices calculated by mean monthly temperatures (T_i, near-surface atmosphere, or ground surface), respectively; D_i is the number of days of the ith month; M_F and M_T represent months with negative and positive mean monthly values of near-surface air temperature or ground surface temperature, respectively.

2.2.2. Surface Frost Index

To estimate the frozen ground distribution in each decade, the surface frost index, which has been validated to capture permafrost distribution reasonably on the QTP [36,43,44], was calculated and used to diagnose permafrost. In this study, the boundary of permafrost is determined by the ground surface freezing index and thawing index:

$$F = \frac{\sqrt{GFI^+}}{\sqrt{GFI^+} + \sqrt{GTI}} \quad (3)$$

where F is a parameter to diagnose permafrost. Using a negative exponential function depending on snow density, depth, heat capacity, and thermal conductivity, the effects of snow cover were adjusted while evaluating the ground freezing index. As a result, the GFI with a superscript (+) represents the underlying ground freezing index under snow cover [45]. The value of F ranging from 0.50 to 0.60 indicates sporadic permafrost, 0.60 to 0.67 indicates extensive permafrost, and above 0.67 indicates continuous permafrost [46]. A threshold value of 0.58 was taken to estimate the absence or presence of permafrost in this study because a minimum frost index value of 0.58 has been confirmed by Slater and Lawrence [46], matching the observed permafrost area. The simulated permafrost and SFG distribution in the 2000s (Figure 1e) were very similar to the new map of the Plateau's frozen ground from 2009–2014 (Figure 1f). It covers a total area of 1.02×10^6 km^2 (excluding glaciers and lakes) in the 2000s, which is similar to but slightly lower than the value of 1.06×10^6 km^2 presented in the new map [5]. Less permafrost was simulated in the northeastern parts and southern regions of the QTP. The discrepancies between the simulations and the new frozen ground map could be partially attributed to possible inaccuracies in the permafrost parameterization schemes in the model and could also be due to possible inaccuracies in the frozen ground map.

2.2.3. Statistical Analysis Method

The modified Mann–Kendall (MMK) trend and Sen's slope estimator methods were applied in this study to detect the tendencies and estimate the trends of variations in freezing and thawing indices across the QTP. The two nonparametric methods have been widely used to conduct statistical diagnosis in climate analysis studies, and a detailed description of these methods can be found in the previous study [47].

2.2.4. Model and Numerical Simulation Design

The Community Land Surface Model version 4.5 (CLM 4.5) was used in this study to obtain the essential monthly air and ground surface temperatures to calculate freezing/thawing indices across the QTP. As the land component of the Community Climate System Model and the Community Earth System Model, CLM 4.5 simulates the partitioning of mass and energy from the atmosphere, redistributes the mass and energy of the land surface, and then exports the fresh water and heat to the oceans [48]. It explicitly considers the coupling interaction between water and heat flows in the frozen soil by modifying the snow cover setting and hydraulic properties parameterization, which reflect the hysteresis of snow between accumulation and melt phases at a given depth [48] during the freezing and thawing processes of frozen ground on the QTP. Previous studies have proven CLM is effective for thermal dynamics research on the QTP at different spatial scales [49,50]. In this study, the CLM4.5 was driven by the BNU dataset, and the numerical simulation was conducted at a spatial resolution of $0.05^\circ \times 0.05^\circ$ across the QTP. The outputs were set to a monthly time step from 1961 to 2010. It is important to state that the reliabilities of the simulated soil temperature and moisture have been validated with station records and in situ observations across the QTP in preceding works [36,39].

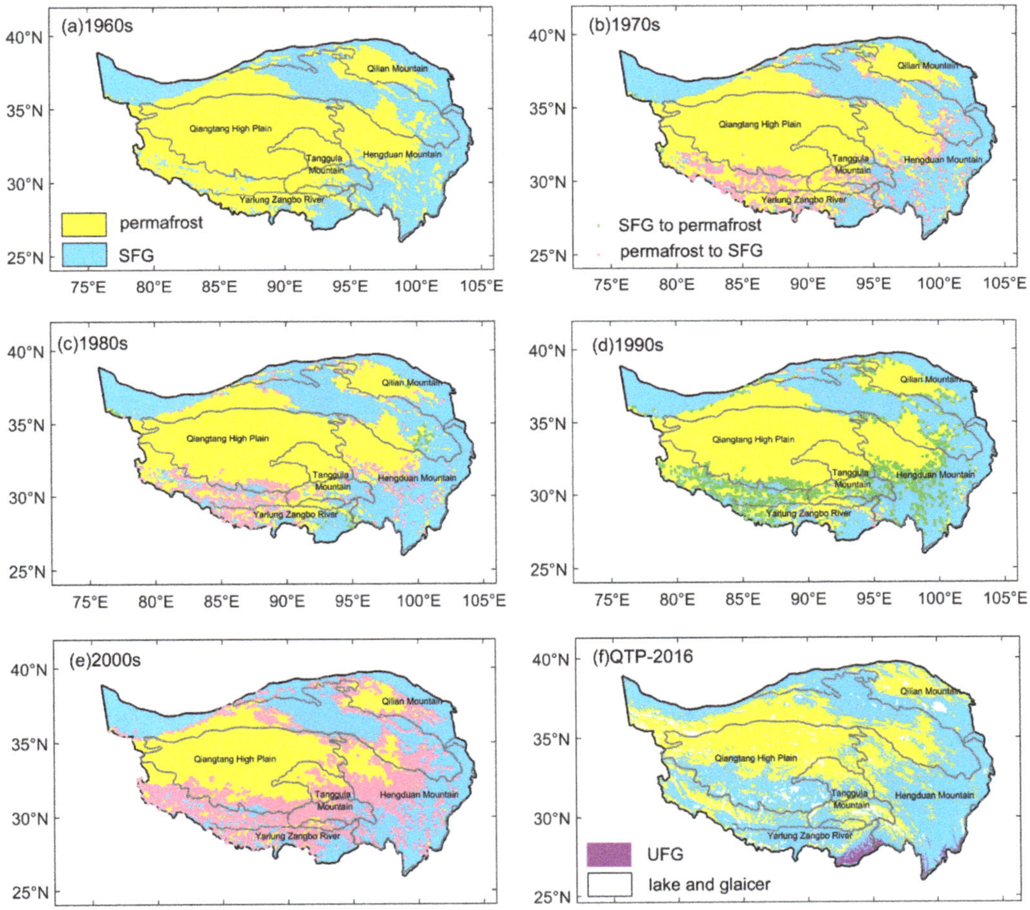

Figure 1. Simulated spatial distribution of frozen ground types in (**a**) the 1960s, (**b**) the 1970s, (**c**) the 1980s, (**d**) the 1990s, and (**e**) the 2000s, against the (**f**) new map of permafrost distribution on the QTP (QTP-2016). The yellow, navy blue, purple, and white colors represent the areas of permafrost, seasonally frozen ground (SFG), unfrozen ground (UFG), glaciers, and lakes, respectively. The spatial distribution pattern of permafrost in the current decade served as the benchmark for determining the type of transition between permafrost and SFG in the next one.

3. Results

3.1. Decadal Changes of Permafrost Distribution on the QTP from 1961 to 2010

The spatial patterns of frozen ground distribution in each decade from 1961 to 2010 across the QTP are presented in Figure 1. A new presentation of permafrost distribution on the QTP, mapping permafrost, SFG, and unfrozen ground based on remote sensing land surface temperature products from 2009 to 2014 (Figure 1f, hereafter as QTP-2016), is also included to provide a contrast benchmark [5]. It is obvious that the conversion between permafrost and SFG occurred mostly around the rims of mountainous permafrost areas on the southwestern QTP and the sporadic permafrost regions on the eastern QTP during the 1970s and 1980s (Figure 1b,c). Statistically, the permafrost area shrank from 1.62×10^6 km^2 to 1.40×10^6 km^2 in the 1960s–1980s (Table 1), broadly consistent with previous studies. A remote sensing-based study estimated that the total area of permafrost was 1.74×10^6 km^2 in the 1960s and turned to 1.58×10^6 km^2 in the 1980s [14]. The significant decreases in the AFI occurred in the 1980s, which implies the occurrence of dramatic warming of air temperature over the entire QTP in the freezing period during this decade. The air warming over the permafrost region was also detected during the thawing period, as evidenced by the increased rate of 0.65 °C·d/decade in the ATI. The prominent atmospheric warming should have exerted important effects on the permafrost degradation (6.67% reduction in area) and the thermal instability in SFG, where the GFI decreased significantly (11.73 °C·d/decade). Furthermore, we noticed that the GTI showed an opposite decreasing trend in permafrost areas, even though the ATI increased (as did air warming in the thawing period) in the 1980s. It suggests that the top permafrost showed fewer thermal responses within the active layer to the rise in air temperature during the thawing process in the 1980s. While the intensified air warming in the cold season (from November to March) [26] was more favorable for the remarkable shrinkage of permafrost areas during this period.

Table 1. Decadal permafrost areas (km^2) and the trends of AFI, GFI, and ATI, GTI (°C·d/decade) in permafrost, SFG, and the entire QTP from 1961 to 2010.

Periods	Permafrost Area (×10^6 km^2)	Region	AFI (°C·d/Decade)	GFI (°C·d/Decade)	ATI (°C·d/Decade)	GTI (°C·d/Decade)	AFI/ATI	GFI/GTI
1960s	1.62	permafrost	−20.75	14.99	2.07	2.25	4.94	1.15
		SFG	−20.87	8.13	−4.79	−9.10 *	0.64	0.28
		QTP	−10.59	16.57	−8.31	−11.09	1.57	0.57
1970s	1.50 (7.41% ↓)	permafrost	0.40	1.72	−0.53	0.25	4.73	1.18
		SFG	2.14	1.16	0.05	2.40	0.64	0.29
		QTP	1.32	0.73	0.45	2.67	1.44	0.57
1980s	1.40 (6.67% ↓)	permafrost	−23.28 *	−13.68 *	0.65	−1.95	4.83	1.28
		SFG	−14.87 *	−11.73 *	−3.79	−5.19	0.68	0.31
		QTP	−22.64 *	−17.13 *	0.81	−0.13	1.45	0.56
1990s	1.53 (9.29% ↑)	permafrost	−6.93	−5.55	1.17	−4.09	4.31	1.44
		SFG	−13.03 *	−14.75 *	16.39	24.33 *	0.60	0.36
		QTP	−19.23	−18.60	20.96	27.21	1.18	0.59
2000s	1.02 (33.33% ↓)	permafrost	13.37	14.73	−6.41	10.98	3.36	0.98
		SFG	25.68	32.66 *	−12.91	−9.57	0.53	0.24
		QTP	38.07	52.42 *	−20.07	−38.16	1.20	0.49

Trends with an asterisk (*) denotes $p < 0.10$. Downward (upward) arrow indicates shrinkage (expansion) permafrost in area.

The QTP has experienced obvious permafrost expansion in the 1990s (0.13×10^6 km^2) after the pronounced shrinkage of permafrost area in the 1980s. Ran et al. [14] also claimed similar thermal condition improvement in a specific permafrost area located primarily east of Lhasa in the southeastern part of QTP, where there are numerous marine glaciers and substantial snow cover. Figure 1d shows that the conversions from SFG to permafrost mostly occurred in the south of the Qiangtang High Plain, including the Yarlung Zangbo River, as well as the Hengduan Mountains in the southeastern QTP. However, the AFI and the ATI in permafrost areas show decreasing and increasing trends with rates of 6.93 °C·d/decade and 1.17 °C·d/decade, respectively (Table 1), implying that the atmosphere over permafrost regions exhibited warming changes in both freezing and thawing periods in the 1990s. These results indicate that air warming does not always decrease the permafrost area because the active layer shortens in freezing (5.55 °C·d/decade) and thawing (4.09 °C·d/decade) periods. Zhang et al. [20] reported that the Qiangtang High Plain has experienced the most intensive precipitation increase since 1998, and it has exerted a cooling effect on the top permafrost regions by reducing their thermal responses to climate change. Numerical tests showed that an increase of 100 mm in summer precipitation causes a mean reduction of 0.35 m in active layer thickness and 0.36 °C in the top of permafrost [20]. More precipitation increases soil wetness and soil moisture, so surface ground heat loss occurs inevitably. It will reduce the heat transferred to the deeper layer while rainfall with higher temperatures percolates into the frozen soil column [36]. Furthermore, abundant heat will also be absorbed by increased liquid water in soil pores due to its higher specific heat capacity than ice, thus diminishing heat transfer to the deeper layer. Intensified rainfall will have a cooling effect on permafrost when heat is consumed more by evaporating and percolating liquid water than by transferring to soils [20]. This phenomenon is more prominent in permafrost regions, which belong to the arid zone on the QTP, as evidenced by the remarkable positive and negative correlation coefficients between the annual precipitation and the GFI and GTI, respectively, since the 1990s [36]. Based on these results, we can conclude that precipitation has played an important role in favoring permafrost development in the 1990s, despite the background of air warming.

Table 1 shows that the permafrost area grew in the 1990s and dropped to a new low in the 2000s, attaining a minimum value of 1.02×10^6 km^2, which is also comparable but slightly lower than the observational-based value of 1.33×10^6 km^2 in the same period [14]. Figure 1e shows that the conversions from permafrost into SFG prevailed across the entire QTP, especially in the southern QTP. The reduction of the freezing and thawing durations in the active layer in the 1990s is likely to cause permafrost to be more vulnerable to climate change, which coincides with previous results [51]. Permafrost monitoring through 10 boreholes on the QTP shows that mean annual temperatures at 6.0 m depth have increased by 0.12 °C to 0.67 °C, with an average increase of 0.43 °C from the mid-1990s to the late 2000s [16]. The increase in active layer thickness was driven mainly by increases in the thawing index and annual precipitation [52]. Meanwhile, a field investigation conducted in the permafrost region from 2005 to 2009 in the interior QTP demonstrated that declines in vegetation cover in alpine meadows led to an increase in the thawing and an advance in the onset of seasonal changes within the active layer [53]. Our results further indicated that, in contrast to the 1990s, a permafrost area of 33.33% thawed into SFG in the 2000s over the entire QTP. While the AFI increased and the ATI decreased at rates of 13.37 °C·d/decade and 6.41 °C·d/decade, respectively, in permafrost regions during this decade (Table 1). It illustrates that the QTP has undergone a warming hiatus during the 2000s as the period of air temperature below (above) 0 °C extended (shortened). SFG also exhibited an increasing trend in GFI and a decreasing one in GTI, following the variations in AFI and ATI. However, the variations in ATI and GTI have shown an opposite direction in the permafrost parts since the 2000s. The temperature of top permafrost above 0 °C lasted longer while the atmospheric temperature exceeding 0 °C lasted shorter, as evidenced by the increasing rate of 10.98 °C·d/decade in the GTI and the decreasing rate of 6.41 °C·d/decade in the ATI (Table 1). It is reported that dramatic wetting over cold and arid permafrost zones tends to induce a cooling of the atmosphere [36], because the surface energy balance responds to variations in annul precipitation by influencing the surface radiation components, which alter the near-surface air temperature [54]. However, the warming hiatus during thawing duration has not restrained the thermal degradation in permafrost regions in the 2000s. Persistent wetting and more heavy rainfall events in summer are favorable for liquid water to rapidly reach and accumulate at the bottom of the active layer, which will serve to warm the permafrost body during the subsequent freezing period [20,37].

As a whole, the QTP witnessed prominent permafrost shrinkage in the 1960s–2000s, accounting for 37.04% of the permafrost area in the 1960s. The most pronounced decline in permafrost area took place in the 2000s, when QTP was undergoing a warming hiatus. From the perspective of the surface freeing/thawing processes, the ratios of the AFI and the ATI fell in the 1970s and decreased again in the 1990s–2000s from 4.31 to 3.36. However, the ratios of the GFI and the GTI rose in the 1990s and dropped sharply to a minimum of 0.98 in the 2000s in response to the atmospheric warming during the freezing period over the permafrost (Table 1). It signifies the QTP has presented more complicated permafrost responses to climate warming and wetting since the 1990s. The frozen ground was thermally unstable in the 1980s owing to the strikingly atmospheric warming in the cold season. The dramatic thermal degradation in the 2000s, which might be due to the accelerated wetting process over the QTP, mainly occurred in permafrost regions in the warm season (from May to October) [26]. This differs from the 1980s. Although recent studies have confirmed that intensified summer precipitation has exerted an important warming effect on permafrost bodies [20,36,37], our results further reveal that thermal degradation also terminated in SFG in the 2000s. Relatively warmer and wetter conditions in SFG regions might experience more heat loss via evaporation when gradually plentiful water occupies soil pores, which alters the thermal regimes of surface ground [36].

3.2. Comparisons of Thermal States between Near-Surface Atmosphere and Ground Surface in Climatology

The mean values of GFI and GTI from 1961 to 2010 across the QTP show spatial patterns similar to the AFI and ATI, respectively (Figure 2). From the perspective of value, the climatology of the GTI (GFI) is higher (lower) than that of the ATI (AFI), especially in the permafrost regions. The maximum value of the ATI is 1505.80 °C·d, which is substantially lower than that of the GTI (2485.06 °C·d). The maximum AFI reaching the value of 2042.24 °C·d is higher than that of the GFI (1463.06 °C·d) for the entire QTP. This implies that, in general, the period of ground temperature above 0 °C lasted longer than air temperature in the thawing period, while air temperature below 0 °C retained longer than ground surface temperature in the freezing period in past decades. The maximum value of the averaged annual ground surface temperature is 3.44 °C, which is more than five times the value of the air temperature (−0.61 °C, Figure 2c,f).

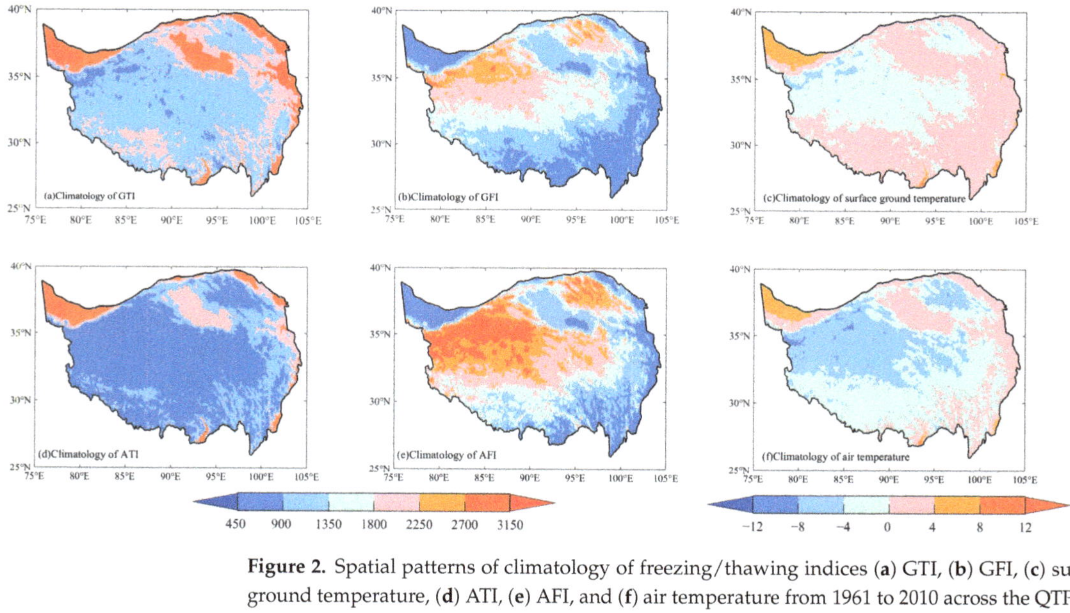

Figure 2. Spatial patterns of climatology of freezing/thawing indices (**a**) GTI, (**b**) GFI, (**c**) surface ground temperature, (**d**) ATI, (**e**) AFI, and (**f**) air temperature from 1961 to 2010 across the QTP.

The spatial patterns for the AFI and GFI, ATI, and GTI climatology show distinctive regional differences (Figure 2). In contrast to the GFI, permafrost and SFG tend to show more pronounced distinctions in the GTI from 1961 to 2010, with a value of 1267.83 °C·d. However, the difference between the AFI in permafrost and SFG is 1298.38 °C·d, which is comparable to the ATI value of 1222.01 °C·d. It shows apparent distinctions between permafrost and SFG responses to climate change in terms of thawing intensities. Dramatic higher differences between the AFI and the GFI in permafrost regions suggest more pronounced freezing intensity. Conversely, the differences between the ATI and GTI are more remarkable in SFG parts, which illustrate that SFG regions are more vulnerable to thawing when air temperature increases. However, owing to the wetting and warming conditions over the QTP, the decreased cold stress in high altitude parts of the QTP encouraged an expansion of climatically suitable areas for plant growth [55]. Air temperature and precipitation changes over time appear to exert substantial impacts on regulating the thermal regimes of the surface permafrost. As one of the typical characteristics of land surface processes at high altitudes in the QTP, alpine permafrost is also not immune to these impacts. The vegetation expansion and snow cover reduction in high-altitude areas can produce positive feedback on permafrost degradation when the underlying snow with high albedo is replaced by vegetation with a lower albedo [26,56].

3.3. Spatial Changes of Freezing and Thawing Indices in Permafrost and Seasonally Frozen Ground

As the 1990s represent the pivotal period of permafrost area variations and freezing/thawing indices dynamics, the spatial patterns of changes in the AFI and GFI, the ATI, and the GTI in permafrost and SFG regions before and after the 1990s are shown in Figures 3 and 4, respectively. Before the 1990s, the changes in AFI in permafrost regions were dominated by negative values, which illustrates the air warming during the freezing period (Figure 3a). Persistent negative trends of the GFI in the Qilian Mountains, Tanggula Mountains, and parts of Qiangtang High Plain in the interior QTP, which indicate these surface permafrost areas have warmed substantially, are broadly consistent with the dynamics of permafrost distribution during these decades (Figure 1b,c). It can be well explained that the increased air temperature, which has been widely recognized to control thermal regime changes in recent studies [19,34,35], has triggered the thermal regime changes of the underlying permafrost. However, it should also be emphasized that the major Qiangtang High Plain in the western QTP exhibited prominent increases in the GFI, where the freezing duration was maintained longer within the surface active layer. It appears that the air warming before the 1990s in the cold season only induced thermally unstable conditions in some small portions of surface permafrost. Statistically, in contrast to the decreasing trend of the AFI (13.59 °C·d/decade) from 1961 to 1990, the annual GFI in permafrost regions was growing at a rate of 67.59 °C·d/decade (Table 2). Meanwhile, positive-dominated changes in the ATI before the 1990s have consistently shown that the QTP has experienced a prevalent warming process in thawing duration, except for some portions in the sporadic permafrost regions in the Hengduan Mountains and Qilian Mountains where negative trends were present (Figure 3c). Table 2 demonstrates that the ATI increased significantly at a rate of 39.86 °C·d/decade during 1961–1990, while the GTI shows pronounced spatial distinctions in terms of permafrost responses to warming. The eastern parts of the Qiangtang High Plain in the interior QTP exhibited decreasing changes in the GTI, which implies that the thawed duration has a tendency to decline to some extent even through an accelerated warming process over the permafrost body. However, the western parts exhibit an extended thawing duration as a response to the warming. Being different from the freezing process, which releases heat to the surroundings, the extension of the thawing process to the surface of the ground should be accompanied by a certain amount of heat input from the surroundings. Positive changes in the GTI and the GFI took place consistently in the western parts of the Qiangtang High Plain (Figure 3b,d). The longer freezing duration and extended thawing times that occurred in the top continuous permafrost regions tend to deepen the active layer, although the permafrost in these areas remained relatively stable in type before the 1990s (Figure 1c).

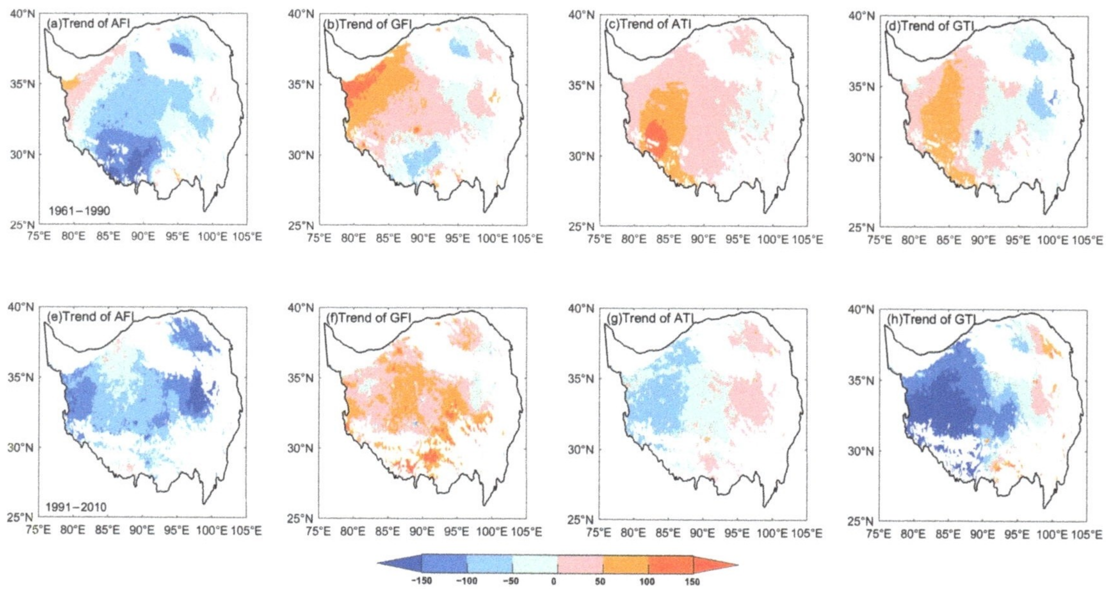

Figure 3. Spatial changes of freezing and thawing indices in permafrost regions between 1961–1990 and 1991–2010. (**a**,**e**) AFI, (**b**,**f**) GFI, (**c**,**g**) ATI, and (**d**,**h**) GTI on the QTP.

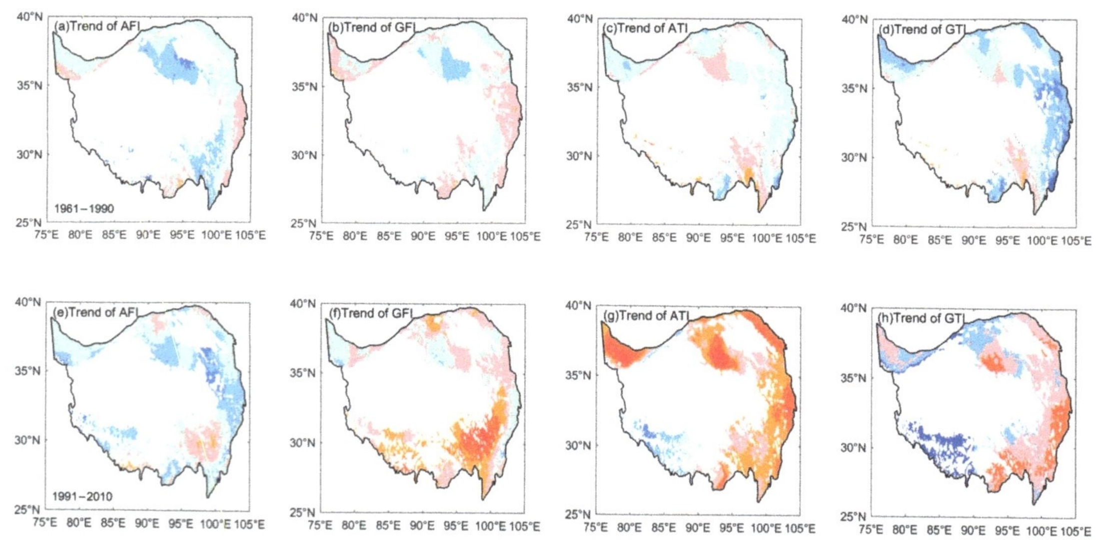

Figure 4. The same as Figure 3, but for seasonally frozen ground on the QTP. (**a**,**e**) AFI, (**b**,**f**) GFI, (**c**,**g**) ATI, and (**d**,**h**) GTI on the QTP.

In contrast to the slight decrease at a rate of 13.59 °C·d/decade before the 1990s, the AFI started to decrease dramatically since the 1990s and reached a magnified value of 202.57 °C·d/decade (Table 2). However, the freezing duration within the surface active layer extended intensively, as evident by the prevalent GFI increase in permafrost regions (Figure 3f). This is in contrast to the changes in AFI (Figure 3e). The warming effect of increased air temperature shows less control over thermal regimes in permafrost in the cold season from 1991 to 2010. Furthermore, spatial changes in the ATI also presented a reverse pattern after 1990. Most areas of the Qiangtang High Plain experienced a warming hiatus during the thawing period. From the perspective of magnitude, the increasing rate of the ATI is 16.94 °C·d/decade after 1990, which is substantially lower than that before 1990 (39.86 °C·d/decade). It is shown that air warming in the cold season eventually exceeded that in the warm season after 1990. This corroborates the findings of previous numerical experiments, which reported that summer warming has slowed and winter warming has begun to accelerate over the QTP around 1998 [34]. Meanwhile, the ground surface, however, presented shortened thawing periods in the active layer of permafrost regions (Figure 3h), with a rate of 83.58 °C·d/decade in the GTI (Table 2). Considering the relative milder increase in the ATI (16.94 °C·d/decade), it can be reasonably concluded that the top permafrost has not been particularly sensitive to the air warming in warm seasons since the 1990s. That is, pronounced air warming in the warm season had only marginal effects on thermal regime changes in permafrost on the western QTP.

Table 2. Comparison of the annual mean (°C·d) and trends (°C·d/decade) of the AFI, GFI, ATI, and GTI over the QTP in permafrost and SFG regions from 1961 to 1990 and 1991 to 2010.

Periods	Frozen Type	AFI		GFI		ATI		GTI	
		Mean	Trend	Mean	Trend	Mean	Trend	Mean	Trend
1961–1990	permafrost	3357.15	−13.59	2427.27	67.59	187.27	**39.86 ***	940.77	29.26
	SFG	1806.63	**−44.27 ***	1258.67	1.48	1165.43	19.61	2120.32	−10.70
	QTP	1991.66	−41.85	1397.33	8.58	1063.34	27.71	2009.92	−0.11
1991–2010	permafrost	3168.49	**−202.57 ***	2393.41	**−85.22 ***	263.83	16.94	918.51	−83.58
	SFG	1640.24	**−148.97 ***	1184.92	−42.39	1297.28	68.33	2162.22	−15.20
	QTP	1796.94	−57.24	1275.56	3.21	1203.37	2.35	2154.49	−45.90
1961–2010	permafrost	3286.73	−72.74	2413.87	−0.50	217.57	**32.73 ***	931.96	−5.38
	SFG	1740.77	−68.74	1229.48	−24.35	1217.62	47.98	2136.69	6.13
	QTP	1914.58	**−71.00 ***	1349.13	**−34.33 ***	1118.77	**48.13 ***	2067.14	**40.37 ***

* Trends with an asterisk (*) denotes $p < 0.10$.

Nevertheless, the QTP has also experienced a dramatic wetting process around 1998 and a remarkable greening process after 2000, which was favored by substantial climate change [20,26,57]. Permafrost monitoring has indicated that intensified rainfall imposed a cooling effect in the freezing duration and a warming effect in the thawing duration in the active layer [58], which has a pivotal impact on deepening the active layer thickness [59]. Meanwhile, an expansion of climatically suitable areas for plant growth might favor the increasing impact of vegetation on regulating the current thermal regimes of permafrost over the QTP [26]. From these results, it can be inferred that variations in freezing/thawing indices after 1990 might also be related to local variations in precipitation and to vegetation cover dynamics in permafrost ranges.

In SFG areas, the changes in AFI decreased significantly with rates of 44.27 °C·d/decade and 148.97 °C·d/decade during the two subperiods of 1961–1990 and 1991–2010 (Table 2). Spatially, the largest decreasing areas of the AFI expanded from the Qaidam Basin to the eastern parts of the QTP (Figure 4a,e). These remarkable declines in the AFI indicate the near-surface atmosphere over the SFG has warmed substantially in the freezing period during the study period. Figure 4b shows that the Qaidam Basin presented a prominent decreasing tendency in the GFI as responses to air temperature increased. However, some portions of the eastern QTP underwent cooling processes where the GFI decreased. As a whole, the significant air temperature rise over the SFG has not exerted prominent warming effects on frozen surface ground before the 1990s because the GFI was increasing at a rate of 1.48 °C·d/decade (Table 2).

From 1991, the ground surface underwent an obvious GFI decrease at a rate of 42.39 °C·d/decade (Table 2). Spatial patterns also indicate that SFG in the southeastern parts of the QTP also presented increasing GFI trends (Figure 4f). It is suggestive of diminishing thermal responses of SFG in these areas to air warming during the freezing period since the 1990s. Moreover, distinct from freezing indices changes, the changing thawing indices show reversed patterns before and after the 1990s. It is clear that the SFG experienced progressive increases in the ATI after 1990 (Figure 4g). Correspondingly, prominent increases in the GTI occurred in the eastern QTP (Figure 4h). While the southwestern QTP, where belong the rims of the transitional areas between permafrost and SFG in the 1990s–2000s (Figure 1d,e), showed strikingly decreasing tendencies in the ATI and GTI (Figure 4g,h). These results suggest that atmospheric warming over SFG after 1990 mainly occurred in the warm season, which differs from that in permafrost parts. The SFG degraded from permafrost in transitional areas shows fewer thermal responses to air warming in the cold season. A warming hiatus in the warm season was also favorable for permafrost development in these regions, as it occurred in the 1990s (Figure 1d).

4. Discussion

To characterize the effects of seasonal warming on frozen ground, we compared the freezing/thawing indices between the near-surface atmosphere and frozen ground surface after presenting the changes in frozen ground distribution in each decade. Air warming, which has shifted between freezing and thawing processes since the 1990s, shows a diminished control role on thermal regimes in both permafrost and SFG areas since the 1990s. More specifically, after 1990, air temperature rise over permafrost was more pronounced in the freezing period, while in the thawing period, air warming over SFG exhibited more intensive increase. The warming hiatus over permafrost in thawing period and the prominent shrinkage of permafrost area in the 2000s implies that air warming appears to play only a marginal controlling role in thermal regime changes, especially in western QTP.

As the occurrence of annual precipitation increases and the vegetation cover extends, especially in the 2000s [6,20,26,55], it is essential to investigate the effects on reregulating surface thermal regimes of frozen ground. Focused on a single precipitation event on the QTP, Li et al. [58] reported that increased rainfall created a heating effect on active layers in the thawing period, which led to a cooling effect in the freezing period. Similar warming effects of rainfall on the thermal regimes of permafrost have also been concluded over the Russian Arctic drainage basin [60]. Moreover, vegetation expansion in the warmer and wetter Arctic coincides with that on the QTP [55,61]. It will partially reduce permafrost degradation in the Siberian tundra [62]. Furthermore, on the QTP, recent observational results reveal that increases in vegetation greenness since the 2000s promote the role of SFG in responding to the intensified wetting process [18]. In this study, we more deeply investigate the changes in ground freezing and thawing indices responding to the atmospheric freezing and thawing indices on the entire QTP. To state this conclusion more reasonably, we also presented the relations between freezing/thawing indices and precipitation, the NDVI, respectively (Figure 5).

It is obvious that the precipitation anomaly over the QTP experienced a first peak in 1998 and rose to a new high of 191.10 mm in 2005. This corroborates the findings of the previous numerical study [20]. A similar breakpoint was also detected in northern Canada [63]. Permafrost borehole records from Canadian stations indicate that air temperatures have generally been higher since the 1980s and then decreased since the peak temperature in 1998. In this study, it was found that the negative AFI and positive ATI anomalies reach the maxima of 453.72 °C·d/decade and 255.90 °C·d/decade, suggesting that the QTP also underwent the most progressive atmospheric warming in 1998. The GFI anomaly generally followed the evolution of the AFI anomaly until 2001. After that, the GFI reaches a new minimum anomaly of 359.79 °C·d/decade in 2004, whereas the AFI anomaly remains stable. Furthermore, the ATI showed weakening but stable positive anomalies during 2001–2005, while the GTI presented strikingly negative ones during this period, especially in the permafrost regions (Figure 5c,d). These results suggest that the ground thawing duration shortened sharply even though the air temperature rose. Instead, the intensified precipitation began to act as a control mechanism for regulating surface thermal regimes. Statistical results indicate that precipitation anomalies yield significant negative correlation relationships with the AFI, GFI, and GTI ($p < 0.05$, Figure 5a,b,d) but a positive one with the ATI ($p < 0.05$, Figure 5c). The increased rainfall has a tendency to promote air temperature rise but restrain freezing and thawing cycles within the active layer of permafrost. This conclusion is coincident with the work of Luo et al. [37], who reported that increased summer precipitation amounts tend to reduce ground thawing indices and exert an obvious cooling effect on top permafrost, based on permafrost monitoring in the northeastern QTP. However, when a large amount of rainfall percolates into the deeper soils, it may cause an opposite warming effect on permafrost [20]. It can be explained that the liquid rainfall that commonly comes with a higher temperature than that in deep soils will effectively transfer heat to the surroundings after percolating through the soil quickly and, as a result, warm the permafrost body.

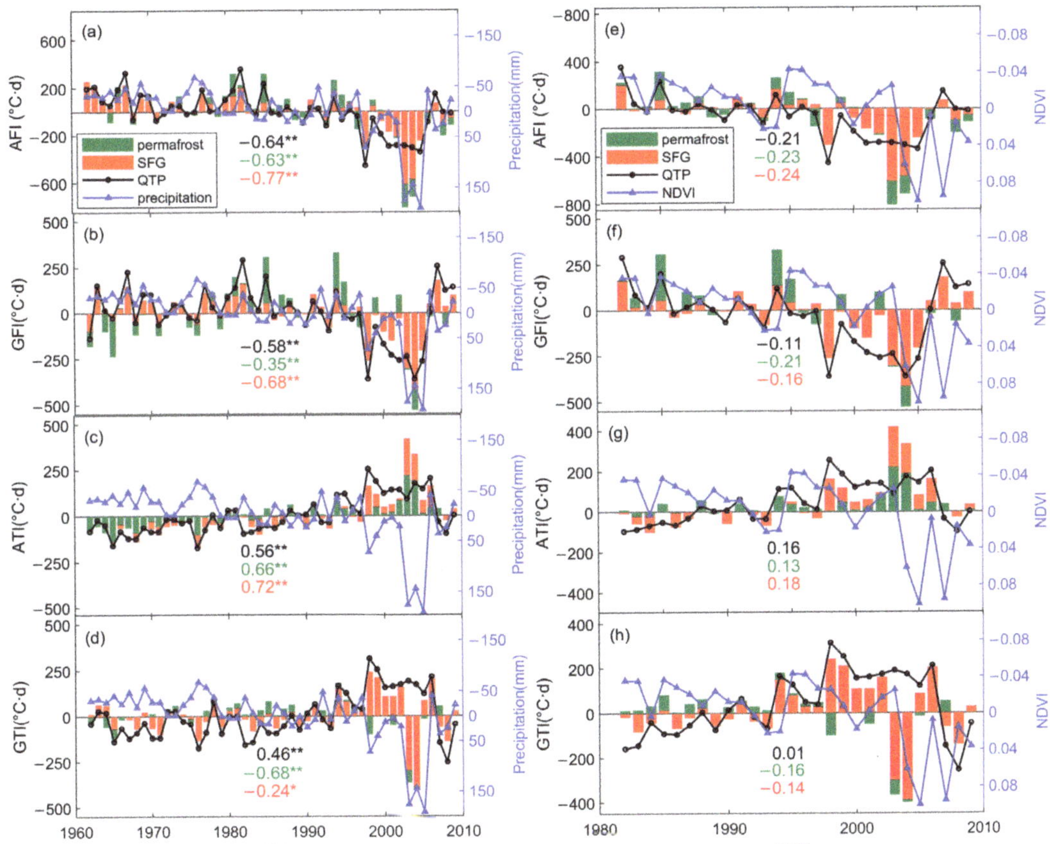

Figure 5. The time series of anomaly comparisons between precipitation (left panels), NDVI (right panels), and freezing/thawing indices during 1961–2010, respectively. (**a**,**e**) AFI; (**b**,**f**) GFI; (**c**,**g**) ATI; and (**d**,**h**) GTI. Two colors of green (red) bars represent index anomalies in permafrost (SFG), and the black lines with circles indicate those across the whole QTP. The blue lines with triangles represent the anomalies in the precipitation and the NDVI. Figures in the same colors as permafrost, SFG, and the entire QTP indices shown in the plots represent their correlation coefficients with precipitation and NDVI anomalies. One asterisk (*) denotes $p < 0.10$ and two asterisks (**) denote $p < 0.05$.

It is important to note that the NDVI started to increase in 2004 and reached a maximum value of a positive anomaly (0.10) in 2005. As shown in Figure 5, the NDVI anomaly yields a negative correlation relationship with the AFI anomaly and a positive one with the ATI anomaly, although they are not statistically significant. It implies that the shortened periods of air temperature below zero in the freezing process and the extended ones above zero in the thawing process are favorable for vegetation cover expansion. For top permafrost, a positive anomaly in the NDVI will also reduce its thermal response to climate warming because of its negative correlation relationship with the GFI and the GTI, respectively. Similar conclusions have also been drawn in a recent study [25]. However, the hiatus in vegetation greening after 2006 (Figure 5) mainly showed warming effects due to nighttime warming [64]. It could accelerate permafrost thawing [64], as evidenced by the increased GFI and GTI (Figure 5f,h). The vegetation dynamics coincide with the persistent wetting over the QTP in the 2000s, which intensified thermal instability in permafrost. As a result, these results suggest that the dramatic permafrost shrinkage and thermal degradation across the QTP are not always only induced by air warming; intensive wetting and greening dynamics over the QTP also play important roles. Further research into the influences of vegetation growth and precipitation anomalies on the thermal regime changes of frozen ground at different altitude levels of the QTP is also required in the near future.

5. Conclusions

Based on the numerical simulation results, this study analyzed the thermal dynamics of frozen ground on the QTP from a new perspective. Freezing and thawing indices between frozen ground and near-surface atmosphere are analyzed in five decades from 1961 to 2010 and related to the changes in frozen ground distribution in each decade. The main conclusions are the following:

The net shrinkage of permafrost area in the 2000s was 37.04% relative to the 1960s across the QTP. The permafrost area rose in the 1990s and dropped to a new low value of 1.02×10^6 km^2 in the 2000s. Decadal permafrost distribution patterns indicate the first occurrence of degradation in the 1970s–1980s lies in the Tanggula Mountain areas on the southwestern QTP, followed by the pronounced permafrost reduction in areas of the Qilian Mountain in the northwestern QTP and the Hengduan Mountain in the interior QTP in the 2000s.

The ratios of the AFI and the ATI presented a steady decrease from 1.57 to 1.44 in the 1960s–1980s, in contrast to the slight decrease in the ratios of the GFI and the GTI from 0.57 to 0.56 during this period. Atmospheric warming over the QTP played a more controlling role in the thermal dynamics of frozen ground before the 1990s. The air temperature reached a peak in 1998 in both freezing and thawing periods. The ratios of the GFI and the GTI in permafrost and SFG also reached their maximum in the 1990s, with values of 1.44 and 0.36, respectively. The ratio of the GFI and the GTI was <1 in the 2000s, indicating their prominent thermal responses to intensified wetting and warming processes.

A warming hiatus was detected in the 2000s, especially for the permafrost regions in the cold season, which favorably restrained the permafrost degradation. In this decade, the QTP has been under the influence of prominently thermally unstable conditions, with the permafrost area shrinking by 33.33%. The occurrence of increasing precipitation over the QTP began to play a more controlling role in permafrost degradation. The hiatus in vegetation greening after 2006 could promote the thermal responses of frozen ground to climate change.

Author Contributions: Conceptualization, X.F.; methodology, X.F.; software, A.W.; validation, X.F.; formal analysis, X.F. and A.W.; investigation, X.F.; resources, X.F. and S.L.; data curation, X.F.; writing—original draft preparation, X.F.; writing—review and editing, X.F. and K.F., visualization, A.W.; supervision, X.F.; project administration, X.F. and S.L.; funding acquisition, S.L. and X.F. All authors have read and agreed to the published version of the manuscript.

Funding: This research was funded jointly by the National Natural Science Foundation of China (41905008, 41975007, 42275080) and the Innovation and Entrepreneurship Training Program for College Students of Chengdu University of Information Technology (202210621006, 202210621003).

Data Availability Statement: The China Dataset of Meteorological Forcing for Land Surface Modeling can be obtained online (http://globalchange.bnu.edu.cn/research/ (accessed on 3 January 2020)). The NDVI product of NOAA Global Inventory Monitoring and Modeling System (GIMMS), version 3g.v1 is accessible on website (https://data.tpdc.ac.cn/en/data/9775f2b4-7370-4e5e-a537-3482c9a83d88/ (accessed on 12 June 2022)).

Acknowledgments: Thanks are due to the reviewers who provide valuable suggestions and supportive comments which are favored to improve the quality of the manuscript. We also thank the Max Planck Institute of Meteorology (Atmosphere in the Earth System) for providing hospitality combined with an excellent working environment.

Conflicts of Interest: The authors declare no conflict of interest.

References

1. Zhang, T.; Zhou, Y.; Guo, D.; Qiu, G.; Li, S. Geocryology in China. *Arct. Antarct. Alp. Res.* **2001**, *33*, 245.
2. Yang, K.; Ye, B.; Zhou, D.; Wu, B.; Foken, T.; Qin, J.; Zhou, Z.; Oppenheimer, M.; Yohe, G. Response of hydrological cycle to recent climate changes in the Tibetan Plateau. *Clim. Chang.* **2011**, *109*, 517–534. [CrossRef]
3. Yao, T.; Thompson, L.; Mosbrugger, V.; Zhang, F.; Ma, Y.; Luo, T.; Xu, B.; Yang, X.; Joswiak, D.; Wang, W.; et al. Third Pole Environment (TPE). *Environ. Dev.* **2012**, *3*, 52–64. [CrossRef]
4. Pan, X.; Guo, X.; Li, X.; Niu, X.; Yang, X.; Feng, M.; Che, T.; Jin, R.; Ran, Y.; Guo, J.; et al. National Tibetan Plateau Data Center: Promoting Earth System Science on the Third Pole. *Bull. Am. Meteorol. Soc.* **2021**, *102*, E2062–E2078.
5. Zou, D.; Zhao, L.; Sheng, Y.; Chen, J.; Hu, G.; Wu, T.; Wu, J.; Xie, C.; Wu, X.; Pang, Q.; et al. A new map of permafrost distribution on the Tibetan Plateau. *Cryosphere* **2017**, *11*, 2527–2542. [CrossRef]
6. Liu, L.; Luo, D.L.; Wang, L.; Huang, Y.D.; Chen, F.F. Dynamics of freezing/thawing indices and frozen ground from 1900 to 2017 in the upper Brahmaputra River Basin, Tibetan Plateau. *Adv. Clim. Chang. Res.* **2021**, *12*, 6–17. [CrossRef]
7. Dobinski, W. Permafrost. *Earth Sci. Rev.* **2011**, *108*, 158–169. [CrossRef]
8. Li, X.; Cheng, G. A GIS-aided response model of high-altitude permafrost to global change. *Sci. China Earth Sci.* **1999**, *42*, 72–79. [CrossRef]
9. Guo, D.; Wang, H. CMIP5 permafrost degradation projection:A comparison among different regions. *J. Geophys. Res. Atmos.* **2016**, *121*, 4499–4517. [CrossRef]
10. Schuur, E.; Vogel, J.G.; Crummer, K.G.; Lee, H.; Sickman, J.O.; Osterkamp, T.E. The effect of permafrost thaw on old carbon release and net carbon exchange from tundra. *Nature* **2009**, *459*, 556–559. [CrossRef]
11. Cheng, G.; Jin, H. Permafrost and groundwater on the Qinghai-Tibet Plateau and in northeast China. *Hydrogeol. J.* **2012**, *21*, 5–23. [CrossRef]
12. Cheng, G.; Wu, T. Responses of permafrost to climate change and their environmental significance, Qinghai-Tibet Plateau. *Geophys. Res. Earth Surf.* **2007**, *112*, F2. [CrossRef]
13. Xue, X.; Guo, J.; Han, B.; Sun, Q.; Liu, L. The effect of climate warming and permafrost thaw on desertification in the Qinghai-Tibetan Plateau. *Gepmorphology* **2009**, *108*, 182–190. [CrossRef]
14. Ran, Y.; Li, X.; Cheng, G. Climate warming over the past half century has led to thermal degradation of permafrost on the Qinghai-Tibet Plateau. *Cryosphere* **2018**, *12*, 595–608. [CrossRef]
15. Wang, B.; Bao, Q.; Hoskins, B.; Wu, G.; Liu, Y. Tibetan Plateau warming and precipitation changes in East Asia. *Geophys. Res. Lett.* **2008**, *35*, L14702. [CrossRef]
16. Wu, Q.; Zhang, T. Recent permafrost warming on the Qinghai-Tibetan Plateau. *J. Geophys. Res. Atmos.* **2008**, *113*, D13108. [CrossRef]
17. Wu, Q.; Zhang, T. Changes in active layer thickness over the Qinghai-Tibetan Plateau from 1995 to 2007. *J. Geophys. Res. Atmos.* **2010**, *115*, D09107. [CrossRef]
18. Frauenfeld, O.W.; Zhang, T.; Mccreight, J.L. Northern Hemisphere freezing/thawing index variations over the twentieth century. *Int. J. Climatol.* **2010**, *27*, 47–63. [CrossRef]
19. Wu, T.; Qin, Y.; Wu, X.; Li, R.; Zou, D.; Xie, C. Spatiotemporal changes of freezing/thawing indices and their response to recent climate change on the Qinghai-Tibet Plateau from 1980 to 2013. *Theor. Appl. Climatol.* **2018**, *132*, 1187–1199. [CrossRef]
20. Zhang, G.; Nan, Z.; Zhao, L.; Liang, Y.; Cheng, G. Qinghai-Tibet Plateau wetting reduces permafrost thermal responses to climate warming. *Earth Planet. Sci. Lett.* **2021**, *31*, 916–930. [CrossRef]
21. Bai, Y.; Guo, C.; Degen, A.A.; Ahmad, A.A.; Wang, W.; Zhang, T.; Li, W.; Ma, L.; Huang, M.; Zeng, H. Climate Warming Benefits Alpine Vegetation Growth in Three-River Headwater Region, China. *Sci. Total Environ.* **2020**, *742*, 140574. [CrossRef]
22. Cai, D.; Fraedrich, K.; Sielmann, F.; Zhang, L.; Zhu, X.; Guo, S.; Guan, Y. Vegetation Dynamics on the Tibetan Plateau (1982–2006); An Attribution by Ecohydrological Diagnostics. *J. Clim.* **2015**, *28*, 4576–4584. [CrossRef]

23. Li, J.; Chen, F.; Zhang, G.; Barlage, M.; Gan, Y.; Xin, X.; Wang, C. Impacts of Land Cover and Soil Texture Uncertainty on Land Model Simulations Over the Central Tibetan Plateau. *J. Adv. Model. Earth Syst.* **2018**, *10*, 2121–2146. [CrossRef]
24. Zhong, L.; Ma, Y.; Salama, M.S.; Su, Z. Assessment of vegetation dynamics and their response to variations in precipitation and temperature in the Tibetan Plateau. *Clim. Chang.* **2010**, *103*, 519–535. [CrossRef]
25. Li, J.; Wu, C.; Wu, M.; Zhang, Y.; Ran, Y. Modeling the start of frozen dates with leaf senescence over Tibetan Plateau. *Remote Sens. Environ.* **2022**, *281*, 113258. [CrossRef]
26. Fang, X.; Chen, Y.; Cheng, C.; Wang, Z.; Lyu, S.; Fraedrich, K. Changes of timing and duration of the ground surface freeze on the Tibetan Plateau in the highly wetting period from 1998 to 2021. *Clim. Chang.* **2023**, *176*, 59. [CrossRef]
27. Meng, X.; Li, R.; Luan, L.; Lyu, S.; Zhang, T.; Ao, Y.; Han, B.; Zhao, L.; Ma, Y. Detecting hydrological consistency between soil moisture and precipitation and changes of soil moisture in summer over the Tibetan Plateau. *Clim. Dyn.* **2018**, *51*, 4157–4168. [CrossRef]
28. Qin, Y.; Wu, T.; Zhao, L.; Wu, X.; Li, R.; Xie, C.; Pang, Q.; Hu, G.; Qiao, Y.; Zhao, G.; et al. Numerical Modeling of the Active Layer Thickness and Permafrost Thermal State Across Qinghai-Tibetan Plateau. *J. Geophys. Res. Atmos.* **2017**, *122*, 11604–11620. [CrossRef]
29. Li, D.; Chen, J.; Meng, Q.; Liu, D.; Fang, J.; Liu, J. Numeric simulation of permafrost degradation in the eastern Tibetan Plateau. *Permafr. Periglac. Process.* **2008**, *19*, 93–99. [CrossRef]
30. Yang, M.; Nelson, F.E.; Shiklomanov, N.I.; Guo, D.; Wan, G. Permafrost degradation and its environmental effects on the Tibetan Plateau: A review of recent research. *Earth Sci. Rev.* **2010**, *103*, 31–44. [CrossRef]
31. Yang, M.; Wang, X.; Pang, G.; Wan, G.; Liu, Z. The Tibetan Plateau cryosphere: Observations and model simulations for current status and recent changes. *Earth Sci. Rev.* **2019**, *190*, 353–369. [CrossRef]
32. Guo, D.; Wang, H. Simulation of permafrost and seasonally frozen ground conditions on the Tibetan Plateau, 1981–2010. *J. Geophys. Res. Atmos.* **2013**, *118*, 5216–5230. [CrossRef]
33. Yin, G.; Niu, F.; Lin, Z.; Luo, J.; Liu, M. Data-driven spatiotemporal projections of shallow permafrost based on CMIP6 across the Qinghai-Tibet Plateau at 1 km^2 scale. *Adv. Clim. Chang. Res.* **2021**, *12*, 814–827. [CrossRef]
34. Zhang, G.; Nan, Z.; Wu, X.; Ji, H.; Zhao, S. The Role of Warming in Permafrost Change over the Qinghai-Tibet Plateau. *Geophys. Res. Lett.* **2019**, *46*, 11261–11269. [CrossRef]
35. Zhao, L.; Hu, G.; Zou, D.; Wu, X.; Ma, L.; Sun, Z.; Yuan, L.; Zhou, H.; Liu, S. Permafrost changes and its effects on hydrological processes on Qinghai-Tibet Plateau. *Bull. Chin. Acad. Sci.* **2019**, *34*, 1233–1246. (In Chinese)
36. Fang, X.; Li, Z.; Cheng, C.; Fraedrich, K.; Wang, A.; Chen, Y.; Xu, Y.; Lyu, S. Response of freezing/thawing indexes to the wetting trend under warming climate conditions over the Qinghai-Tibetan Plateau during 1961–2010: A Numerical Simulations. *Adv. Atmos. Sci.* **2023**, *40*, 211–222. [CrossRef]
37. Luo, D.; Jin, H.; Bense, V.F.; Jin, X.; Li, X. Hydrothermal processes of near-surface warm permafrost in response to strong precipitation events in the Headwater Area of the Yellow River, Tibetan Plateau. *Geoderma* **2020**, *376*, 114531. [CrossRef]
38. Li, T.; Zheng, X.; Dai, Y.; Yang, C.; Chen, Z.; Zhang, S.; Wu, G.; Wang, Z.; Huang, C.; Shen, Y.; et al. Mapping near-surface air temperature, pressure, relative humidity and wind speed over Mainland China with high spatiotemporal resolution. *Adv. Atmos. Sci.* **2014**, *31*, 1127–1135. [CrossRef]
39. Fang, X.; Luo, S.; Lyu, S.; Cheng, C.; Li, Z.; Zhang, S. Numerical modeling of the responses of soil temperature and soil moisture to climate change over the Tibetan Plateau. 1961–2010. *Int. J. Climatol.* **2021**, *41*, 4134–4150. [CrossRef]
40. Pinzon, J.E.; Tucker, C.J. A Non-Stationary 1981–2012 AVHRR NDVI3g Time Series. *Remote Sens.* **2014**, *6*, 6929–6960. [CrossRef]
41. Klene, A.E.; Nelson, F.E.; Shiklomanov, N.I.; Hinkel, K.M. The N-factor in natural landscapes: Variability of air and soil-surface temperatures, Kuparuk River Basin, Alaska, USA. *Arct. Antarct. Alp. Res.* **2001**, *38*, 140–148. [CrossRef]
42. Luo, D.; Jin, H.; Marchenko, S.S.; Romanovsky, V.E. Difference between near-surface air, land surface and ground surface temperatures and their influences on the frozen ground on the Qinghai-Tibe Plateau. *Geoderma* **2018**, *312*, 74–85. [CrossRef]
43. Wang, C.; Jin, S.; Wu, Z.; Cui, Y. Evaluation and Application of the Estimation Methods of Frozen (Thawing) Depth over China. *Adv. Earth Sci.* **2009**, *24*, 132–140.
44. Change, Y.; Lyu, S.; Luo, S.; Li, Z.; Fang, X.; Chen, B.; Li, R.; Chen, S. Estimation of permafrost on the Tibetan Plateau under current and future climate conditions using the CMIP5 data. *Int. J. Climatol.* **2018**, *38*, 5659–5676. [CrossRef]
45. Anisinov, O.A.; Nelson, F.E. Permafrost distribution in the Northern Hemishpere under scenarios of climatic change. *Glob. Planet. Chang.* **1996**, *14*, 59–72. [CrossRef]
46. Slater, A.G.; Lawrence, D.M. Diagnosing Present and Future Permafrost from Climate Models. *J. Clim.* **2013**, *26*, 5608–5623. [CrossRef]
47. Fang, X.; Luo, S.; Lyu, S. Observed soil temperature trends associated with climate change in the Tibetan Plateau, 1960–2014. *Theor. Appl. Climatol.* **2019**, *135*, 169–181. [CrossRef]
48. Swenson, S.; Lawrence, D. A new fractional snow-covered area parameterization for the Community Land Model and its effect on the surface energy balance. *J. Geophys. Res. Atmos.* **2012**, *117*, D21107. [CrossRef]
49. Guo, D.; Wang, H. Simulated change in the near-surface soil freeze/thaw cycle on the Tibetan Plateau from 1981 to 2010. *Chin. Sci. Bull.* **2014**, *59*, 2439–2448. [CrossRef]
50. Fang, X.; Luo, S.; Lyu, S.; Chen, B.; Zhang, Y.; Ma, D.; Chang, Y. A Simulation and Validation of CLM during Freeze-Thaw on the Tibetan Plateau. *Adv. Meteorol.* **2016**, *2016*, 9476098. [CrossRef]

51. Yang, K.; Wang, C. Water storage effect of soil freeze-thaw process and its impacts on soil hydro-thermal regime variations. *Agric. For. Meteorol.* **2019**, *265*, 280–294. [CrossRef]
52. Wu, Q.; Hou, Y.; Yun, H.; Liu, Y. Changes in active-layer thickness and near-surface permafrost between 2002 and 2012 in alpine ecosystems, Qinghai-Xizang (Tibet) Plateau, China. *Glob. Planet. Chang.* **2015**, *124*, 149–155. [CrossRef]
53. Wang, G.; Liu, G.; Li, C.; Yang, Y. The variability of soil thermal and hydrological dynamics with vegetation cover in a permafrost region. *Agric. For. Meteorol.* **2012**, *162–163*, 44–57.
54. Yin, D.; Roderick, M.L. A framework to quantify the inter-annual variation in near-surface air temperature due to change in precipitation in snow-free regions. *Environ. Res. Lett.* **2020**, *15*, 114028. [CrossRef]
55. Lamsal, P.; Kumar, L.; Shabani, F.; Atreya, K. The greening of the Himalayas and Tibetan Plateau under climate change. *Glob. Planet. Chang.* **2017**, *159*, 77–92. [CrossRef]
56. Alessandri, A.; Catalano, F.; Felice, M.D.; Hurk, B.; Balsamo, G. Varying snow and vegetation signatures of surface-albedo feedback on the Northern Hemisphere land warming. *Environ. Res. Lett.* **2021**, *16*, 034023. [CrossRef]
57. Gao, Y.; Li, X.; Leung, L.R.; Chen, D.; Xu, J. Aridity changes in the Tibetan Plateau in a warming climate. *Environ. Res. Lett.* **2015**, *10*, 034013. [CrossRef]
58. Li, D.; Wen, Z.; Cheng, Q.; Xing, A.; Zhang, M.; Li, A. Thermal dynamics of the permafrost active layer under increased precipitation at the Qinghai-Tibet Plateau. *J. Mt. Sci.* **2019**, *16*, 309–322. [CrossRef]
59. Wu, T.; Zhao, L.; Li, R.; Wang, Q.; Xie, C.; Pang, Q. Recent ground surface warming and its effects on permafrost on the central Qinghai-Tibet Plateau. *Int. J. Climatol.* **2013**, *33*, 920–930. [CrossRef]
60. Wang, K.; Zhang, T.; Yang, D. Permafrost dynamics and their hydrologic impacts over the Russian Arctic drainage basin. *Adv. Clim. Chang. Res.* **2021**, *12*, 482–498. [CrossRef]
61. Pattison, R.R.; Jorgenson, J.C.; Raynolds, M.K.; Welker, J.M. Trend in NDVI and tundra community composition in the arctic of NE Alaska between 1984 and 2009. *Ecosystems* **2015**, *18*, 707–719. [CrossRef]
62. Blok, D.; Heijmans, M.M.P.D.; Schaepman-Strub, G.; Kononov, A.V.; Maximov, T.C.; Berendse, F. Shrub expansion may reduce summer permafrost thaw in Siberian tundra. *Glob. Chang. Biol.* **2010**, *16*, 1296–1305. [CrossRef]
63. Smith, S.L.; Lewkowicz, A.G.; Duchesne, C.; Ednie, M. Variability and change in permafrost thermal state in northern Canada. In Proceedings of the 68th Canadian Geotechnical Conference and Seventh Canadian Conference on Permafrost, Québec, QC, Canada, 20–23 September 2015.
64. Lu, Y.; Yang, Y.; Wang, L.; Liu, J. Recent hiatus of Tibetan Plateau vegetation greening and the consequence impact on climate. In Proceedings of the EGU General Assembly 2023, Vienna, Austria, 24–28 April 2023.

remote sensing

MDPI

Article

Creation and Verification of a High-Resolution Multi-Parameter Surface Meteorological Assimilation Dataset for the Tibetan Plateau for 2010–2020 Available Online

Xiaohang Wen [1,2,*], Xian Zhu [3], Maoshan Li [1], Mei Chen [1], Shaobo Zhang [1], Xianyu Yang [1], Zhiyuan Zheng [4,5,6], Yikun Qin [7], Yu Zhang [1] and Shihua Lv [1]

1 Plateau Atmosphere and Environment Key Laboratory of Sichuan Province, School of Atmospheric Sciences, Chengdu University of Information Technology, Chengdu 610225, China; 3220101022@stu.cuit.edu.cn (M.C.)
2 State Key Laboratory of Severe Weather, Chinese Academy of Meteorological Sciences, Beijing 100081, China
3 School of Atmospheric Sciences, Sun Yat-Sen University, Zhuhai 519082, China
4 Zhuhai Branch of State Key Laboratory of Earth Surface Process and Resource Ecology, Advanced Institute of Natural Sciences, Beijing Normal University, Zhuhai 519087, China
5 State Key Laboratory of Earth Surface Processes and Resource Ecology, Beijing Normal University, Beijing 100875, China
6 Southern Marine Science and Engineering Guangdong Laboratory (Zhuhai), Zhuhai 519082, China
7 Beijing XTD Digital Mapping Information Technology Company, Beijing 102200, China
* Correspondence: wxh@cuit.edu.cn

Abstract: The Qinghai–Tibet Plateau (QTP) is a crucial component of the global climate system, influencing the regional and global climate through complex thermal and dynamic mechanisms. The high-altitude region, which is the largest part of the extra-polar cryosphere, encompasses extensive mountain glaciers, permafrost, and seasonally frozen land, making it highly sensitive to global climate change. However, the challenging environmental conditions, such as the harsh terrain and high altitude, coupled with sparse weather station distribution and weak observatory representation, make it difficult to accurately quantify the atmospheric conditions and land–atmosphere coupling systems and their effects on the surrounding areas. To address these challenges, we utilized the Weather Research and Forecasting (WRF) model and a three-dimensional variational (3DVAR) assimilation method to create a high-resolution assimilated dataset (HRAD). The QTP-HRAD, covering the spatial range of 70 to 110°E and 25 to 40°N, was validated using both surface weather station observations and the European Center for Medium-Range Weather Forecasts Reanalysis V5, and can now be utilized for further studies on land–atmosphere interactions, water cycling and radiation energy transfer processes, and extreme weather events in the region.

Keywords: Tibetan Plateau; assimilation dataset; land–atmosphere interaction; WRF model

Citation: Wen, X.; Zhu, X.; Li, M.; Chen, M.; Zhang, S.; Yang, X.; Zheng, Z.; Qin, Y.; Zhang, Y.; Lv, S. Creation and Verification of a High-Resolution Multi-Parameter Surface Meteorological Assimilation Dataset for the Tibetan Plateau for 2010–2020 Available Online. *Remote Sens.* **2023**, *15*, 2906. https://doi.org/10.3390/rs15112906

Academic Editor: Pradeep Wagle

Received: 30 March 2023
Revised: 31 May 2023
Accepted: 31 May 2023
Published: 2 June 2023

1. Introduction

The Qinghai–Tibet Plateau (QTP) is widely recognized as the world's highest and largest plateau, with a highly complex terrain that includes mountain ranges, vast plains, and numerous lakes and rivers. This unique geographical region, sometimes referred to as the "Roof of the World," is situated in the heart of Asia and covers an area of over 2.5 million square kilometers. Due to its high elevation, harsh climate, and unique ecology, the QTP is often referred to as the "third pole" of the Earth, after the North and South Poles. The region plays a crucial role in global climate, as it is a major source of freshwater and a key driver of atmospheric circulation patterns that affect weather patterns around the world [1]. It serves as the origin of numerous major Asian rivers, making it a vital "hydrological tower" in Asia [2,3]. However, the plateau is particularly susceptible to the impacts of global climatic alteration, exhibiting a heightened rate of temperature amplification in comparison to other regions [4,5]. The impacts of global warming on the Tibetan Plateau have become a subject of great scientific interest in recent years [6–8]. Multiple observational and monitoring projects were implemented on the QTP. To better understand the mechanism of the land–atmosphere interaction and land surface processes on the Plateau, researchers have conducted numerical simulation studies [6].

The Tibetan Plateau's thermal impact on the atmosphere significantly impacts the land–atmosphere circulation of China, East Asia, and the entire world [9–16]. The Plateau's cryospheric processes, such as the absorption of solar radiation and seasonal variations in surface heat and water fluxes, lead to complex interactions between the land surface and atmosphere that make it challenging to fully understand and simulate the energy and water cycle of the QTP and its impact on global climate change [10,12,17]. Although there has been a growing amount of research on land–atmosphere interactions over the QTP, observational experiments to determine terrestrial and atmospheric parameters are limited by the harsh environmental conditions, high altitudes, and sparse distribution of weather stations. Additionally, most observational experiments are conducted only in summer at a few locations, leaving gaps in our quantitative understanding of the local land–atmosphere coupling systems and their effects on the surrounding areas [18].

A series of atmospheric field experiments have been conducted over the Tibetan Plateau (TP) since the 1970s. Notable examples include the first Qinghai–Xizang Plateau Meteorology Experiment (QXPMEX) [19], the Global Energy and Water Cycle Exchanges (GEWEX) Asian Monsoon Experiment (GAME)/Tibet intensive observation [20], and the Coordinated Enhanced Observing Period (CEOP) Asia–Australia Monsoon Project on the Tibetan Plateau (CAMP/Tibet) [21]. Through these experiments, several field observational stations were established and kept in operation. After decades of effort and with an optimized scientific design and layout, the level of atmospheric observation has greatly improved with respect to observation infrastructure, technology, and meteorological elements observed. Since 2014, the third Tibetan Plateau Atmospheric Science Experiment (TIPEX-III) has been underway. As part of this experiment, routine automatic sounding systems were deployed in the western plateau region, filling the gap of routine operational sounding stations previously lacking in this area. Additionally, observational networks for soil temperature and moisture in the central and western TP were established. TIPEX-III also conducted plateau and regional-scale boundary layer observations, measured cloud–precipitation microphysical characteristics through the use of multiple radars and aircraft campaigns, and collected tropospheric–stratospheric atmospheric compositions at multiple sites [22].

With the development of numerical simulations and data assimilation techniques, regional numerical models have become widely used for studying climate change over the QTP [23–26]. However, existing numerical models have defects in reflecting the earth–atmosphere coupling process, particularly the physical process of clouds and precipitation, and their influence on complex terrain conditions [27–29]. The simulation of the 2 m temperature over the QTP using regional climate and global circulation models tends to have a consistent underestimation or cold bias. This systematic error affects the accuracy of the climate model simulations for the QTP region [30], and the cold bias of different regional models over the QTP is between −11 °C and −0.3 °C [31–35]. At present, there are large uncertainties in the numerical simulation data of the precipitation in high-altitude mountainous areas. The rugged and steep terrain, location of observation sites, wind-blown snow, and low-temperature environments hinder the establishment of high-quality grid precipitation data [36]. Because of the coarse resolution, model physics, and dynamics, the global climate model showed a substantial bias over the QTP, and the simulated precipitation was significantly overestimated over this area. The dynamic downscaling simulation of precipitation is better than that of the global climate model, but the overestimation of precipitation still exists, with the highest value of 35% [34]. Atmospheric reanalysis data and various remote sensing products in the QTP region also have significant uncertainties [37–40]. These factors cause a low accuracy of regional numerical models in the QTP region, which seriously affects the assessment of climate change in this region [41].

With advances in computational resources, dynamic downscaling has become a candidate for climate studies with small-scale information, which can now be simulated over decades at a 1–10 km resolution [42,43]. The regional model can generate high spatial and temporal resolutions and climate-relevant long-term series of local climatic factors. In addition, the consistency of model physical processes can be used to generate datasets across different climatic scales [44]. Therefore, with the continuous improvement of model resolution and the establishment of a rapid assimilation update cycle system [45], ground, radiosonde, satellite, and aerial observation data with high spatial and temporal resolutions are playing an increasingly important role in assimilation systems [46,47].

In the assimilation process, the assimilation of large amounts of ground observation data not only makes the ground elements of the model closer to the actual ground observations, but also affects the simulation of the boundary layer and above the boundary layer, as well as the vertical motion and structure related to the surface features observed by these ground elements [46,48]. Therefore, effectively integrating multi-source observational data into the numerical model system improves the numerical simulation results of the complex underlying surface of the QTP, enhances the simulation accuracy of the numerical model, and uses the assimilation data results to analyze the occurrence and evolution of extreme temperature and precipitation events. Understanding and exploring the formation mechanism and response to global warming of extreme weather and climate events in the southeastern plateau of the QTP is of significant importance.

For the first time, we generated high spatial and temporal resolution (horizontal 5 × 5 km, once every 1 h) assimilation data in the QTP (QTP-HRAD) over 11 years (2010–2020). The assimilation data include 28 near-surface meteorological variables, including the radiation and energy field variables. The hourly values of the 2 m temperature, land surface temperature, relative humidity, surface pressure, water vapor mixing ratio, 10 m U and V wind components, precipitation, and dew point temperature at 5 × 5 km are available online [49].

Our compiled QTP-HRAD is expected to help more users to understand the spatial–temporal characteristics of key attributes in the QTP region of China and be applied in different fields. In addition, the QTP-HRAD can provide data references for the comprehensive evaluation of plateau ecological environment, extreme weather, and climatology. The flowchart for constructing the QTP-HRAD is shown in Figure 1.

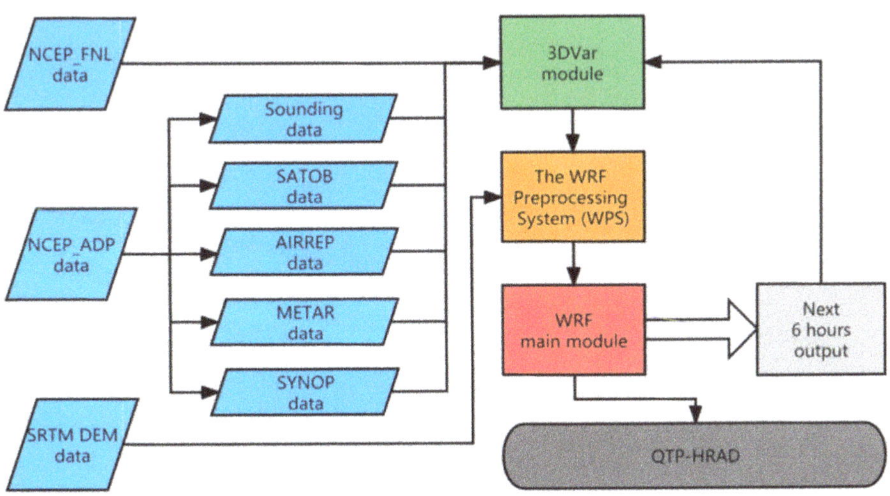

Figure 1. Flowchart for generating the QTP-HRAD.

2. Methods

2.1. Introduction to WRF Model Configuration

This dataset is generated and produced by the Weather Research and Forecasting (WRF) model version 4.2.2 and the three-dimensional variation (3DVAR) module. The atmospheric vertical layer has 35 layers, the atmospheric pressure at the upper boundary of the WRF model is 50 hPa, and the "CONUS" physics suite was selected as the parameterization series scheme of the simulation and assimilation. The "CONUS" series scheme is a combination of physical options that are highly tested and show reasonable results. The configuration is as follows: The microphysics was characterized using the Thompson scheme [50], and the Rapid Radiative Transfer Model for General Circulation Models (RRTMG) [51,52] was utilized to determine the calculation of longwave and shortwave radiations and their atmospheric transfer. The cumulus clouds were modeled using the "Tiedtke and Zhang" convection scheme [53,54]. The Noah land surface model (Noah-LSM) is particularly effective in simulating soil temperature and moisture in multiple layers, which allows for a more precise representation of the vertical structure of the land surface within the WRF, taking into account soil temperature and moisture in four layers, fractional snow cover, and frozen soil physics. This helps to accurately represent processes over ice sheets and snow-covered areas [55]. The Mellor–Yamada–Janjic (MYJ) scheme was utilized to resolve planetary boundary layer processes [56]. The thermal roughness length and standard similarity functions were obtained from look-up tables in the surface layer scheme, which is based on the Monin–Obukhov theory with Zilitinkevich and used in the Eta model [57]. During long simulations, the model offers the option to input time-varying data and continuously update these fields [58,59]. The 12-month values of vegetation fraction and albedo were obtained from the geogrid program to be used in the WRF model, but it does not have the capability of predicting sea surface temperature, vegetation fraction, albedo, or sea ice. To overcome this limitation, time-varying sea surface temperature (SST) and sea ice fields must be read into the model, allowing for updates of these fields during long simulations.

The WRF configuration is presented in Table 1. In this study, a double-nested grid system was utilized, with a horizontal grid spacing of 25 km in domain 1 and 5 km in domain 2 (Figure 2). The two domains used the geographic data resolution of the "usgs_lakes + default" option. The central latitude and longitude of domain 2 are 87.76°E and 33.41°N, and the latitude and longitude of domain 2 ranges from 72.3 to 103.2°E and from 25.9 to 40.3°N, respectively. The National Centers for Environmental Prediction (NCEP) Final Analysis with a 6-h resolution and 1 degree $\times$ 1 degree spatial resolution were used as the initial and lateral boundary conditions for the WRF model. The data were obtained from the Global Forecast System (GFS). The simulation covered a period of 10 years, starting on 1 January 2010, at 00:00 UTC and ending on 31 December 2020, at 23:00 UTC.

2.2. Introduction to NCEP_ADP Data and Assimilation Methods

For the assimilation data source, global observation data from the National Centers for Environmental Prediction (NCEP) Automated Data Processing (ADP) was utilized. The timeframe of this data spans from 1 January 2010 to 31 December 2020, and was incorporated into the WRF-3DVAR system for assimilation.

The NCEP-ADP data contain surface synoptic observations (SYNOP), radio-sounding data (SOUNDING), satellite observations (SATOBs), and a few aircraft reports (AIRREPs) and the Meteorological Terminal Air Report (METAR). Satellite observations (SATOBs) track cloud and wind data using cloud motion obtained using space-borne infrared derivations or water vapor imagers. These data and reports include air pressure, geopotential height, air temperature, dew point temperature, wind direction, and wind speed, with up to 20 layers of upper-air observations available ranging from 1000 to 10 hPa. The reporting intervals range from hourly to 12 h (https://rda.ucar.edu/datasets/ds461.0/, accessed on 31 December 2020).

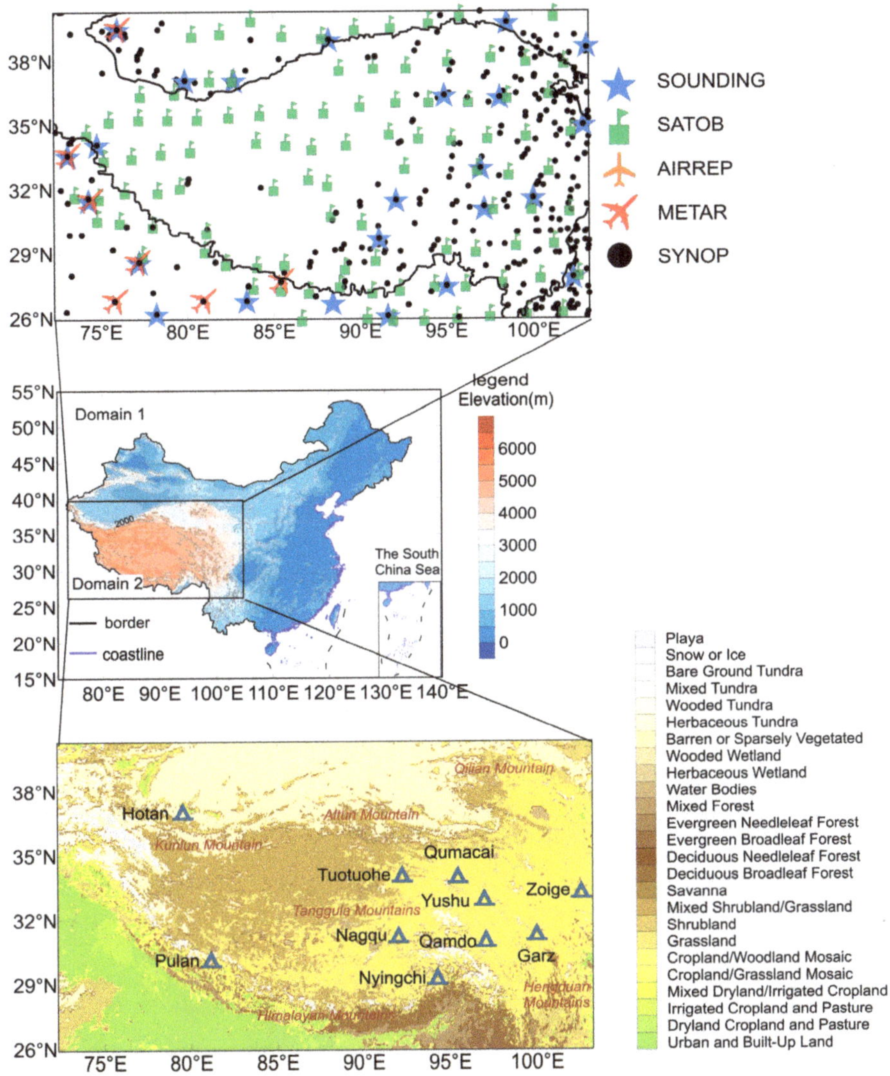

Figure 2. The schematic diagram of simulated region, underlying land use/vegetation types, and the distribution of the NCEP_ADP observation data. There are a total of 10 stations (Qamdo, Garz, Hotan, Nyingchi, Nagqu, Pulan, Qumacai, Zoige, Tuotuohe, and Yushu) at different locations With the blue triangle in and around the QTP as representative stations. The Hotan station stands on the northwest side of the QTP, located in the northwest arid region, and represents the simulated situation in the arid region north of the plateau. Nyingchi and Qamdo stations are located in the semi-humid and humid regions of the southeastern plateau, representing a simulated situation in the humid region of the southeastern plateau. Except for the Hotan station, which has a lower altitude of 1370 m, all other stations are located in the hinterland of the QTP at an altitude of over 3000 m.

Table 1. Summary of the WRF configuration.

Domain Configuration	Domain 1	Domain 2
Horizontal grid spacing	25 km	5 km
Grid dimensions	370 × 270	544 × 304
Time step	180 s	30 s
Central longitude–latitude	103°E, 38°N	87.76°E, 33.41°N
Geog data resolution	usgs_lakes + default	usgs_lakes + default
Model physics		
Microphysics	Thompson scheme	
Cumulus parameterization	Tiedtke scheme	
Longwave and shortwave radiation	RRTMG scheme	
Surface layer	Eta-similarity scheme	
Land surface	Noah land surface model	
Planetary boundary layer	Mellor–Yamada–Janjic scheme	
Other physics options for long simulations	tmn_update: update deep soil temperature sst_skin: calculate skin SST based on Zeng and Beljaars [60] sst_update: update sea-surface temperature, vegetation fraction, albedo	

The NCEP-ADP was used in the NCEP-FNL analysis. However, due to the resolution of FNL data being 1 × 1 degree, it is difficult to reconcile multiple observation data within a grid when assimilating these dense observation data. Therefore, when using the WRF model with 3DVAR method for assimilation, we not only increase the grid resolution, but also consider some control and constraint options when setting assimilation parameters to better iterate and converge. For example:

(1) The multi-level q adjustment should be made while ensuring the conservation of column-integrated water vapor, considering the need to verify the supersaturation (RH > 95%) and the minimum relative humidity (RH < 11%) requirements.

(2) The assimilation of surface observations will rely on the surface similarity theory within the planetary boundary layer instead of using the lowest model level's first guess.

(3) Any stations where the model's interpolated height differs from the actual observation station's height by more than 100 m will be deemed invalid and excluded.

(4) This specifies the method used to interpolate the background into the observation space. The method involves selecting the neighboring model grid point with the smallest height difference, while disregarding the grid points over water.

Figure 2 shows a schematic diagram of the simulated region, underlying land use/vegetation types, and distribution of the NCEP_ADP observation data. It can be seen that the surface SYNOP and SOUNDING sites are mostly distributed in the eastern part of the QTP, whereas there are only a few ground and radio-sounding sites in the western part of the plateau. SATOBs were evenly distributed in the entire simulated region. A few AIRREPs and the METAR are available for the areas south of the Himalayas.

The three-dimensional variational assimilation theory entails the minimization of the cost function as follows [61,62]:

$$J = \frac{1}{2}\left\{\left[H(x) - y^0\right]^T R^{-1}\left[H(x) - y^0\right] + \left(x - x^b\right)^T B^{-1}\left(x - x^b\right)\right\} \quad (1)$$

where y^0 is the observed data, x is the model state, x^b is the background guess, and B is the background error covariance matrix. The first term of the cost function represents the difference between the observed data and the model state, and the second term represents the difference between the model state and the background. The purpose of the assimilation process is to minimize the cost function, which means that the difference between the observed data and the model state is reduced as much as possible, and the background error is controlled [63]. The last stage involves updating the lower boundary and inserting the analysis obtained from the WRF data assimilation system into the WRF model. This study utilizes a cycling assimilation approach that incorporates a quasi-isentropic coordinate for the analysis increment. By utilizing the potential temperature structure around the observation, this method dynamically adjusts the impact of observations. The 6 h update cycle guarantees that the analysis and short-term forecast remain current [46].

Every 6 h, the cycling 3DVAR technique assimilates recent observations by utilizing the previous 6 h model output as a reference to generate a fresh estimate of 3D atmospheric fields. The approach involves analyzing the observation-minus-simulation residuals, or innovations, to obtain a 3D multivariate error field estimate known as the analysis increment. This increment is then combined with the 6 h forecast background to produce the updated analysis. Through the model's filter, the 6 h forecast incorporates insights from previous observations into the current analysis [46]. The 3DVAR system operates in cycling mode, which allows the background cloud water and rainwater from the previous cycle to be incorporated into the current cycle. By including these variables in the 3DVAR system, their information can be transferred to the subsequent cycle, reducing the time required for the model's cloud water and rainwater spin-up during the integration from the analysis stage [64]. At the same time, the "SST update" option is turned on to update the lower boundary conditions in the WRF model, including sea surface temperature (SST), background albedo (ALBBCK), green vegetation fraction (GVF), and underlying soil temperature (TMN) once every 6 h. These variables are updated to the forecast file via "warm start" after the assimilation.

A schematic diagram of the cycling assimilation process is shown in Figure 3.

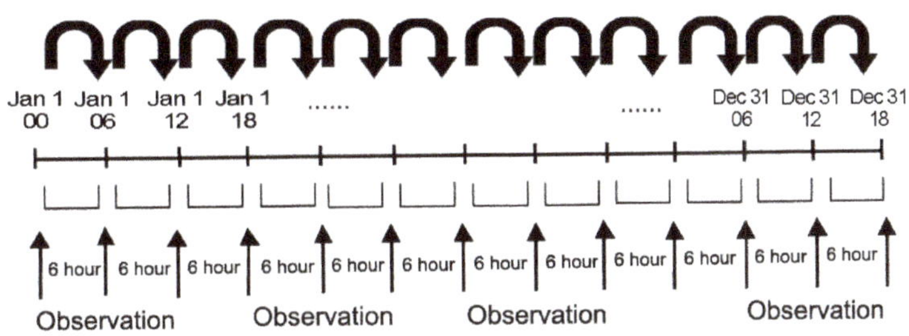

Figure 3. The schematic diagram of cycling assimilation data in the WRF model. The bottom brackets mean every cycle of assimilation, and the bold arrows mean the observations are assimilated from the time window "00", "06", "12", and "18".

2.3. Introduction to Observational Data and ERA-5 Data

The National Meteorological Information Center (NMIC) of the China Meteorological Administration (CMA) published the China surface climatic daily dataset (V3.0) on http://data.cma.cn/, accessed on 29 March 2023. This dataset comprises daily meteorological observations from 91 stations located in the simulated QTP region. This dataset includes the daily averaged surface pressure (hPa), 2 m temperature (°C), 2 m relative humidity (%), precipitation (mm), evaporation (mm), 10 m wind direction (°), wind speed (m/s), and ground surface temperature (°C) from 1 January 2010 to 31 December 2019. Before being utilized, the NMIC conducts quality control and inspection on the data. We used these observation data for the model evaluation.

The European Center for Medium-Range Weather Forecasts Reanalysis V5 (ERA-5) reanalysis data were used to provide additional information on atmospheric variables such as temperature, humidity, wind speed, wind direction, and precipitation in the QTP region [65,66]. These data were used in conjunction with the China surface climatic daily dataset to evaluate the performance of the WRF model simulations. To encompass the timeframe from January 2010 to December 2020, monthly mean data from ERA-5 were employed in this investigation. These data cover a 30 km grid and span 137 atmospheric levels, up to 80 km in altitude from the surface [67].

The QTP-HRAD from the WRF model was averaged to a daily time scale for comparison and error analysis with observations from meteorological stations. We also compared the QTP-HRAD with the gridded dataset of ERA-5, comparing variables such as temperature, 2 m temperature, surface temperature, surface pressure, 2 m specific humidity, 10 m wind speed, and total precipitation. We used patch interpolation and the Earth System Modeling Framework (ESMF) re-gridding toolbox in the National Center for Atmospheric Research (NCAR) Command Language (NCL) (https://www.ncl.ucar.edu/Document/Functions/ESMF/ESMF_regrid.shtml, accessed on 8 June 2022) and calculated the difference between the two datasets.

3. Results

3.1. Comparison of Observations and QTP-HRAD Simulations from 10 Weather Stations

We selected 10 stations (Qamdo, Garz, Hotan, Nyingchi, Nagqu, Pulan, Qumacai, Zoige, Tuotuohe, and Yushu) at different locations in and around the QTP as representative stations to compare the daily mean values of the 2 m temperature, relative humidity, precipitation, and evaporation.

Figure 4 shows the time series comparison of the 2 m temperature and relative humidity observations and the QTP-HRAD data at 10 stations from 2010 to 2019. It can be seen that the observed values of the 2 m temperature at the Garz, Nyingchi, and Pulan stations are all higher than the simulated values throughout the year, and the observed values at the other stations are in good agreement with the QTP-HRAD values. It can also be seen from the error analysis table (Table 2) that the RMSE between the observed and simulated daily mean 2 m temperatures of the Garz, Nyingchi, and Pulan stations are higher than 6.8 °C, and the bias is below −6.6 °C. The RMSE and bias of the other stations between the observations and simulations were lower in the Hotan, Nagqu, Zoige, and Tuotuohe stations, at approximately 2 °C (RMSE)and between −1 and 1 °C (bias), respectively. The correlation coefficients of the 2 m temperature between the observed and simulated values were all from 0.95 to 0.99 at these 10 stations, which all passed the 95% significance test.

The simulated value of the relative humidity also varied annually. Except for the Hotan station, the simulated values of the other stations showed that the relative humidity peaks in the summer and valleys in the winter, mainly due to the higher temperature in the summer and more water vapor evaporating into the air. The Hotan station is located at the edge of the Taklimakan Desert. Owing to the lack of adequate water supply on the underlying surface, less water vapor evaporates into the atmosphere in the summer. The relative humidity increases in the winter when the temperatures are lower, and therefore, the annual variation in the relative humidity and temperature is reversed. Both the observed and simulated results reflect this trend. The average annual relative humidity and specific humidity at the Hotan station are 37% and 4.1 g/kg, respectively, which are much lower than those at the other stations. The statistical error results (Table 2) show that the RMSE and bias of the specific humidity in the Hotan station are the lowest among the 10 observation stations (0.75 and 0.26 g/kg, respectively). The Garz station, located in southeast Tibet, has the highest annual average relative humidity of 79% and a specific humidity of 5.1 g/kg.

In general, the WRF model showed a good simulation effect on the 2 m temperature and relative humidity in the QTP region. The daily averaged simulated values were consistent with the observed values, and accurately reflected the annual and interannual variations in the temperature and humidity. For some stations located in the complex terrain area, the simulation errors for the temperature and humidity were slightly larger than those in the flat terrain area. As the Garz and Nyingchi stations are located in the Hengduan Mountains in the southeast of the QTP, and the Pulan station is located in the Ali Plateau in the southwest of the QTP, the complex and steep terrain caused a large 2 m temperature simulation error. Other stations such as Hotan are located in the Tarim Basin north of the Kunlun Mountains on the northern side of the QTP. The Zoige station is located in the flat Zoige Wetland in the east of the QTP, and the Nagqu station is positioned in the northern QTP on hilly terrain that lies between the Tanggula Mountains, Nyainqentanglha Mountains, and Gangdise Mountains. The terrain of the area gently slopes.

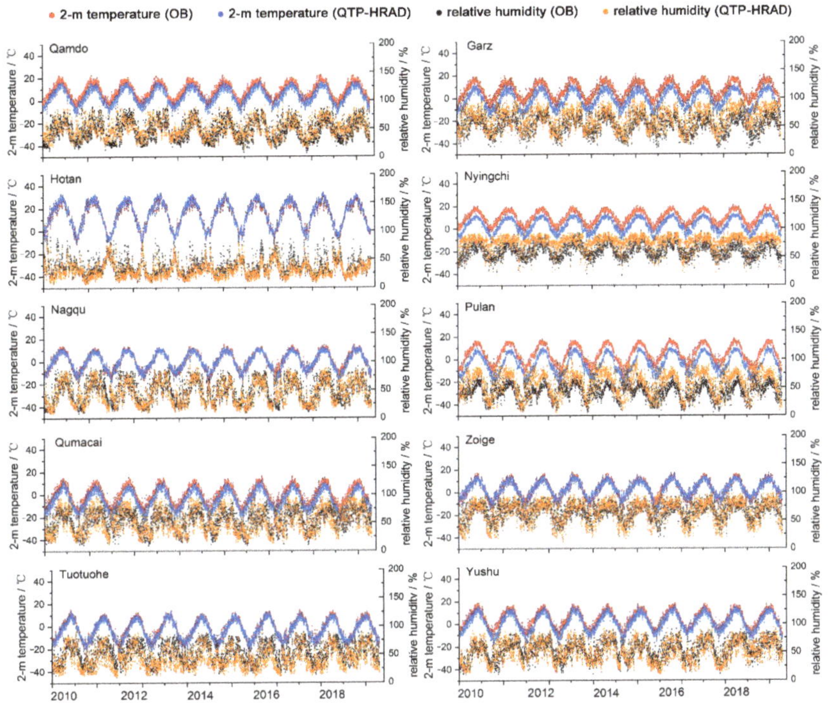

Figure 4. Comparison of observations and QTP-HRAD simulations for daily averaged 2 m temperature (°C) and relative humidity (%) of 10 weather stations in QTP region from 2010 to 2019. The left vertical axis shows the range of 2 m temperature value and the right vertical axis shows the range of relative humidity value.

Table 2. RMSE, bias, and correlation coefficient of the daily averaged 2 m temperature, specific humidity, surface pressure, and 10 m wind speed values between observations and QTP-HRAD from 10 weather stations; * means the correlation coefficient passed the significance *t*-test level of 0.05.

Weather Stations	2 m Temperature (°C)			Specific Humidity (g/kg)			Surface Pressure (hPa)			10 m Wind Speed (m/s)		
	RMSE	Bias	R	RMSE	Bias	R	RMSE	Bias	R	RMSE	Bias	R
Qamdo	4.88	−4.51	0.96 *	1.07	−0.51	0.95 *	59.9	−59.86	0.89 *	0.22	1.65	0.22 *
Garz	8.39	−8.11	0.96 *	1.14	−0.82	0.96 *	93.9	−93.84	0.85 *	1.72	1.29	0.22 *
Hotan	1.35	0.36	0.99 *	0.75	−0.26	0.96 *	11.56	−11.45	0.96 *	1.73	1.34	0.51 *
Nyingchi	6.84	−6.61	0.95 *	0.90	−0.48	0.97 *	74.93	−74.89	0.79 *	1.151	0.536	−0.15
Nagqu	2.01	−0.54	0.97 *	0.77	−0.39	0.97 *	9.08	−9.03	0.97 *	1.54	1.16	0.54 *
Pulan	8.66	−8.24	0.95 *	0.81	−0.46	0.97 *	62.48	−62.44	0.85 *	1.259	0.117	0.003
Qumacai	4.97	−4.43	0.96 *	1.18	−0.89	0.96 *	52.21	−52.16	0.95 *	2.59	2.0	0.23 *
Zoige	2.24	−0.31	0.96 *	0.73	−0.22	0.97 *	8.24	−8.153	0.96 *	1.67	1.25	0.33 *
Tuotuohe	2.19	−1.00	0.97 *	1.24	−0.93	0.95 *	20.48	−20.46	0.97 *	1.38	0.44	0.56 *
Yushu	3.68	−2.95	0.96 *	1.08	−0.78	0.97 *	39.50	−39.47	0.90 *	1.67	1.41	0.39 *

Figure 5 shows a comparison of the daily total precipitation and the daily mean actual evaporation at 10 stations in the QTP region. The red and blue column lines represent the observed and simulated daily total precipitation, respectively, and the gray line represents the actual simulated daily evaporation. As no actual evaporation was observed, only the simulated evaporation was plotted on the graph. It can be seen that the annual variations of the simulated and observed precipitation were relatively consistent, with more precipitation in the summer and less in the winter. Except for the Hotan station, the maximum daily precipitation of the other stations can reach between 30 and 40 mm in the summer, or even 50 mm, which indicates heavy rain and rainstorm. The simulated value reflects the peak value of summer precipitation. At the Pulan station, more precipitation days were observed than those that were simulated. At the Nyingchi station, the observed precipitation in the winter was slightly greater than the simulated value. The annual evaporation was positively correlated with the precipitation. The daily average value of evaporation in the summer is approximately 2–4 mm/d, and the evaporation in the winter tends to be zero. In general, the stations in the eastern and southern parts of the QTP, such as the Qamdo, Garz, Nyingchi, and Zoige stations, have more precipitation and evaporation, while the western stations and the stations in the arid region on the north side of the plateau, such as Hotan and Tuotuohe, have less precipitation and evaporation. The variation trend of precipitation was simulated well by the WRF model, which was in good agreement with the observed values.

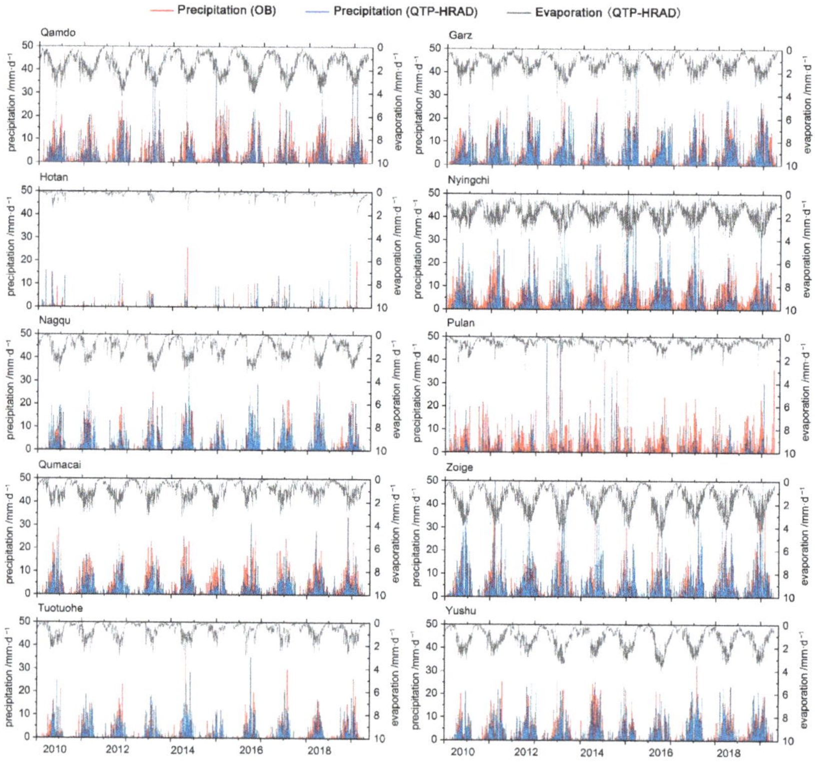

Figure 5. Comparison of observations and QTP-HRAD simulations for daily precipitation (mm) and evaporation (mm) from 10 weather stations in the QTP region from 2010 to 2019. The left vertical axis shows the range of precipitation values, and the right vertical axis shows the range of evaporation values.

The observed and simulated hourly values of 2 m air temperature, dew point temperature, and precipitation at 10 representative stations on the plateau were also compared (figure omitted). The hourly simulated values of some stations are closer to the observed values, such as in the Hotan, Nagqu, Zoige, Tuotuohe, and Yushu stations, but the simulated values of some stations are higher than the observed values by about 3 to 5 °C, such as the Garz, Nyingchi, and Pulan stations. The simulated hourly precipitation can also reflect sporadic precipitation events over the plateau in the summer. At the Hotan and Pulan stations, precipitation events rarely occur, and the simulated value is almost zero.

The large bias between the hourly simulation values and the observation values at stations such as Garz, Nyingchi, and Pulan is caused by the large difference between the model grid altitude and the station altitude under complex terrain. Therefore, it is recommended to make corrections to the "pressure-altitude" height difference in order to make better comparisons between the assimilation data and the station observation data.

It can be seen from Table 2 that the RMSE between the observed and simulated 2 m temperatures at the Garz, Pulan, and Nyingchi stations are higher at 8.39, 8.66, and 6.84 °C, respectively, and the bias is negative, indicating that the simulated values are lower than the observed values. Nevertheless, the correlation coefficient was above 0.95 at these 10 stations and passed the 95% confidence test, indicating that the daily variations in the simulated and observed values are in good agreement. The RMSE between the observed and simulated values of the specific humidity at the 10 stations was all below 1.24 g/kg, and the bias was also negative. The minimum bias value appeared in Zoige (−0.22 g/kg). The error statistics of the surface pressure show that the observed and simulated values of the RMSE and bias in the Nagqu, Hotan, and Zoige stations are lower than those at the other stations, which are approximately 10 hPa, and the RMSE and bias between the simulated and observed values of the 2 m temperature and specific humidity were also lower in these three stations. The RMSE between the simulated and observed surface pressure in the Garz, Pulan, and Nyingchi stations were higher, which was above 60 hPa, and the simulated 2 m temperature and specific humidity also showed a larger bias compared with the observations. This may be due to the large difference in altitude between the observation site and the grid corresponding to the model, which increases the simulation errors of other meteorological elements, such as temperature and humidity. The simulated value can be corrected using the actual altitude of the observation site to reduce the errors. From the simulation error results of the 10 m wind speed, the correlation coefficient of only two sites failed the 95% confidence test, and the RMSE and bias of the simulated and observed values of the 10 sites were between 1 and 2 m/s. For high altitude complex terrain mountain areas, the location and orientation of the weather stations, as well as the type of underlying surface, can significantly impact the simulation accuracy of the temperature, humidity, wind direction, and wind speed. Table 2 shows that the root mean square error (RMSE) and bias of the temperature, humidity, and wind speed are also significant when there is a large difference between the simulated and observed atmospheric pressure values at the stations. For instance, the bias of the observed and simulated surface pressure values at the Garz, Nyingchi, and Pulan stations reaches −93.84, −74.89, and −62.44 hPa, respectively. Meanwhile, the bias of the observed and simulated 2 m temperature values at the three stations also reached −8.11, −6.61, and −8.24 °C, respectively, with the RMSE values of 8.39, 6.84, and 8.66 °C. This may be due to the significant difference in elevation between the simulated grid and the observation site. Despite generating 5 × 5 km horizontal simulation data in the QTP region, the grid distance of 5 × 5 km is still relatively coarse compared to these sites. At the Hotan, Nagqu, and Zoige stations, there was little difference between the simulated surface pressure and observed values, and the simulated 2 m temperature, relative humidity, and wind speed were also consistent with the observed values. The RMSE and bias of the 2 m temperature and humidity were the lowest among the ten representative stations. This is likely because the three stations are not obstructed by high mountains and the terrain is relatively flat, and the elevation of the observation stations is close to that of the model grid points. When using these simulation data, it is better to correct the altitude to reduce the error.

3.2. Comparison of Observations and QTP-HRAD over QTP Region

We selected 91 meteorological stations located in the QTP from 2010 to 2019 to compare the statistical errors between the simulated and observed daily mean values of the 2 m temperature, surface temperature, surface pressure, relative humidity, and 10 m wind speed, and plotted the regional distribution of the correlation coefficients, RMSE, and bias. Since most observation stations are located in the eastern part of the plateau, with fewer in the north of the Kunlun Mountain and Altun Mountain, there are almost no meteorological observation stations in the central and western parts of the plateau.

It can be seen from Figure 6a that the correlation coefficients of the 2 m temperature between the observed and simulated values are all above 0.9, and those in the north of the plateau are above 0.97. The RMSE ranged from 1.4 to 12 °C. The RMSE of some stations in the eastern and northern parts of the QTP was lower, while the RMSE of stations along the Himalayan Mountains, Tanggula Mountains, and Qilian Mountains was higher. This may be caused by the large difference between the station altitude and the model grid altitude owing to the complex topography of the mountains. A total of 70% of the stations (64 stations) had negative bias values, indicating that the simulated values were lower than the observed values, and only some stations in the southeastern plateau and north of the Kunlun Mountains had a positive bias. The statistical error distribution of the land surface temperature was similar to the 2 m temperature (Figure 6b), and the correlation coefficient between the simulations and the observations was slightly lower than that of the 2 m temperature. However, the correlation coefficient for almost all stations exceeded 0.9. The RMSE and bias of the surface temperature were larger than those of the 2 m temperature. The RMSE of 70% of the stations (64 stations) was above 4 °C, and the bias of 85% of the stations (76 stations) was negative. This may be due to the underestimation of the air and surface temperatures in the QTP by the NCEP data of the forcing field in the WRF model. Studies have shown that the temperature in the QTP is underestimated in the NCEP data [68,69]. According to the research conducted by Frauenfeld et al. [70], there was an approximate difference of 7 °C between the ERA-40 reanalysis data and the observations. Similarly, Baolin et al. [71] discovered that the NCEP reanalysis data significantly underestimated the observational data for the QTP.

As shown in Figure 6c, the correlation coefficient between the observed and simulated values of the surface pressure was above 0.9. There were 17 stations with the RMSE exceeding 80 hPa, accounting for 19% of all stations, and only 15 stations had an RMSE between 1 and 10 hPa, accounting for 16.7% of the total. The distribution of bias also showed that the values of the simulated surface pressure at 68 stations were lower than the observed values, and 50% of the stations had a bias between −20 and −75 hPa. The low simulated surface temperature values can be explained by the simulation error of the surface pressure. The average elevation of the model grid cell was higher than the elevation of the observation station point, which led to a low simulated surface pressure and low simulated 2 m and surface temperatures.

The correlation coefficients between the observed and simulated values of relative humidity between 0.35 and 0.9, all passing the 95% significance level test, are illustrated in Figure 6d. The RMSE with 60 stations (66.7%) was less than 15%, and only 14 stations had an RMSE between 17% and 30%. The bias of stations along the Himalayan Mountains and near the Hengduan Mountains was positive (30 stations, accounting for 33%), indicating that the simulated value was higher than the observed value. The simulation error of the 10 m wind speed showed that the simulated value of the wind speed at most stations was approximately 2 m/s higher, and 21 stations had a correlation coefficient higher than 0.5, accounting for 24% of the total (Figure 6e). The RMSE ranged between 0.5 and 3.2 m/s, and the bias error statistics showed that the bias in most of the stations was positive (60 sites), and the simulated value was higher than the observed value. The simulated wind speed in the WRF model was high, possibly because the model underestimated the roughness of the underlying surface. This is because the simulation of the wind speed requires the consideration of many parameterization schemes, such as turbulence parameterization schemes and boundary layer parameterization schemes, and the inaccurate estimation of parameters, such as surface roughness, within these schemes can cause errors. Due to the uncertainty and randomness of atmospheric turbulence, the current mesoscale meteorological models have a lower simulation accuracy for the wind speed compared to other variables such as temperature, humidity, and atmospheric pressure [72]. Due to the complex terrain and the limited data input into the meteorological model, the deviation of the average wind speed between the simulation and observation can reach 2 m/s, and the RMSE can be as high as 6 m/s. Even with better models, reducing uncertainties in the simulations of near-surface wind speed and direction can be challenging, as they stem from random or turbulent fluctuations. Furthermore, the randomness is compounded by variations in sub-grid topography and land use [72–74].

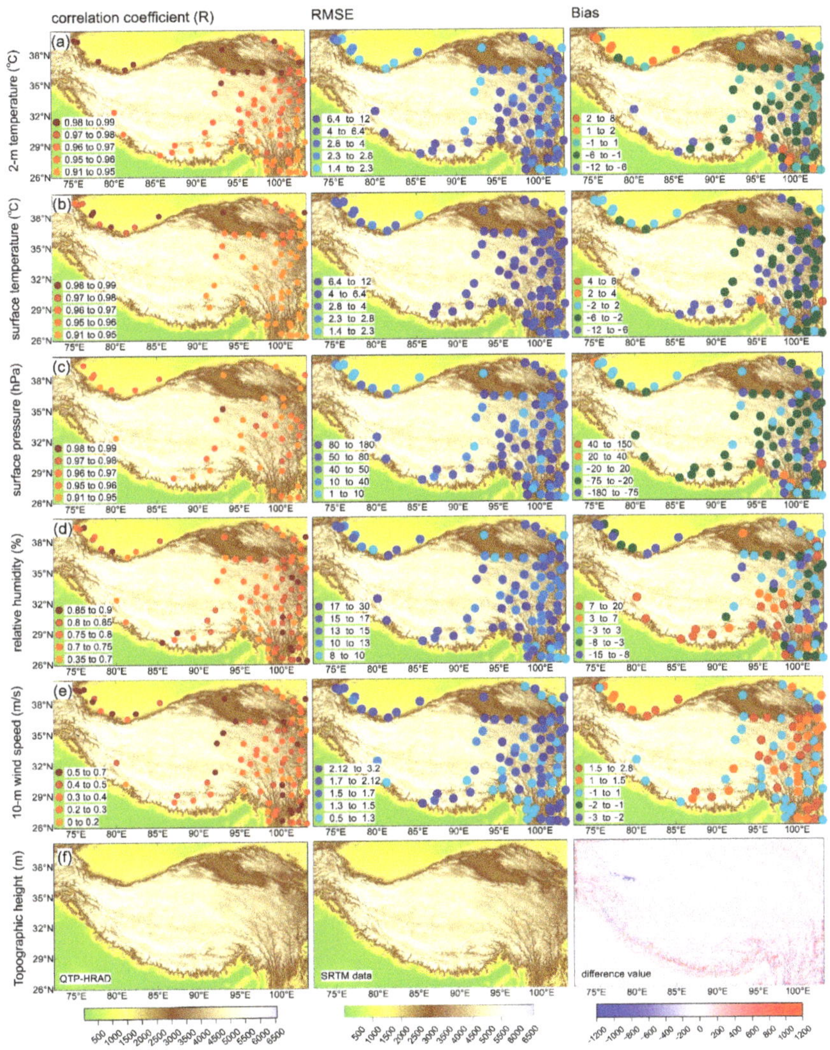

Figure 6. Correlation coefficient, RMSE, and bias between observation and the QTP-HRAD daily mean (**a**) 2 m temperature (°C), (**b**) land surface temperature (°C), (**c**) surface pressure (hPa), (**d**) relative humidity (%), and (**e**) 10 m wind speed (m/s) from 2010 to 2019. (**f**) Topographic height from QTP-HRAD, SRTM data, and the difference value.

Overall, the QTP-HRAD showed good results for the QTP. Except for the wind speed, the correlation coefficients between the simulated and observed values of the temperature, humidity, and surface pressure were higher than 0.8. However, the QTP-HRAD underestimated the air temperature and surface temperature in the QTP region, and underestimated the surface pressure at most observation stations, with simulated biases ranging from

−20 to −180 hPa. The model overestimated the relative humidity at the southern and southeastern stations and underestimated the relative humidity in the northern part of the plateau. The simulated wind speed value of the QTP-HRAD was higher, and the bias was between 1 and 2.8 m/s.

3.3. Comparison of QTP-HRAD and ERA-5 Data over QTP Region

We compared the QTP-HRAD data with the ERA-5 data for 2020 and calculated the difference between the QTP-HRAD and ERA-5 data in the QTP. To calculate the difference, we used bilinear interpolation to scale up the QTP-HRAD data from 5 × 5 km to 0.25° × 0.25°, corresponding to the grid size of the ERA-5 data, and then subtracted. Six meteorological elements, including 2 m temperature, land surface temperature, surface pressure, 2 m specific humidity, 10 m wind speed, and annual accumulated precipitation were compared.

Figure 7 shows the spatial distribution of the QTP-HRAD and ERA-5. As can be seen from Figure 7, the spatial distribution characteristics of the meteorological elements of the QTP-HRAD are consistent with those of the ERA-5, which reflects the distribution of the 2 m temperature, surface pressure, specific humidity, and 10 m wind speed over the QTP. As the spatial resolution of the QTP-HRAD is 5 × 5 km, the distribution of near-surface meteorological elements is finer than that of the ERA-5, which can better demonstrate the differences caused by large topographic fluctuations.

The spatial distribution of the 2 m temperature data is depicted in Figure 7a. In the plateau, the annual average temperature at this height is below 0 °C, with the western part of the plateau having a temperature of around −10 °C, and the eastern part having a temperature of around 0 °C. The 2 m temperature of the ERA-5 is more consistent with the spatial distribution of the QTP-HRAD, but is not as fine as that of the QTP-HRAD. The difference diagram reveals that, with the exception of the central and western regions of the plateau, the difference value is around −4 °C and negative, while the rest of the regions show positive values. The spatial distribution of the surface temperature is comparable to that of the 2 m air temperature. Both the QTP-HRAD and ERA-5 are capable of capturing the extremely low surface temperatures in the western part of the plateau, near the Karakoram Mountains, which are around −15 °C. As the average altitude of the Karakoram Mountains is above 5500 m, the annual 0 °C isotherm is approximately the same as the 4200 m contour line, which is caused by low temperatures throughout the year in the vast mountainous area. From the difference map of the surface temperature, it can be seen that the difference between the central and western parts of the plateau and the Qaidam Basin in Qinghai Province was also found to be negative, and the lowest value was under −5 °C. This value was positive in the central and northern parts of the plateau and the southeastern Hengduan Mountains (6–10 °C).

The regional distribution of the annual average surface pressure is as follows: the surface pressure in the plateau area is low (below 600 hPa), where the lowest in the central and western parts of the plateau can reach 500 hPa, and the surface pressure in the Qaidam Basin on the north side of the plateau is approximately 700 hPa. The areas with significant discrepancies between the QTP-HRAD and ERA-5 are primarily located along the southern Himalayas and the northern Kunlun Mountains, Altun Mountains, and Qilian Mountains. The difference in surface pressure is between −50 and 50 hPa (Figure 7c). The 2 m specific humidity of the QTP-HRAD is a high value area in the south of the Himalayan Mountains, above 10 g/kg, while the annual average specific humidity in the plateau area is approximately 5 g/kg. The specific humidity of the ERA-5 is lower than that of the QTP-HRAD in the plateau area at approximately 3 g/kg, and the difference value is positive (Figure 7d). The 10 m wind speed of the QTP-HRAD in the eastern part of the plateau is below 3 m/s, and the wind speed in other parts of the plateau is higher. The ERA-5 data only reflect that the wind speed in the central and western regions of the plateau is higher than 3 m/s, whereas the wind speed in the Pamir Plateau and Hengduan Mountains is lower than 1 m/s (Figure 7e). As there are only a few surface observation stations in the central and western parts of the plateau, the QTP-HRAD and ERA-5 data have large differences in the 10 m wind speed simulation of the Pamir region, and more observational data need to be obtained in the future for verification. The spatial distribution of the total annual precipitation is shown in Figure 7f. Along the Himalayan Mountains, the precipitation on the southern side can reach more than 2000 mm/year, which occurs on the windward slope and Hengduan Mountains in the southern QTP, while the precipitation in the hinterland of the plateau is less than 200 mm/year. There is slightly more precipitation in the eastern part of the plateau, which is between 500 and 1000 mm/year, and the difference between these two datasets is positive in the southeastern part of the plateau, and negative in other areas, indicating that the total annual precipitation simulated by the QTP-HRAD is higher than that of the ERA-5 (Figure 7f).

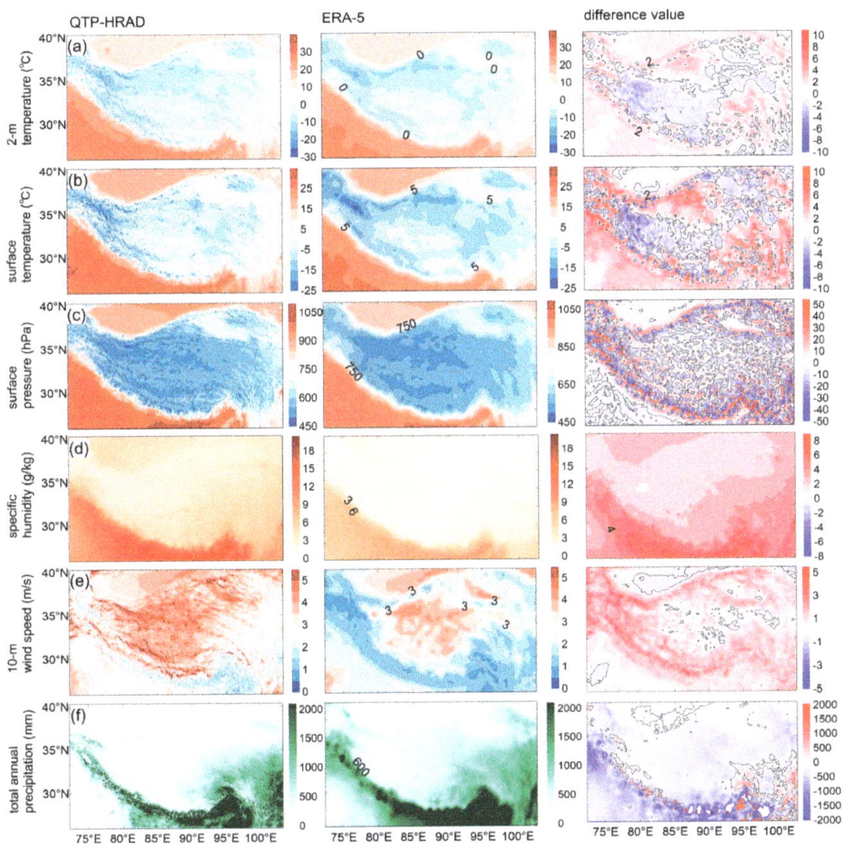

Figure 7. Spatial distribution of the QTP-HRAD (left column), ERA-5 (middle column), and their difference values (right column) over QTP region in 2020. The bold black line is the zero contour line. (**a**) Annual mean 2 m temperature (°C); (**b**) annual mean land surface temperature (°C); (**c**) surface pressure (hPa); (**d**) specific humidity (g/kg); (**e**) 10 m wind speed (m/s); and (**f**) total annual precipitation (mm).

4. Data Records and Availability

The assimilation data in the QTP were generated with a spatial resolution of 5×5 km at a temporal resolution of once every hour from 2010 to 2020. The assimilation data include 28 near-surface meteorological variables, including the radiation and energy field variables. The latitude and longitude of the dataset area range from 72.3°E to 103.2°E and from 25.9°N to 40.3°N. This dataset began at 00:00 (UTC) on 1 January 2010 and ended at 23:00 (UTC) on 31 December 2020, which includes the 2 m temperature (K), land surface temperature (K), relative humidity (%), surface pressure (Pa), 2 m water vapor mixing ratio (kg/kg), 10 m U and V wind components (m/s), 10 m wind speed (m/s), wind direction (°), precipitation (mm), and dew point temperature (°C), are available online (https://doi.org/10.57760/sciencedb.01840, accessed on 31 December 2020).

The structure of the dataset is as follows:

<Year-Short_name>.nc. Here, "Year" represents the data of a certain year, and "Short_name" represents the short name of the variable as presented in Table 3. The "Short_Name", "Long_name", "Missing_value", and the "Unit" of the variables are also summarized in Table 3. From 2010 to 2020, there are 8760 times per variable in each year, and 8784 times in leap years (2012, 2016, and 2020), and the variables are double-precision floating-point data (float type).

Table 3. The description and spatiotemporal variables of QTP-HRAD.

Data Description				
Data type	Gridded			
Projection	Mercator projection			
Horizontal coverage	The QTP region, 72.3°E to 103.2°E and 25.9°N to 40.3°N			
Horizontal resolution	0.05° × 0.05°, about 5 × 5 km			
Vertical coverage	Land surface, 2 m and 10 m above the surface level			
Temporal coverage	Began at 00:00 (UTC) 1 January 2010 and ended at 23:00 (UTC) 31 December 2020			
Temporal resolution	Hourly			
File format	NETCDF			
Variables of QTP-HRAD				
Variables	Short_name	Long_name	Missing_value	Unit
2 m temperature	t2	TEMP at 2 m	1×10^{30}	K
Land surface temperature	tsk	Surface skin temperature	1×10^{30}	K
2 m relative humidity	rh2	Relative Humidity at 2 m	1×10^{30}	%
Surface pressure	prs	Sfc pressure	1×10^{30}	Pa
2 m water vapor mixing ratio	q2	QV at 2 m	1×10^{30}	$kg \cdot kg^{-1}$
10 m U wind components	u10	U at 10 m	1×10^{30}	$m \cdot s^{-1}$
10 m V wind components	v10	V at 10 m	1×10^{30}	$m \cdot s^{-1}$
10 m wind speed	ws10	Wind speed at 10 m	1×10^{30}	$m \cdot s^{-1}$
10 m wind direction	wd10	Wind direction at 10 m	1×10^{30}	Degrees
Precipitation	pre	Accumulated total cumulus precipitation, accumulated total grid scale precipitation	1×10^{30}, 1×10^{30}	mm, mm
2 m dew point temperature	td2	Dewpoint temperature at 2 m	1×10^{30}	°C

5. Conclusions

The WRF model and its assimilation system can precisely replicate the near-surface meteorological parameters in the QTP region, and the simulation data time series demonstrated good correspondence with the observed data. The simulation results reflect the interannual variations and regional distribution characteristics of meteorological field elements such as temperature, humidity, precipitation, evaporation, air pressure, and wind speed. A comparison between the observations and simulations shows that the simulated values are more consistent with the observations in flat areas, whereas the statistical error is larger in areas with complex terrain conditions. The main conclusions of this study are as follows:

1. We selected 10 representative weather stations in the QTP region and compared the observed and simulated values of the 2 m temperature, relative humidity, and precipitation for the period 2011–2020. It was found that the simulated surface pressure bias and RMSEs of the stations located in complex mountainous and terrain areas on the plateau were significantly different from the observed values. The bias between the observed and simulated surface pressure values at the Garz, Nyingchi, and Pulan stations were −93.84 hPa, −74.89 hPa, and −62.44 hPa, respectively, and the temperature and relative humidity bias of these sites were also large. However, at the stations with relatively flat terrain on the plateau such as Hotan, Nagqu, and Zoige, the disparities between the simulated and observed surface pressure values were minor, and the simulated values of the 2 m temperature, relative humidity, and wind speed were mostly in agreement with the observed values.
2. The annual variations in the simulated and observed precipitation are relatively consistent. The simulated values can reflect the peak periods and magnitudes of summer precipitation on the plateau. The daily average evaporation and precipitation at the 10 stations show a significant positive correlation. The daily average evaporation in the summer is about 2–4 mm/d, while in the winter, it tends to be zero. Overall, there is more precipitation and evaporation at the stations in the eastern and southern parts of the QTP, such as Qamdo, Garz, Nyingchi, and Zoige, while the stations in the western and northern arid areas of the plateau, such as Hotan and Tuotuohe, have less precipitation and evaporation.
3. An error statistical analysis was conducted on the observed and simulated values of the 2 m temperature, surface temperature, surface pressure, relative humidity, and 10 m wind speed at all 91 meteorological stations in the QTP region. It was found that the correlation coefficients of the 2 m temperature, surface temperature, and surface pressure were relatively high, all above 0.9. The correlation coefficient between the observed and simulated values of relative humidity was around 0.7 for most stations on the plateau, with positive bias at the stations in the southern part of the plateau and negative bias at the stations in the northern part. The correlation coefficient between the observed and simulated values of wind speed was relatively low, around 0.5, with positive bias at most stations on the plateau, indicating that the simulated values of the near-surface wind speed were higher than the observed values.
4. We also compared the QTP-HRAD with ERA-5 on a regional scale and calculated their differences. The spatial distribution patterns of the two datasets for the 2 m temperature and surface temperature are quite consistent, with the QTP-HRAD values slightly lower than the ERA-5 in the southwest region of the plateau. The difference in the surface pressure between the two datasets shows clear positive and negative differences along the plateau boundary. The QTP-HRAD simulated specific humidity is about 2–5 g/kg higher than the ERA-5. The simulated values of the 10 m wind speed near the ground are also higher than the ERA-5, except for the central plateau area, with a difference of 1–3 m/s. The simulated precipitation amount shows the largest difference along the southern side of the plateau, particularly along the Himalayan range, and the annual total simulated precipitation amount by the QTP-HRAD is higher than that of the ERA-5.

When using the QTP-HRAD, it should be noted that the grid data near the Himalayan Mountains in the south, the Hengduan Mountains in the southeast, and the Kunlun Mountains in the northwest of the plateau may be inaccurate, which is the limitation of using this dataset.

Author Contributions: Conceptualization, X.W. and S.L.; methodology, M.L.; software, S.Z.; validation, X.Y. and Z.Z.; formal analysis, M.C.; investigation, Y.Z.; writing—original draft preparation, X.W.; writing—review and editing, X.Z. and Y.Q. All authors have read and agreed to the published version of the manuscript.

Funding: The Second Tibetan Plateau Scientific Expedition and Research Program (STEP) (grant number 2019QZKK010304), the National Natural Science Foundation of China (grant number 41975096 and 41975130), the Innovation Team Fund of Southwest Regional Meteorological Center, the China Meteorological Administration (unnumbered), the Science and Technology Program of Sichuan Province (grant number 2022YFS0536), The Open Grants of the State Key Laboratory of Severe Weather (2023LASW-B08) and the Technological Innovation Capacity Enhancement Program of the Chengdu University of Information Technology (grant number KYQN202203) provided financial support for this study.

Data Availability Statement: The hourly values of 2 m temperature, land surface temperature, relative humidity, surface pressure, water vapor mixing ratio, 10 m U and V wind components, precipitation, and dew point temperature at 5 × 5 km are available online at https://doi.org/10.57760/sciencedb.01840 (accessed on 31 December 2020). Due to the extensive amount of work required to validate and compare the dataset with other reanalyzed datasets, this will be our main focus for future research. After publishing this dataset online, we welcome researchers to validate and compare the dataset, and provide us with feedback for further improvement. If there are any questions during the use of the dataset, please contact wenzerg@126.com.

Acknowledgments: We express our gratitude to the computing resources and assistance provided by the School of Atmospheric Sciences, Chengdu University of Information and Technology (SAS, CUIT).

Conflicts of Interest: The authors declare no conflict of interest.

References

1. Qiu, J. The third pole: Climate change is coming fast and furious to the Tibetan plateau. Jane Qiu reports on the changes atop the roof of the world. *Nature* **2008**, *454*, 393–397. [CrossRef] [PubMed]
2. Immerzeel, W.W.; van Beek, L.P.H.; Bierkens, M.F.P. Climate change will affect the Asian water towers. *Science* **2010**, *328*, 1382–1385. [CrossRef] [PubMed]
3. Immerzeel, W.W. Importance and vulnerability of the world's water towers. *Nature* **2020**, *577*, 364–369. [CrossRef] [PubMed]
4. Pepin, N. Elevation-dependent warming in mountain regions of the world. *Nat. Clim. Chang.* **2015**, *5*, 424–430. [CrossRef]
5. Liu, X.; Cheng, Z.; Yan, L.; Yin, Z.Y. Elevation dependency of recent and future minimum surface air temperature trends in the Tibetan Plateau and its surroundings. *Glob. Planet Change* **2009**, *68*, 164–174. [CrossRef]
6. Yao, T. Tackling on environmental changes in Tibetan Plateau with focus on water, ecosystem and adaptation. *Sci. Bull.* **2019**, *64*, 417. [CrossRef]
7. Yao, T. Recent Third Pole's rapid warming accompanies cryospheric melt and water cycle intensification and interactions between monsoon and environment: Multidisciplinary approach with observations, modeling, and analysis. *Bull. Amer. Meteor. Soc.* **2019**, *100*, 423–444. [CrossRef]
8. Chen, F.H. The Tibetan Plateau as the engine for Asian environmental change: The Tibetan Plateau Earth system research into a new era. *Sci. Bull.* **2021**, *66*, 1263–1266. [CrossRef]
9. Gao, Y.-X.; Tang, M.-C.; Luo, S.-W.; Shen, Z.-B.; Li, C. Some aspects of recent research on the Qinghai-Xizang Plateau meteorology. *Bull. Am. Meteorol. Soc.* **1981**, *62*, 31–35. [CrossRef]
10. Ye, D. Some characteristics of the summer circulation over the Qinghai-Xizang (Tibet) Plateau and its neighborhood. *Bull. Am. Meteorol. Soc.* **1981**, *62*, 14–19. [CrossRef]
11. Yanai, M.; Li, C.; Song, Z. Seasonal heating of the Tibetan Plateau and its effects on the evolution of the Asian summer monsoon. *J. Meteorol. Soc. Japan. Ser. II* **1992**, *70*, 319–351. [CrossRef]
12. Ye, D.-Z.; Wu, G.-X. The role of the heat source of the Tibetan Plateau in the general circulation. *Meteorol. Atmos. Phys.* **1998**, *67*, 181–198. [CrossRef]
13. Tsukamoto, O.; Ishikawa, H.; Tamagawa, I. Analysis of aerodynamic and thermodynamic parameters on the grassy marshland surface of Tibetan Plateau. *Prog. Nat. Sci.* **2002**, *12*, 36–40.
14. Ma, Y.; Su, Z.; Koike, T.; Yao, T.; Ishikawa, H.; Ueno, K.i.; Menenti, M. On measuring and remote sensing surface energy partitioning over the Tibetan Plateau—-From GAME/Tibet to CAMP/Tibet. *Phys. Chem. Earth Parts A/B/C* **2003**, *28*, 63–74. [CrossRef]
15. Ma, Y.; Fan, S.; Ishikawa, H.; Tsukamoto, O.; Yao, T.; Koike, T.; Zuo, H.; Hu, Z.; Su, Z. Diurnal and inter-monthly variation of land surface heat fluxes over the central Tibetan Plateau area. *Theor. Appl. Climatol.* **2005**, *80*, 259–273. [CrossRef]

16. Ma, Y.; Zhong, L.; Su, Z.; Ishikawa, H.; Menenti, M.; Koike, T. Determination of regional distributions and seasonal variations of land surface heat fluxes from Landsat-7 Enhanced Thematic Mapper data over the central Tibetan Plateau area. *J. Geophys. Res. Atmos.* **2006**, *111*, D10305. [CrossRef]
17. Qie, X.; Wu, X.; Yuan, T.; Bian, J.; Lu, D. Comprehensive pattern of deep convective systems over the Tibetan Plateau–South Asian monsoon region based on TRMM data. *J. Clim.* **2014**, *27*, 6612–6626. [CrossRef]
18. Zhao, P.; Li, Y.; Guo, X.; Xu, X.; Liu, Y.; Tang, S.; Xiao, W.; Shi, C.; Ma, Y.; Yu, X. The Tibetan Plateau surface-atmosphere coupling system and its weather and climate effects: The Third Tibetan Plateau Atmospheric Science Experiment. *J. Meteorol. Res.* **2019**, *33*, 375–399. [CrossRef]
19. Tao, S.; Luo, S.; Zhang, H. The Qinghai-Xizang Plateau Meteorological Experiment (Qxpmex) May–August 1979. In Proceedings of the International Symposium on the Qinghai-Xizang Plateau and Mountain Meteorology, Beijing, China, 20–24 March 1984; American Meteorological Society: Boston, MA, USA, 1986; pp. 3–13. Available online: https://link.springer.com/chapter/10.1007/978-1-935704-19-5_1 (accessed on 31 December 2020).
20. Wang, J. Land surface process experiments and interaction study in China: From HEIFE to IMGRASS and GAME-Tibet/TIPEX. *Plateau Meteorol.* **1999**, *18*, 280–294.
21. Ma, Y.-M.; Yao, T.-D.; Wang, J.-M. Experimental Study of Energy and Water Cycle in Tibetan Plateau—The Progress Introduction on the Study of GAME/Tibet and CAMP/Tibet. *Plateau Meteorol.* **2006**, *25*, 344–351.
22. Zhao, P.; Xu, X.; Chen, F.; Guo, X.; Zheng, X.; Liu, L.; Hong, Y.; Li, Y.; La, Z.; Peng, H. The third atmospheric scientific experiment for understanding the earth–atmosphere coupled system over the Tibetan Plateau and its effects. *Bull. Am. Meteorol. Soc.* **2018**, *99*, 757–776. [CrossRef]
23. Wang, H.-J.; Sun, J.-Q.; Chen, H.-P.; Zhu, Y.-L.; Zhang, Y.; Jiang, D.-B.; Lang, X.-M.; Fan, K.; Yu, E.-T.; Yang, S. Extreme climate in China: Facts, simulation and projection. *Meteorol. Z.* **2012**, *21*, 279. [CrossRef]
24. Dasari, H.P.; Salgado, R.; Perdigao, J.; Challa, V.S. A regional climate simulation study using WRF-ARW model over Europe and evaluation for extreme temperature weather events. *Int. J. Atmos. Sci.* **2014**, *2014*, 704079. [CrossRef]
25. Sun, Q.; Zwiers, F.; Zhang, X.; Li, G. A comparison of intra-annual and long-term trend scaling of extreme precipitation with temperature in a large-ensemble regional climate simulation. *J. Clim.* **2020**, *33*, 9233–9245. [CrossRef]
26. Sasai, T.; Kawase, H.; Kanno, Y.; Yamaguchi, J.; Sugimoto, S.; Yamazaki, T.; Sasaki, H.; Fujita, M.; Iwasaki, T. Future projection of extreme heavy snowfall events with a 5-km large ensemble regional climate simulation. *J. Geophys. Res. Atmos.* **2019**, *124*, 13975–13990. [CrossRef]
27. Altaratz, O.; Koren, I.; Remer, L.; Hirsch, E. Cloud invigoration by aerosols—Coupling between microphysics and dynamics. *Atmos. Res.* **2014**, *140*, 38–60. [CrossRef]
28. Morrison, H.; van Lier-Walqui, M.; Fridlind, A.M.; Grabowski, W.W.; Harrington, J.Y.; Hoose, C.; Korolev, A.; Kumjian, M.R.; Milbrandt, J.A.; Pawlowska, H. Confronting the challenge of modeling cloud and precipitation microphysics. *J. Adv. Model. Earth Syst.* **2020**, *12*, e2019MS001689. [CrossRef]
29. Zhuo, H.; Liu, Y.; Jin, J. Improvement of land surface temperature simulation over the Tibetan Plateau and the associated impact on circulation in East Asia. *Atmos. Sci. Lett.* **2016**, *17*, 162–168. [CrossRef]
30. Meng, X.; Lyu, S.; Zhang, T.; Zhao, L.; Li, Z.; Han, B.; Li, S.; Ma, D.; Chen, H.; Ao, Y. Simulated cold bias being improved by using MODIS time-varying albedo in the Tibetan Plateau in WRF model. *Environ. Res. Lett.* **2018**, *13*, 044028. [CrossRef]
31. Peng, Q.; Meixue, Y.; Donglin, G.; Chu-xiong, C. Simulation of summer air temperature and precipitation over Tibetan Plateau with Regional Climate Model (RegCM3). *Plateau Meteor.* **2009**, *28*, 738–744.
32. Zhang, D.; Gao, X.; Bai, H.; Li, D. Simulation of climate over Qinghai-Xizang Plateau utilizing RegCM3. *Plateau Meteor.* **2005**, *24*, 714–720.
33. Gao, X.; Shi, Y.; Zhang, D.; Giorgi, F. Climate change in China in the 21st century as simulated by a high resolution regional climate model. *Chin. Sci. Bull.* **2012**, *57*, 1188–1195. [CrossRef]
34. Gao, Y.; Xu, J.; Chen, D. Evaluation of WRF mesoscale climate simulations over the Tibetan Plateau during 1979–2011. *J. Clim.* **2015**, *28*, 2823–2841. [CrossRef]
35. Yu, E.; Sun, J.; Chen, H.; Xiang, W. Evaluation of a high-resolution historical simulation over China: Climatology and extremes. *Clim. Dyn.* **2015**, *45*, 2013–2031. [CrossRef]
36. Gao, Y.; Chen, F.; Jiang, Y. Evaluation of a convection-permitting modeling of precipitation over the Tibetan Plateau and its influences on the simulation of snow-cover fraction. *J. Hydrometeorol.* **2020**, *21*, 1531–1548. [CrossRef]
37. You, Q.; Min, J.; Zhang, W.; Pepin, N.; Kang, S. Comparison of multiple datasets with gridded precipitation observations over the Tibetan Plateau. *Clim. Dyn.* **2015**, *45*, 791–806. [CrossRef]
38. Zhu, X.; Liu, Y.; Wu, G. An assessment of summer sensible heat flux on the Tibetan Plateau from eight data sets. *Sci. China Earth Sci.* **2012**, *55*, 779–786. [CrossRef]
39. Su, Z.; De Rosnay, P.; Wen, J.; Wang, L.; Zeng, Y. Evaluation of ECMWF's soil moisture analyses using observations on the Tibetan Plateau. *J. Geophys. Res. Atmos.* **2013**, *118*, 5304–5318. [CrossRef]
40. Zeng, Y.; Su, Z.; Van der Velde, R.; Wang, L.; Xu, K.; Wang, X.; Wen, J. Blending satellite observed, model simulated, and in situ measured soil moisture over Tibetan Plateau. *Remote Sens.* **2016**, *8*, 268. [CrossRef]
41. Grose, M.R.; Narsey, S.; Delage, F.; Dowdy, A.J.; Bador, M.; Boschat, G.; Chung, C.; Kajtar, J.; Rauniyar, S.; Freund, M. Insights from CMIP6 for Australia's future climate. *Earth's Future* **2020**, *8*, e2019EF001469. [CrossRef]

42. Ban, N.; Schmidli, J.; Schär, C. Evaluation of the convection-resolving regional climate modeling approach in decade-long simulations. *J. Geophys. Res. Atmos.* **2014**, *119*, 7889–7907. [CrossRef]
43. Collier, E.; Mölg, T.; Sauter, T. Recent atmospheric variability at Kibo Summit, Kilimanjaro, and its relation to climate mode activity. *J. Clim.* **2018**, *31*, 3875–3891. [CrossRef]
44. Ekström, M.; Grose, M.R.; Whetton, P.H. An appraisal of downscaling methods used in climate change research. *Wiley Interdiscip. Rev. Clim. Chang.* **2015**, *6*, 301–319. [CrossRef]
45. Smith, T.L.; Benjamin, S.G.; Gutman, S.I.; Sahm, S. Short-range forecast impact from assimilation of GPS-IPW observations into the Rapid Update Cycle. *Mon. Weather Rev.* **2007**, *135*, 2914–2930. [CrossRef]
46. Benjamin, S.G.; Dévényi, D.; Weygandt, S.S.; Brundage, K.J.; Brown, J.M.; Grell, G.A.; Kim, D.; Schwartz, B.E.; Smirnova, T.G.; Smith, T.L. An hourly assimilation–forecast cycle: The RUC. *Mon. Weather Rev.* **2004**, *132*, 495–518. [CrossRef]
47. James, E.P.; Benjamin, S.G. Observation system experiments with the hourly updating Rapid Refresh model using GSI hybrid ensemble–variational data assimilation. *Mon. Weather Rev.* **2017**, *145*, 2897–2918. [CrossRef]
48. Ha, S.-Y.; Snyder, C. Influence of surface observations in mesoscale data assimilation using an ensemble Kalman filter. *Mon. Weather Rev.* **2014**, *142*, 1489–1508. [CrossRef]
49. Wen, X.; Zhu, X.; Li, M.; Guo, H.; Zhang, S.; Yang, X.; Zheng, Z.; Yan, D.; Zhang, Y.; Lv, S. QTP-HRAD[DS/OL]. V2. Science Data Bank. 2022. Available online: https://www.scidb.cn/en/detail?dataSetId=f0860df1bd8e4fee842baa70fe6d9b94# (accessed on 29 March 2023).
50. Thompson, G.; Field, P.R.; Rasmussen, R.M.; Hall, W.D. Explicit forecasts of winter precipitation using an improved bulk microphysics scheme. Part II: Implementation of a new snow parameterization. *Mon. Weather Rev.* **2008**, *136*, 5095–5115. [CrossRef]
51. Iacono, M.J.; Delamere, J.S.; Mlawer, E.J.; Shephard, M.W.; Clough, S.A.; Collins, W.D. Radiative forcing by long-lived greenhouse gases: Calculations with the AER radiative transfer models. *J. Geophys. Res. Atmos.* **2008**, *113*, D13103. [CrossRef]
52. Bae, S.Y.; Hong, S.-Y.; Lim, K.-S.S. Coupling WRF double-moment 6-class microphysics schemes to RRTMG radiation scheme in weather research forecasting model. *Adv. Meteorol.* **2016**, *2016*, 5070154. [CrossRef]
53. Tiedtke, M. A comprehensive mass flux scheme for cumulus parameterization in large-scale models. *Mon. Weather Rev.* **1989**, *117*, 1779–1800. [CrossRef]
54. Zhang, C.; Wang, Y.; Hamilton, K. Improved representation of boundary layer clouds over the southeast Pacific in ARW-WRF using a modified Tiedtke cumulus parameterization scheme. *Mon. Weather Rev.* **2011**, *139*, 3489–3513. [CrossRef]
55. Ek, M.; Mitchell, K.; Lin, Y.; Rogers, E.; Grunmann, P.; Koren, V.; Gayno, G.; Tarpley, J. Implementation of Noah land surface model advances in the National Centers for Environmental Prediction operational mesoscale Eta model. *J. Geophys. Res. Atmos.* **2003**, *108*, GCP12-1. [CrossRef]
56. Janjić, Z.I. The step-mountain eta coordinate model: Further developments of the convection, viscous sublayer, and turbulence closure schemes. *Mon. Weather. Rev.* **1994**, *122*, 927–945. [CrossRef]
57. Chen, F.; Janjić, Z.; Mitchell, K. Impact of atmospheric surface-layer parameterizations in the new land-surface scheme of the NCEP mesoscale Eta model. *Bound.-Layer Meteorol.* **1997**, *85*, 391–421. [CrossRef]
58. Bassett, R.; Young, P.; Blair, G.; Samreen, F.; Simm, W. A large ensemble approach to quantifying internal model variability within the WRF numerical model. *J. Geophys. Res. Atmos.* **2020**, *125*, e2019JD031286. [CrossRef]
59. Zhang, X.; Cornuelle, B.D.; Martin, A.; Weihs, R.R.; Ralph, M. West-WRF Sensitivity to Sea Surface Temperature Boundary Condition in California Precipitation Forecasts of AR Related Events. In Proceedings of the AGU Fall Meeting Abstracts, New Orleans, LA, USA, 11–15 December 2017; pp. A33H–2496.
60. Zeng, X.; Beljaars, A. A prognostic scheme of sea surface skin temperature for modeling and data assimilation. *Geophys. Res. Lett.* **2005**, *32*. [CrossRef]
61. Ide, K.; Courtier, P.; Ghil, M.; Lorenc, A.C. Unified notation for data assimilation: Operational, sequential and variational (gtspecial issueltdata assimilation in meteology and oceanography: Theory and practice). *J. Meteorol. Soc. Japan. Ser. II* **1997**, *75*, 181–189. [CrossRef]
62. Lorenc, A.C. Analysis methods for numerical weather prediction. *Q. J. R. Meteorol. Soc.* **1986**, *112*, 1177–1194. [CrossRef]
63. Barker, D.M.; Huang, W.; Guo, Y.-R.; Bourgeois, A.; Xiao, Q. A three-dimensional variational data assimilation system for MM5: Implementation and initial results. *Mon. Weather. Rev.* **2004**, *132*, 897–914. [CrossRef]
64. Routray, A.; Mohanty, U.; Osuri, K.K.; Kiran Prasad, S. Improvement of monsoon depressions forecast with assimilation of Indian DWR data using WRF-3DVAR analysis system. *Pure Appl. Geophys.* **2013**, *170*, 2329–2350. [CrossRef]
65. Dragani, R.; Hersbach, H.; Poli, P.; Pebeuy, C.; Hirahara, S.; Simmons, A.; Dee, D. Recent Reanalysis Activities at ECMWF: Results from ERA-20C and Plans for ERA5. *AGU Fall Meet. Abstr.* **2015**, *2015*, A11R-02.
66. Urraca, R.; Huld, T.; Gracia-Amillo, A.; Martinez-De-Pison, F.J.; Kaspar, F.; Sanz-Garcia, A. Evaluation of global horizontal irradiance estimates from ERA5 and COSMO-REA6 reanalyses using ground and satellite-based data. *Sol. Energy* **2018**, *164*, 339–354. [CrossRef]
67. Hersbach, H.; Bell, B.; Berrisford, P.; Hirahara, S.; Horányi, A.; Muñoz-Sabater, J.; Nicolas, J.; Peubey, C.; Radu, R.; Schepers, D. The ERA5 global reanalysis. *Q. J. R. Meteorol. Soc.* **2020**, *146*, 1999–2049. [CrossRef]
68. Bao, X.; Zhang, F. Evaluation of NCEP–CFSR, NCEP–NCAR, ERA-Interim, and ERA-40 reanalysis datasets against independent sounding observations over the Tibetan Plateau. *J. Clim.* **2013**, *26*, 206–214. [CrossRef]

69. You, Q.; Kang, S.; Pepin, N.; Flügel, W.-A.; Yan, Y.; Behrawan, H.; Huang, J. Relationship between temperature trend magnitude, elevation and mean temperature in the Tibetan Plateau from homogenized surface stations and reanalysis data. *Glob. Planet. Change* **2010**, *71*, 124–133. [CrossRef]
70. Frauenfeld, O.W.; Zhang, T.; Serreze, M.C. Climate change and variability using European Centre for Medium-Range Weather Forecasts reanalysis (ERA-40) temperatures on the Tibetan Plateau. *J. Geophys. Res. Atmos.* **2005**, *110*, D02101.1–D02101.9. [CrossRef]
71. Baolin, L.; Yichi, Z.; Chenghu, Z. Remote sensing detection of glacier changes in Tianshan Mountains for the past 40 years. *J. Geogr. Sci.* **2004**, *14*, 296–302. [CrossRef]
72. Hanna, S.R.; Yang, R. Evaluations of mesoscale models' simulations of near-surface winds, temperature gradients, and mixing depths. *J. Appl. Meteorol.* **2001**, *40*, 1095–1104. [CrossRef]
73. Horvath, K.; Koracin, D.; Vellore, R.; Jiang, J.; Belu, R. Sub-kilometer dynamical downscaling of near-surface winds in complex terrain using WRF and MM5 mesoscale models. *J. Geophys. Res. Atmos.* **2012**, *117*, D11111. [CrossRef]
74. Wen, X.; Lu, S.; Jin, J. Integrating remote sensing data with WRF for improved simulations of oasis effects on local weather processes over an arid region in northwestern China. *J. Hydrometeorol.* **2012**, *13*, 573–587. [CrossRef]

remote sensing

MDPI

Article

Annual and Seasonal Trends of Vegetation Responses and Feedback to Temperature on the Tibetan Plateau since the 1980s

Fangfang Wang [1,2,3], Yaoming Ma [1,2,4,5,6,7,*], Roshanak Darvishzadeh [3] and Cunbo Han [1]

1 Land-Atmosphere Interaction and Its Climatic Effects Group, State Key Laboratory of Tibetan Plateau Earth System, Environment and Resources (TPESER), Institute of Tibetan Plateau Research, Chinese Academy of Sciences, Beijing 100101, China
2 College of Earth and Planetary Sciences, University of Chinese Academy of Sciences, Beijing 100049, China
3 Faculty of Geo-Information Science and Earth Observation (ITC), University of Twente, 7500 AE Enschede, The Netherlands
4 College of Atmospheric Science, Lanzhou University, Lanzhou 730000, China
5 National Observation and Research Station for Qomolangma Special Atmospheric Processes and Environmental Changes, Dingri 858200, China
6 Kathmandu Center of Research and Education, Chinese Academy of Sciences, Beijing 100101, China
7 China-Pakistan Joint Research Center on Earth Sciences, Chinese Academy of Sciences, Islamabad 45320, Pakistan
* Correspondence: ymma@itpcas.ac.cn

Abstract: The vegetation–temperature relationship is crucial in understanding land–atmosphere interactions on the Tibetan Plateau. Although many studies have investigated the connections between vegetation and climate variables in this region using remote sensing technology, there remain notable gaps in our understanding of vegetation–temperature interactions over different timescales. Here, we combined site-level air temperature observations, information from the global inventory modeling and mapping studies (GIMMS) dataset, and moderate-resolution imaging spectroradiometer (MODIS) products to analyze the spatial and temporal patterns of air temperature, vegetation, and land surface temperature (LST) on the Tibetan Plateau at annual and seasonal scales. We achieved these spatiotemporal patterns by using Sen's slope, sequential Mann–Kendall tests, and partial correlation analysis. The timescale differences of vegetation-induced LST were subsequently discussed. Our results demonstrate that a breakpoint of air temperature change occurred on the Tibetan Plateau during 1994–1998, dividing the study period (1982–2013) into two phases. A more significant greening response of NDVI occurred in the spring and autumn with earlier breakpoints and a more sensitive NDVI response occurred in recent warming phase. Both MODIS and GIMMS data showed a common increase in the normalized difference vegetation index (NDVI) on the Tibetan Plateau for all timescales, while the former had a larger greening area since 2000. The most prominent trends in NDVI and LST were identified in spring and autumn, respectively, and the largest areas of significant variation in NDVI and LST mostly occurred in winter and autumn, respectively. The partial correlation analysis revealed a significant negative impact of NDVI on LST during the annual scale and autumn, and it had a significant positive impact during spring. Our findings improve the general understanding of vegetation–climate relationships at annual and seasonal scales.

Keywords: vegetation change; near-surface air temperature; annual and seasonal variations; land–atmosphere interactions; Tibetan Plateau

Citation: Wang, F.; Ma, Y.; Darvishzadeh, R.; Han, C. Annual and Seasonal Trends of Vegetation Responses and Feedback to Temperature on the Tibetan Plateau since the 1980s. *Remote Sens.* **2023**, *15*, 2475. https://doi.org/10.3390/rs15092475

Academic Editor: Conghe Song

Received: 6 April 2023
Revised: 2 May 2023
Accepted: 5 May 2023
Published: 8 May 2023

1. Introduction

Vegetation is an important part of the global terrestrial ecosystem and can significantly impact global physical energy cycles, carbon balance regulations, and regional climate [1,2]. Global vegetation cover generally increases with the warming of the Earth's climate [3], and the surface albedo changes caused by vegetation change affect both the net amount of solar radiation absorbed by the earth surface and the evapotranspiration rate. Consequently, vegetation-related evapotranspiration and albedo affect vegetation–temperature relationships [4,5]. The Tibetan Plateau (TP) is the highest plateau on earth and possesses a complex topography, which generates a unique relationship between vegetation and temperature that is extremely sensitive to global climate change [6]. Investigating the responses and feedbacks of vegetation to the TP's temperature is crucial for understanding land–atmosphere interactions.

Numerous researchers have studied the vegetation–temperature relationships on the TP through different approaches, including field observations [7,8], numerical simulations, and remote sensing observations [9,10]. However, given the challenging topography and extreme climate of the TP [11], fieldwork and in situ analysis can be logistically difficult to perform. It is not easy to consistently assess vegetation–temperature relationships at different timescales. As such, many studies have employed numerical models to simulate and analyze their relationships [12], and significant research progress has been made using these techniques. However, since models often have considerable uncertainties and coarse spatial resolution, it is unfeasible to numerically simulate the vegetation–temperature relationships at different timescales for long periods [13]. Satellite Earth observation is an effective instrument for monitoring bio-geophysical variables of vegetation at large scales [14]. The vegetation index and albedo derived from satellite observations can capture vegetation greening and browning and surface albedo change. These changes alter surface bio-geophysical properties and near-surface aerodynamics, leading to an effect on local temperature through bio-geophysical feedbacks [15–17]. As satellite remote sensing technology has developed, the products of widespread vegetation–temperature interactions that accumulate on the TP can be measured remotely, allowing a quantitative investigation of vegetation coverage at large scales and over extended periods [6,18]. Consequently, remote sensing observations are becoming a valuable instrument for investigating vegetation and climate interactions.

Using the available remote sensing data, previous papers report the trends and the spatial variability of vegetation, surface albedo, and land surface temperature (LST) on the TP at annual scales. The satellite-derived normalized difference vegetation index (NDVI) indicates the status of plant growth. This indicator can be quantified by the difference between the near-infrared (representing vegetation reflection) and red bands (representing vegetation absorption) [19]. Global inventory modeling and mapping studies (GIMMS) data suggest an increasing NDVI trend on the TP during 1985–1999, which is likely due to the shift from an arid to a warm-humid climate and the reduction in human activities in this region [20]. Similar studies have found an increasing NDVI trend during the latter two decades of the 20th century [21], which has been confirmed by researchers since the year 2000. Statistical analysis of SPOT (Satellite Pour l' Observation de la Terre) data collected during 1999–2014 shows an overall increasing trend in NDVI on the TP coupled with a moderate increase in air temperature [6]. Studies based on MODIS data collected over the TP during 2001–2019 also support this, as they show an increasing trend in NDVI and LST but a decreasing trend in albedo during that period. Increased forest coverage and decreased snow coverage are considered to be the dominant factors that drove these changes [22]. Further studies have shown that increased snowfall induced an increase in albedo on the southwestern TP due to anomalous water vapor transport [23]. Indeed, the LST warmed at a significantly faster rate than the air temperature, with the annual temperature increase during 1987–2008 in the former showing 0.78 ± 0.0631 K/decade, but in the latter, it showed 0.275 ± 0.0216 K/decade [24]. LST on the TP is influenced by various factors such as elevation, surface radiation, subsurface temperature, and surface properties [25–27]. Nevertheless, vegetation and albedo are becoming the hot spots for LST warming studies. This can be attributed to the significant linear relationship between vegetation and LST [28], the direct contribution of surface albedo to LST [25], and satellite advantages [29]. Despite different spatial patterns being inferred from different datasets, they show consistent support for the overall observed trends and patterns of vegetation and temperature on the TP.

Most of the studies mentioned above focused on the annual scale, which leaves two important issues unresolved: (1) no systematic analysis has been performed to analyze the differences in both the warming phases (well known as a period of increasing temperature) and the vegetation response to temperature across different timescales over the TP; and (2) the spatial–temporal pattern of vegetation and its effects on LST across the TP remains unclear. We addressed these problems by combining site-level observations with GIMMS data to (1) analyze spatial and temporal patterns at the annual and seasonal scales of air temperature and NDVI during different warming phases in 1982–2013 and (2) examine the annual and seasonal scales of NDVI and LST changes during 2000–2021 using MODIS data products and thus discuss the possible impacts of vegetation on LST at the annual and seasonal scales. Our results have revealed systematic annual and seasonal characteristics of air temperature, LST, and vegetation changes on the TP, and this can be used to form a better understanding of land–atmosphere interaction patterns across the TP.

2. Study Area and Datasets

2.1. Study Area

With an area of 257 × 10^4 km^2 and an average altitude above 4000 m, the TP is the largest and highest plateau in the world, and it exhibits a complex topography and diverse underlying surface conditions (Figure 1). The climate and ecological environment on the TP have both changed dramatically since the middle of the 20th century due to intensified human activity [30]. Significant warm and wet trends on the TP have occurred since the 1960s as documented by temperature and precipitation data [31]. Due to its distinctive geographic location and susceptibility to climate change, the TP serves as a natural laboratory for investigating the intricate relationship between vegetation and temperature.

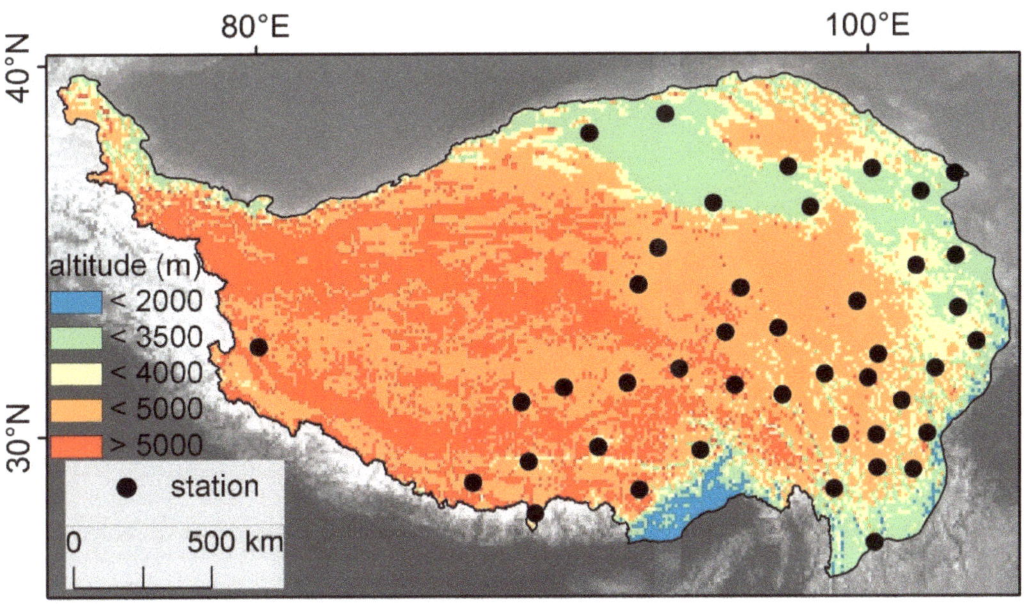

Figure 1. Location of meteorological stations on the Tibetan Plateau.

2.2. Data Sources and Processing

The data used in this study were obtained from meteorological station observations (Figure 1) (near-surface air temperature) and remote sensing (Table 1). Remote sensing data comprised NDVI products derived from GIMMS and MODIS, albedo, land cover type, and

LST products derived from MODIS. In this study, winter, spring, summer, and autumn are defined as extending from December to the following February (DJF), March to May (MAM), June to August (JJA), and September to November (SON), respectively.

Table 1. Data sources used in this study.

Parameter	Dataset	Spatial Resolution	Temporal Resolution	Download Links
GIMMS NDVI	Version 5	8 km	15 days	https://climatedataguide.ucar.edu/climate-data/ndvi-normalized-difference-vegetation-index-3rd-generation-nasagfsc-gimms, accessed on 3 October 2022
Normalized difference vegetation index (NDVI)	MOD13A2	1000 m	16 days	https://lpdaac.usgs.gov/products/mod13a2v061, accessed on 3 October 2022
Land surface temperature (LST)	MOD11A1	1000 m	8 days	https://lpdaac.usgs.gov/products/mod11a2v061, accessed on 3 October 2022
Albedo	MCD43A3	500 m	daily	https://lpdaac.usgs.gov/products/mcd43a3v061, accessed on 3 October 2022
Land Cover Type	MCD12Q1	500 m	yearly	https://lpdaac.usgs.gov/products/mcd12q1v061, accessed on 3 October 2022
Air temperature	/	/	monthly	https://www.ncdc.noaa.gov/cdo-web, accessed on 3 October 2022

2.2.1. Meteorological Station Data

Air temperature data were collected from the National Climate Data Center, which is affiliated with the National Oceanic and Atmosphere Administration (NOAA) (Figure 1). These data were used to produce the average air temperature at different timescales from 1982 to 2013. First, we acquired 44 high-quality meteorological stations considering the geographical scope of the TP and the requirement of high-density temperature data. Second, we collected the daily mean temperature data for each station during 1982–2013 and replaced the missing values for a few of stations using the average air temperature of the neighboring days [32,33]. Finally, the mean temperature of annual and seasonal scales was calculated and used to detect trends and breakpoints in different warming phases of the TP.

2.2.2. Remote Sensing Products

The remote sensing data used in this study included the GIMMS and MODIS datasets. The GIMMS NDVI product (1982–2013) was derived from the AVHRR sensor of the National Oceanic and Atmospheric Administration (http://www.ncdc.noaa.gov/cdo-web, accessed on 3 October 2022), and it has a temporal resolution of 15 days and a spatial resolution of 8 km. The initial version of this dataset is not ideal for capturing vegetation dynamics, and therefore, the latest version mitigates this problem by correcting sensors, aerosols, and view geometry [34–36]. A maximum value composite procedure was used to remove some sources of interference, such as clouds, the atmosphere, and variation in solar altitude angle; after that, annual and seasonal NDVI values were obtained [37].

Compared to the AVHRR instrument, the updated MODIS instrument has a better sensitivity to chlorophyll with higher spatial resolution. We employed MODIS datasets of 2000–2021 including NDVI (MOD13A2), LST (MOD11A1), albedo (MCD43A3), and land cover types (MCD12Q1). Specifically, MOD13A2 and MOD11A1 have a 1 km spatial resolution and a 16-day temporal resolution. MOD11A1 and MCD43A3 were used in this study to better understand the relationships between LST and vegetation change. MCD12Q1 (version 6.1), providing annual land cover types (2001–2021), was also obtained to analyze the effect of land cover change on NDVI change trends. Additionally, all of the MODIS datasets were aggregated to 1 km to ensure the consistent spatial resolution of these datasets. We adopted different strategies in processing the MODIS-derived surface parameters by referring to previous studies. For each timescale, we processed the NDVI data using the maximum value composite method [37], and we processed the LST and albedo data using the mean value composite method [29,38].

2.3. Methods

Our analyses were performed in three major steps. Firstly, to analyze the NDVI responses at different warming phases, we performed breakpoint detection of air temperature data. Considering previous artificial temporal segmentation on NDVI variation [17,39–41], the Sequential Mann–Kendall (SQMK) [32–34,42] was employed to estimate the breakpoints of temperature change, and it was used as a proxy for classifying the different phases of NDVI responses. Please see the Supporting Information S1 for the detailed introduction to the SQMK test.

Secondly, we analyzed trends and performed a significance analysis [43,44] on key surface parameters. Sen's slope estimator, a nonparametric test [45], was employed to determine trends in NDVI, LST, and albedo for different warming phases of all timescales. Please see the Supporting Information S2 for the detailed introduction to the Sen's slope.

Thirdly, to identify the impacts of the NDVI on LST, both the detrending method and partial correlation analysis were performed among the NDVI, albedo, and LST. The purpose of using the detrending method was to eliminate any spurious correlations among the three parameters that may have been caused by temporal variations. As for the partial correlation analysis, it was used to better quantify the individual impact of NDVI or albedo on LST. The specific procedures included two aspects: (1) Using the first-difference detrending method (i.e., the difference of values in one year to the previous year), we examined and filtered the temporal trends of NDVI and albedo. (2) For partial correlation analysis, the partial correlation coefficient is an assessment of the net correlation between a single factor and the target value, provided that the impact of other factors is fixed or deducted. Considering the significant co-impact of vegetation and albedo on LST, the partial correlation coefficient is a good indicator for analyzing the relationship between them.

3. Results

3.1. Air Temperature and Vegetation Trends during 1982–2013 at Annual and Seasonal Scales

3.1.1. Annual Trends in Air Temperature and Vegetation

The air temperature and vegetation coverage on the TP generally increased on an annual scale (Figure 2a), while vegetation changes during each warming phase showed significant spatial and temporal variation. A significant abrupt change ($p < 0.05$) occurred in 1996 for air temperature trends on the TP (Figure 2b), and the warming trend during 1996–2013 (0.043 °C/year) was notably higher than that during 1982–1996 (0.042 °C/year), suggesting that the warming rate on the TP accelerated after 1996. The significant result ($p < 0.05$) revealed that the annual NDVI is generally increasing (Figure 2a) and the clustered NDVI increase and decrease in the second warming phase are greater than that in the first warming phase (Figure 2d,f). Specifically, the NDVI greened more than it browned during the first warming phase (Figure 2d), and there was clustered NDVI greening and browning in the eastern and western plateau (Figure 2f), respectively.

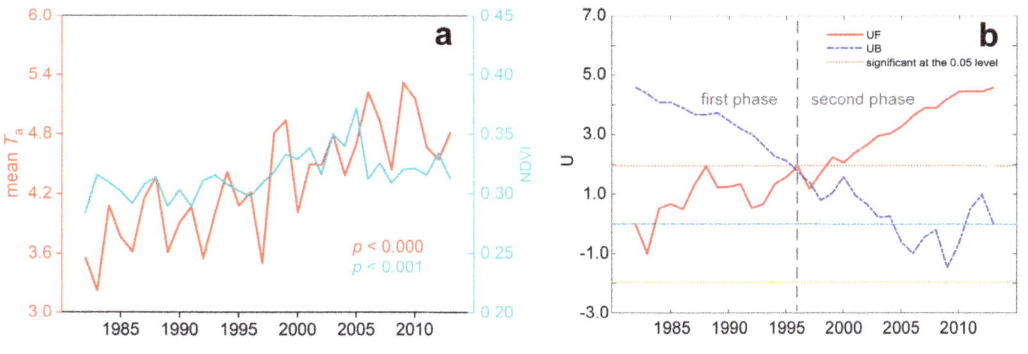

Figure 2. *Cont.*

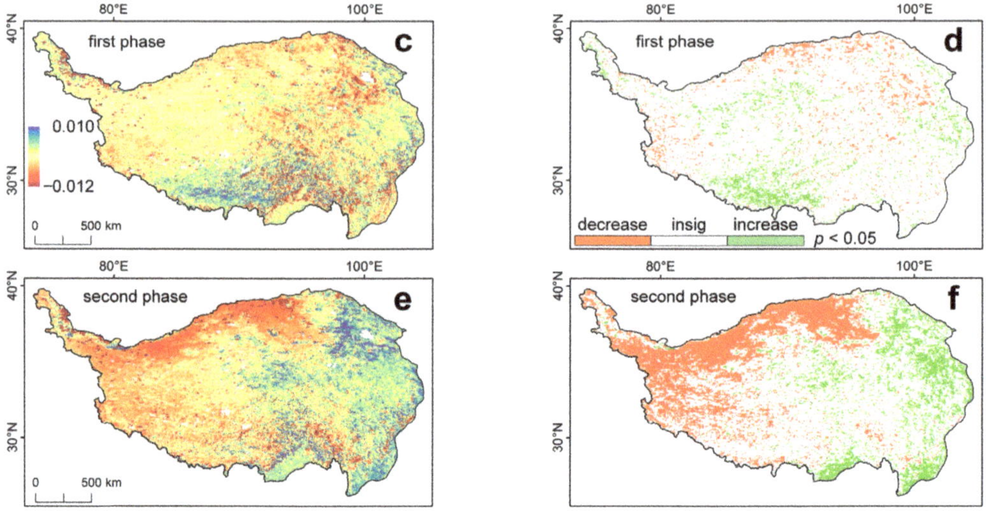

Figure 2. Vegetation and air temperature trends on the annual scale. GIMMS-based NDVI and air temperature trends during 1982–2013 (**a**); air temperature breakpoint detection based on the MK method (**b**). In (**b**), the gray dashed line denotes the detected temperature breakpoint, and UF and UB refer to the statistics of forward and backward sequence, respectively; GIMMS-based NDVI trends (**c**) and regions of significant change (**d**) during the first warming phase; GIMMS-based NDVI trends (**e**) and regions of significant change (**f**) during the second warming phase.

3.1.2. Seasonal Trends in Air Temperature and Vegetation

The air temperature and vegetation coverage on the TP during the spring seasons generally showed an increasing trend throughout the study period (Figure 3a). The breakpoint and warming rates were very different from those of the annual scale. The breakpoint for spring's air temperature trend occurred in 1994, with a slope of 0.040 °C/year in the first warming phase and 0.034 °C/year in the second warming phase (Figure 3b), indicating that warming on the TP during the spring has slowed down. The NDVI showed a general increase over time, although the trend was more significant during the second warming phase (Figure 3c,e). Furthermore, the NDVI showed a significant increasing trend ($p < 0.05$) in the eastern TP and a significant decreasing trend in the northwestern part (Figure 3e). The significance statistics ($p < 0.05$) suggested that the clustered NDVI changes occurred in the second phase rather than in the first phase (Figure 3d,f). For example, TP NDVI decreased in the western part and increased in the eastern part (Figure 3f).

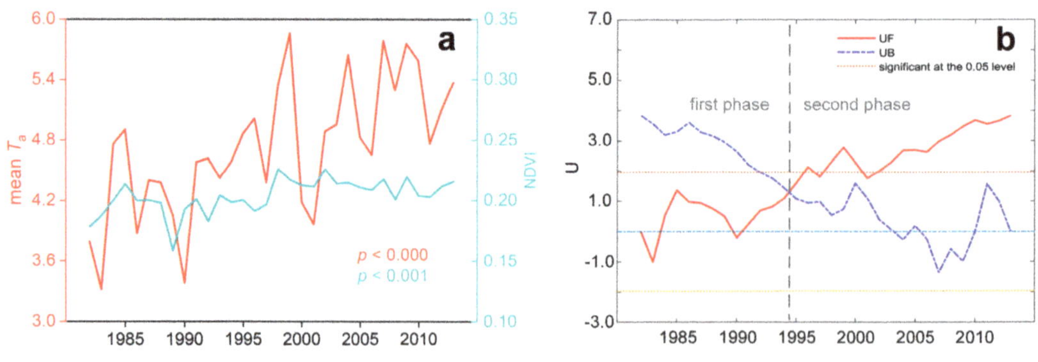

Figure 3. *Cont.*

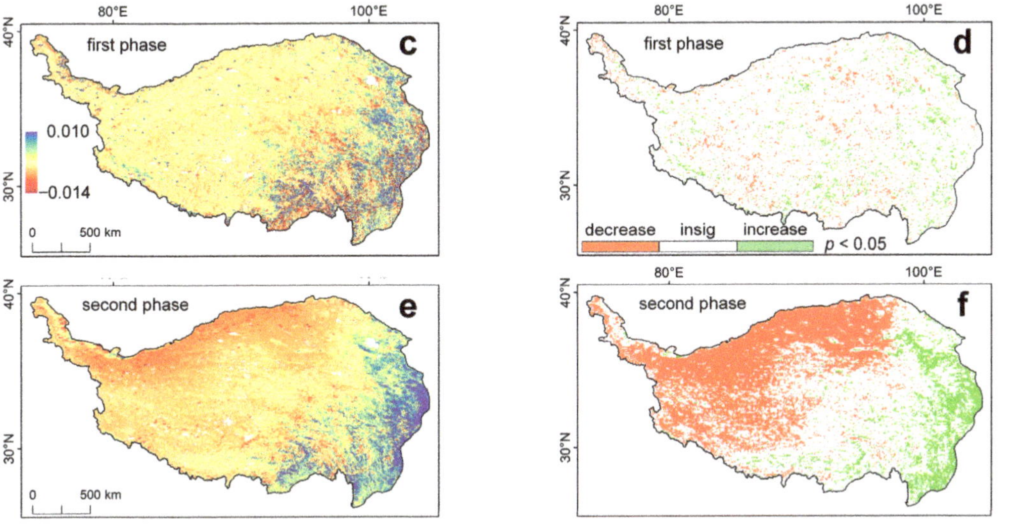

Figure 3. Vegetation and air temperature trends in spring. GIMMS-based NDVI and air temperature trends during 1982–2013 (**a**); air temperature breakpoint detection based on the MK method (**b**). In (**b**), the gray dashed line denotes the detected temperature breakpoint, and UF and UB refer to the statistics of forward and backward sequence, respectively; GIMMS-based NDVI trends (**c**) and regions of significant change (**d**) during the first warming phase. GIMMS-based NDVI trends (**e**) and regions of significant change (**f**) during the second warming phase.

For summer, the TP exhibited both the largest NDVI value and the most significant warming trend (Figure 4a, $p < 0.05$), as did the largest pre- and post-phase difference in air temperature (Figure 4b). The breakpoint for the summer temperature on the TP occurred in 1998, which is slightly later than for the spring and annual scales. The trend for the second warming phase reached a rate of 0.047 °C/year, which is notably higher than that of the first warming phase. This implied that the warming rate in summer on the TP is much greater than that in other timescales. The significance analysis ($p < 0.05$) indicated that the NDVI increase is greater in the first warming phase than that in the second warming phase (Figure 4d,f), particularly in the western TP.

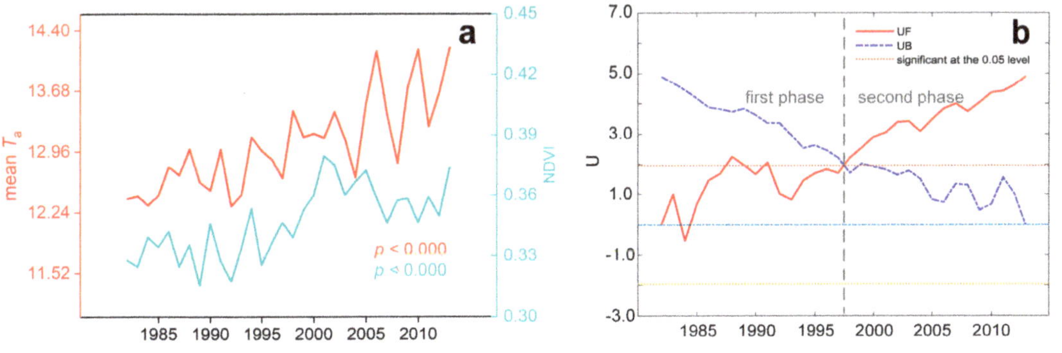

Figure 4. *Cont.*

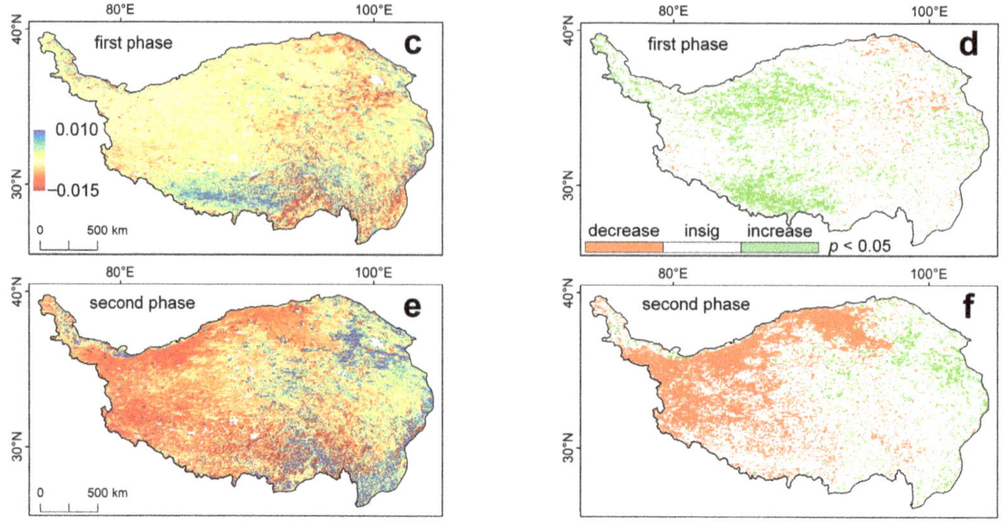

Figure 4. Vegetation and air temperature trends in summer. GIMMS-based NDVI and air temperature trends during 1982–2013 (**a**); air temperature breakpoint detection based on the MK method (**b**). In (**b**), the gray dashed line denotes the detected temperature breakpoint, and UF and UB refer to the statistics of forward and backward sequence, respectively; GIMMS-based NDVI trends (**c**) and regions of significant change (**d**) during the first warming phase. GIMMS-based NDVI trends (**e**) and regions of significant change (**f**) during the second warming phase.

Autumn vegetation trends and warming both showed a lower value compared to the summer (Figure 5a). The breakpoint of temperature during the autumn occurred in 1994 (Figure 5b), similar to the spring. The first and second phases during autumn recorded trends of 0.029 °C/year and 0.018 °C/year, respectively. As such, the trend for the second phase was significantly weaker than that of the first phase, demonstrating that the warming rate on the TP becomes slow during autumn. The NDVI showed a non-significant fluctuation in general, and the significance statistics ($p < 0.05$) showed that the increase or decrease in the TP's NDVI is less significant in the first phase than that in the second phase (Figure 5d,f), which is similar to the vegetation change pattern in spring.

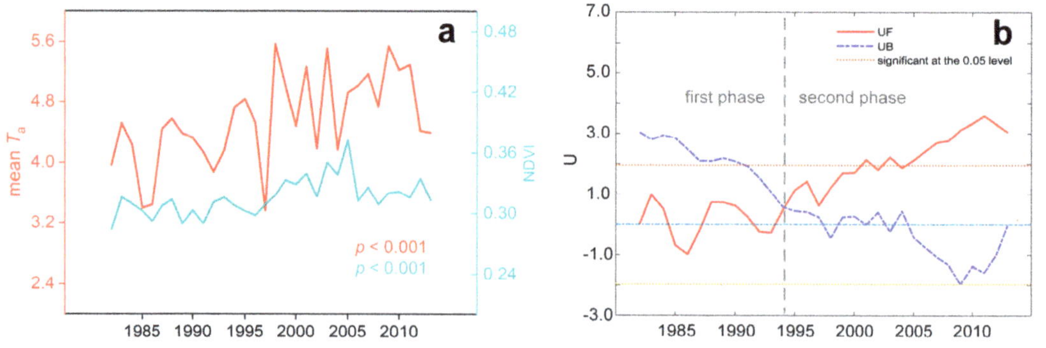

Figure 5. *Cont.*

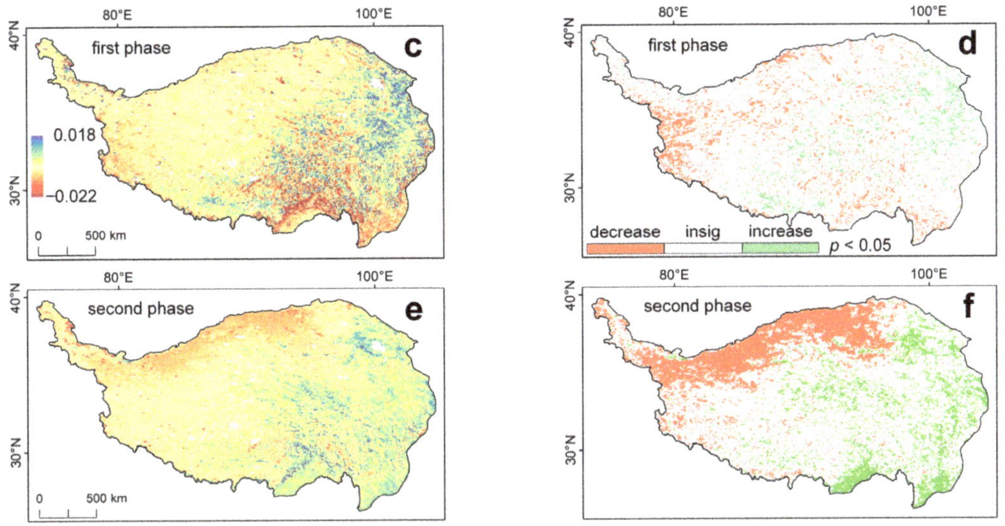

Figure 5. Vegetation and air temperature trends in autumn. GIMMS-based NDVI and air temperature trends during 1982–2013 (**a**); air temperature breakpoint detection based on the MK method (**b**). In (**b**), the gray dashed line denotes the detected temperature breakpoint, and UF and UB refer to the statistics of forward and backward sequence, respectively; GIMMS-based NDVI trends (**c**) and regions of significant change (**d**) during the first warming phase. GIMMS-based NDVI trends (**e**) and regions of significant change (**f**) during the second warming phase.

The vegetation and air temperature changes on the TP documented during the winter were similar to those during the autumn, although they showed much smaller magnitudes than other timescales (Figure 6a,b). The breakpoint in the winter temperature between warming phases on the TP occurred in 1998; the trends of the first and second warming phases were 0.017 °C/year and 0.021 °C/year, respectively, indicating that the winter warming on the TP is accelerating (Figure 6b). Winter changes in the NDVI values were generally small. A fragmented NDVI increase occurred in the eastern TP of the first phase, but a massive NDVI decrease occurred in the second phase, which was not found for any other timescales (Figure 6d,f).

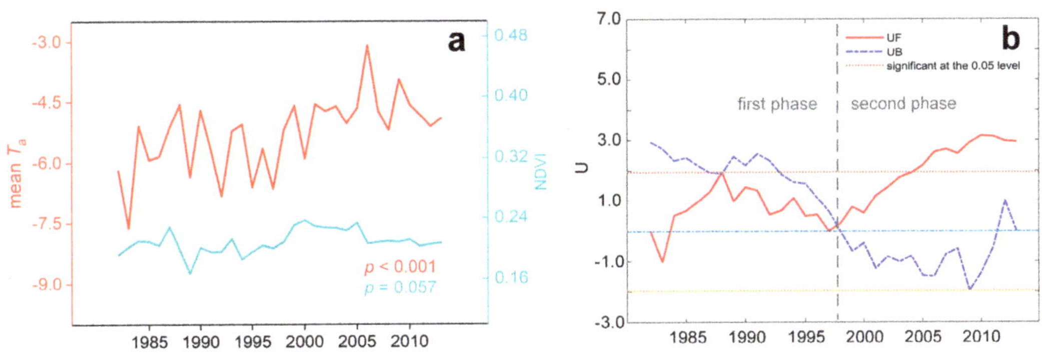

Figure 6. *Cont.*

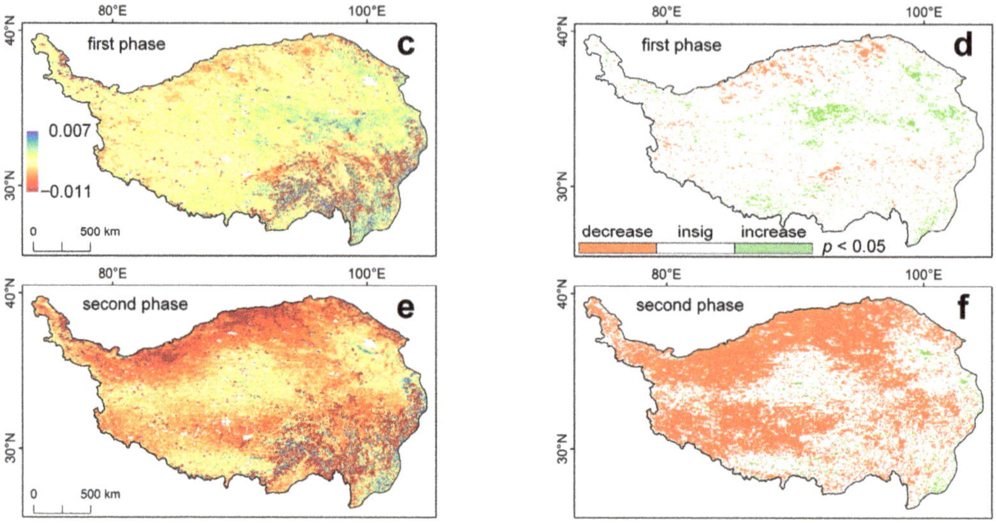

Figure 6. Vegetation and air temperature trends in winter. GIMMS-based NDVI and air temperature trends during 1982–2013 (**a**); air temperature breakpoint detection based on the MK method (**b**). In (**b**), the gray dashed line denotes the detected temperature breakpoint, and UF and UB refer to the statistics of forward and backward sequence, respectively; GIMMS-based NDVI trends (**c**) and regions of significant change (**d**) during the first warming phase. GIMMS-based NDVI trends (**e**) and regions of significant change (**f**) during the second warming phase.

3.1.3. The Spatial and Temporal Responses of NDVI to Air Temperature

Our study identified consistent breakpoints in air temperature for spring and autumn as well as for summer and winter. We observed that the first warming phase showed a smaller trend than the second warming phase on most timescales. This suggests that while warming continues on the TP, the later warming trend is slightly decreasing.

We also investigated the spatial trends of NDVI at different warming phases. During the first warming phase, we found that the area of NDVI greening is smaller in spring and autumn with earlier breakpoints compared to summer and winter with later breakpoints. However, in the second warming phase, we observed that the area of NDVI increase is larger in spring and autumn with earlier breakpoints compared to summer and winter with later breakpoints. This indicates that the timescale with the earlier breakpoints exhibits a greater NDVI increase in the second warming phase. For instance, the air temperature breakpoint at the annual scale earlier than summer and winter showed a significantly larger area of NDVI increase in the second warming phase.

Furthermore, we found that in the relatively weaker second warming phase, there is more NDVI decrease at all timescales across the sparsely vegetated northwestern TP. This may be attributed to two factors: (1) limitations in the ability of the AVHRR sensor to capture detailed NDVI changes in sparsely vegetated areas, and (2) an increased vegetation sensitivity to air temperature during the second warming phase.

3.2. LST, Vegetation, and Albedo Trends during 2000–2021 at Annual and Seasonal Scales

3.2.1. LST Trends at Different Timescales

The spatial pattern of LST trends recorded at different timescales was more concentrated than that for the NDVI data. Alongside being common in winter, the LST warming trend on the TP from 2000 to 2021 was most prominent in the southern TP (Figure 7a–e). The LST warming rate was significantly higher in summer and autumn than during winter and spring, and it was mainly concentrated in the southwestern TP (Figure 7c,d). The LST cooling trend was most prominent on the northern TP on an annual scale and during spring, summer, and autumn (Figure 7a–d). During the winter, cooling mostly occurred in the southwestern TP. The spatial distribution of significance trends ($p < 0.05$) showed that the most significant decrease in LST occurs at annual scale and in spring (Figure 7f,j), with 7.06% and 7.93% of all pixels recording these changes, respectively. Significant LST increases ($p < 0.05$) in autumn were represented by 10.60% of all pixels (Figure 7i), which covered a much larger area than that for spring (5.01%; Figure 7g), summer (6.71%; Figure 7h), and winter (1.42%; Figure 7j).

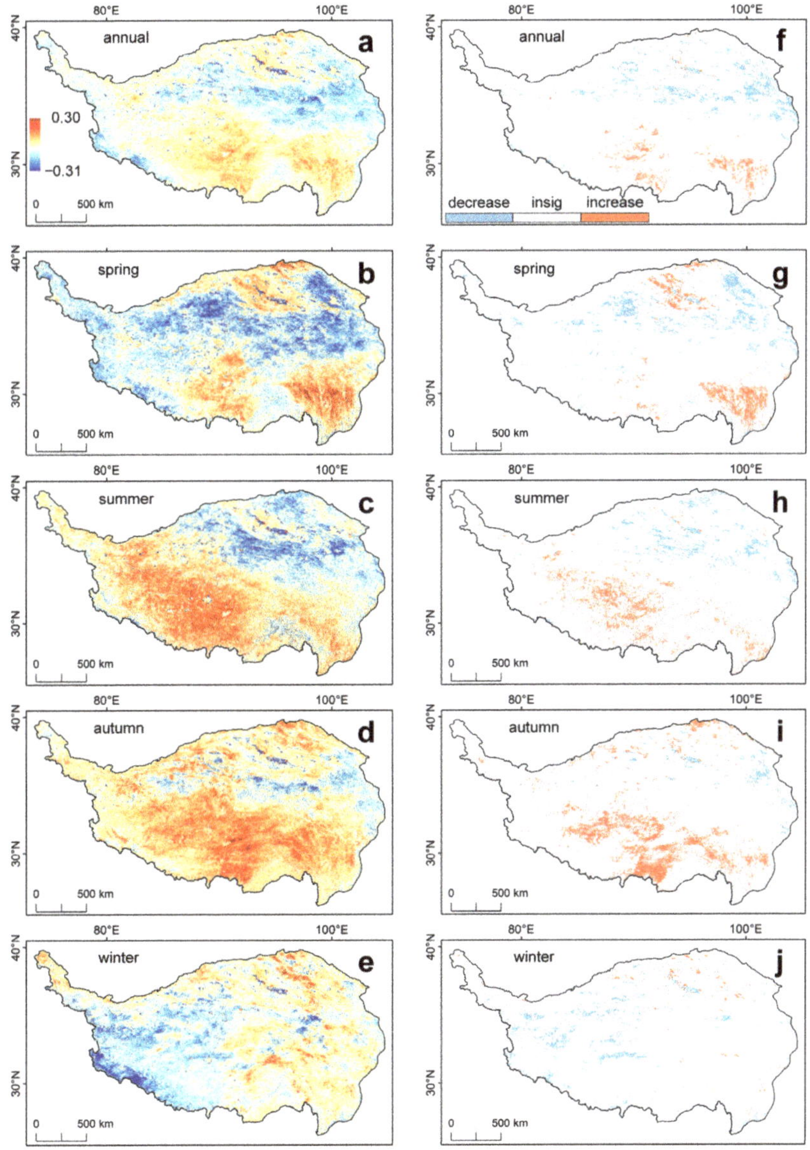

Figure 7. MODIS-based LST trends during 2000–2021 at annual and seasonal scales. (**a**–**e**) denote the LST trends for 2000–2021 in annual, spring, summer, autumn, and winter, respectively; (**f**–**j**) denote the LST trends with $p < 0.05$ for 2000–2021 in annual, spring, summer, autumn, and winter, respectively.

3.2.2. NDVI Trends at Different Timescales

Spatial variations in the surface parameters across the study region revealed that MODIS-based NDVI data show significantly more greening trends than GIMMS-based NDVI data (Figure 8). These NDVI data showed that the eastern TP records the largest greening area during 2000–2021 at all timescales (Figure 8a–e), and the greening rate in the spring is significantly higher than for the other timescales (Figure 8b). The southern TP showed the highest NDVI browning rate (Figure 8a–d), which was highest in autumn out of all the considered timescales (Figure 8d). The NDVI data recorded an increasing trend on an annual scale, with 40.16% of pixels showing a significant increase and only 3.56% showing a significant decrease trend (Figure 8f, $p < 0.05$). Approximately 44.62% of the NDVI pixels showed a significant increase in the spring, and 2.53% significantly decreased (Figure 8g). Furthermore, 38.84% of the NDVI pixels showed significant greening ($p < 0.05$), and 3.49% showed a considerable browning during the summer (Figure 8h). An approximately equal proportion of NDVI pixels showed browning (21.13%) and greening (21.35%) in autumn (Figure 8i). Finally, 52.56% and 2.24% of the NDVI pixels recorded greening and browning in winter, respectively, with these values being significantly higher than their equivalents during the other seasons (Figure 8j).

3.2.3. Albedo Trends at Different Timescales

Albedo trends at all timescales showed relatively few spatial hotspots except for in winter (Figure 9a–e). The fastest albedo growth rates during 2000–2021 occurred in spring and winter (Figure 9b,e); however, the albedo trends in summer and autumn were less pronounced (Figure 9c,d). The decreasing rate of albedo was similar to its increasing rate in the temporal pattern. A significant increase in albedo was observed in the western plateau region at all timescales, and a significant decrease is found in the south and northeast plateau (Figure 9f–j, $p < 0.05$). In terms of areas distribution, a relatively high percentage of pixels showed a significant increase during spring (7.78%; Figure 9g) and winter (4.92%; Figure 9j), and they were notably higher than the values of 0.49% in summer and 0.38% in autumn. This effect may have been due to snow accumulation in spring and winter. The largest areas with significant decreasing trends in albedo were monitored in summer and autumn (Figure 9h,i, $p < 0.05$). These areas comprised 26.30% and 16.36% of the total number of pixels in the study region, respectively, and these changes might potentially be linked to increased vegetation coverage during the growing season.

3.2.4. Vegetation Impacts on LST at Different Time Scales

A statistical method was used to quantify the individual effect of vegetation on LST, and we simultaneously considered the varying albedo (see Section 3.2.3) due to its significant role in the vegetation–LST relationship [46,47]. Specifically, we employed the detrending method and partial correlation analysis to analyze the individual effects of the NDVI and albedo on LST. First, we examined the linear trends of the NDVI and albedo (see the Supporting Information S3 for details). The results suggested that the NDVI of the TP shows significant temporal trends at different timescales, and the summer albedo also shows significant temporal trends (Table 2). Consequently, we filtered the temporal trends of the indicated six variables using the detrending method. Second, we fixed NDVI (albedo) to analyze the correlation between albedo (NDVI) and LST. The results showed that the partial correlation coefficients of the NDVI and albedo with LST are significant except in summer (Table 3). This may be due to the saturation effect of the summer vegetation and albedo contributions on the LST [48], resulting in their insignificant trend of contribution. The individual contribution of albedo to LST was larger than that of NDVI at all timescales, especially in spring and autumn. This may be attributed to the fact that (1) the individual contribution of albedo to LST includes both the indirect contribution of vegetation altering albedo [49] and the direct contribution of albedo; and (2) an advanced vegetation growing season and a delayed vegetation ending season can alter surface albedo strongly [50,51], causing the contribution of surface albedo to be larger in spring and autumn. In terms of winter, faded vegetation and deepened snow cover would reduce the contribution of vegetation but increase the contribution of albedo [47].

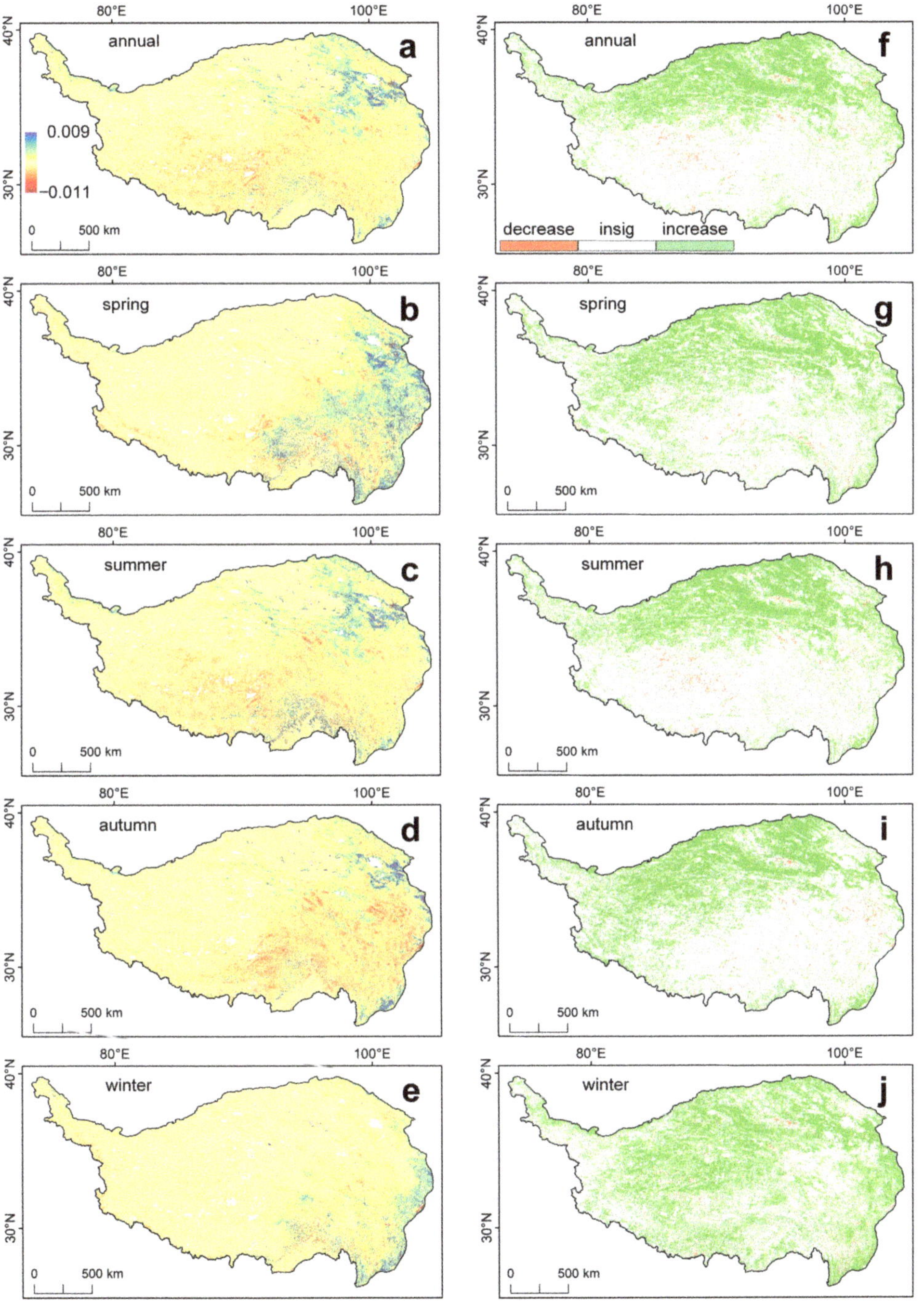

Figure 8. MODIS-based NDVI trends during 2000–2021 at annual and seasonal scales. (**a**–**e**) denote the NDVI trends for 2000–2021 in annual, spring, summer, autumn, and winter, respectively; (**f**–**j**) denote the NDVI trends with $p < 0.05$ for 2000–2021 in annual, spring, summer, autumn, and winter, respectively.

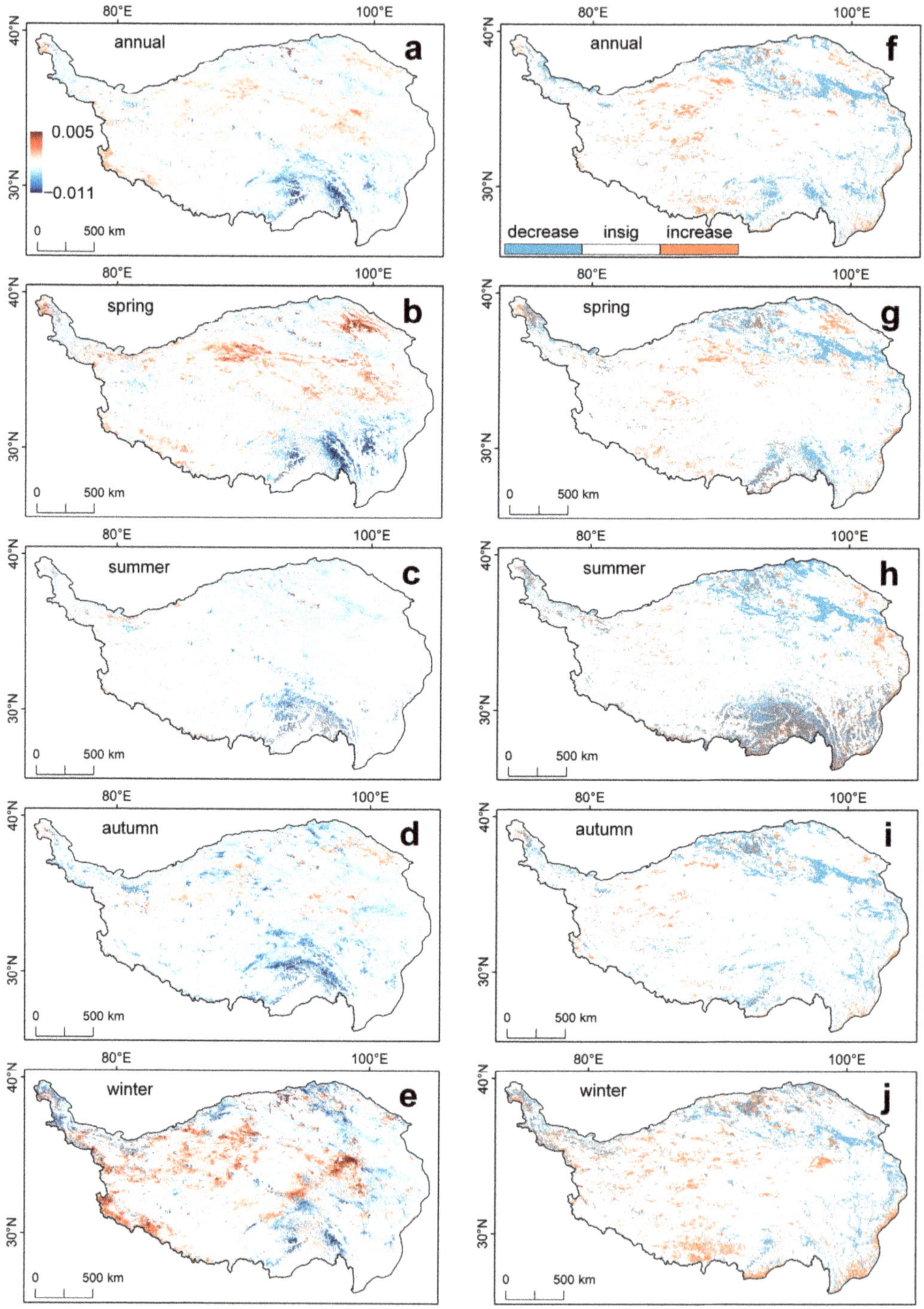

Figure 9. MODIS-based albedo trends during 2000–2021 at annual and seasonal scales. (**a**–**e**) denote the albedo trends for 2000–2021 in annual, spring, summer, autumn, and winter, respectively; (**f**–**j**) denote the albedo trends with $p < 0.05$ for 2000–2021 in annual, spring, summer, autumn, and winter, respectively.

Table 2. Significance of temporal trends in NDVI and albedo over 2000–2021.

Independent Variables	R^2	F-Test	Sig.
MODIS-based annual NDVI	0.628	33.729	0.000
MODIS-based spring NDVI	0.437	15.509	0.001
MODIS-based summer NDVI	0.581	27.779	0.000
MODIS-based autumn NDVI	0.412	14.040	0.001
MODIS-based winter NDVI	0.567	26.174	0.000
MODIS-based annual albedo	0.000	0.004	0.950
MODIS-based spring albedo	0.008	0.156	0.697
MODIS-based summer albedo	0.490	19.188	0.000
MODIS-based autumn albedo	0.030	0.608	0.445
MODIS-based winter albedo	0.050	1.061	0.315
MODIS-based annual NDVI	0.628	33.729	0.000

Table 3. Partial correlation coefficients and significance of NDVI (albedo) with LST after fixing for albedo (NDVI).

MODIS-Based LST	MODIS-Based NDVI		MODIS-Based Albedo	
	Partial Correlation Coefficient	Two-Tailed Test	Partial Correlation Coefficient	Two-Tailed Test
annual	−0.478	0.028	−0.603	0.004
spring	0.467	0.033	−0.856	0.000
summer	−0.283	0.214	−0.277	0.223
autumn	−0.436	0.048	−0.883	0.000
winter	0.041	0.858	−0.803	0.000

Note that we utilized a statistical method to elucidate the relationship between vegetation, albedo, and LST, but we acknowledge its limitations due to the absence of a detailed surface radiation balance analysis. Achieving an accurate decomposition of surface radiation components would require further discussion beyond the scope and word limit of this paper. Therefore, for the purpose of this study, we considered the statistical method as a reasonable proxy for interpreting the individual impacts of vegetation and albedo on LST at various timescales. The individual impacts of vegetation and albedo to LST using surface radiative balance methods is expected in the near future. In addition, vegetation data are considered in producing LST products, yet it is difficult to estimate the contribution of vegetation data due to a lack of robust methods. Therefore, the vegetation contribution to LST estimation may have some uncertainties in the dense vegetation growth of the southeastern TP.

4. Discussion

Vegetation–temperature relationships can be affected by estimation methods, data sources, and land cover types. We therefore discussed the breakpoint method for vegetation trends, the differences between GIMMS and MODIS NDVI datasets, and the impacts of land cover types on annual NDVI trends.

Firstly, we analyzed vegetation–temperature relationships during different warming phases. In terms of methodology, the relationship between vegetation and climate variables on the TP presented remarkable phase changes. However, the phase division of such a relationship was usually subject to the limitation of the time span of remote sensing observed vegetation datasets and the study period. For example, a 5-year time step was used to investigate the relationship between climatic and non-climatic factors and vegetation on the TP during 1980–2010 [40], and a visual graphical approach was used to determine the breakpoints in TP vegetation during 1982–2002 [41] as well as artificial time segmentation to investigate the TP vegetation feedback to climatic factors [39,46]. These methods may lead to bias in the vegetation response analysis. We therefore used the breakpoints in temperature change as the reference time points when analyzing the TP vegetation changes at different phases. We found that the breakpoints of TP temperature during 1982–2013 at different time scales largely occurred between 1994 and 1998 (Figures 2–6). This result is not only closer to direct NDVI segmentation investigation on the TP [52], but it is also closer to the artificial breakpoint of 2000 years in some investigations of TP NDVI response analysis [39,46]. Nevertheless, our results strongly imply that the artificial abrupt points later than 2000 may bring greater bias in TP NDVI response analysis. Our study also found that a common warming trend occurred at all timescales during 1982–2013, and the breakpoints of the TP were earlier in spring and autumn than those annually and in the summer and winter (Figures 2–6). This indicates that the growth and end season of vegetation may advance the warming breakpoint of the TP [50]. With the TP warming, the second phase of NDVI increase was significant at all timescales except winter, implying potential temperature accumulation induced by the first warming phase contributing to the vegetation growth in the second phase.

Secondly, our investigation demonstrated a significant NDVI trend difference between the GIMMS and MODIS datasets. Overall, NDVI trends were greater in MODIS data than in GIMMS, and this finding was consistent with [46,53]. The GIMMS NDVI data did not show a significant increase in 2000–2013, which was also found in [46,54]. This may be attributed to the wider NIR band [55] and the imperfect atmospheric effects processing technique for GIMMS data [56]. Fortunately, MODIS data are usually more reliable than GIMMS data and fill the GIMMS data gap well [57]. For example, our results indicated a significant increase in MODIS NDVI trends during 2000–2021 (Figure 8). Nevertheless, both GIMMS and MODIS products showed a common trend of vegetation changes on the TP (Figures 3a and 8a). However, given the advantages of the long time series of GIMMS data and the moderate accuracy of MODIS data, it is of great significance to integrate multiple datasets to analyze the future vegetation and climatic factors on the TP.

Thirdly, given the recent warm and wet trends on the TP [31,58], we further estimated the relative impact of NDVI changes in different land cover types on annual NDVI trend on the TP. Based on the MODIS land cover product, we detected annual land cover data in 2001 and 2021 and produced change pixels and unchanged land cover types on the TP. The results showed that 89% of the land cover types on the TP have remained unchanged over the 2001–2021 period (Figure 10). Given that the unchanged pixels cover the majority of the TP, a stepwise backward method based on these areas was employed to filter the intended land cover types and then evaluate their relative impact on the estimation of NDVI trends of TP. The results suggested that only bare land has a relative impact of 9.746% on the TP NDVI trend, but the relative impact of other land cover types on the TP NDVI trend does not exceed 5% (Table 4). This implied that the NDVI growth of bare land contributes significantly to the TP vegetation. In addition, we also found that the mean NDVI value of the TP reaches the highest after filtering the bare land compared to other cover types (Figure 11). Accordingly, we suggest that future NDVI trend and intensity studies on the TP should pay more attention to vegetation changes in bare land.

Table 4. Relative impact of land cover on annual NDVI trends on the Tibetan Plateau.

Filtered Land Cover	Trend	R^2	Sig.	Relative Impact (%)
Evergreen Needleleaf Forests	10.839	0.625	0.000	0.324
Evergreen Broadleaf Forests	10.826	0.624	0.000	0.204
Deciduous Needleleaf Forests	10.804	0.628	0.000	0.000
Deciduous Broadleaf Forests	10.798	0.627	0.000	0.056
Mixed Forests	10.828	0.622	0.000	0.222
Closed Shrublands	10.804	0.628	0.000	0.000
Open Shrublands	10.814	0.628	0.000	0.093
Woody Savannas	10.785	0.621	0.000	0.176
Savannas	10.782	0.626	0.000	0.204
Grasslands	10.986	0.883	0.000	1.685
Permanent Wetlands	10.802	0.628	0.000	0.019
Croplands	10.787	0.628	0.000	0.157
Urban and Built-Up Lands	10.809	0.628	0.000	0.046
Cropland/Natural Vegetation Mosaics	10.804	0.628	0.000	0.000

Table 4. *Cont.*

Filtered Land Cover	Trend	R^2	Sig.	Relative Impact (%)
Permanent Snow and Ice	10.834	0.625	0.000	0.278
Barren	11.857	0.518	0.000	9.746
Water Bodies	10.293	0.599	0.000	4.730
Null	10.804	0.628	0.000	/

Notes: Null means no filtered land cover type.

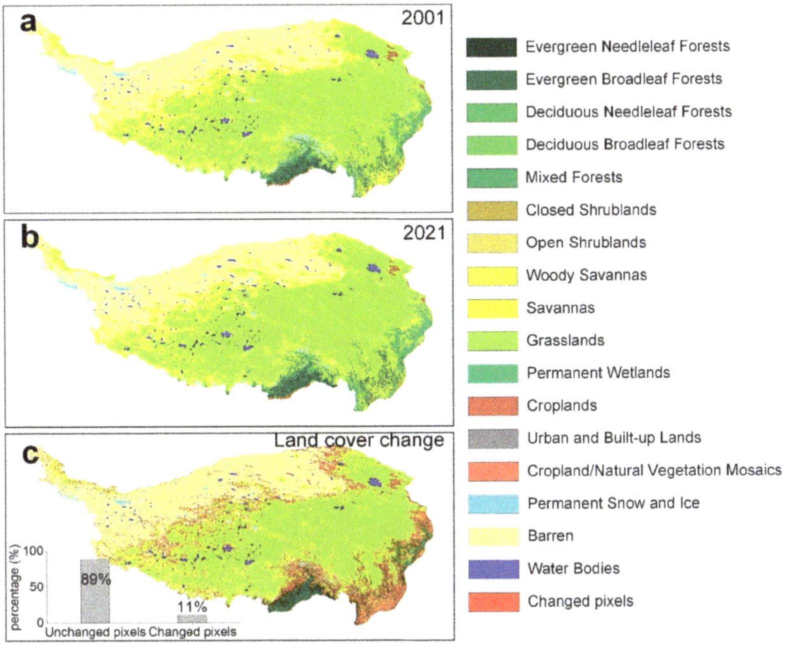

Figure 10. Land cover change detection on the Tibetan Plateau during 2001–2021.

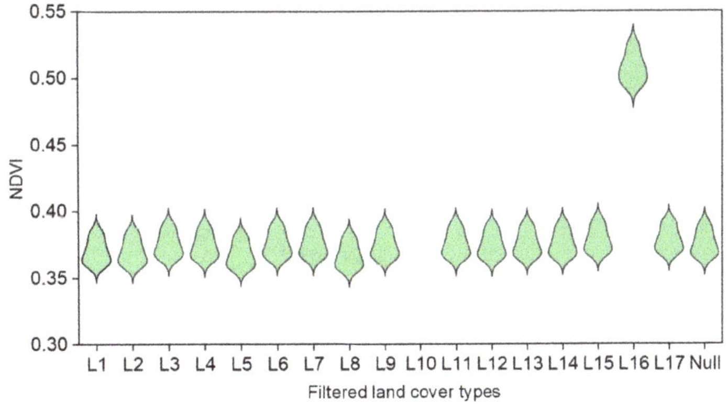

Figure 11. Mean NDVI distribution for 2000–2021 after removing different land cover types. L1, L2, L3, L4, L5, L6, L7, L8, L9, L10, L11, L12, L13, L14, L15, L16, and L17 represent evergreen needleleaf forests, evergreen broadleaf forests, deciduous needleleaf forests, deciduous broadleaf forests, mixed forests, closed shrublands, open shrublands, woody savannas, savannas, grasslands, permanent wetlands, croplands, urban and built-up lands, cropland/natural vegetation, mosaics, permanent snow and ice, barren, and water bodies, respectively. Null means no filtered land cover type.

5. Conclusions

This study aimed to investigate the differences in vegetation changes during various warming phases on the Tibetan Plateau (TP) as well as the trends and patterns of NDVI at different timescales and its influence on LST over the past two decades. First, the warming trend on the TP occurred at different timescales, corresponding to breakpoints that took place in 1994–1998, with earlier occurrences observed in spring and autumn. Secondly, as the warming of the TP continued, the vegetation in the second warming phase exhibited a larger greening area. These findings suggested that the time of temperature breakpoints may have an impact on the area of vegetation increase, and NDVI changes are more sensitive to air temperature during the recent warming phase. Third, MODIS data highlighted more greening compared to GIMMS data. Using MODIS data, we also found the fastest NDVI increase trend in spring and the fastest LST warming in autumn. The partial correlation analysis indicated that NDVI has a significant negative impact on LST during the annual scale and autumn while also having a significant positive impact on LST during spring. This suggests that the contribution of NDVI to LST varies across different timescales since 2000. To summarize, our findings systematically uncover the spatiotemporal patterns of air temperature, LST, and NDVI on the TP across different timescales. These results provide significant insights into the annual and seasonal patterns of vegetation responses and feedback to climate change on the TP. Furthermore, our analysis reveals distinct seasonal trends between NDVI and LST, which can be leveraged to enhance the accuracy of numerical simulations that aim to replicate the relationships between vegetation and climate over the TP.

Supplementary Materials: The following supporting information can be downloaded at: https://www.mdpi.com/article/10.3390/rs15092475/s1, Section S1: Detailed procedures of SQMK test; Section S2: Detailed procedures of Sen's slope; Section S3: Overall trends of NDVI, LST, and albedo over the Tibetan Plateau (Figure S1: MODIS-based NDVI, LST, and albedo trends over the Tibetan Plateau at annual and seasonal scales during 2000–2021) [59,60].

Author Contributions: Conceptualization, F.W. and Y.M.; methodology, F.W.; manuscript, F.W. and Y.M.; funding acquisition, Y.M. Writing—review and editing, F.W., Y.M., R.D. and C.H.; All authors have read and agreed to the published version of the manuscript.

Funding: This research was funded by the Second Tibetan Plateau Scientific Expedition and Research (STEP) program (2019QZKK0103) and the National Natural Science Foundation of China (42230610 and 91837208).

Data Availability Statement: All input data for this research are publicly available: global inventory modeling and mapping studies (GIMMS) NDVI data (https://climatedataguide.ucar.edu/climate-data/ndvi-normalized-difference-vegetation-index-3rd-generation-nasagfsc-gimms), moderate-resolution imaging spectroradiometer (MODIS) NDVI data (https://lpdaac.usgs.gov/products/mod13a2v061), moderate-resolution imaging spectroradiometer (MODIS) LST data (https://lpdaac.usgs.gov/products/mod11a2v061), moderate-resolution imaging spectroradiometer (MODIS) albedo data (https://lpdaac.usgs.gov/products/mcd43a3v061), moderate-resolution imaging spectroradiometer (MODIS) Land Cover Type data (https://lpdaac.usgs.gov/products/mcd12q1v061), and surface-near air temperature (https://www.ncdc.noaa.gov/cdo-web).

Acknowledgments: The author would like to acknowledge all the members of the Land–Atmosphere Interaction and Its Climatic Effects Group. We also thank reviewers for their valuable comments and the China Scholarship Council (CSC).

Conflicts of Interest: The authors declare no conflict of interest.

References

1. Gerten, D.; Schaphoff, S.; Haberlandt, U.; Lucht, W.; Sitch, S. Terrestrial vegetation and water balance—Hydrological evaluation of a dynamic global vegetation model. *J. Hydrol.* **2004**, *286*, 249–270. [CrossRef]
2. Zuo, Z.; Zhang, R.; Zhao, P. The relation of vegetation over the Tibetan Plateau to rainfall in China during the boreal summer. *Clim. Dyn.* **2011**, *36*, 1207–1219. [CrossRef]

3. Piao, S.; Wang, X.; Park, T.; Chen, C.; Lian, X.; He, Y.; Bjerke, J.W.; Chen, A.; Ciais, P.; Tømmervik, H.; et al. Characteristics, drivers and feedbacks of global greening. *Nat. Rev. Earth Environ.* **2020**, *1*, 14–27. [CrossRef]
4. Forzieri, G.; Alkama, R.; Miralles, D.G.; Cescatti, A. Satellites reveal contrasting responses of regional climate to the widespread greening of Earth. *Science* **2017**, *356*, 1180–1184. [CrossRef] [PubMed]
5. Zeng, Z.; Piao, S.; Li, L.Z.X.; Zhou, L.; Ciais, P.; Wang, T.; Li, Y.; Lian, X.; Wood, E.F.; Friedlingstein, P.; et al. Climate mitigation from vegetation biophysical feedbacks during the past three decades. *Nat. Clim. Chang.* **2017**, *7*, 432–436. [CrossRef]
6. Zhong, L.; Ma, Y.; Xue, Y.; Piao, S. Climate Change Trends and Impacts on Vegetation Greening over the Tibetan Plateau. *J. Geophys. Res. Atmos.* **2019**, *124*, 7540–7552. [CrossRef]
7. Yu, H.; Luedeling, E.; Xu, J. Winter and spring warming result in delayed spring phenology on the Tibetan Plateau. *Proc. Natl. Acad. Sci. USA* **2010**, *107*, 22151–22156. [CrossRef]
8. Zhou, H.K.; Yao, B.Q.; Xu, W.X.; Ye, X.; Fu, J.J.; Jin, Y.X.; Zhao, X.Q. Field evidence for earlier leaf-out dates in alpine grassland on the eastern Tibetan Plateau from 1990 to 2006. *Biol. Lett.* **2014**, *10*, 20140291. [CrossRef]
9. Ding, M.; Zhang, Y.; Liu, L.; Zhang, W.; Wang, Z.; Bai, W. The relationship between NDVI and precipitation on the Tibetan Plateau. *J. Geogr. Sci.* **2007**, *17*, 259–268. [CrossRef]
10. Pang, G.; Wang, X.; Yang, M. Using the NDVI to identify variations in, and responses of, vegetation to climate change on the Tibetan Plateau from 1982 to 2012. *Quat. Int.* **2017**, *444*, 87–96. [CrossRef]
11. Duan, A.; Wu, G.; Liu, Y.; Ma, Y.; Zhao, P. Weather and climate effects of the Tibetan Plateau. *Adv. Atmos. Sci.* **2012**, *29*, 978–992. [CrossRef]
12. Latif, A.; Ilyas, S.; Zhang, Y.; Xin, Y.; Zhou, L.; Zhou, Q. Review on global change status and its impacts on the Tibetan Plateau environment. *J. Plant Ecol.* **2019**, *12*, 917–930. [CrossRef]
13. Li, Y.; Zhao, M.; Motesharrẽi, S.; Mu, Q.; Kalnay, E.; Li, S. Local cooling and warming effects of forests based on satellite observations. *Nat. Commun.* **2015**, *6*, 6603. [CrossRef] [PubMed]
14. Verrelst, J.; Camps-Valls, G.; Muñoz-Marí, J.; Rivera, J.P.; Veroustraete, F.; Clevers, J.G.P.W.; Moreno, J. Optical remote sensing and the retrieval of terrestrial vegetation bio-geophysical properties—A review. *ISPRS J. Photogramm. Remote Sens.* **2015**, *108*, 273–290. [CrossRef]
15. Bright, R.M.; Davin, E.; O'halloran, T.; Pongratz, J.; Zhao, K.; Cescatti, A. Local temperature response to land cover and management change driven by non-radiative processes. *Nat. Clim. Chang.* **2017**, *7*, 296–302. [CrossRef]
16. Ge, J.; Guo, W.; Pitman, A.J.; De Kauwe, M.G.; Chen, X.; Fu, C. The Nonradiative Effect Dominates Local Surface Temperature Change Caused by Afforestation in China. *J. Clim.* **2019**, *32*, 4445–4471. [CrossRef]
17. Shen, M.; Piao, S.; Jeong, S.-J.; Zhou, L.; Zeng, Z.; Ciais, P.; Chen, D.; Huang, M.; Jin, C.-S.; Li, L.Z.X.; et al. Evaporative cooling over the Tibetan Plateau induced by vegetation growth. *Proc. Natl. Acad. Sci. USA* **2015**, *112*, 9299–9304. [CrossRef]
18. Zhong, L.; Ma, Y.; Salama, M.S.; Su, Z. Assessment of vegetation dynamics and their response to variations in precipitation and temperature in the Tibetan Plateau. *Clim. Chang.* **2010**, *103*, 519–535. [CrossRef]
19. Huete, A.R.; Liu, H.Q.; Batchily, K.V.; van Leeuwen, W. A comparison of vegetation indices over a global set of TM images for EOS-MODIS. *Remote Sens. Environ.* **1997**, *59*, 440–451. [CrossRef]
20. Piao, S.; Fang, J.; Liu, H.; Zhu, B. NDVI-indicated decline in desertification in China in the past two decades. *Geophys. Res. Lett.* **2005**, *32*, L06402. [CrossRef]
21. Yuan-He, Y.; Shi-Long, P. Variations in grassland vegetation cover in relation to climatic factors on the Tibetan Plateau. *Chin. J. Plant Ecol.* **2006**, *30*, 1–8. [CrossRef]
22. Lin, X.; Wen, J.; Liu, Q.; You, D.; Wu, S.; Hao, D.; Xiao, Q.; Zhang, Z.; Zhang, Z. Spatiotemporal Variability of Land Surface Albedo over the Tibet Plateau from 2001 to 2019. *Remote Sens.* **2020**, *12*, 1188. [CrossRef]
23. Guo, D.; Sun, J.; Yang, K.; Pepin, N.; Xu, Y.; Xu, Z.; Wang, H. Satellite data reveal southwestern Tibetan plateau cooling since 2001 due to snow-albedo feedback. *Int. J. Clim.* **2020**, *40*, 1644–1655. [CrossRef]
24. Salama, M.S.; Van Der Velde, R.; Zhong, L.; Ma, Y.; Ofwono, M.; Su, Z. Decadal variations of land surface temperature anomalies observed over the Tibetan Plateau by the Special Sensor Microwave Imager (SSM/I) from 1987 to 2008. *Clim. Chang.* **2012**, *114*, 769–781. [CrossRef]
25. Luo, D.L.; Jin, H.J.; He, R.X.; Wang, X.F.; Muskett, R.R.; Marchenko, S.S.; Romanovsky, V.E. Characteristics of Water-Heat Exchanges and Inconsistent Surface Temperature Changes at an Elevational Permafrost Site on the Qinghai-Tibet Plateau. *J. Geophys. Res. Atmos.* **2018**, *123*, 10057–10075. [CrossRef]
26. Stroppiana, D.; Antoninetti, M.; Brivio, P.A. Seasonality of MODIS LST over Southern Italy and correlation with land cover, topography and solar radiation. *Eur. J. Remote Sens.* **2014**, *47*, 133–152. [CrossRef]
27. Zhou, Y.; Ran, Y.; Li, X. The contributions of different variables to elevation-dependent land surface temperature changes over the Tibetan Plateau and surrounding regions. *Glob. Planet. Chang.* **2023**, *220*, 104010. [CrossRef]
28. Zhu, W.; Lű, A.; Jia, S. Estimation of daily maximum and minimum air temperature using MODIS land surface temperature products. *Remote Sens. Environ.* **2013**, *130*, 62–73. [CrossRef]
29. Yang, M.; Zhao, W.; Zhan, Q.; Xiong, D. Spatiotemporal Patterns of Land Surface Temperature Change in the Tibetan Plateau Based on MODIS/Terra Daily Product From 2000 to 2018. *IEEE J. Sel. Top. Appl. Earth Obs. Remote Sens.* **2021**, *14*, 6501–6514. [CrossRef]

30. Yang, K.; Wu, H.; Qin, J.; Lin, C.; Tang, W.; Chen, Y. Recent climate changes over the Tibetan Plateau and their impacts on energy and water cycle: A review. *Glob. Planet. Chang.* **2014**, *112*, 79–91. [CrossRef]
31. Li, L.; Yang, S.; Wang, Z.; Zhu, X.; Tang, H. Evidence of Warming and Wetting Climate over the Qinghai-Tibet Plateau. *Arct. Antarct. Alp. Res.* **2010**, *42*, 449–457. [CrossRef]
32. Zhang, Q.; Singh, V.P.; Li, J.; Chen, X. Analysis of the periods of maximum consecutive wet days in China. *J. Geophys. Res. Atmos.* **2011**, *116*, D23106. [CrossRef]
33. Guan, Y.; Zheng, F.; Zhang, X.; Wang, B. Trends and variability of daily precipitation and extremes during 1960–2012 in the Yangtze River Basin, China. *Int. J. Clim.* **2017**, *37*, 1282–1298. [CrossRef]
34. Bachelet, D.; Neilson, R.P.; Lenihan, J.M.; Drapek, R.J. Climate Change Effects on Vegetation Distribution and Carbon Budget in the United States. *Ecosystems* **2001**, *4*, 164–185. [CrossRef]
35. Ma, M.; Frank, V. Interannual variability of vegetation cover in the Chinese Heihe River Basin and its relation to meteorological parameters. *Int. J. Remote Sens.* **2006**, *27*, 3473–3486. [CrossRef]
36. Anyamba, A.; Small, J.L.; Tucker, C.J.; Pak, E.W. Thirty-two Years of Sahelian Zone Growing Season Non-Stationary NDVI3g Patterns and Trends. *Remote Sens.* **2014**, *6*, 3101–3122. [CrossRef]
37. Holben, B.N. Characteristics of maximum-value composite images from temporal AVHRR data. *Int. J. Remote Sens.* **1986**, *7*, 1417–1434. [CrossRef]
38. Zheng, L.; Qi, Y.; Qin, Z.; Xu, X.; Dong, J. Assessing albedo dynamics and its environmental controls of grasslands over the Tibetan Plateau. *Agric. For. Meteorol.* **2021**, *307*, 108479. [CrossRef]
39. Du, J.; Zhao, C.; Shu, J.; Jiaerheng, A.; Yuan, X.; Yin, J.; Fang, S.; He, P. Spatiotemporal changes of vegetation on the Tibetan Plateau and relationship to climatic variables during multiyear periods from 1982–2012. *Environ. Earth Sci.* **2015**, *75*, 77. [CrossRef]
40. Pan, T.; Zou, X.; Liu, Y.; Wu, S.; He, G. Contributions of climatic and non-climatic drivers to grassland variations on the Tibetan Plateau. *Ecol. Eng.* **2017**, *108*, 307–317. [CrossRef]
41. Zhou, D.; Fan, G.; Huang, R.; Fang, Z.; Liu, Y.; Li, H. Interannual variability of the normalized difference vegetation index on the Tibetan Plateau and its relationship with climate change. *Adv. Atmos. Sci.* **2007**, *24*, 474–484. [CrossRef]
42. Wang, X.; Xiao, J.; Li, X.; Cheng, G.; Ma, M.; Che, T.; Dai, L.; Wang, S.; Wu, J. No Consistent Evidence for Advancing or Delaying Trends in Spring Phenology on the Tibetan Plateau. *J. Geophys. Res. Biogeosci.* **2017**, *122*, 3288–3305. [CrossRef]
43. McLeod, A.I. Kendall rank correlation and Mann-Kendall trend test. *R Package Kendall* **2005**, *602*, 1–10. Available online: https://cran.r-project.org/web/packages/Kendall/index.html (accessed on 5 April 2023).
44. Mann, H.B. Nonparametric tests against trend. *Econometrica* **1945**, *13*, 245–259. [CrossRef]
45. Sen, P.K. Estimates of the regression coefficient based on Kendall's Tau. *J. Am. Stat. Assoc.* **1968**, *63*, 1379–1389. [CrossRef]
46. Querin, C.A.S.; Beneditti, C.A.; Machado, N.G.; da Silva, M.J.G.; da Silva Querino, K.A.; dos Santos Neto, L.A.; Biudes, M.S. Spatiotemporal NDVI, LAI, albedo, and surface temperature dynamics in the southwest of the Brazilian Amazon forest. *J. Appl. Remote Sens.* **2016**, *10*, 026007. [CrossRef]
47. Pang, G.; Chen, D.; Wang, X.; Lai, H.-W. Spatiotemporal variations of land surface albedo and associated influencing factors on the Tibetan Plateau. *Sci. Total. Environ.* **2022**, *804*, 150100. [CrossRef]
48. Lian, X.; Jiao, L.; Liu, Z. Saturation response of enhanced vegetation productivity attributes to intricate interactions. *Glob. Chang. Biol.* **2023**, *29*, 1080–1095. [CrossRef]
49. Tian, L.; Zhang, Y.; Zhu, J. Decreased surface albedo driven by denser vegetation on the Tibetan Plateau. *Environ. Res. Lett.* **2014**, *9*, 104001. [CrossRef]
50. Shen, M.; Zhang, G.; Cong, N.; Wang, S.; Kong, W.; Piao, S. Increasing altitudinal gradient of spring vegetation phenology during the last decade on the Qinghai-Tibetan Plateau. *Agric. For. Meteorol.* **2014**, *189–190*, 71–80. [CrossRef]
51. Du, J.; He, P.; Fang, S.; Liu, W.; Yuan, X.; Yin, J. Autumn NDVI contributes more and more to vegetation improvement in the growing season across the Tibetan Plateau. *Int. J. Digit. Earth* **2017**, *10*, 1098–1117. [CrossRef]
52. Wang, Z.; Wang, Q.; Wu, X.; Zhao, L.; Yue, G.; Nan, Z.; Wang, P.; Yi, S.; Zou, D.; Qin, Y.; et al. Vegetation Changes in the Permafrost Regions of the Qinghai-Tibetan Plateau from 1982–2012: Different Responses Related to Geographical Locations and Vegetation Types in High-Altitude Areas. *PLoS ONE* **2017**, *12*, e0169732. [CrossRef] [PubMed]
53. Liu, L.; Wang, Y.; Wang, Z.; Li, D.; Zhang, Y.; Qin, D.; Li, S. Elevation-dependent decline in vegetation greening rate driven by increasing dryness based on three satellite NDVI datasets on the Tibetan Plateau. *Ecol. Indic.* **2019**, *107*, 105569. [CrossRef]
54. Zhang, G.; Zhang, Y.; Dong, J.; Xiao, X. Green-up dates in the Tibetan Plateau have continuously advanced from 1982 to 2011. *Proc. Natl. Acad. Sci. USA* **2013**, *110*, 4309–4314. [CrossRef] [PubMed]
55. Guo, X.; Zhang, H.; Wu, Z.; Zhao, J.; Zhang, Z. Comparison and Evaluation of Annual NDVI Time Series in China Derived from the NOAA AVHRR LTDR and Terra MODIS MOD13C1 Products. *Sensors* **2017**, *17*, 1298. [CrossRef] [PubMed]
56. Donohue, R.J.; Roderick, M.L.; McVicar, T.R. Deriving consistent long-term vegetation information from AVHRR reflectance data using a cover-triangle-based framework. *Remote Sens. Environ.* **2008**, *112*, 2938–2949. [CrossRef]
57. Beck, P.S.; Goetz, S.J. Satellite observations of high northern latitude vegetation productivity changes between 1982 and 2008: Ecological variability and regional differences. *Environ. Res. Lett.* **2011**, *6*, 045501. [CrossRef]
58. Wang, J.; Chen, X.; Hu, Q.; Liu, J. Responses of terrestrial water storage to climate variation in the Tibetan Plateau. *J. Hydrol.* **2020**, *584*, 124652. [CrossRef]

59. Hussain, A.; Cao, J.; Ali, S.; Ullah, W.; Muhammad, S.; Hussain, I.; Rezaei, A.; Hamal, K.; Akhtar, M.; Abbas, H.; et al. Variability in Runoff and Responses to Land and Oceanic Parameters in the Source Region of the Indus River. *Ecol. Indicators* **2022**, *140*, 109014. [CrossRef]
60. Ullah, S.; You, Q.; Ullah, W.; Ali, A. Observed Changes in Precipitation in China-Pakistan Economic Corridor during 1980–2016. *Atmos.* **2018**, *210*, 1–14. [CrossRef]

remote sensing MDPI

Article

Spatiotemporal Variation Characteristics of Groundwater Storage and Its Driving Factors and Ecological Effects in Tibetan Plateau

Wenhao Ren [1,2], Yanyan Gao [1,2], Hui Qian [1,2,*], Yaoming Ma [3,4,5,6], Zhongbo Su [7], Weiqiang Ma [3,4], Yu Liu [8,9] and Panpan Xu [1,2]

1 School of Water and Environment, Chang'an University, Xi'an 710054, China
2 Key Laboratory of Subsurface Hydrology and Ecological Effect in Arid Region of Ministry of Education, Chang'an University, Xi'an 710054, China
3 Land-Atmoshere Interaction and Its Climatic Effects Group, State Key Laboratory of Tibetan Plateau Earth System, Resources and Environment (TPESRE), Institute of Tibetan Plateau Research, Chinese Academy of Sciences, Beijing 100101, China
4 Research Centre of Environment Change and Land Surface Processes, Institute of Tibetan Plateau Research, Chinese Academy of Sciences, Beijing 100101, China
5 University of Chinese Academy of Sciences, Beijing 100101, China
6 College of Atmospheric Science, Lanzhou University, Lanzhou 730000, China
7 Faculty of Geo-Information Science and Earth Observation (ITC), University of Twente, 7500 AE Enschede, The Netherlands
8 College of Water Resources and Architectural Engineering, Northwest A&F University, Xianyang 712100, China
9 Key Laboratory of Agricultural Soil and Water Engineering in Arid and Semiarid Areas, Ministry of Education, Northwest A&F University, Xianyang 712100, China
* Correspondence: qianhui@chd.edu.cn

Abstract: Known as the "Asian Water Tower", the Tibetan Plateau (TP) is a rich water resource and serves an important ecological function. Climate change may cause changes to the water cycle, and these changes may affect the alpine vegetation growth. However, the variation characteristics of groundwater storage (GWS) and its driving factors and associated ecological effects in the TP are poorly understood. In this study, terrestrial water storage changes retrieved by GRACE (Gravity Recovery and Climate Experiment) were combined with GLDAS (Global Land Data Assimilation System) to estimate the GWS changes in the TP. The temporal and spatial variation characteristics of GWS were identified using linear regression and the modified Mann–Kendall (MMK) test, respectively. The analyses showed that the GWS of the TP decreased at an average rate of −0.89 mm/a from January 2003 to December 2021, but since January 2016, it gradually recovered at a rate of 1.47 mm/a. This shows that the GWS in the eastern and northern parts of the TP is decreasing, while the GWS in the western and southern parts is increasing. The influence of climate change on GWS in time and space was determined using the correlation analysis method. Decreased precipitation and permafrost degradation caused by increasing temperatures will lead to a decrease in GWS. On the other hand, rising temperatures may result in an increase in GWS in regions where glaciers are distributed. In this study, the ecological effects were represented by the relationship between GWS and vegetation change. A decline in GWS means that the vegetation will not receive enough water, leading to a decrease in the NDVI and the eventual degradation of grassland to sand, desert, or other kinds of unused land on the TP. On the other hand, an increase in GWS would promote vegetation restoration. The results of this study offer a new opportunity to reveal the groundwater changes in a cryosphere region and to assess the impact of changes in hydrological conditions on ecology.

Keywords: groundwater storage; GRACE; GLDAS; climate change; vegetation response

Citation: Ren, W.; Gao, Y.; Qian, H.; Ma, Y.; Su, Z.; Ma, W.; Liu, Y.; Xu, P. Spatiotemporal Variation Characteristics of Groundwater Storage and Its Driving Factors and Ecological Effects in Tibetan Plateau. *Remote Sens.* **2023**, *15*, 2418. https://doi.org/10.3390/rs15092418

Academic Editor: Marouane Temimi

Received: 16 March 2023
Revised: 19 April 2023
Accepted: 3 May 2023
Published: 5 May 2023

1. Introduction

The Tibetan Plateau (TP) has important ecological functions such as global water circulation, ecological security, and protection [1–3]. It is one of the most sensitive regions to climate change due to its unique geographic location [4–6]. Due to its rich water resources, the TP is also called the "Asian Water Tower"; it has a profound impact on the survival and development of about two billion people downstream and important implications in the protection and sustainability of water resources [7]. Unlike surface water, groundwater is invisible, and its distribution and change are difficult to understand. Compared with extensive and well-developed studies on other surface-water resources (glaciers [8,9], snow [10,11], lakes [12,13], rivers [14,15], etc.), there are relatively few studies on the TP groundwater [16]. Therefore, groundwater is one of the most challenging but most important components of the "Asian Water Tower".

The traditional groundwater monitoring method is to regularly monitor the changes to the groundwater levels by monitoring wells. However, considering its high altitude, extensive size, remote geographical location, harsh climate, and difficult working conditions, the adoption of this method in the TP is unrealistic. Therefore, a new method must urgently be adopted to enable long-term and large-scale monitoring. The development of satellites means that remote sensing technology can be effectively used as a method to monitor water storage changes. The Gravity Recovery and Climate Experiment (GRACE) gravity satellite and GRACE-FO, which is GRACE's follow-up satellite, launched in 2002 and 2018, respectively, were shown to have significant advantages in monitoring the changes in regional terrestrial water storage (TWS). The TWS in the TP have been confirmed to have undergone significant changes [17,18]. Additionally, the changes in GWS can be obtained by removing known contributors (soil moisture, accumulated snow, and plant canopy surface water) from the changes in TWS which were observed by GRACE/FO [18–23]. This method has been widely used, and relatively accurate results were obtained. The difference between the results obtained using the well-based and GRACE model-based GWS trends was not more than 1.5 cm/year in Poland [24]. Xiang et al. [16] quantitated the GWS changes in the TP and the surrounding area from 2003 to 2009 and showed increasing trend rates in eight basins. Li et al. [7] pointed out that the GWS in the endorheic and exorheic TP basins decreased during 2002–2017, with a rate of 1.17 Gt/a and 4.89 Gt/a, respectively. However, alongside continual climate change, GWS has changed significantly. The exploration of the latest changes in GWS in the TP remains challenging.

Changes in groundwater storage are mainly influenced by climate change and human activities. Different levels of climate change in different regions have different impacts on the changes in GWS. Compared with decreased precipitation, the influence of increased temperatures on the decrease in GWS is much more pronounced in Turkey [19]. However, in arid Central Asia, precipitation in mountainous areas is considered the main factor affecting the water storage in the piedmont area, while human activities may have a significant impact on the water storage in the Turgay Valley [25]. Human activities do have a great impact on GWS in some small basins. In the Shiyang River Basin, there are many large reservoirs, and irrigation agriculture has been developed in several large oases. Therefore, groundwater storage has been decreasing in recent years due to human activities [26]. The TP covers a vast area but is sparsely populated, accounting for one quarter of China's total area but less than 1% of China's total population, and it contains large uninhabited areas. Therefore, from the perspective of the entire TP, human activities have a relatively low impact on GWS [7,18]. It is widely acknowledged that the climate in the TP has changed significantly in the past half century, mainly due to climate warming and wetting [27]. Rising temperatures are not only accelerating the melting of glaciers [28], but also permafrost degradation [29], both of which, together with precipitation, directly or indirectly affect GWS. Therefore, the elicidation of the influence of climate factors on GWS is key.

As an indicator of ecological environment change, vegetation is highly dependent on groundwater [30–32], especially when the groundwater level is lower than the root depth [31]. However, in a large number of studies, the effect of conventional climatic indicators on ecological systems has been studied [32–35]. Zhang and Zhou [36] found that grassland has undergone the largest decrease in area, which has decreased by 9.47%, while the LUCC that has undergone the largest increase in area is unused land, which increased by 7.25% in the TP from 1980 to 2018. However, they attributed this difference to temporal and spatial variations in precipitation. Xu et al. [32] pointed out that the effect of the changes to soil water storage on vegetation in the Three Rivers Source Region was considerably greater than the effects of precipitation and temperature. The response of vegetation to groundwater changes has been poorly studied, and the impact of groundwater changes has been ignored in the TP [37].

Therefore, the objectives of the present study are to (1) identify the spatial–temporal characteristics of groundwater storage in the TP and its ten sub-regions during 2002–2021; (2) clarify the spatial–temporal characteristics of climate factors and the influence of climate factors on GWS; and (3) illustrate the spatial–temporal characteristics of vegetation changes and assess the vegetation responses to GWS changes. The results of the present study can act as a reference for the management of groundwater resources in different sub-regions in the TP.

2. Methodology

2.1. Study Area

The study area was in the geographic domain of the Tibetan Plateau (TP) in China and was composed of ten sub-regions with diverse geographical environments (Figure 1), including the Hexi Corridor (HC), the Qaidam Basin (QB), the Yellow River Basin (YRB), the Yangtze River Basin (YB), the Lancang River (upper Mekong River) Basin (L-MRB), the Nu River (upper Salween River) Basin (N-SRB), the Yarlung Zangbo River (upper Brahmaputra River) Basin (YZ-BRB), the Inner Basin (IB), the Sengezangbu River (upper Indus River) Basin (S-IRB), and the Tarim Basin (TB) [38].

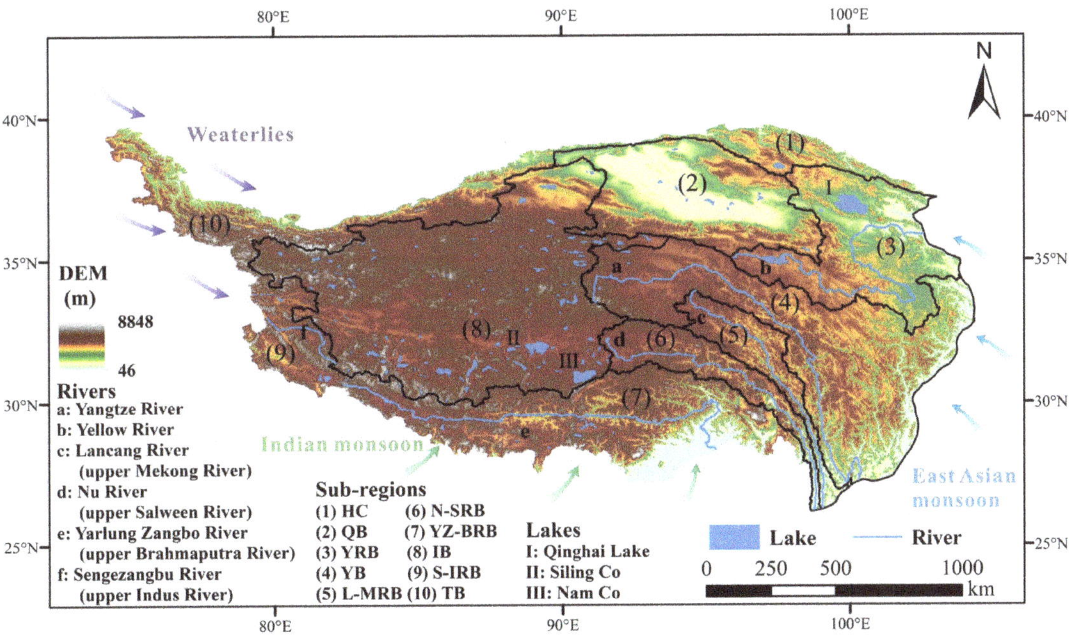

Figure 1. Location map of the TP.

These different sub-regions are characterized by different climate and hydrogeological conditions [3,39]. The western, eastern, and southern TP are affected by the westerlies, the East Asian monsoon, and the Indian monsoon, respectively, and other areas are generally controlled by a combination of two or three of these conditions. Therefore, the water cycle patterns in the TP will be significantly affected by large-scale atmospheric circulation [3]. The abundant precipitation caused by monsoons can greatly replenish large rivers on the edge of the TP. During 1980−2018, the annual runoff of these rivers showed different change characteristics, such as a significant increase in the Sengezangbu River (+3.9 Gt per decade), but a stable status in the Yangtze River and Nu River, while the Yellow River recorded a decline (−1.5 Gt per decade) during the same period [3]. In the center of the TP, there are numerous endorheic lakes, rather than large exorheic rivers, because of the lower annual precipitation level [39].

2.2. Data Sources

2.2.1. GRACE/FO

The CSR GRACE/FO RL06 mascon dataset has been frequently used in the study of terrestrial water storage changes due to its highest spatial resolution at this stage compared with other products [40–42]. The spatial resolution was 0.25 degrees. We used the GRACE data for 237 consecutive months (from April 2002 to December 2021). The cubic spline interpolation method was used to interpolate the short-term missing data of GRACE in the time series. For the long-term data interruption caused by the replacement of the GRACE and GRACE-FO satellites, the reconstruction method based on precipitation, which was invented in previous research, was used, and more detailed information regarding this method can be found in previous studies [43,44]. Additionally, all the grids used in this study were relative to the 2004–2009 mean baseline. The TWS was recorded in terms of equivalent water height (EWH) in cm.

2.2.2. GLDAS

The Global Land Data Assimilation System (GLDAS) uses data assimilation technology to fuse satellite-based data and in situ ground observation data to generate surface state quantities and fluxes that are closest to the observation data [45,46]. GLDAS includes four land surface models, NOAH, VIC, CLM, and MOSAIC. Compared with the other models, the NOAH model has the advantages of being an advanced model, having a high spatial resolution, and having a stable driving field. Thus, the model selected in this study was the GLDAS Noah Land Surface Model L4 monthly 0.25 × 0.25-degree V2.1 [47,48]. It has been widely used worldwide and is a highly recognized data source [24,45,46,49]. The spatial resolution is 0.25 degrees and is consistent with GRACE.

2.2.3. Temperature and Precipitation

Temperature (Tmp) and precipitation (Pre) are the most widely studied and important meteorological factors. Previous studies also showed that Tmp and Pre have significant effects on the variation in GWS [23,50].

The Climatic Research Unit gridded Time Series (CRU TS) can provide 0.5-degree-resolution monthly data covering the global land surface [51]. The dataset of Tmp and Pre from the CRU TS from April 2002 to December 2021 was used in this study. It is worth noting that the spatial resolution was interpolated to 0.25 degrees using the bilinear interpolation method to match the resolution of GRACE.

2.3. Vegetation Response

The vegetation response to the GWS changes was represented by the Normalized Difference Vegetation Index (NDVI) and Land Use and Cover Change (LUCC). The satellite-based NDVI, which has become the most popular index to reflect the state of regional climates and environments, is obtained by monitoring the vegetation growth status and estimating the vegetation coverage without damaging or altering the vegetation [52,53].

The satellite-based LUCC is one of the essential driving factors for regional climate variability, and its impact on hydrological cycle has become an important aspect of water resources [54,55]. The monthly data of the NDVI from April 2002 to December 2019 and the LUCC data of five periods in 2000, 2005, 2010, 2015, and 2020, were used in this study. The NDVI dataset is generated by calculating the maximum values of the first, middle, and last three ten days of each month. The LUCC types are divided into six first-level groups, which include cropland, forest, grassland, water, urban land, and unused land.

2.4. Method

2.4.1. Groundwater Storage Anomalies (GWSA)

The Terrestrial Water Storage Anomaly (TWSA) consists of four parts, namely, groundwater storage anomalies (GWSA), soil storage anomalies (SMSA), snow water equivalent anomalies (SWEA), and canopy water storage anomalies (CWSA) [26]. The GWSA could be isolated using the following formula [26]:

$$\mathrm{GWSA} = \mathrm{TWSA} - (\mathrm{SMSA} + \mathrm{SWEA} + \mathrm{CWSA}) \tag{1}$$

where TWSAs were derived from gravity anomalies observed by GRACE/FO and the other three components were provided by GLDAS.

It should be emphasized that all data are also relative to the 2004–2009 mean baseline, meaning the value of the GWSA can be positive, negative, or zero. If the GWSA calculated for a certain pixel in a certain month are positive, this indicates that the GWSA in this area increase compared with the 2004–2009 mean baseline; a negative value indicates a decrease, and zero indicates no change.

2.4.2. Trend Test and Significance Analysis

Temporally, unitary linear regression analysis was used to construct a linear regression equation, and the k was used to quantify the trends as follows [52,56]:

$$k = \frac{n \times \sum_{i=1}^{n} (i \times x_i) - \sum_{i=1}^{n} i \times \sum_{i=1}^{n} x_i}{n \times \sum_{i=1}^{n} i^2 - (\sum_{i=1}^{n} i)^2} \tag{2}$$

where n is the number of months and x_i is variable in month i. If $k > 0$, the variables show a positive trend; otherwise, they would show a negative trend. $k = 0$ means no change.

Spatially, the change trend for various factors in each pixel was analyzed using the modified Mann–Kendall (MMK) test and represented by *Slope* [52,56–59]. Additionally, whether this trend has statistical significance was determined by t-tests. The corresponding formulas were as follows:

$$Slope = \mathrm{median}\left(\frac{x_j - x_k}{j - k}\right) \quad (1 \le k < j \le n) \tag{3}$$

$$\mathrm{sgn}(x_j - x_k) = \begin{cases} 1 & x_j > x_k \\ 0 & x_j = x_k \\ -1 & x_j < x_k \end{cases} \tag{4}$$

$$S = \sum_{k=1}^{n-1} \sum_{j=k+1}^{n} \mathrm{sgn}(x_j - x_k) \tag{5}$$

$$Var(S) = \frac{n(n-1)(2n+5)}{18} \tag{6}$$

$$\eta = 1 + \frac{2}{n(n-1)(n-2)} \times \sum_{k=1}^{n-1} (n-k)(n-k-1)(n-k)\rho_{ACF}(k) \tag{7}$$

$$Z = \begin{cases} \frac{S-1}{\sqrt{\eta \times Var(S)}} & \text{for } S > 0 \\ 0 & \text{for } S = 0 \\ \frac{S+1}{\sqrt{\eta \times Var(S)}} & \text{for } S < 0 \end{cases} \tag{8}$$

Here, x_j is the parameter value at time j, and x_k at time k. *Slope* > 0 represents an upward trend; *Slope* < 0 represents a downward trend; and *Slope* = 0 means no change. $\rho_{ACF}(k)$ is the autocorrelation function (ACF) of the ranks of the observations [58]. The Z value indicates the trend of factors. In this study, the given significance levels were 99%, 95%, and 90%, when p = 0.01, 0.05, and 0.1, and $|Z| - p/2$ was equal to 2.58, 1.96, and 1.64, respectively [52,56]. Based on the factors trend, the nine levels were further divided and are presented in Table 1.

Table 1. Index classification.

Level	Description of the Trend or Correlation	*Slope* or *CC*	p	$\mid Z \mid$
1	Extremely significant increase or positive correlation	>0	$p < 0.01$	$\mid Z \mid > 2.58$
2	Significant increase or positive correlation		$0.01 < p \leq 0.05$	$1.96 < \mid Z \mid \leq 2.58$
3	Weakly significant increase or positive correlation		$0.05 < p \leq 0.1$	$1.64 < \mid Z \mid \leq 1.96$
4	Insignificant increase or positive correlation		$p > 0.1$	$\mid Z \mid \leq 1.64$
5	Extremely significant decrease or negative correlation	<0	$p < 0.01$	$\mid Z \mid > 2.58$
6	Significant decrease or negative correlation		$0.01 < p \leq 0.05$	$1.96 < \mid Z \mid \leq 2.58$
7	Weakly significant decrease or negative correlation		$0.05 < p \leq 0.1$	$1.64 < \mid Z \mid \leq 1.96$
8	Insignificant decrease or negative correlation		$p > 0.1$	$\mid Z \mid \leq 1.64$
9	Unchanged or uncorrelated	=0	-	-

2.4.3. Correlation Analysis

The correlation analysis method was used to analyze the relationship between two variables in each pixel [52]. The corresponding formula was as follows:

$$CC_{xy} = \frac{\sum_{i=1}^{n} [(x_i - \overline{x})(y_i - \overline{y})]}{\sqrt{\sum_{i=1}^{n} (x_i - \overline{x})^2 \sum_{i=1}^{n} (y_i - \overline{y})^2}} \tag{9}$$

where CC_{xy} represents the correlation coefficient (CC) between variables x and y; x_i and y_i represent the value of the variables x and y in month I, respectively; and $\overline{x}$ and $\overline{y}$ are the mean value of the variables x and y, respectively. The *CC* ranged from −1 to 1. There was a positive correlation when the *CC* was greater than 0 and a negative correlation when it was less than 0. Additionally, the significance of the results was evaluated using a t-test, and it could be divided into four groups, extremely significant, significant, weakly significant, and insignificant. Thus, the relationships between the dependent variable and independent variable were further divided into nine levels, as detailed in Table 1.

3. Results and Discussion

3.1. Spatial–Temporal Patterns of GWSA

The GWSA across the TP had obvious spatial heterogeneity, with a mean of 0.17 cm (Figure 2a) and a decreasing rate of −0.25 mm/a (Figure 2b). The higher values were mainly distributed in the QB, the IB, the source of the Yangtze River, and the southeastern edge of the TP. The difference in the GWSA reached 9.26 mm. The value of the GWSA in the QB was the highest, 3.42 cm, ranking first in the sub-regions. However, it had been decreasing with the highest decreasing rate of −1.24 mm/a (Figure 2b) and in a large area of 79.3% of the region (Figure 2c). On the contrary, the YZ-BRB had the lowest GWSA of −5.84 cm, with the highest increasing rate at 2.12 mm/a and a large increasing area of 80.2% of the basin. Furthermore, it revealed an increasing trend in 42.0% of the plateau area, while a decreasing trend was present in 58.0% of the plateau area (Figure 2c,d). The areas where the slope of the GWSA was positive were mainly HC, S-IRB, YZ-BRB, and south and southwest of the IB. In the HC and S-IRB, 53.9% and 33.3% of the regions showed a significant increasing trend, and a significant decrease was observed in <2% of these two basins, which indicated that the GWS levels in these sub-regions were increasing. In the YB and the YRB, 83.0% and 63.7% showed a decreasing trend, while 31.2% and 17.0% showed a significant decreasing trend, respectively.

The GWSA varied seasonally, with an increase in rainy seasons (May–October) and a decrease in dry seasons (November to April in the next year) (Figure 2e). The difference in the GWSA in the YB reached 1.06 cm, where the GWSA in rainy seasons and dry seasons were 1.15 cm and 0.09 cm, respectively. The variation in the GWSA in the YB was consistent with the variation in rainfall, indicating that the annual variation in groundwater may be affected by the recharge of rainfall. The GWSA of the IB in rainy seasons and dry seasons were 3.67 cm and 3.09 cm, respectively. As the amount of precipitation in this basin is still less in the rainy season, it is possible that permafrost degradation and glacier melting caused by increasing temperature have a stronger impact on the GWS than precipitation. The minimum negative anomaly of the interannual variation in the GWSA was −2.16 cm in 2009, while the maximum positive anomaly of 2.11 cm occurred in 2012 (Figure 2f). Although the GWS increased rapidly from 2009 to 2012, it was still declining overall from 2003 to 2021, which was similar to the result inferred from the monthly GWSA from April 2002 to December 2021 (Figure 2g). The latter was divided into five periods, April 2002 to June 2005, July 2005 to June 2010, June 2010 to May 2013, June 2013 to December 2015, and January 2016 to December 2021. Overall, the GWS in the TP was decreasing at an average rate of −0.89 mm/a from January 2003 to December 2021. Despite the lowest GWSA value of −2.66 cm occurring in October 2004, the GWS was still rising at a speed of 15.23 mm/a from April 2002 to June 2005. Then, there was a 60-month decrease at −7.31 mm/a from July 2005 to June 2010. The third period, from June 2010 to May 2013, was an increasing period. The increasing speed was 11.29 mm/a, and the GWSA reached the maximum value of 5.82 cm in May 2013. The largest decrease rate of −21.29 mm/a occurred from June 2013 to December 2015. It had been slowly rising since January 2016, with a speed of 1.47 mm/a. The results of the study by Liu et al. [22] were the same. Liu et al. [22] also came to a similar conclusion that GWS in the Qinghai province and the Tibet Autonomous Region reached a split point in 2016, and the groundwater storage changed from decreasing with a mean rate of −0.2 mm/a from 2003 to 2015 to increasing with a mean rate of 3.28 mm/a from 2016 to 2019. The GWSA in the TP showed an overall downward trend, but increases after 2016. This shows that the amount of groundwater storage is gradually recovering.

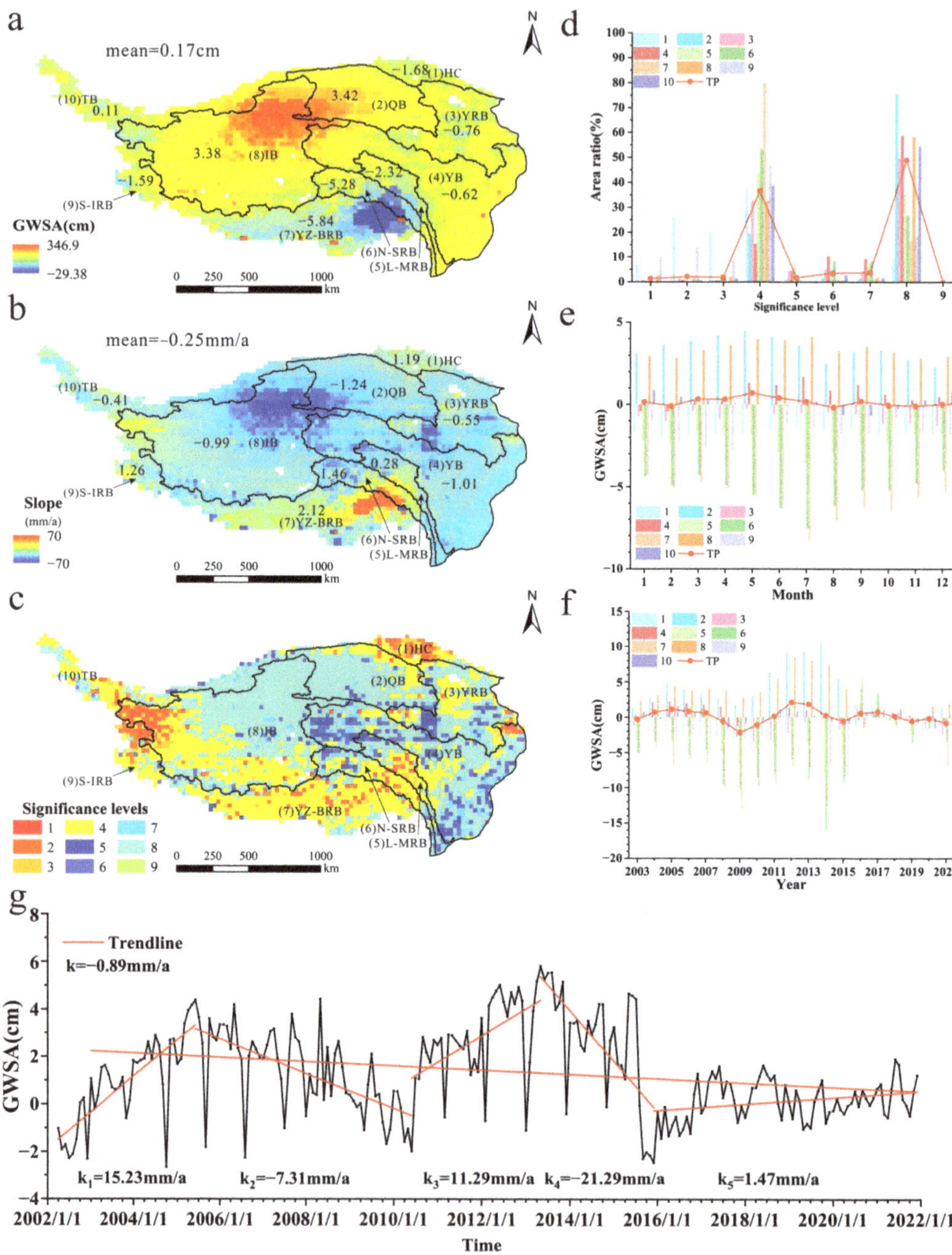

Figure 2. Characteristics of GWS in the TP and ten sub-regions. (**a**) Spatial distribution of monthly mean GWSA from January 2003 to December 2021; (**b**) spatial distribution of slope of annual mean GWSA from 2003 to 2021; (**c**) spatial distribution of significant changes in annual mean GWSA from 2003 to 2021; (**d**) significance analysis of changes in annual mean GWSA from 2003 to 2021; (**e**) variation in monthly mean GWSA; (**f**) variation in annual mean GWSA; (**g**) temporal variations in monthly GWSA from April 2002 to December 2021.

3.2. The Influence of Climate Change on GWS in the TP

3.2.1. Fluctuation Characteristics of Regional Climate

The changes in temperature and precipitation were calculated, and the spatial characteristics of climate change are shown in Figure 3.

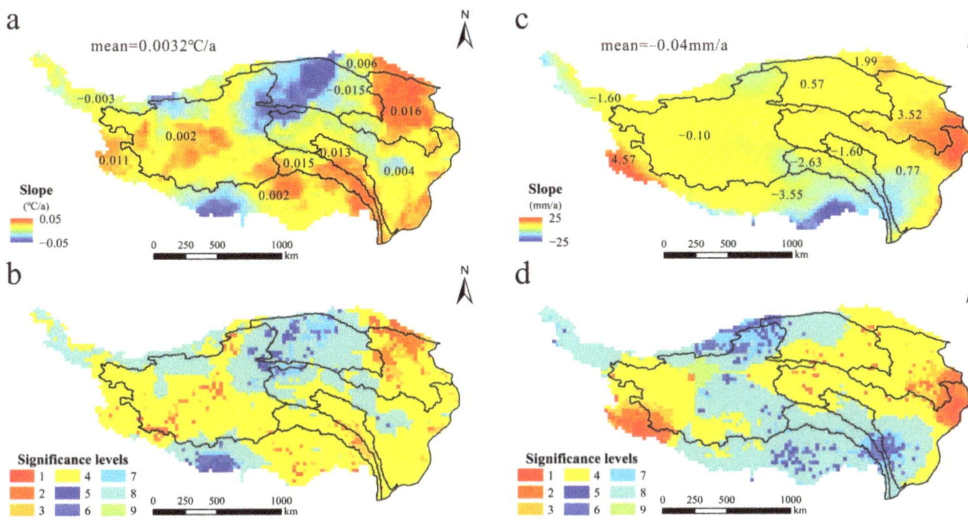

Figure 3. Spatial characteristics of climate change from 2003 to 2021. (**a**) Spatial distribution of slope of annual mean temperature; (**b**) spatial distribution of significant changes in annual mean temperature; (**c**) spatial distribution of slope of annual precipitation; (**d**) spatial distribution of significant changes in annual precipitation.

As shown in Figure 3a,b, the changes indicate that the temperature increased in 64.1% of the TP and decreased in 35.1% of the TP. The areas where the temperature increased were mainly distributed in the YRB, the L-MRB, the N-SRB, the S-IRB, the east of the YB, the east of the YZ-BRB, and the south of the IB. The increase rate of the temperature in the YRB was 0.016 °C/a. Additionally, 89.0% of the YRB showed an increasing trend. The decrease rate of the temperature in the QB, in which 88.3% showed a decreasing trend, was −0.015 °C/a.

The precipitation showed a decreasing trend at −0.04 mm/a and in 49.6% of the TP from 2003 to 2021, mainly over the YZ-BRB, the N-SRB, the TB, the south of the YB, the south of the L-MRB, the center of the IB, and the northwest of the QB (Figure 3c,d). The Brahmaputra River Basin had the highest decreasing rate of −3.55 mm/a and a decreasing area of 90.5%. Meanwhile, the largest increase in precipitation was shown to be in the Indus River Basin, with an increasing rate of 4.57 mm/a and an increasing area of 92.4%. These results indicate that there have been different or even opposite trends in the rate of climate change in different sub-regions of the TP in recent years.

3.2.2. Relationship between Climate Change and GWS

The correlation coefficients between the GWSA and Tmp and Pre were calculated to analyze the relationships between the GWS and climate change. Then, significance tests were carried out. The results are shown in Figure 4.

Overall, the GWSA were positively correlated with Tmp, and the mean correlation coefficient was 0.0492 (Figure 4a). The correlation coefficient in the YB was 0.143, which was the highest among the ten sub-regions. The HC, the QB, the YRB, the YB, the IB, and the S-IRB were the main regions that showed positive correlations. This shows that the areas that had positive correlations accounted for about 74.94% of the study area. In these areas, 15.23% showed extremely significant positive correlations. There were mostly insignificant positive correlations in 47.51% of the study area, which were mainly the HC, the QB, and the YRB (Figure 4b). The negatively correlated area covered 25.06% of the study area, with insignificant negative correlation occupying 13.66% of the area (Figure 4a,b). In the L-MRB, the N-SRB, the YZ-BRB, and the TB, the GWSA were insignificantly negatively correlated with the temperature in 25.87% of the area.

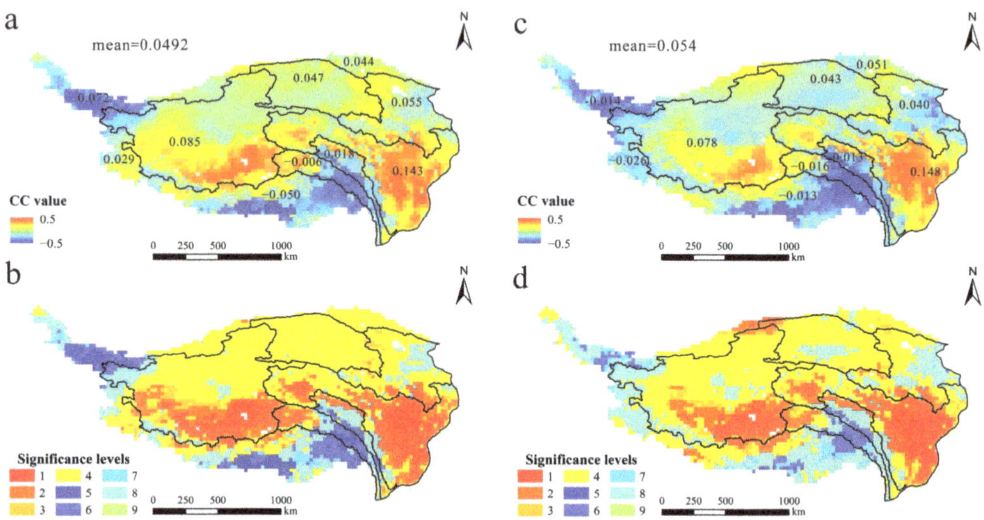

Figure 4. Relationship between climate change and GWS. (**a**) The correlation coefficients between GWSA and temperature; (**b**) the significance test between GWSA and temperature; (**c**) the correlation coefficients between GWSA and precipitation; (**d**) the significance test between GWSA and precipitation.

As shown in Figure 4c, the mean correlation coefficient was 0.054, which was greater than 0, indicating a positive correlation between the GWSA and Pre in the TP. The variables were positively correlated in 70.81% of the study area and negatively correlated in 29.19% of the study area. The mean correlation coefficient was 0.148 in the YB, which was the highest among the ten sub-regions. The L-MRB, the N-SRB, the YZ-BRB, the IB, and the TB were the main regions showing negative correlations, accounting for 68.24% of the negatively correlated areas in total. The results of the significance tests in Figure 4d indicate that the GWSA were extremely significantly positively correlated with precipitation in 13.96% of the TP, mainly in the YB and south of the IB.

Increasing precipitation will lead to more groundwater recharge, and thus, an increase in the GWS. Precipitation in the TP has been slightly increasing since the 1960s and is projected to further increase in the future in most areas of the TP [60]. The direct manifestation of permafrost degradation caused by increased temperature is the increase in the number and area of thermal karst lakes, which leads to the reduction in GWS in the form of runoff or evaporation [61,62]. However, when a large number of glaciers are distributed in the basin, the increase in glacial meltwater caused by the increase in temperature will recharge the groundwater, weaken the decrease in GWS, and even cause the increase in GWS. This is the reason why the GWS in the north and south of the IB showed different trends. There are many glaciers in the southern part of the IB and almost no glaciers in the northern part of the IB. Similar results were obtained in previous studies in the central Qiangtang Nature Reserve and the Upper Indus Basin [16]. The GWS in the L-MRB, N-SRB, and YZ-BRB increased in the context of decreased precipitation and increased temperature. As decreased precipitation and increased glacial melt due to temperature increases may cause decreased GWS, these three regions may require more groundwater recharge. The decreased GWS in the YRB and YB, where the temperature and precipitation increased, may indirectly support this hypothesis that the groundwater in the YRB and YB may seep to the L-MRB, N-SRB, and YZ-BRB [16].

3.3. Ecological Effect of Groundwater Storage

3.3.1. Vegetation Change

As shown in Figure 5, grassland has the largest area of 84.285×10^4 km^2 of the TP (Table 2), mostly distributed in the IB, YB, YRB, and YZ-BRB. The area of unused land has the second largest area, which is 59.736×10^4 km^2, mainly covering the IB and QB. The forest land that is mainly distributed in the YB and YZ-BRB covers 23.457×10^4 km^2. The area of water, mainly lakes and glaciers, is 9.577×10^4 km^2. The cropland and urban land types cover 1.994×10^4 km^2 and 0.210×10^4 km^2, respectively, which means there is a low level of human activity on the TP. However, these areas have increased by 0.19×10^4 km^2 and 0.10×10^4 km^2 from 2000 to 2020, respectively, indicating that human activities cannot be ignored.

The highest decrease and increase in land cover changes from 2000 to 2020 were grassland and unused land, with changed areas of -17.19×10^4 km^2 and 12.18×10^4 km^2, respectively. This means that large areas of grassland have degraded to sand, desert, and other kinds of unused land, especially in the IB, where the area of grassland decreased to 14.926×10^4 km^2 and the unused land increased to 12.284×10^4 km^2. It should be noted that different trends occurred in the LUCC of the YRB and YB. The area of unused land decreased while the area of forest and grassland increased, which proved the effectiveness of ecological protection and construction projects, such as eco-migration, grazing bans, forest and wetland reservations, etc. [16,63].

As LUCC is discrete data, it is difficult to quantify the vegetation response to GWS. Therefore, the influences of groundwater change on vegetation in time and space were discussed using the NDVI, an indicator of vegetation change.

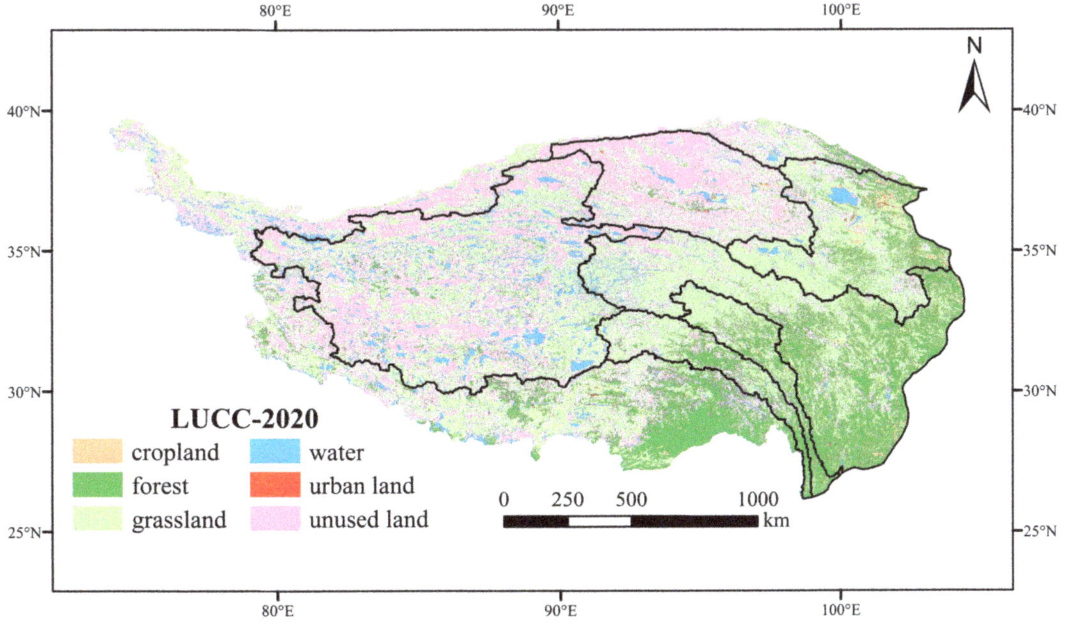

Figure 5. Land use map of the TP in 2020.

Table 2. Area of different land use types in 2020 and changed area from 2000 to 2020 (10^4 km^2).

Basin		Cropland	Forest	Grassland	Water	Urban Land	Unused Land
HC	Area	0.009	0.382	2.181	0.200	0.002	2.027
	Changed area	0.002	0.001	0.017	0.080	0.001	−0.125
QB	Area	0.061	0.169	5.730	0.522	0.035	12.090
	Changed area	0.009	−0.006	0.106	0.171	0.015	−0.294
YRB	Area	0.634	2.022	12.367	0.696	0.093	2.121
	Changed area	−0.010	0.015	0.654	0.066	0.033	−0.762
YB	Area	0.745	9.313	18.935	0.723	0.041	3.197
	Changed area	0.044	0.388	0.210	0.059	0.023	−0.694
L-MRB	Area	0.150	1.720	3.870	0.063	0.003	0.398
	Changed area	0.046	0.242	−0.060	0.028	0.002	−0.247
N-SRB	Area	0.046	1.664	4.429	0.166	0.005	1.266
	Changed area	−0.001	0.355	0.128	−0.032	0.004	−0.447
YZ-BRB	Area	0.339	7.097	11.342	1.093	0.027	5.920
	Changed area	0.090	0.920	−2.278	−0.158	0.019	1.486
IB	Area	0.001	0.692	21.507	4.752	0.002	22.990
	Changed area	0.001	0.681	−14.926	2.022	0.002	12.284
S-IRB	Area	0.001	0.376	3.036	0.375	0.002	2.962
	Changed area	0.001	0.374	−1.840	0.164	0.002	1.321
TB	Area	0.007	0.023	5.889	0.989	0.001	6.765
	Changed area	0.006	−0.001	0.800	−0.449	0.001	−0.342
The TP	Area	1.994	23.457	89.285	9.577	0.210	59.736
	Changed area	0.19	2.97	−17.19	1.96	0.10	12.18

3.3.2. Vegetation Responses to GWS Changes

Overall, the NDVI increased with the mean rate of 0.0035 per ten years, but the increase was lower in the west and faster in the east from 2003 to 2019 (Figure 6a). The YB showed the fastest increasing rate of 0.0144 per ten years, while the lowest was in the QB, with a rate of −0.0071 of per ten years. The areas where the NDVI increased significantly are mainly distributed over the east edge of the TP (Figure 6b). In Figure 6c, it can be seen that the GWS and the NDVI were positively correlated, and the correlation coefficient was 0.068. Additionally, they had a positive correlation relationship in 73.17% of the study area. The highest correlation coefficient was 0.113 in the YB, while the lowest was −0.016 in the L-MRB. The weakly to extremely significant positive correlations were in 36.91% of the aforementioned areas, mostly in the QB, YB, and YZ-BRB (Figure 6d). The regions where there was a negative correlation between the GWS and the NDVI covered 26.83% of the TP, of which only 23.59% were significantly or extremely significantly negatively correlated (Figure 6d).

The GWS was positively correlated with the NDVI, which indicated that the vegetation was dependent on the groundwater condition. In the IB, large areas of grassland have degenerated into unused land because of the decline in the groundwater level in the context of climate change [64,65]. This is analogous to the earlier findings reported in the TP. Peng et al. [66] also found that the decline of the groundwater level in the source area of the Yellow River was closely related to the deterioration of the ecological environment in the permafrost-degraded area. This may have been due to the decrease in the groundwater level, which would have caused a great loss to occur in the shallow soil moisture, which would have meant that the root system of some vegetation in cold areas could not effectively use soil water, eventually leading to the degradation or disappearance of vegetation [67–69].

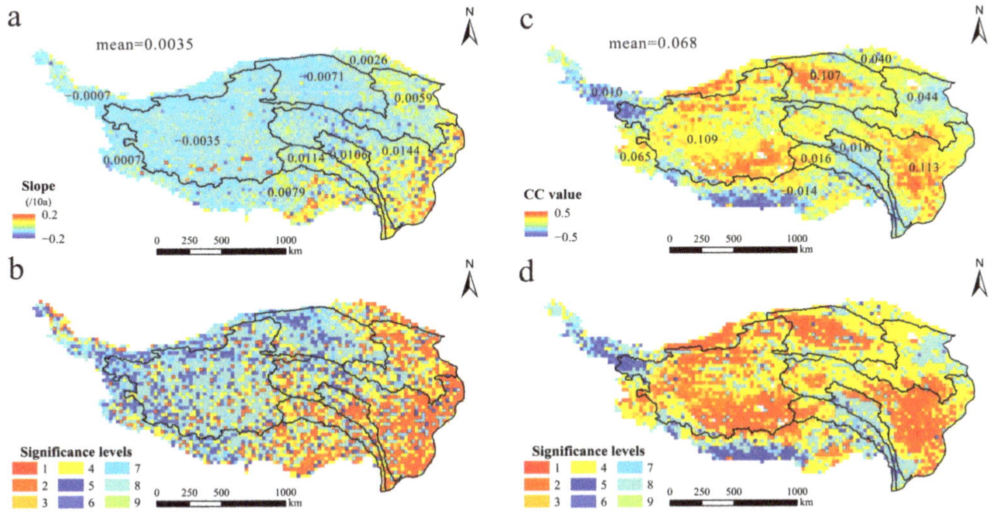

Figure 6. Relationship between NDVI and GWS. (**a**) Spatial distribution of slope of annual mean NDVI; (**b**) spatial distribution of significant changes in NDVI; (**c**) the correlation coefficients between GWS and NDVI; (**d**) the significance test between GWS and NDVI.

4. Conclusions

Based on the linear regression and modified Mann–Kendall test analysis methods, the temporal and spatial variations in GWS were analyzed using the GRACE and GLDAS data from April 2002 to December 2021. Then, the driving factors and ecological effects of the changes in GWS in the Tibetan Plateau were discussed, using the correlation analysis model and multi-source remote sensing data. This study of groundwater storage changes provides a useful reference for ecological protection and the high-quality development of the TP. The conclusions are as follows:

1. The higher values were mainly distributed in the QB, the IB, the source of the Yangtze River, and the southeastern edge of the TP. An increasing trend in the GWS was revealed in 42.0% of the plateau area and a decreasing trend was revealed in 58.0% of the plateau area. The areas where the GWS increased were mainly the HC, S-IRB, YZ-BRB, and the south and southwest of the IB.
2. Overall, the GWS in the TP was decreasing at an average rate of −0.89 mm/a from January 2003 to December 2021. However, the GWS has been slowly rising at a rate of 1.47 mm/a since January 2016. This shows that the level of groundwater storage is gradually recovering.
3. The different climate conditions in the different sub-regions had different impacts on the change in the GWS. The change in precipitation may be the main reason for the change in the GWS in the YB. Rising temperatures have a two-sided effect on the groundwater storage. On the one hand, the melting of ice and snow caused by rising temperatures will replenish the groundwater, which will increase the groundwater storage in some areas, such as the south of the IB. On the other hand, permafrost degradation caused by climate change will lead to a decrease in the GWS in other regions, such as the north of the IB.
4. The potential ecological effects were investigated, with the results showing that the reduction in the GWS is an important cause of vegetation degradation. The decrease in the GWS reduced the efficiency of plant roots in absorbing and utilizing groundwater

and eventually led to the degradation of 17.19×10^4 km^2 of grassland to sand, desert, or other kinds of unused land on the TP.

The results of this study could offer a new opportunity to reveal the groundwater changes in a cryosphere region and to assess the impact of the changes in hydrological conditions on ecology. However, this study only provided a preliminary investigation of the changes to GWS, and more in-depth studies should focus on the following specific aspects: (1) in situ groundwater monitoring with high temporal and spatial resolution at the regional scale to gain insight into the accurate variation mechanism of groundwater storage on the TP; (2) the integration of multivariate satellite data and improving modeling capacity to deal with the quality change signals caused by rapid structural uplift; (3) the quantitative analysis of the response of groundwater to climate change and the evaluation of the effects of groundwater changes on vegetation evolution under climate change.

Author Contributions: W.R., conceptualization, methodology, formal analysis, and writing—original draft; Y.G., conceptualization and supervision; H.Q., conceptualization, supervision, resources, and writing—review and editing; Y.M., Z.S. and W.M., supervision and resources; Y.L. and P.X., methodology. All authors have read and agreed to the published version of the manuscript.

Funding: This research was supported by the Second Tibetan Plateau Scientific Expedition and Research (STEP) program (Grant No. 2019QZKK0103), the National Natural Science Foundation of China (41931285, 42007184), the Fundamental Research Funds for the Central Universities, CHD (Grant No. 300102292901), and the Program of Introducing Talents of Discipline to Universities (Grant No. B08039).

Data Availability Statement: The CSR GRACE/FO RL06 mascon dataset is provided by the Center for Space Research at the University of Texas, Austin and downloaded at http://www2.csr.utexas.edu/grace (accessed on 27 March 2022). The GLDAS Noah model is developed by the National Aeronautics and Space Administration and the National Oceanic and Atmospheric Administration and downloaded at https://disc.gsfc.nasa.gov/datasets (accessed on 27 March 2022). The CRU TS data is produced by the UK's National Centre for Atmospheric Science (NCAS) and downloaded at https://crudata.uea.ac.uk/ (accessed on 21 May 2022). The NDVI and LUCC data are both obtained from the Resources and Environment Science and Data Center, Chinese Academy of Sciences at http://www.resdc.cn/ (accessed on 1 May 2022).

Acknowledgments: The authors gratefully acknowledge the supports of various foundations. The authors are grateful to the editor and anonymous reviewers whose very insightful comments have contributed to improving the quality of this paper.

Conflicts of Interest: The authors declare that they have no known competing financial interest or personal relationships that could have appeared to influence the work reported in this paper.

References

1. Yao, T.; Masson-Delmotte, V.; Gao, J.; Yu, W.; Yang, X.; Risi, C.; Sturm, C.; Werner, M.; Zhao, H.; He, Y.; et al. A review of climatic controls on δ18O in precipitation over the Tibetan Plateau: Observations and simulations. *Rev. Geophys.* **2013**, *51*, 525–548. [CrossRef]
2. Wang, J.; Huang, L.; Ju, J.; Daut, G.; Ma, Q.; Zhu, L.; Haberzettl, T.; Baade, J.; Mäusbacher, R.; Hamilton, A.; et al. Seasonal stratification of a deep, high-altitude, dimictic lake: Nam Co, Tibetan Plateau. *J. Hydrol.* **2020**, *584*, 124668. [CrossRef]
3. Yao, T.; Bolch, T.; Chen, D.; Gao, J.; Immerzeel, W.; Piao, S.; Su, F.; Thompson, L.; Wada, Y.; Wang, L.; et al. The imbalance of the Asian water tower. *Nat. Rev. Earth Environ.* **2022**, *3*, 618–632. [CrossRef]
4. Chen, D.; Tan-Dong, Y. Climate change and its impact on the Third Pole and beyond. *Adv. Clim. Chang. Res.* **2021**, *12*, 297–298. [CrossRef]
5. Xiaodong, L.; Baode, C. Climatic warming in the Tibetan Plateau during recent decades. *Int. J. Climatol.* **2000**, *20*, 1729–1742. [CrossRef]
6. Kang, S.; Xu, Y.; You, Q.; Fluegel, W.-A.; Pepin, N.; Yao, T. Review of climate and cryospheric change in the Tibetan Plateau. *Environ. Res. Lett.* **2010**, *5*, 015101. [CrossRef]
7. Li, X.; Long, D.; Scanlon, B.R.; Mann, M.E.; Li, X.; Tian, F.; Sun, Z.; Wang, G. Climate change threatens terrestrial water storage over the Tibetan Plateau. *Nat. Clim. Chang.* **2022**, *12*, 801–807. [CrossRef]
8. Yao, T.D.; Thompson, L.; Yang, W.; Yu, W.S.; Gao, Y.; Guo, X.J.; Yang, X.X.; Duan, K.Q.; Zhao, H.B.; Xu, B.Q.; et al. Different glacier status with atmospheric circulations in Tibetan Plateau and surroundings. *Nat. Clim. Chang.* **2012**, *2*, 663–667. [CrossRef]

9. He, Z.; Yang, W.; Wang, Y.J.; Zhao, C.X.; Ren, S.T.; Li, C.H. Dynamic Changes of a Thick Debris-Covered Glacier in the Southeastern Tibetan Plateau. *Remote Sens.* **2023**, *15*, 357. [CrossRef]
10. Yongsheng, Z.; Li, T.; Bin, W. Decadal change of the spring snow depth over the Tibetan Plateau: The associated circulation and influence on the East Asian summer monsoon. *J. Clim.* **2004**, *17*, 2780–2793. [CrossRef]
11. Zhaoxia, P.; Xu, L.; Salomonson, V.V. MODIS/Terra observed seasonal variations of snow cover over the Tibetan Plateau. *Geophys. Res. Lett.* **2007**, *34*, 137–161. [CrossRef]
12. Zhang, G.Q.; Xie, H.J.; Kang, S.C.; Yi, D.H.; Ackley, S.F. Monitoring lake level changes on the Tibetan Plateau using ICESat altimetry data (2003-2009). *Remote Sens. Environ.* **2011**, *115*, 1733–1742. [CrossRef]
13. Zhang, G.; Yao, T.; Xie, H.; Yang, K.; Zhu, L.; Shum, C.K.; Bolch, T.; Yi, S.; Allen, S.; Jiang, L.; et al. Response of Tibetan Plateau lakes to climate change: Trends, patterns, and mechanisms. *Earth-Sci. Rev.* **2020**, *208*, 103269. [CrossRef]
14. Ren, W.; Tian, L.D.; Shao, L.L. Temperature and precipitation control the seasonal patterns of discharge and water isotopic signals of the Nyang River on the southeastern Tibetan Plateau. *J. Hydrol.* **2023**, *617*, 129064. [CrossRef]
15. Immerzeel, W.W.; Droogers, P.; de Jong, S.M.; Bierkens, M.F.P. Large-scale monitoring of snow cover and runoff simulation in Himalayan river basins using remote sensing. *Remote Sens. Environ.* **2009**, *113*, 40–49. [CrossRef]
16. Xiang, L.; Wang, H.; Steffen, H.; Wu, P.; Jia, L.; Jiang, L.; Shen, Q. Groundwater storage changes in the Tibetan Plateau and adjacent areas revealed from GRACE satellite gravity data. *Earth Planet. Sci. Lett.* **2016**, *452*, 309. [CrossRef]
17. Deng, H.J.; Pepin, N.C.; Liu, Q.; Chen, Y.N. Understanding the spatial differences in terrestrial water storage variations in the Tibetan Plateau from 2002 to 2016. *Clim. Chang.* **2018**, *151*, 379–393. [CrossRef]
18. Gao, G.; Zhao, J.; Wang, J.; Zhao, G.; Chen, J.; Li, Z. Spatiotemporal Variation and Driving Analysis of Groundwater in the Tibetan Plateau Based on GRACE Downscaling Data. *Water* **2022**, *14*, 3302. [CrossRef]
19. Khorrami, B.; Gunduz, O. Evaluation of the temporal variations of groundwater storage and its interactions with climatic variables using GRACE data and hydrological models: A study from Turkey. *Hydrol. Process.* **2021**, *35*, 3406–3425. [CrossRef]
20. Han, Z.; Huang, S.; Huang, Q.; Bai, Q.; Leng, G.; Wang, H.; Zhao, J.; Wei, X.; Zheng, X. Effects of vegetation restoration on groundwater drought in the Loess Plateau, China. *J. Hydrol.* **2020**, *591*, 125566. [CrossRef]
21. Zhang, H.; Ding, J.; Wang, Y.; Zhou, D.; Zhu, Q. Investigation about the correlation and propagation among meteorological, agricultural and groundwater droughts over humid and arid/semi-arid basins in China. *J. Hydrol.* **2021**, *603*, 127007. [CrossRef]
22. Liu, F.; Kang, P.; Zhu, H.; Han, J.; Huang, Y. Analysis of Spatiotemporal Groundwater-Storage Variations in China from GRACE. *Water* **2021**, *13*, 2378. [CrossRef]
23. Ahmed, M.; Aqnouy, M.; El Messari, J.S. Sustainability of Morocco's groundwater resources in response to natural and anthropogenic forces. *J. Hydrol.* **2021**, *603*, 126866. [CrossRef]
24. Sliwinska, J.; Birylo, M.; Rzepecka, Z.; Nastula, J. Analysis of Groundwater and Total Water Storage Changes in Poland Using GRACE Observations, In-situ Data, and Various Assimilation and Climate Models. *Remote Sens.* **2019**, *11*, 2949. [CrossRef]
25. Hu, W.; Liu, H.; Bao, A.; El-Tantawi, A.M. Influences of environmental changes on water storage variations in Central Asia. *J. Geogr. Sci.* **2018**, *28*, 985–1000. [CrossRef]
26. Liu, X.; Hu, L.; Sun, K.; Yang, Z.; Sun, J.; Yin, W. Improved Understanding of Groundwater Storage Changes under the Influence of River Basin Governance in Northwestern China Using GRACE Data. *Remote Sens.* **2021**, *13*, 2672. [CrossRef]
27. Yang, K.; Wu, H.; Qin, J.; Lin, C.; Tang, W.; Chen, Y. Recent climate changes over the Tibetan Plateau and their impacts on energy and water cycle: A review. *Glob. Planet. Chang.* **2014**, *112*, 79–91. [CrossRef]
28. Luo, J.; Ke, C.-Q.; Seehaus, T. The West Kunlun Glacier Anomaly and Its Response to Climate Forcing during 2002–2020. *Remote Sens.* **2022**, *14*, 3465. [CrossRef]
29. Xu, L.; Gao, B. Understanding the Effects of Cold and Warm Season Air Warming on the Permafrost Hydrology Changes in the Source Region of the Lancang River, the Qinghai-Tibetan Plateau. *J. Geophys. Res.-Atmos.* **2022**, *127*, e2022JD036551. [CrossRef]
30. Li, F.W.; Wang, Y.; Zhao, Y.; Qiao, J.L. Modelling the response of vegetation restoration to changes in groundwater level, based on ecologically suitable groundwater depth. *Hydrogeol. J.* **2018**, *26*, 2189–2204. [CrossRef]
31. Zhang, H.; Wang, X.-S. The impact of groundwater depth on the spatial variance of vegetation index in the Ordos Plateau, China: A semivariogram analysis. *J. Hydrol.* **2020**, *588*, 125096. [CrossRef]
32. Xu, M.; Kang, S.; Chen, X.; Wu, H.; Wang, X.; Su, Z. Detection of hydrological variations and their impacts on vegetation from multiple satellite observations in the Three-River Source Region of the Tibetan Plateau. *Sci. Total Environ.* **2018**, *639*, 1220–1232. [CrossRef]
33. Wei, Y.Y.; Sun, S.G.; Liang, D.; Jia, Z.H. Spatial-temporal variations of NDVI and its response to climate in China from 2001 to 2020. *Int. J. Digit. Earth* **2022**, *15*, 1463–1484. [CrossRef]
34. Gao, Y.; Zhou, X.; Wang, Q.; Wang, C.; Zhan, Z.; Chen, L.; Yan, J.; Qu, R. Vegetation net primary productivity and its response to climate change during 2001–2008 in the Tibetan Plateau. *Sci. Total Environ.* **2013**, *444*, 356–362. [CrossRef] [PubMed]
35. Piao, S.; Tan, K.; Nan, H.; Ciais, P.; Fang, J.; Wang, T.; Vuichard, N.; Zhu, B. Impacts of climate and CO_2 changes on the vegetation growth and carbon balance of Qinghai–Tibetan grasslands over the past five decades. *Glob. Planet. Chang.* **2012**, *98–99*, 73–80. [CrossRef]
36. Zhang, B.; Zhou, W. Spatial-Temporal Characteristics of Precipitation and Its Relationship with Land Use/Cover Change on the Qinghai-Tibet Plateau, China. *Land* **2021**, *10*, 269. [CrossRef]

37. Xu, W.; Su, X. Challenges and impacts of climate change and human activities on groundwater-dependent ecosystems in arid areas–A case study of the Nalenggele alluvial fan in NW China. *J. Hydrol.* **2019**, *573*, 376–385. [CrossRef]
38. Zhang, G.Q.; Yao, T.D.; Xie, H.J.; Kang, S.C.; Lei, Y.B. Increased mass over the Tibetan Plateau: From lakes or glaciers? *Geophys. Res. Lett.* **2013**, *40*, 2125–2130. [CrossRef]
39. Li, Y.; Su, F.; Chen, D.; Tang, Q. Atmospheric Water Transport to the Endorheic Tibetan Plateau and Its Effect on the Hydrological Status in the Region. *J. Geophys. Res.-Atmos.* **2019**, *124*, 12864–12881. [CrossRef]
40. Wei, L.; Jiang, S.; Ren, L.; Tan, H.; Ta, W.; Liu, Y.; Yang, X.; Zhang, L.; Duan, Z. Spatiotemporal changes of terrestrial water storage and possible causes in the closed Qaidam Basin, China using GRACE and GRACE Follow-On data. *J. Hydrol.* **2021**, *598*, 126274. [CrossRef]
41. Abhishek; Kinouchi, T.; Sayama, T. A comprehensive assessment of water storage dynamics and hydroclimatic extremes in the Chao Phraya River Basin during 2002-2020. *J. Hydrol.* **2021**, *603*, 126868. [CrossRef]
42. Save, H.; Bettadpur, S.; Tapley, B.D. High-resolution CSR GRACE RL05 mascons. *J. Geophys. Res.-Solid Earth* **2016**, *121*, 7547–7569. [CrossRef]
43. Yulong, Z.; Wei, F.; Min, Z.; Zutao, M. *Dataset of Reconstructed Terrestrial Water Storage in China Based on Precipitation (2002–2019)*; National Tibetan Plateau Data Center: Beijing, China, 2020. [CrossRef]
44. Zhong, Y.; Feng, W.; Humphrey, V.; Zhong, M. Human-Induced and Climate-Driven Contributions to Water Storage Variations in the Haihe River Basin, China. *Remote Sens.* **2019**, *11*, 3050. [CrossRef]
45. Sun, Z.; Long, D.; Yang, W.; Li, X.; Pan, Y. Reconstruction of GRACE Data on Changes in Total Water Storage Over the Global Land Surface and 60 Basins. *Water Resour. Res.* **2020**, *56*, e2019WR026250. [CrossRef]
46. Meng, F.; Su, F.; Li, Y.; Tong, K. Changes in Terrestrial Water Storage During 2003-2014 and Possible Causes in Tibetan Plateau. *J. Geophys. Res.-Atmos.* **2019**, *124*, 2909–2931. [CrossRef]
47. Rodell, M.; Houser, P.R.; Jambor, U.; Gottschalc, J.; Mitchell, K.; Meng, C.-J.; Arsenault, K.; Cosgrove, A.; Radakovich, J.; Bosilovich, M.; et al. The Global Land Data Assimilation System. *Bull. Amer. Meteor. Soc.* **2004**, *85*, 381–394. [CrossRef]
48. Beaudoing, H.; Rodell, M. *NASA/GSFC/HSL. GLDAS Noah Land Surface Model L4 monthly 0.25 x 0.25 Degree V2.1*; Goddard Earth Sciences Data and Information Services Center (GES DISC): Greenbelt, MD, USA, 2020. [CrossRef]
49. Qu, W.; Jin, Z.; Zhang, Q.; Gao, Y.; Zhang, P.; Chen, P. Estimation of Evapotranspiration in the Yellow River Basin from 2002 to 2020 Based on GRACE and GRACE Follow-On Observations. *Remote Sens.* **2022**, *14*, 730. [CrossRef]
50. Ramjeawon, M.; Demlie, M.; Toucher, M. Analyses of groundwater storage change using GRACE satellite data in the Usutu-Mhlatuze drainage region, north-eastern South Africa. *J. Hydrol.-Reg. Stud.* **2022**, *42*, 101118. [CrossRef]
51. Harris, I.; Osborn, T.J.; Jones, P.; Lister, D. Version 4 of the CRU TS monthly high-resolution gridded multivariate climate dataset. *Sci. Data* **2020**, *7*, 109. [CrossRef]
52. Liu, Y.; Tian, J.; Liu, R.; Ding, L. Influences of Climate Change and Human Activities on NDVI Changes in China. *Remote Sens.* **2021**, *13*, 4326. [CrossRef]
53. Jin, H.; Chen, X.; Wang, Y.; Zhong, R.; Zhao, T.; Liu, Z.; Tu, X. Spatio-temporal distribution of NDVI and its influencing factors in China. *J. Hydrol.* **2021**, *603*, 127129. [CrossRef]
54. Chang, X.; Xing, Y.; Wang, J.; Yang, H.; Gong, W. Effects of land use and cover change (LUCC) on terrestrial carbon stocks in China between 2000 and 2018. *Resour. Conserv. Recycl.* **2022**, *182*, 106333. [CrossRef]
55. Guan, Q.; Yang, L.; Pan, N.; Lin, J.; Xu, C.; Wang, F.; Liu, Z. Greening and Browning of the Hexi Corridor in Northwest China: Spatial Patterns and Responses to Climatic Variability and Anthropogenic Drivers. *Remote Sens.* **2018**, *10*, 1270. [CrossRef]
56. Hamed, K.H. Exact distribution of the Mann–Kendall trend test statistic for persistent data. *J. Hydrol* **2008**, *365*, 86–94. [CrossRef]
57. Guclu, Y.S. Improved visualization for trend analysis by comparing with classical Mann-Kendall test and ITA. *J. Hydrol.* **2020**, *584*, 124674. [CrossRef]
58. Hamed, K.H.; Ramachandra Rao, A. A modified Mann-Kendall trend test for autocorrelated data. *J. Hydrol.* **1998**, *204*, 182–196. [CrossRef]
59. Liu, Y.; Jia, Z.F.; Ma, X.Y.; Wang, Y.Q.; Guan, R.H.; Guan, Z.L.; Gu, Y.H.; Zhao, W. Analysis of Drought Characteristics Projections for the Tibetan Plateau Based on the GFDL-ESM2M Climate Model. *Remote Sens.* **2022**, *14*, 5084. [CrossRef]
60. Kuang, X.X.; Jiao, J.J. Review on climate change on the Tibetan Plateau during the last half century. *J. Geophys. Res.-Atmos.* **2016**, *121*, 3979–4007. [CrossRef]
61. Lei, Y.B.; Yang, K.; Wang, B.; Sheng, Y.W.; Bird, B.W.; Zhang, G.Q.; Tian, L.D. Response of inland lake dynamics over the Tibetan Plateau to climate change. *Clim. Chang.* **2014**, *125*, 281–290. [CrossRef]
62. Ge, S.; McKenzie, J.; Voss, C.; Wu, Q. Exchange of groundwater and surface-water mediated by permafrost response to seasonal and long term air temperature variation. *Geophys. Res. Lett.* **2011**, *38*. [CrossRef]
63. Shao, Q.; Liu, J.; Huang, L.; Fan, J.; Xu, X.; Wang, J. Integrated assessment on the effectiveness of ecological conservation in Sanjiangyuan National Nature Reserve (in Chinese). *Geogr. Res.* **2013**, *32*, 1645–1656.
64. Zhao, L.; Zou, D.; Hu, G.; Wu, T.; Du, E.; Liu, G.; Xiao, Y.; Li, R.; Pang, Q.; Qiao, Y.; et al. A synthesis dataset of permafrost thermal state for the Qinghai-Tibet (Xizang) Plateau, China. *Earth Syst. Sci. Data* **2021**, *13*, 4207–4218. [CrossRef]
65. Li, R.; Zhao, L.; Ding, Y.J.; Wu, T.H.; Xiao, Y.; Du, E.J.; Liu, G.Y.; Qiao, Y.P. Temporal and spatial variations of the active layer along the Qinghai-Tibet Highway in a permafrost region. *Chin. Sci. Bull.* **2012**, *57*, 4609–4616. [CrossRef]

66. Peng, X.; Wu, Q.; Tian, M. The Effect of Groundwater Table Lowering on Ecological Environment in the Headwaters for the Yellow River. *J. Glaciol. Geocryol.* **2003**, *25*, 667–671. (In Chinese)
67. Chang, J.; Ye, R.; Wang, G. Review: Progress in permafrost hydrogeology in China. *Hydrogeol. J.* **2018**, *26*, 1387–1399. [CrossRef]
68. Cheng, G.D.; Jin, H.J. Permafrost and groundwater on the Qinghai-Tibet Plateau and in northeast China. *Hydrogeol. J.* **2013**, *21*, 5–23. [CrossRef]
69. Huijun, J.; Ruixia, H.; Guodong, C.; Qingbai, W.; Shaoling, W.; Lanzhi, L.; Xiaoli, C. Changes in frozen ground in the Source Area of the Yellow River on the Qinghai-Tibet Plateau, China, and their eco-environmental impacts. *Environ. Res. Lett.* **2009**, *4*, 045206. [CrossRef]

remote sensing

MDPI

Article

Relationships between Landscape Patterns and Hydrological Processes in the Subtropical Monsoon Climate Zone of Southeastern China

Chong Wei [1,2], Xiaohua Dong [1,2,*,†], Yaoming Ma [3,4,5,6,7,8,†], Menghui Leng [1,2], Wenyi Zhao [1,2], Chengyan Zhang [1,2], Dan Yu [1,2] and Bob Su [9]

1 College of Hydraulic & Environmental Engineering, China Three Gorges University, Yichang 443002, China
2 Engineering Research Center for the Ecological Environment of the Three Gorges Reservoir Area, Ministry of Education, Yichang 430072, China
3 State Key Laboratory of Tibetan Plateau Earth System, Environment and Resources (TPESER), Institute of Tibetan Plateau Research, Chinese Academy of Sciences, Beijing 100101, China
4 College of Earth and Planetary Sciences, University of Chinese Academy of Sciences, Beijing 100049, China
5 College of Atmospheric Science, Lanzhou University, Lanzhou 730000, China
6 National Observation and Research Station for Qomolongma Special Atmospheric Processes and Environmental Changes, Zhikatsé 858200, China
7 Kathmandu Center of Research and Education, Chinese Academy of Sciences, Beijing 100101, China
8 China-Pakistan Joint Research Center on Earth Sciences, Chinese Academy of Sciences, Islamabad 45320, Pakistan
9 Faculty of Geo-Information Science and Earth Observation, University of Twente, 7500 AE Enschede, The Netherlands
* Correspondence: xhdong@ctgu.edu.cn
† These authors contributed equally to this work.

Abstract: With rapid economic development, extensive human activity has changed landscape patterns (LPs) dramatically, which has further influenced hydrological processes. However, the effects of LPs changes on hydrological processes, especially for the streamflow–sediment relationship in the subtropical monsoon climate zone, have not been reported. In our study, 10 watersheds with different sizes in the subtropical monsoon climate zone of southeastern China were chosen as the study area, and the effect of the 14 most commonly used landscape metrics (LMs) on 4 typical hydrological indices (water yields (WY), the runoff coefficient (RC), the soil erosion modulus (SEM), and the suspended sediment concentration (SSC)) were analyzed based on land use maps and hydrological data from 1990 to 2019. The results reveal that the LP characteristics within the study area have changed significantly. The number of patches and landscape shape indices were significantly positively correlated with watershed size ($p < 0.01$). For most watersheds, the largest patch index was negatively correlated with WY, RC, and SEM, and the perimeter area fractal dimension was positively correlated with WY, RC, SEM, and SSC. The effects of several LMs on the hydrological indices had scale effects. WY/RC and the interspersion and juxtaposition index were negatively correlated in most larger watersheds but were positively correlated in most smaller watersheds. Similar results were found for Shannon's diversity/evenness index and SEM. In general, an increase in a small patch of landscape and in landscape diversity would increase WY, the fragmentation of LPs would result in more soil erosion, and LPs would affect the relationship between streamflow and sediment yield. As a result, a proper decrease in landscape fragmentation and physical connectivity in the subtropical monsoon climate zone of southeastern China would benefit soil erosion prevention. These results enhance the knowledge about the relationship between LPs and hydrological processes in the subtropical monsoon climate zone of southeastern China and benefit local water and soil conservation efforts.

Keywords: landscape pattern; runoff coefficient; soil erosion modulus; suspended sediment concentration; subtropical monsoon climate zone; southeastern China

Citation: Wei, C.; Dong, X.; Ma, Y.; Leng, M.; Zhao, W.; Zhang, C.; Yu, D.; Su, B. Relationships between Landscape Patterns and Hydrological Processes in the Subtropical Monsoon Climate Zone of Southeastern China. *Remote Sens.* **2023**, *15*, 2290. https://doi.org/10.3390/rs15092290

Academic Editor: Konstantinos X. Soulis

Received: 6 March 2023
Revised: 18 April 2023
Accepted: 25 April 2023
Published: 26 April 2023

1. Introduction

With the rapid development of human society, land use has changed dramatically [1], which has altered ecosystem structures, functions, and services [2], further influencing eco-hydrological processes [3,4]. The characteristics of land use changes mainly contain land use quantity and landscape patterns (LPs). Scholars have investigated the differences in hydrological effects for various land uses [5–8] to better determine a proper quantity for each land use type for these study regions. Afterward, some scholars further explored the evolution of land use spatial distribution based on various land use prediction models such as the CLUE-S [9], CA [10], and FLUS [11] models. In addition, the spatial distribution and configuration of various land use types could also affect hydrological processes, even with the same quantity. In recent decades, more and more scholars have focused on the effect of LPs on hydrological processes such as runoff and soil erosion [12–14].

LPs are the spatial configuration characteristics of various landscape units with different sizes and shapes [15]. Various landscape metrics (LMs) have been developed to describe the characteristics of LPs [16–18]. Over recent decades, scholars have indicated that LPs have significant correlations with runoff [12], soil erosion [13,14], water quality [19,20], and organic matter decomposition [21,22]. Sadeghi et al. [23] investigated the relationships between LMs and hydrographic components within the Galazchai watershed in Iran. The results show that the number of disjunct core areas (NDCA) was positively correlated with flood volume, and patch density (PD) was positively correlated with peak discharge. Zhao and Huang [24] examined the effects of LPs on runoff within a small watershed in southern China. The results indicate that PD and landscape shape index (LSI) were negatively correlated with runoff, whereas Shannon's diversity index (SHDI) and the landscape division index (DIVISION) were positively correlated with runoff. In addition, PD, SHDI, and largest patch index (LPI) had the largest impacts on variations in runoff. Zhang et al. [25] attempted to reveal the effect of changes in LPs on soil erosion in a small watershed. The results suggest that soil erosion was positively correlated with the patch cohesion index (COHESION) but negatively correlated with LPI, modified Simpson's evenness index (MSIEI), and aggregation index (AI). Another study conducted on the island of Crete in Greece [26] found that there was a significant correlation between average soil erosion and PD, edge density (ED), LPI, and percentage of landscape (PLAND). The topic of the effects of LPs variations on various hydrological processes has been a research hotspot over the last decade.

The most popular methods used to investigate the relationships between LPs and hydrological processes are various correlation analysis (the correlation coefficient method [27], stepwise regression analysis [28], multiple linear regression [25], etc.) methods. In addition, hydrological series have mainly been simulated using multiple hydrological or soil erosion models such as the SWAT [29], InVEST [30], IUH [23], RUSLE [26], WaTEM/SEDEM [31] models with various land use maps as inputs, which are further used for analysis with LMs calculated from the corresponding land use maps. The number of land use maps ranges from 3 to 8 in the previously mentioned works, which is insufficient and possibly brought uncertainty into the results. In addition, most of them used simulated hydrological series from multiple eco-hydrological models that ignored the effect of different LPs on hydrological processes. For instance, the SWAT model separates the watershed into several sub-watersheds and further divides them into more hydrological response units (HRUs) based on different combinations of land use, soil, and slope belt [32]. The hydrological processes were simulated at an HRU scale and then aggregated into a sub-watershed scale. The SWAT model is used to calculate the runoff and sediment generation based on specific combinations of land use, soil, and slope belt, but it does not consider the effects of LPs on the streamflow convergence process and the sediment transportation process. As reported by previous studies, the impacts of LPs on hydrological processes are mainly concentrated in the streamflow converge process and sediment transportation process [33]. Hence, simulated hydrological processes based on the SWAT model ignore the major effects of LPs on these hydrological processes. The RUSLE model simulates soil erosion using an experience equation containing six factors [26], which also does not take LPs into consideration. In addition, the simulation results also contain errors compared to the observed data to some extent.

The existing relevant studies mainly focus on specific watersheds all over the world, and the areas of these watersheds range from 16.6 to 320,000 km^2 [17,23,24,26]. As scale is a fundamental concept, analyses at various scales may have resulted in different results or even reverse results [34]. In addition, differences in soil type, topography, and meteorology within various regions could also have been a possible driver for the different results. For example, SHDI was positively correlated with runoff in some regions [17,35] and negatively correlated with runoff in other regions [36,37]; PD was positively correlated with the sediment yields load (SYL) in some regions [34,38] and negatively correlated with SYL in other regions [35,36]. Similar reverse results were also observed for COHESION, LPI, SHEI, AI, CONTAG, LSI, etc. in various regions [25,31,33,39]. These phenomena may suggest that it is insufficient to reveal the relationships between LPs and hydrological processes using relevant data from only one watershed. To investigate the impacts of LPs on water quality across Taiwan, Chiang et al. [40] revealed relationships between 12 water quality parameters and 12 LMs based on water quality records for 10 watersheds. The results show that temperature, PH, NO_3-N, and TN are significantly correlated with most LMs in these watersheds. A similar relevant study has not been reported to have been conducted for runoff/SYL and LMs. In addition, the relationship between LMs and suspended sediment concentration (SSC) has not been reported until now. As SSC is an important hydrological index that could reflect the relationship between streamflow and suspended sediment in a watershed, understanding the driving factors behind SSC would benefit not only soil erosion prevention, but also the prevention and control of river reservoir siltation. As the existing studies report, LPs influence runoff and SYL, mainly during the transportation process. It is necessary to conduct a relevant study to investigate whether LPs changes affect SSC to better understand the driving factors behind SSC.

In previous studies, scholars investigated the relationships between LPs and hydrological processes in specific watersheds, but relevant studies focused on climate zones are still rare. Soil erosion brought by rainfall and runoff is strongly controlled by climate [41], and it is important to investigate the effects of LPs on hydrological processes in various climate zones. The subtropical monsoon climate zone is characterized by high annual rainfall and concentrated summer rainfall, which would result in severe soil erosion [42]. The subtropical monsoon climate zone covers ~29% of China's total area, is mainly distributed in southeastern China, and has experienced rapid economic development and land use changes. In addition, southeastern China contains a variety of landforms and shows a general conversion from mountains and hills to plains from south to north, which are more likely to suffer soil erosion. As pronounced by the Resource and Environment Science and Data Center, Chinese Academy of Sciences (RESDC), ~56.39% of Chinese land suffers from water erosion, with this percentage accounting for almost all of southeastern China [43]. However, how LPs changes affect hydrological processes within the subtropical monsoon climate zone of southeastern China is still unclear. Under the background of global climate change and land use change resulting from extensive human activity, it is significant to reveal the relationships between LPs and representative hydrological processes for better land use management for water and soil conservation. To settle this issue, 10 watersheds with area ranging from 1700 to 80,900 km^2 were chosen as the study area, and water yields (WY), runoff coefficient (RC), soil erosion modulus (SEM), and SSC were chosen as the representative hydrological indices to determine the correlations between the chosen hydrological indices and the most studied LMs from 1990 to 2019 in the subtropical monsoon climate zone of southeastern China. The results enhance the knowledge about the effects of LPs on hydrological processes in the subtropical monsoon climate zone of southeastern China for better land use management with an objective of soil and water conservation.

2. Study Area

A total of 10 watersheds (Zhuji (ZJ), Dufengkeng (DFK), Hushan (HS), Lijiadu (LJD), Lanxi (LX), Boluo (BL), Chaoan (CA), Shijiao (SJ), Zhuqi (ZQ), and Waizhou (WZ)) of different sizes in southeastern China, which were dominated by the subtropical monsoon climate, were chosen as the study area (Figure 1). The study area covered a region between 112.1°~120.7°E, and 22.5°~30.0°N, which is mainly dominated by hills and mountainous terrain, with its highest elevation being 2191 masl and its lowest elevation being −136 masl. The main climatological characteristics in the subtropical monsoon climate zone are temperature and precipitation, which are higher in the summer and lower in the winter, and annual precipitation, which is high but unevenly distributed throughout the year. The annual average temperature and precipitation in southeastern China range from 15 to 22 °C and from 1500 to 2200 mm, respectively. The region is rich with river networks and has a developed water system, which is mainly replenished by precipitation. The main geographical and hydrological characteristics of the chosen watersheds are shown in Table 1. The dominant land use of the study area is forest and agricultural land, and LPs have changed rapidly in the last 30 years due to intensive human activity. The main LP changes were an expansion of urban land and water areas and a reduction in forest and agricultural land from 1990 to 2019.

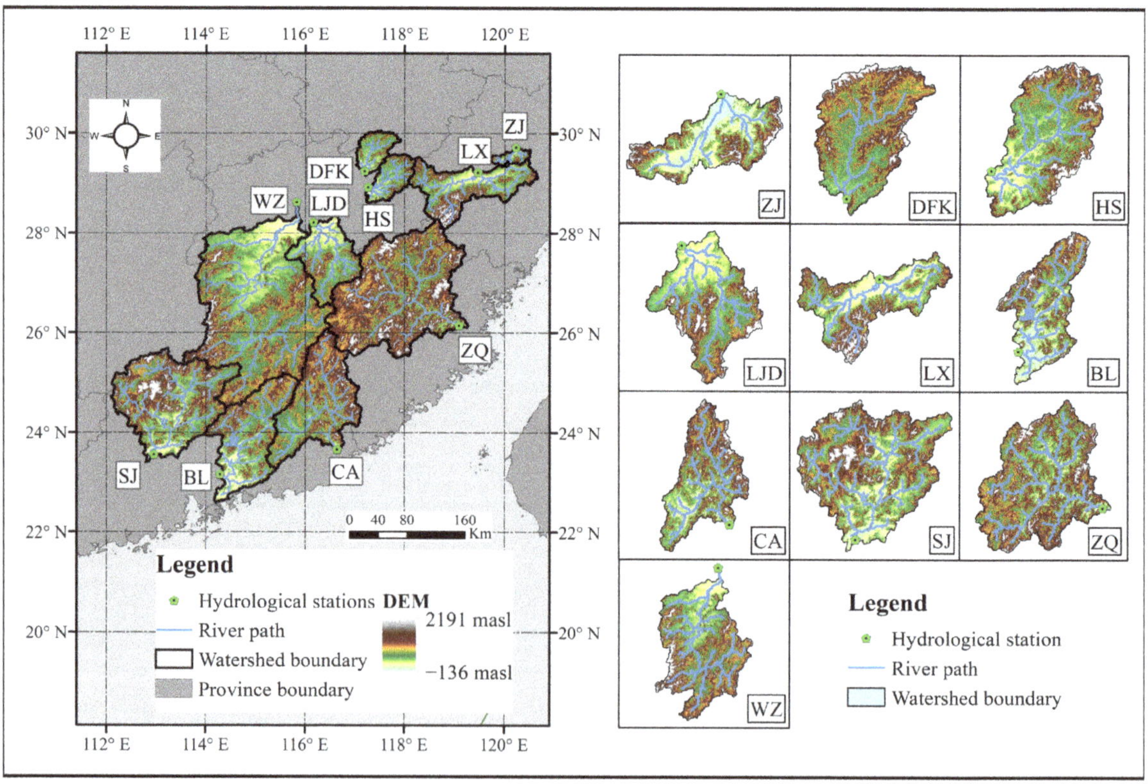

Figure 1. The distribution of the study regions. "ZJ", "DFK", "HS", "LJD", "LX", "BL", "CA", "SJ", "ZQ", and "QZ" represent the "Zhuji", "Dufengkeng", "Hushan", "Lijiadu", "Lanxi", "Boluo", "Chaoan", "Shijiao", "Zhuqi", and "Waizhou" watersheds, respectively.

Table 1. Hydrological and geographical characteristics of the chosen watersheds from 1990 to 2021.

Stations	DA (km^2)	AE (m)	PRE (mm)	WY (10^8 m^3)	RC	SYL (10^4 t)	SEM (t/km^2)	SSC (mg/L)
ZJ	1700	232	1686	12.17	0.41	8.90	52.38	0.07
DFK	5000	243	2054	50.44	0.48	52.97	105.94	0.09
HS	6400	255	2162	71.01	0.52	104.16	162.75	0.13
LJD	15,800	222	2003	124.99	0.39	120.76	76.43	0.09
LX	18,200	352	1942	180.89	0.50	232.69	127.85	0.12
BL	25,300	288	2006	222.21	0.44	131.96	52.16	0.05
CA	29,100	397	1825	235.32	0.45	321.79	110.58	0.12
SJ	38,400	384	1994	416.69	0.55	463.70	120.75	0.11
ZQ	54,500	533	1982	544.55	0.50	278.93	51.18	0.04
WZ	80,900	300	1812	713.52	0.49	382.66	47.30	0.05

Note: DA, AE, and PRE represent drainage area, average elevation, and precipitation, respectively.

3. Data and Methods

3.1. Data Description

The main data used in this study consist of land use maps, precipitation records, and annual runoff and sediment yield records. The land use maps from 1990 to 2019 were obtained from Wuhan University [44] with a spatial resolution of 30 m. The land use maps were generated using the random forest classifier and the visual interpretation method, and they were validated with three sources of test samples. They divided the land use type into 9 categories: cropland, forest, shrub, grassland, water, snow and ice, barren, imperious, and wetland. They can be obtained at https://zenodo.org/record/4417810 (accessed on 28 December 2022), and they have been employed in many studies [19,45,46]. The precipitation records were obtained from China Scientific Data with a temporal resolution of monthly (1960–2020) and a spatial resolution of 1 km [47]. The precipitation records were interpreted with ANUSPLIN 4.4 based on ~2400 precipitation monitoring stations within China and were validated by comparing them with the precipitation records from precipitation gauges and the Chinese hydrological yearbook. The precipitation records that were used can be downloaded at http://doi.org/10.11922/sciencedb.01607 (accessed on 25 December 2022). The measured annual runoff and sediment yield records from 1990 to 2019 of the DFK, LJD, LX, BL, CA, SJ, ZQ, and WZ watersheds and those from 2002 to 2019 of the ZJ and HS watersheds were obtained from the Bulletin of River Sediment in China and were downloaded at http://www.mwr.gov.cn/sj/tjgb/zghlnsgb/ (accessed on 30 December 2022). The bulletins were compiled jointly by the Department of Hydrology of the Ministry of Water Resources, the Monitoring and Forecasting Center for Hydrology and Water Resources of the Ministry of Water Resources, the International Sediment Research and Training Center, and the hydrology bureaus of the river basin institutions, which are very credible. The annual runoff and sediment yield records were used to represent the historical hydrology conditions within the chosen watersheds.

3.2. LMs Selection

LMs were used to represent the spatial distribution characteristics of the various land use patches, and they have been widely used to investigate the relationships between LPs and eco-hydrological processes [21,25,39,48]. The LMs were divided into three categories based on different levels: patch-level, class-level, and landscape-level metrics [39]. The LMs on a landscape level integrated all patch types or classes over the entire study region, and they were separated into four categories reflecting different types of characteristics of landscapes: edge area, shape, aggregation, and diversity metrics [49]. The main purpose of this study was to reveal the relationships between landscape patterns and representative hydrological indices in the subtropical monsoon climate zone. Hence, 14 widely used LMs, which were found to have different relationships with runoff/sediment in various regions, were chosen for this study [17,25,27,31,33,35,36,38,39,50,51]. The selected LMs were LPI, ED, perimeter area fractal dimension (PAFRAC), number of patches (NP), DIVI-

SION, AI, interspersion and juxtaposition index (IJI), contiguity index (CONTAG), PD, LSI, COHESION, SHDI, Shannon's evenness index (SHEI), and MSIEI. Their definitions and relevant literature are shown in Table A1, and these LMs were calculated using Fragstats v4.2.1 software for all land use maps from 1990 to 2019 for all chosen watersheds. Before calculation, the land use types were reclassified into seven categories: agricultural land, forest land, shrubland, grassland, water area, urban land, and bare land.

3.3. Hydrological Indices

To better evaluate the relationships between hydrological processes and LPs, five fundamental hydrological indices (WY (m^3); RC; SYL (t); SEM (t/km^2); SSC (mg/L)) relevant to runoff, sediment yields, and streamflow–sediment relationships were chosen for the study and obtained on an annual scale. These hydrological indices have been employed in previous studies to investigate the hydrological characteristics of various regions [29,52–55]. In this paper, WY, SYL, and SSC values were collected from the Bulletin of River Sediment in China. RC and SEM were calculated using Equations (1) and (2).

$$RC = \frac{WY}{1000 \cdot DA \cdot PRE} \tag{1}$$

$$SEM = \frac{SYL}{DA} \tag{2}$$

where WY represents annual water yield (m^3); DA represents drainage area (km^2); PRE represents annual precipitation (mm); SYL represents the annual sediment yield load (t).

3.4. Analysis Methods

Linear regression [56] was used to analyze the change trends of various hydrological indices and of LMs from 1990 to 2019 for the chosen watersheds. This type of analysis has been widely employed in previous relevant works and has shown reasonable results [43,57,58]. In addition, the F-test [59] was adopted to test the significance of the trend for each series at confidence levels of 0.05 and 0.01. The equation form of the linear regression is as follows:

$$y = ax + b \tag{3}$$

where y represents the dependent variable (WY, RC, SYL, SEM, and SSC); a represents the change trend per year; x represents year; b represents the constant.

To investigate the exchanges between different land use types from 1990 to 2019, the period was separated into two stages: 1990–2005 and 2005–2019. Land use conversion analysis was conducted for land use maps in 1990, 2005, and 2019 in the ArcGIS 10.2 platform. To better show the exchanges between different land use types, land with no land use type exchanges was ignored when conducting relevant analyses. With this method, the exchanges between each land use type from 1990 to 2019 were analyzed for the chosen watersheds.

The effects of LPs changes that propagate to hydrological processes have lag times that vary according to differences in watershed size, climate, landscape, topography, soil type, etc. [60–62]. In previous studies, lag times of 0–4 years were needed before significant responses of hydrological processes to landscape changes could be observed [63,64]. Hence, a slip correlation analysis [43] was conducted with a lag time of 0–4 years to investigate the relationships between LPs and hydrological indices. A description of the slip correlation analysis is as follows:

(1) The origin series X and Y are: $X(x_1, x_2, \ldots, x_{n-1}, x_n)$; $Y(y_1, y_2, \ldots, y_{n-1}, y_n)$. The X and Y series are the landscape series and hydrological indices, respectively; n is the series number, which corresponds to the year.

(2) The clip correlation series for X and Y were established. For the lag time of i years, the X′ and Y′ series are $X'(x_1, x_2, \ldots, x_{n-i-1}, x_{n-i})$ and $Y'(y_i, y_{i+1}, \ldots, y_{n-1}, y_n)$, respectively. The value of *i* ranges from 0 to 4, and it represents lag times of 0–4 years.

(3) The correlation coefficient (CC_i) between X′ and Y′ was calculated, and the result was the lag correlation between X and Y with a lag time of i years. The highest value of CC_i with a significance of 0.05 was regarded as the lag time between X and Y.

As SEM was calculated by dividing SYL by DA, the SEM series was linearly correlated with SYL and removed the effect of watershed size on soil erosion. In this paper, a slip correlation analysis was conducted between the 14 LMs and 4 hydrological indices (WY, RC, SEM, and SSC) in the chosen watersheds.

4. Results

4.1. Variations in SY, RC, SYL, SEM, and SSC

Table 1 shows the annual average values of various hydrological indices, and Figure 2 shows variations in WY, RC, SYL, SEM, and SSC. The annual average WY, SYL, RC, SEM, and SSC values for the chosen watersheds ranged from 12.17×10^8 to 713.52×10^8 m^3, from 8.9×10^4 to 463.7×10^4 t, from 0.39 to 0.55, from 47.3 to 162.75 t/km^2, and from 0.05 to 0.13 mg/L, respectively. The variations in WY were higher within ZQ and WZ, followed by LJD, LX, BL, CA, and SJ, and the variations in SYL were obviously higher within CA, SJ, ZQ, and WZ compared to the other watersheds. In addition, the variations in SEM and SSC were higher within DFK, HS, LX, and CA compared to the other watersheds, which indicates that soil erosion in these watersheds was much more serious compared to the other watersheds.

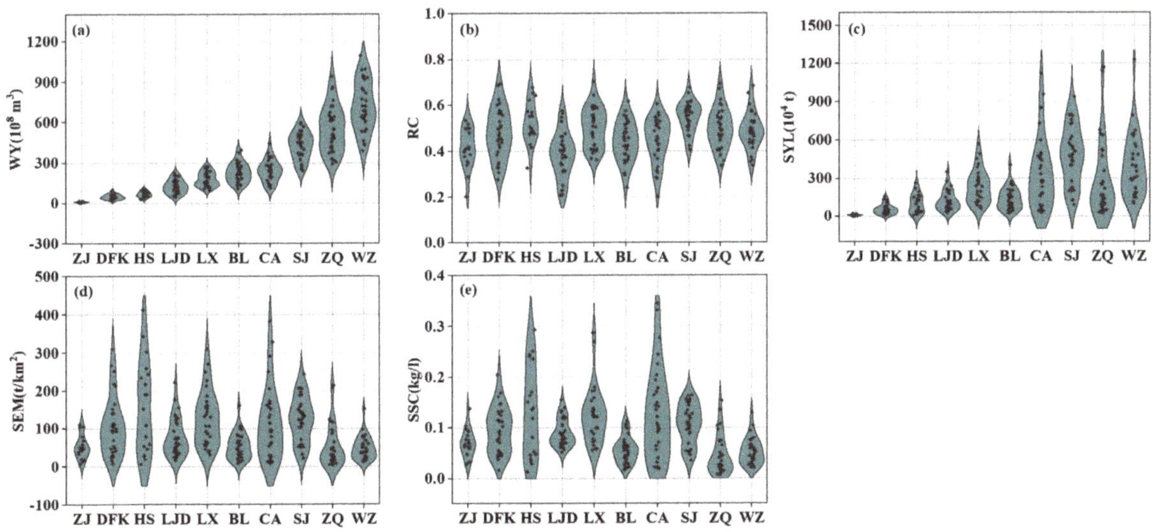

Figure 2. Variations in hydrological indices from 1990 to 2019. (**a**–**e**) represent WY, RC, SYL, SEM, and SSC, respectively.

The change trends of the hydrological indices are shown in Table 2, and their temporal variations are shown in Figure A1. WY showed a decreased trend in the DFK, BL, CA, SJ, and WZ watersheds while exhibiting an increased trend in the ZJ, HS, LJD, LX, and ZQ watersheds. RC increased in all watersheds except for CA and SJ. In particular, DFK, BL, and WZ suffered decreases in WY, whereas their RC values increased. As a whole, the change trends of WY and RC were not significant for all watersheds. Differing from WY and RC, SYL and SEM decreased in all watersheds except for the ZJ, HS, and LX watersheds. SYL and SEM values increased significantly ($p < 0.01$) in HS and decreased significantly ($p < 0.01$) in CA and WZ. Interestingly, LJD and ZQ increased in WY and RC but decreased in SYL and SEM. This phenomenon may have been caused by LP changes and human activity in these regions. As for SSC, it significantly increased in HS ($p < 0.01$) and significantly ($p < 0.05$) decreased in LJD, ZQ, BL, CA, and WZ. In general, soil erosion was mitigated in most watersheds, whereas water resources decreased in the larger watersheds and increased in the smaller watersheds.

Table 2. Change trend statistics of hydrological indices from 1990 to 2019.

Indices	ZJ	DFK	HS	LJD	LX	BL	CA	SJ	ZQ	WZ
WY	0.25	−0.003	1.53	0.16	0.64	−0.12	−1.22	−1.00	0.75	−1.13
RC	0.0048	0.0012	0.0064	0.0008	0.0016	0.0013	−0.0009	−0.0001	0.0008	0.0004
SYL	0.04	−0.02	10.90 **	−1.19	4.43	−3.04	−19.01 **	−4.72	−7.63	−19.76 **
SEM	0.23	−0.04	17.03 **	−0.75	2.43	−1.20	−6.53 **	−1.23	−1.40	−2.44 **
SSC	−0.0005	0.0002	0.0128 **	−0.0011 *	0.0019	−0.0015 **	−0.0073 **	−0.0008	−0.0016 *	−0.0026 **

Note: The units of WY, SYL, SEM, and SSC are 10^8 m^3, 10^4 t, t/km^2, and mg/l, respectively, and RC is a dimensionless quantity. The change trends of these indices are shown as the average variations for each year. * and ** represent significant correlations at confidence levels of 95% and 99%, respectively.

4.2. Temporal Variations in LPs in the Chosen Watersheds

4.2.1. Land Use Changes from 1990 to 2019

As shown in Figure 3, the dominant land use type of the chosen watersheds was forest land, followed by agricultural land. The ratio of agricultural land was higher in ZJ, LJD, LX, BL, and WZ compared to the other watersheds, which were also more intensively distributed. In addition, from 1990 to 2019, urban land expanded drastically in all watersheds, especially in ZJ, LJD, LX, and WZ. The regions in which the land use types for different watersheds were altered occupied about 3.01–8.46% (average of 6.03%), 4.22–9.78% (average of 6.81%), and 5.11–11.47% (average of 8.83%) of the total area from 1990 to 2005, from 2005 to 2019, and from 1990 to 2019, respectively. Interestingly, the average area of the regions whose land use types had been altered from 1990 to 2019 was much lower than the sum of these regions from 1990 to 2005 and from 2005 to 2019. This may suggest that some unreasonable land use conversions from 1990 to 2005 had been corrected from 2005 to 2019. From 1990 to 2019, urban expansion was the only long-lasting land use conversion process.

Figure 4 shows the details of land use conversions among different land use types from 1990 to 2005 and from 2005 to 2019. We found that land use conversions mainly happened on agricultural land, forest land, and urban land. From 1990 to 2005, the main land use conversions were forest land–agricultural land, agricultural land–urban land, and agricultural land–forest land; part of the agricultural land had also been converted into water areas within ZJ, HS, LJD, LX, BL, and SJ. From 2005 to 2019, the main land use conversions were agricultural land–urban land, agricultural land–forest land, and forest land–agricultural land; some urban land was converted from forest land within DFK, HS, LX, CA, ZQ, and WZ. These results reveal that the urban expansion in southeastern China was mainly facilitated by occupying agricultural land, especially from 1990 to 2005. During recent decades, urban expansion also happened in some forest regions for most watersheds. This was reasonable, as urban development had extended to the surrounding rural areas first in most regions, which had large areas of agricultural land and forest land [65–67]. In addition, conversions between agricultural land and forest land were relatively greater. These results may have been caused by land use exchanges. For better soil erosion prevention, some sloped agricultural land was abandoned and converted into forest land, whereas some flat forest land was converted into agricultural land to preserve the amount of agricultural land and promote guaranteed crop production.

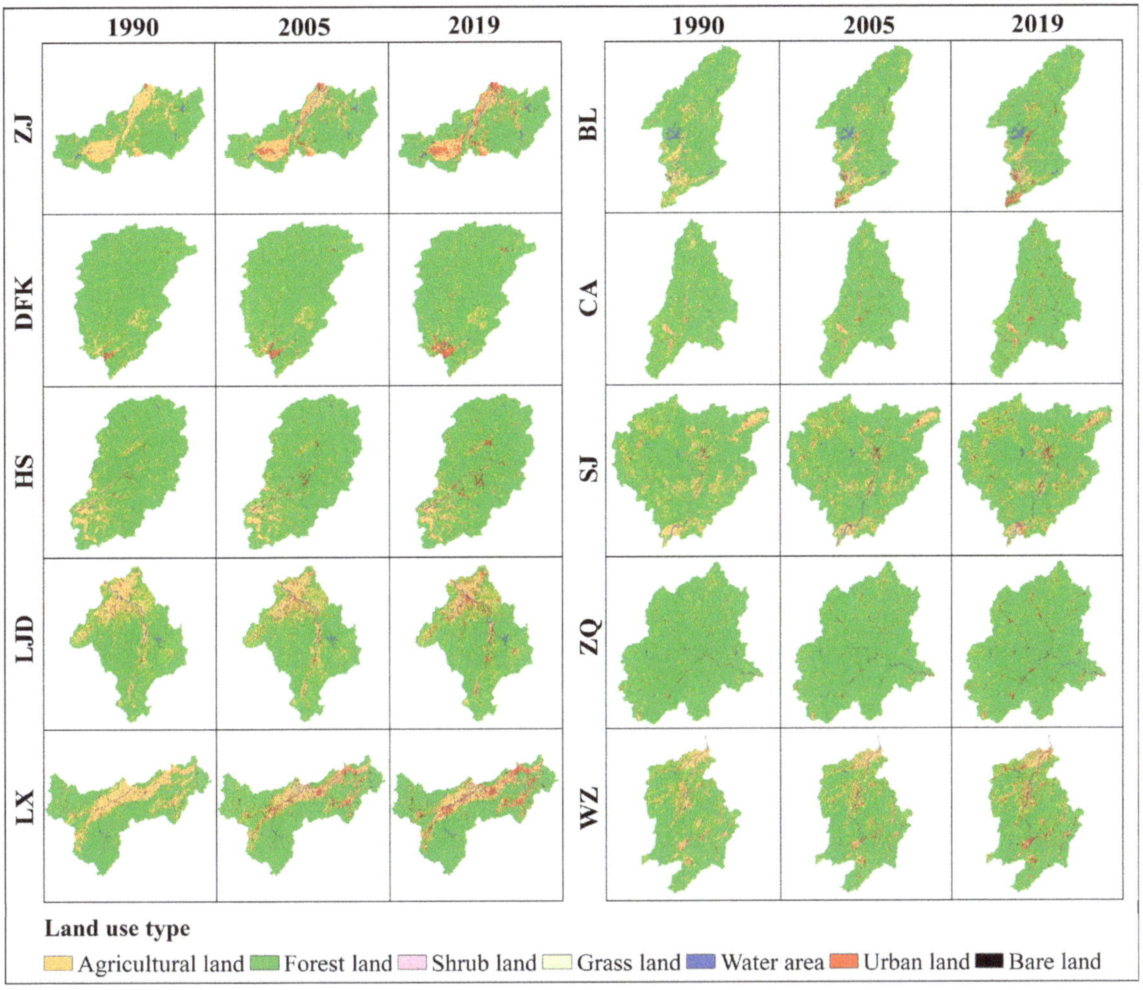

Figure 3. Land use maps of ZJ, DFK, HS, LJD, LX, BL, CA, SJ, ZQ, and WZ in 1990, 2005, and 2019.

4.2.2. LM Changes from 1990 to 2019

As shown in Figure 5, the variations in all LMs of ZJ were the largest among all watersheds. In addition, several LMs of LX, ZQ, and WZ also showed larger variations compared to the other watersheds. The average values of LPI and PAFRAC had similar trends with increases in watershed size. The analogical character was also shown for DIVISION, IJI, PD, SHDI, SHEI, and MSIEI. It is notable that the LPs within ZJ changed intensively in 1998, 2001, 2007, and 2017, with obvious variations in LPI, DIVISION, and COHESION. The LPs within LX changed intensively in 1993, 2007, and 2016 (Figure A2). Different from the other LMs, COHESION changed little in all watersheds except ZJ compared to the other LMs. This may suggest that COHESION was not as sensitive as other LMs to LPs changes.

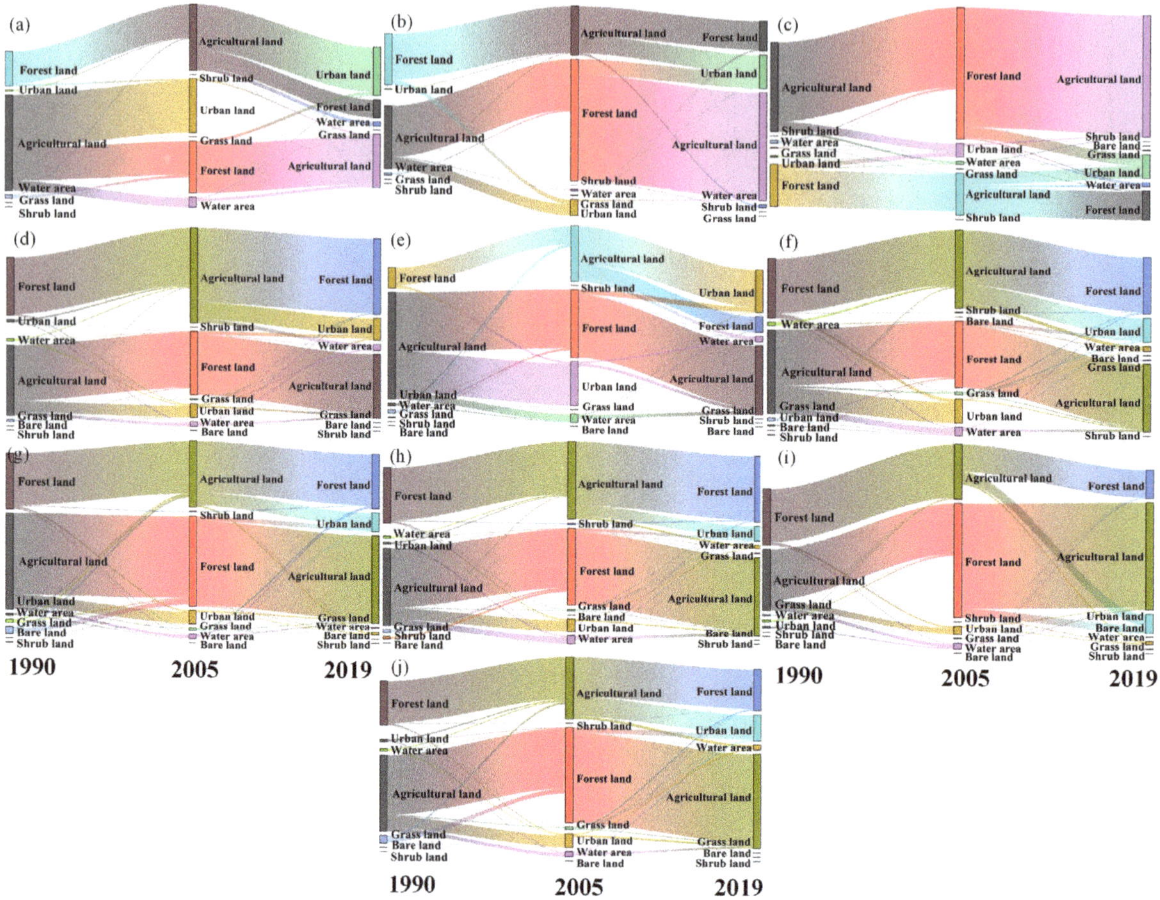

Figure 4. Land use conversion statistics between 1990, 2005, and 2019 for the chosen watersheds. Land whose land use type did not change is not included in this figure. (**a–j**) represent the ZJ, DFK, HS, LJD, LX, BO, CA, SJ, ZQ, and WZ watersheds, respectively.

The change trend statistic results of the chosen LMs from 1990 to 2019 are shown in Table 3. For most watersheds, most LMs changed significantly ($p < 0.05$), which demonstrates that LPs characteristics changed significantly during this period. Among all watersheds, the LPs within ZQ was the most intensively changed, as all LMs of ZQ changed significantly ($p < 0.01$). In detail, COHESION (ED, PAFRAC, AI, IJI, CONTAG, LSI, and SHEI) and SHDI (LPI and DIVISION) significantly ($p < 0.05$) changed in nine (individually, eight and seven, respectively) watersheds; NP and PD (MSIEI) significantly ($p < 0.01$) changed in six (four) watersheds. Among these significant trends, ED, DIVISION, IJI, LSI, SHDI, SHEI, and MSIEI (AI, CONTAG, COHESION, LPI, and PAFRAC) all significantly ($p < 0.05$) increased (or decreased) in most watersheds. These uniformities in the change trends of the LMs may suggest that the development directions in land use planning were similar across the regions of southeastern China. In addition, COHESION significantly decreased in most watersheds, which did not show obvious variations. The change trends of SHDI and SHEI were similar in all watersheds, and the change trend of SHDI was more obvious than that of SHEI (Figure A2). In general, the LPs within the chosen watersheds significantly changed from a landscape perspective.

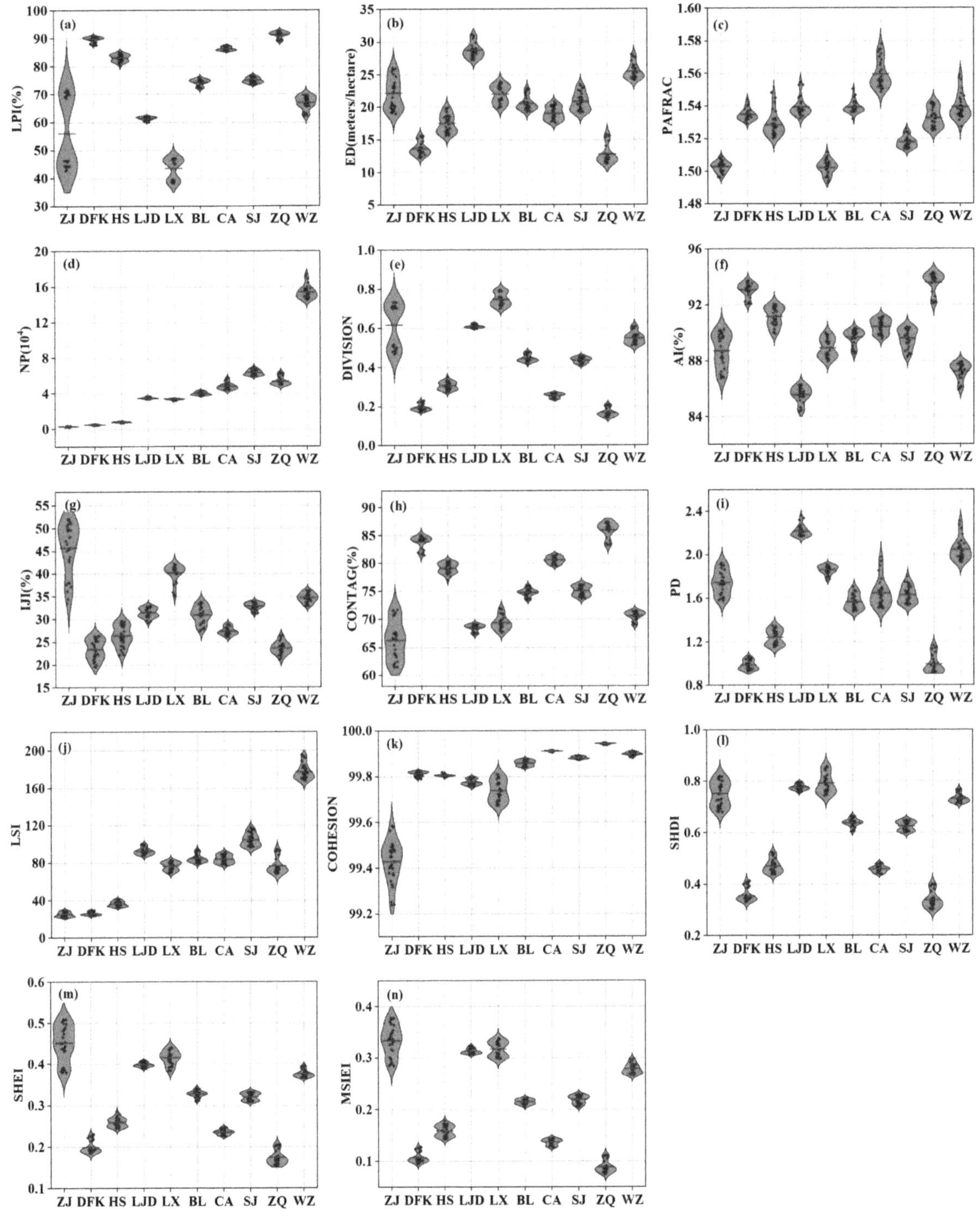

Figure 5. Variations of LMs from 1990 to 2019. (**a**–**n**) represent LPI, ED, PAFRAC, NP, DIVISION, AI, IJI, CONTAG, PD, LSI, COHESION, SHDI, SHEI, and MSIEI, respectively.

Table 3. Change trend statistics of LMs within the chosen watersheds from 1990 to 2019.

Metrics	ZJ	DFK	HS	LJD	LX	BL	CA	SJ	ZQ	WZ
LPI	0.62 *	−0.06 **	−0.01	−0.04 *	−0.09	−0.08 **	0.02	−0.07 **	−0.11 **	−0.11 *
ED	0.26 **	0.05 *	0.03	0.12 **	0.15 **	0.11 **	−0.01	0.14 **	0.13 **	0.06 *
PAFRAC	0.0001	0.0002 **	−0.0005 **	−0.0005 **	−0.0004 **	−0.0002 **	−0.0009 **	0.00005	−0.0004 **	−0.0008 **
NP	0.0013 **	0.0004	−0.0006	−0.0045 **	0.0045 **	0.0072	−0.0264 **	0.0108	0.0263 **	−0.0436 **
DIVISION	−0.0043	0.0010 **	0.0002	0.0007 **	0.0021 **	0.0013 **	−0.0003	0.0011 **	0.020 **	0.0014 *
AI	−0.129 **	−0.025 *	−0.015	−0.058 **	−0.074 **	−0.054 **	0.007	−0.068 **	−0.065 **	−0.029 *
IJI	0.55 **	0.24 **	0.18 **	0.11 **	0.08 *	0.16 **	0.05 *	−0.03	0.12 **	−0.02
CONTAG	−0.376 **	−0.082 **	−0.14	−0.063 **	−0.097 **	−0.073 **	−0.001	−0.087 **	−0.137 **	−0.049 **
PD	0.008 **	0.001	−0.001	−0.003 **	0.002 **	0.003	−0.009 **	0.003	0.005 **	−0.006 **
LSI	0.27 **	0.09 *	0.06	0.36 **	0.50 **	0.44 **	−0.06	0.67 **	0.76 **	0.39 *
COHESION	−0.0034	−0.0011 **	−0.0003 **	−0.0013 **	−0.0045 **	−0.0013 **	−0.0001 **	−0.0005 **	−0.0003 **	−0.0005 **
SHDI	0.005 **	0.002 **	0.001 *	0.001 **	0.004 **	0.002 **	0.001	0.002 **	0.003 **	0.001 **
SHEI	0.005 **	0.001 **	0.001	0.001 **	0.001 **	0.001 **	0.001	0.001 **	0.002 **	0.001 **
MSIEI	0.0031 **	0.0006 **	0	−0.0001	0.0005	0	−0.0001	0.0006 **	0.0011 **	0.0003

Note: The units of LPI, ED, DIVISION, AI, IJI, CONTAG, and PD are %, meters/hectare, proportion, %, %, %, and number/100 hectares, respectively, and PAFRAC, NP, LSI, COHESION, SHDI, SHEI, and MSIEI are dimensionless quantities. The change trends of these indicators are shown as the average variations each year. * and ** represent significant correlations at confidence levels of 95% and 99%, respectively.

4.3. Relationships between Hydrological Indices and LPs

The CC between the hydrological indices (WY, RC, SEM, and SSC) and LMs (LPI, ED, PAFRAC, NP, DIVISION, AI, IJI, CONTAG, PD, LSI, COHESION, SHDI, SHEI, and MSIEI) of the chosen watersheds are shown in Tables 4 and 5. There were discrepancies in the lag times of the correlations between the hydrological indices and LMs in different watersheds in the subtropical monsoon climate zone of southeastern China. In addition, the lag times between the hydrological indices and LMs did not show an obvious relationship with watershed size.

Table 4. CC values between the landscape metrics and the WY and RC.

Hydrological Indices	LMs	ZJ	DFK	HS	LJD	LX	BL	CA	SJ	ZQ	WZ
WY	LPI	**0.57** ^{1}	−0.28 ^{0}	−0.46 ^{0}	−0.29 ^{1}	**−0.46** ^{3}	−0.22 ^{1}	−0.21 ^{0}	0.28 ^{1}	−0.08 ^{0}	**−0.40** ^{4}
	ED	0.48 ^{1}	0.27 ^{0}	0.39 ^{0}	−0.03 ^{1}	0.14 ^{4}	0.13 ^{0}	0.25 ^{2}	−0.13 ^{1}	0.10 ^{0}	0.29 ^{2}
	PAFRAC	−0.12 ^{2}	0.12 ^{0}	**0.50** ^{0}	0.16 ^{3}	−0.33 ^{1}	0.21 ^{2}	0.24 ^{2}	0.34 ^{4}	0.19 ^{2}	0.26 ^{2}
	NP	**0.53** ^{1}	0.33 ^{1}	0.33 ^{0}	0.16 ^{2}	0.37 ^{4}	0.20 ^{3}	0.27 ^{2}	0.30 ^{4}	0.19 ^{2}	0.35 ^{2}
	DIVISION	**−0.56** ^{1}	0.28 ^{0}	0.46 ^{0}	0.16 ^{1}	0.36 ^{3}	0.22 ^{1}	0.21 ^{0}	−0.28 ^{1}	0.08 ^{0}	**0.39** ^{4}
	AI	−0.48 ^{1}	−0.27 ^{0}	−0.39 ^{0}	0.03 ^{1}	−0.14 ^{4}	−0.13 ^{0}	−0.25 ^{2}	0.13 ^{1}	−0.10 ^{0}	−0.29 ^{2}
	IJI	0.47 ^{1}	−0.11 ^{0}	**0.60** ^{4}	0.08 ^{4}	0.11 ^{0}	−0.17 ^{2}	−0.26 ^{0}	−0.26 ^{3}	0.14 ^{4}	−0.35 ^{0}
	CONTAG	−0.47 ^{1}	−0.23 ^{0}	**−0.53** ^{1}	−0.12 ^{0}	−0.24 ^{0}	−0.12 ^{0}	−0.22 ^{2}	0.18 ^{1}	−0.09 ^{0}	−0.31 ^{4}
	PD	**0.53** ^{1}	0.33 ^{1}	0.33 ^{0}	0.16 ^{2}	0.37 ^{4}	0.20 ^{3}	0.27 ^{2}	0.30 ^{4}	0.19 ^{2}	0.35 ^{2}
	LSI	0.48 ^{1}	0.27 ^{0}	0.39 ^{0}	−0.03 ^{1}	0.14 ^{4}	0.13 ^{0}	0.25 ^{2}	−0.13 ^{1}	0.10 ^{0}	0.29 ^{2}
	COHESION	**0.59** ^{3}	−0.16 ^{4}	**−0.51** ^{0}	0.16 ^{2}	−0.21 ^{4}	−0.08 ^{4}	−0.09 ^{4}	0.30 ^{2}	−0.11 ^{3}	−0.33 ^{4}
	SHDI	**0.50** ^{1}	0.20 ^{0}	0.41 ^{0}	0.26 ^{1}	0.17 ^{4}	−0.16 ^{3}	0.18 ^{0}	−0.21 ^{1}	0.08 ^{0}	0.32 ^{4}
	SHEI	0.46 ^{1}	0.20 ^{0}	**0.54** ^{1}	0.26 ^{1}	0.27 ^{0}	−0.16 ^{3}	0.18 ^{0}	−0.21 ^{1}	0.08 ^{0}	0.32 ^{4}
	MSIEI	0.34 ^{0}	0.25 ^{0}	**0.48** ^{0}	0.32 ^{1}	0.34 ^{0}	−0.13 ^{3}	0.21 ^{0}	−0.23 ^{1}	0.08 ^{1}	0.30 ^{2}
RC	LPI	**0.58** ^{1}	**−0.44** ^{0}	**−0.55** ^{0}	−0.28 ^{0}	**−0.52** ^{3}	−0.34 ^{1}	−0.27 ^{2}	0.19 ^{1}	−0.18 ^{0}	**−0.41** ^{1}
	ED	0.45 ^{1}	**0.43** ^{0}	**0.50** ^{0}	0.11 ^{3}	0.18 ^{4}	0.30 ^{0}	0.30 ^{2}	0.13 ^{4}	0.19 ^{0}	**0.39** ^{2}
	PAFRAC	−0.17 ^{3}	0.28 ^{0}	**0.53** ^{1}	−0.12 ^{0}	−0.33 ^{0}	0.14 ^{2}	0.21 ^{2}	0.37 ^{4}	0.14 ^{2}	0.16 ^{2}
	NP	**0.49** ^{1}	**0.44** ^{1}	0.48 ^{2}	0.17 ^{2}	0.33 ^{4}	0.27 ^{3}	0.25 ^{2}	0.31 ^{4}	0.25 ^{2}	0.31 ^{2}
	DIVISION	**−0.57** ^{1}	**0.44** ^{0}	**0.55** ^{0}	0.22 ^{0}	**0.41** ^{3}	0.34 ^{1}	0.27 ^{2}	−0.19 ^{1}	0.18 ^{0}	**0.41** ^{1}
	AI	−0.45 ^{1}	**−0.43** ^{0}	**−0.50** ^{0}	−0.11 ^{3}	−0.18 ^{4}	−0.30 ^{0}	−0.30 ^{2}	−0.13 ^{4}	−0.19 ^{0}	**−0.39** ^{2}
	IJI	0.48 ^{1}	0.18 ^{4}	**0.63** ^{4}	0.13 ^{4}	−0.19 ^{3}	−0.09 ^{2}	−0.28 ^{0}	−0.25 ^{3}	0.18 ^{4}	**−0.42** ^{0}
	CONTAG	−0.46 ^{1}	**−0.40** ^{0}	**−0.57** ^{1}	−0.19 ^{0}	−0.22 ^{0}	−0.28 ^{0}	−0.29 ^{2}	0.09 ^{1}	−0.18 ^{0}	**−0.40** ^{3}
	PD	**0.49** ^{1}	**0.44** ^{1}	0.48 ^{2}	0.18 ^{2}	0.33 ^{4}	0.27 ^{3}	0.25 ^{2}	0.31 ^{4}	0.25 ^{2}	0.31 ^{2}
	LSI	0.45 ^{1}	**0.43** ^{0}	**0.50** ^{0}	0.11 ^{3}	0.18 ^{4}	0.30 ^{0}	0.30 ^{2}	0.13 ^{4}	0.19 ^{0}	**0.39** ^{2}
	COHESION	**0.59** ^{3}	−0.34 ^{4}	**−0.54** ^{0}	0.12 ^{1}	−0.24 ^{4}	−0.19 ^{1}	−0.17 ^{4}	0.22 ^{2}	−0.18 ^{3}	−0.37 ^{4}
	SHDI	0.47 ^{1}	**0.38** ^{0}	**0.51** ^{0}	0.28 ^{0}	0.24 ^{4}	0.25 ^{0}	0.28 ^{2}	−0.11 ^{1}	0.18 ^{0}	**0.40** ^{3}
	SHEI	0.45 ^{1}	**0.38** ^{0}	**0.58** ^{1}	0.28 ^{0}	0.27 ^{0}	0.24 ^{0}	0.28 ^{2}	−0.11 ^{1}	0.18 ^{0}	**0.40** ^{3}
	MSIEI	0.33 ^{1}	**0.42** ^{0}	**0.56** ^{1}	0.28 ^{0}	0.35 ^{0}	0.18 ^{0}	0.27 ^{2}	−0.13 ^{1}	0.18 ^{1}	**0.39** ^{2}

Note: bold font indicates a correlation was significant at a confidence level of 95%, bold and underlined font indicates a correlation was significant at a confidence level of 99%, and a superscript number indicates the lag time of the correlation between the LMs and the WY or RC.

Table 5. CC values between LMs and the SEM and SSC.

Hydrological Indices	LMs	ZJ	DFK	HS	LJD	LX	BL	CA	SJ	ZQ	WZ
SEM	LPI	**0.62** ^{1}	−0.30 ^{2}	**−0.72** ^{0}	−0.18 ^{1}	**−0.63** ^{3}	0.16 ^{0}	**−0.44** ^{3}	−0.35 ^{3}	0.13 ^{0}	−0.21 ^{4}
	ED	0.24 ^{1}	0.27 ^{3}	**0.69** ^{0}	−0.15 ^{0}	0.34 ^{4}	−0.25 ^{0}	**0.42** ^{2}	−0.17 ^{0}	−0.12 ^{0}	0.35 ^{4}
	PAFRAC	−0.33 ^{0}	−0.10 ^{2}	**0.76** ^{0}	0.29 ^{2}	**−0.46** ^{0}	0.30 ^{2}	**0.63** ^{0}	0.34 ^{3}	0.17 ^{2}	**0.82** ^{2}
	NP	0.30 ^{1}	**0.41** ^{3}	**0.66** ^{0}	0.27 ^{2}	**0.59** ^{4}	−0.16 ^{0}	**0.58** ^{0}	0.32 ^{3}	0.13 ^{2}	**0.71** ^{2}
	DIVISION	**−0.63** ^{1}	0.30 ^{2}	**0.72** ^{0}	−0.13 ^{2}	**0.59** ^{3}	−0.17 ^{0}	**0.44** ^{3}	0.34 ^{3}	−0.13 ^{0}	0.20 ^{4}
	AI	−0.24 ^{1}	−0.27 ^{3}	**−0.69** ^{0}	0.15 ^{0}	**−0.34** ^{4}	0.25 ^{0}	**−0.42** ^{2}	0.17 ^{0}	0.12 ^{0}	−0.35 ^{4}
	IJI	**−0.75** ^{4}	−0.16 ^{3}	**0.76** ^{4}	−0.16 ^{2}	0.33 ^{0}	−0.36 ^{2}	**−0.45** ^{2}	0.12 ^{1}	−0.29 ^{2}	−0.30 ^{4}
	CONTAG	**0.64** ^{4}	−0.23 ^{2}	**−0.76** ^{0}	0.07 ^{2}	**−0.44** ^{0}	0.30 ^{2}	−0.38 ^{3}	0.18 ^{0}	0.14 ^{0}	0.21 ^{0}
	PD	0.30 ^{1}	**0.41** ^{3}	**0.66** ^{0}	0.27 ^{2}	**0.59** ^{4}	−0.16 ^{0}	**0.58** ^{0}	0.32 ^{3}	0.13 ^{2}	**0.71** ^{2}
	LSI	0.24 ^{1}	0.27 ^{3}	**0.69** ^{0}	−0.15 ^{0}	0.34 ^{4}	−0.25 ^{0}	**0.42** ^{2}	−0.17 ^{0}	−0.12 ^{0}	0.35 ^{4}
	COHESION	**0.53** ^{1}	−0.09 ^{1}	**−0.73** ^{0}	0.28 ^{2}	**−0.48** ^{4}	0.33 ^{2}	**0.47** ^{0}	0.21 ^{2}	0.14 ^{0}	0.35 ^{2}
	SHDI	0.27 ^{1}	0.20 ^{2}	**0.72** ^{0}	0.14 ^{1}	**0.36** ^{0}	−0.31 ^{2}	0.31 ^{3}	−0.18 ^{0}	−0.14 ^{0}	−0.24 ^{0}
	SHEI	**−0.69** ^{4}	0.20 ^{2}	**0.77** ^{0}	0.14 ^{1}	**0.47** ^{0}	−0.31 ^{2}	0.31 ^{3}	−0.18 ^{0}	−0.14 ^{0}	−0.24 ^{0}
	MSIEI	**−0.74** ^{4}	0.26 ^{2}	**0.75** ^{0}	0.35 ^{1}	**0.42** ^{0}	0.15 ^{4}	**0.41** ^{3}	0.25 ^{3}	−0.25 ^{0}	0.25 ^{4}
SSC	LPI	**0.52** ^{1}	−0.34 ^{2}	**−0.65** ^{0}	0.18 ^{2}	**−0.58** ^{2}	0.30 ^{0}	**−0.50** ^{3}	**−0.45** ^{3}	0.21 ^{0}	0.23 ^{0}
	ED	−0.37 ^{4}	0.34 ^{3}	**0.68** ^{0}	−0.35 ^{0}	0.34 ^{4}	**−0.44** ^{0}	**0.48** ^{4}	−0.15 ^{0}	−0.17 ^{0}	0.26 ^{4}
	PAFRAC	**−0.49** ^{0}	0.16 ^{0}	**0.78** ^{0}	**0.54** ^{2}	**−0.41** ^{0}	0.35 ^{2}	**0.70** ^{0}	0.27 ^{3}	0.26 ^{0}	**0.89** ^{2}
	NP	−0.41 ^{4}	**0.45** ^{3}	**0.66** ^{0}	**0.51** ^{4}	**0.61** ^{3}	−0.22 ^{0}	**0.62** ^{0}	0.26 ^{3}	0.12 ^{2}	**0.74** ^{4}
	DIVISION	**−0.53** ^{1}	0.34 ^{2}	**0.65** ^{0}	−0.34 ^{2}	**0.55** ^{2}	−0.32 ^{0}	**0.50** ^{3}	**0.44** ^{3}	−0.21 ^{0}	−0.23 ^{0}
	AI	0.37 ^{4}	−0.34 ^{3}	**−0.68** ^{0}	0.35 ^{0}	−0.34 ^{4}	**0.44** ^{0}	**−0.48** ^{4}	0.15 ^{0}	0.17 ^{0}	−0.27 ^{4}
	IJI	**−0.71** ^{4}	−0.09 ^{3}	**0.73** ^{4}	−0.36 ^{2}	**0.39** ^{0}	**−0.49** ^{2}	−0.51 ^{2}	0.19 ^{1}	**−0.38** ^{2}	−0.36 ^{4}
	CONTAG	**0.73** ^{4}	−0.30 ^{3}	**−0.70** ^{0}	0.24 ^{2}	**−0.46** ^{0}	**0.46** ^{0}	**−0.45** ^{4}	−0.16 ^{4}	0.22 ^{0}	0.33 ^{0}
	PD	−0.41 ^{4}	**0.45** ^{3}	**0.66** ^{0}	**0.51** ^{4}	**0.61** ^{3}	−0.22 ^{0}	**0.62** ^{0}	0.26 ^{3}	0.12 ^{2}	**0.74** ^{4}
	LSI	−0.37 ^{4}	0.34 ^{3}	**0.68** ^{0}	−0.35 ^{0}	0.34 ^{4}	**−0.44** ^{0}	**0.48** ^{4}	−0.15 ^{0}	−0.17 ^{0}	0.26 ^{4}
	COHESION	**0.53** ^{0}	−0.20 ^{1}	**−0.66** ^{0}	**0.46** ^{0}	**−0.47** ^{4}	**0.49** ^{0}	**0.59** ^{0}	0.15 ^{0}	0.26 ^{0}	**0.49** ^{2}
	SHDI	−0.36 ^{4}	0.28 ^{3}	**0.72** ^{0}	−0.20 ^{4}	**0.37** ^{0}	**−0.44** ^{0}	0.38 ^{4}	0.19 ^{4}	−0.23 ^{0}	**−0.37** ^{0}
	SHEI	**−0.75** ^{4}	0.28 ^{3}	**0.70** ^{0}	−0.20 ^{4}	**0.47** ^{0}	**−0.44** ^{0}	0.38 ^{4}	0.19 ^{4}	−0.23 ^{0}	**−0.37** ^{0}
	MSIEI	**−0.74** ^{4}	0.32 ^{3}	**0.70** ^{0}	**0.43** ^{0}	**0.38** ^{0}	0.32 ^{4}	**0.47** ^{3}	0.33 ^{3}	−0.35 ^{0}	−0.15 ^{0}

Note: bold font indicates a correlation was significant at a confidence level of 95%, bold and underlined font indicates a correlation was significant at a confidence level of 99%, and a superscript number indicates the lag time of the correlation between the LMs and the WY or RC.

In most watersheds, WY and RC were negatively correlated with LPI, AI, and COHESION and positively correlated with ED, PAFRAC, NP, DIVISION, PD, LSI, SHDI, SHEI, and MSIEI; SEM was negatively correlated with LPI and IJI and positively correlated with PAFRAC, NP, PD, COHESION, and MSIEI; SSC was negatively correlated with IJI and positively correlated with PAFRAC, NP, PD, COHESION, and MSIEI. In addition, the correlation between WY/RC and IJI varied in watersheds with different sizes. In general, runoff decreases when a landscape becomes more fragmented or its shape gains complexity [51], and sediment transportation is disturbed when a landscape patch becomes larger or fragmented, further reducing sediment yields [35]. In addition, the results show that the effect of LPs on SSC and SEM were similar, which indicates that the LPs affect SSC by affecting SEM. It can be seen in Table 4 that most watersheds with larger areas had negative correlations between WY/RC and IJI, whereas most watersheds with smaller areas had positive correlations between WY/RC and IJI. A similar phenomenon was found between SHDI/SHEI and SEM. Most watersheds with larger areas had negative correlations between SHDI/SHEI and SEM, whereas most watersheds with smaller areas had positive correlations between SHDI/SHEI and SEM.

The effects of LMs on different hydrological indices varied, even in the same watersheds. For instance, for DKF, no significant ($p < 0.05$) correlations were found between WY and any LMs, but all LMs except for PAFRAC, IJI, and COHESION were significantly ($p < 0.05$) correlated with RC; for WZ, all LMs except for LPI and DIVISION did not show significant ($p < 0.05$) correlations with WY, whereas all LMs except for PAFRAC, NP, PD, and COHESION were significantly ($p < 0.05$) correlated with RC. In general, the effects of the LMs on SEM and SSC were more significant than their effects on WY and RC. Among the chosen watersheds, only four watersheds (LJD, BL, SJ, and ZQ) did not show significant ($p < 0.05$) correlations between SEM and the LMs. In particular, all LMs were significantly ($p < 0.01$) correlated with the SEM in HS; seven, ten, and eleven LMs were significantly ($p < 0.05$) correlated with the SEMs in ZJ, LX, and CA, respectively. Based on the CC between SSC and all LMs, it seems that LPs changes could closely affect the relationship between streamflow and sediment yields. SSC was found to be significantly ($p < 0.05$) correlated with at least one LM in all of the chosen watersheds. In particular, all LMs were significantly ($p < 0.01$) correlated with the SSC in HS, and eight, eleven, eight, and twelve LMs were significantly ($p < 0.05$) correlated with the SSCs in ZJ, LX, BL, and CA, respectively.

5. Discussion

5.1. Correlations between LMs and Watershed Size

Different LMs represent different LPs characteristics, which may be different in different regions due to the varied landscape. In our study, the average LMs were correlated with watershed size (Figure 6). In detail, ED, DIVISION, PD, SHDI, SHEI, and MSIEI had similar correlations to watershed size. These LMs were negatively correlated with watershed size when it ranged from 20,000 to 60,000 km^2 and positively correlated with watershed size when it was in other ranges. LPI, AI, and CONTAG had similar correlations with watershed size. These LMs showed reverse correlations with watershed size compared to ED, DIVISION, PD, SHDI, SHEI, and MSIEI. Generally, ED, DIVISION, PD, SHDI, SHEI, and MSIEI represent the fragmentation of the landscape, whereas LPI, AI, and CONTAG represent the aggregation of the landscape. This result indicates that the LMs that represent similar landscape characteristics had similar correlations with watershed size, whereas the LMs with opposite landscape characteristics had reverse correlations with watershed size. The thresholds of 20,000 and 60,000 km^2 may suggest that fragmentation of the landscape would be decreased in southeastern China where the area ranges from 20,000 to 60,000 km^2. PAFRAC and COHESION showed similar correlations with watersheds, with slow increasing trends where the watershed size ranged from 0 to 30,000 km^2. Interestingly, NP and LSI were significantly ($p < 0.01$) positively correlated with watershed size. The increasing trend was smaller where the watershed size ranged from 25,000 to 55,000 km^2 compared to other ranges. NP represents the number of patches in the watershed, and LSI represents the ratio of edge and area of landscape. Generally, NP and LSI would increase with the increase in watershed size, but the slight increasing trend of NP and LSI where the watershed size ranges from 25,000 to 55,000 km^2 may suggest that the number of patches and the ratio of the edge to the area of a landscape would not be changed obviously where the area of watershed ranges from 25,000 to 55,000 km^2. When the watershed area fell in this scope, the increased area mainly had less fragmentation. These correlations between LMs and watershed size may suggest that the effects of LMs on hydrological processes also have scale effects, as suggested by Xiao, Cao, Liu, and Lu [34]. More LMs should be taken into consideration within more watersheds with wider size ranges to further verify this interesting hypothesis.

5.2. Effects of Various LMs on Hydrological Processes

The relationships between four hydrological indices (WY, RC, SEM, and SSC) and fourteen LMs (LPI, ED, PAFRAC, NP, DIVISION, AI, IJI, CONTAG, PD, LSI, COHESION, SHDI, SHEI, and MSIEI) within ten representative watersheds with various area sizes in the subtropical monsoon climate zone of southeastern China were evaluated using the slip correlation analysis method. In our study, for most watersheds, the land use maps and hydrological series for 30 consecutive years were used to investigate the relationships between landscapes and hydrological processes. Compared to previous studies, the amount of land use maps was much greater, and the hydrological series were measured rather than simulated with various models, which could greatly reduce the uncertainty brought by various models. In addition, the measured hydrological series under different landscapes could directly reflect the effect of landscape on hydrological processes. Taking the SWAT model as an example, as mentioned in the Introduction section, it simulates hydrological processes based on different combinations of land use, soil types, and slope belt. The SWAT model considers the land use amount but ignores the effects of land use spatial variations on hydrological processes. For example, an HRU in a subwatershed would distribute in several places within the subwatershed, which makes it hard to represent the effect of land use spatial distributions on hydrological processes. Hence, it is more reliable to investigate the relationships between landscapes and hydrological processes using measured data compared to using simulated data from various models.

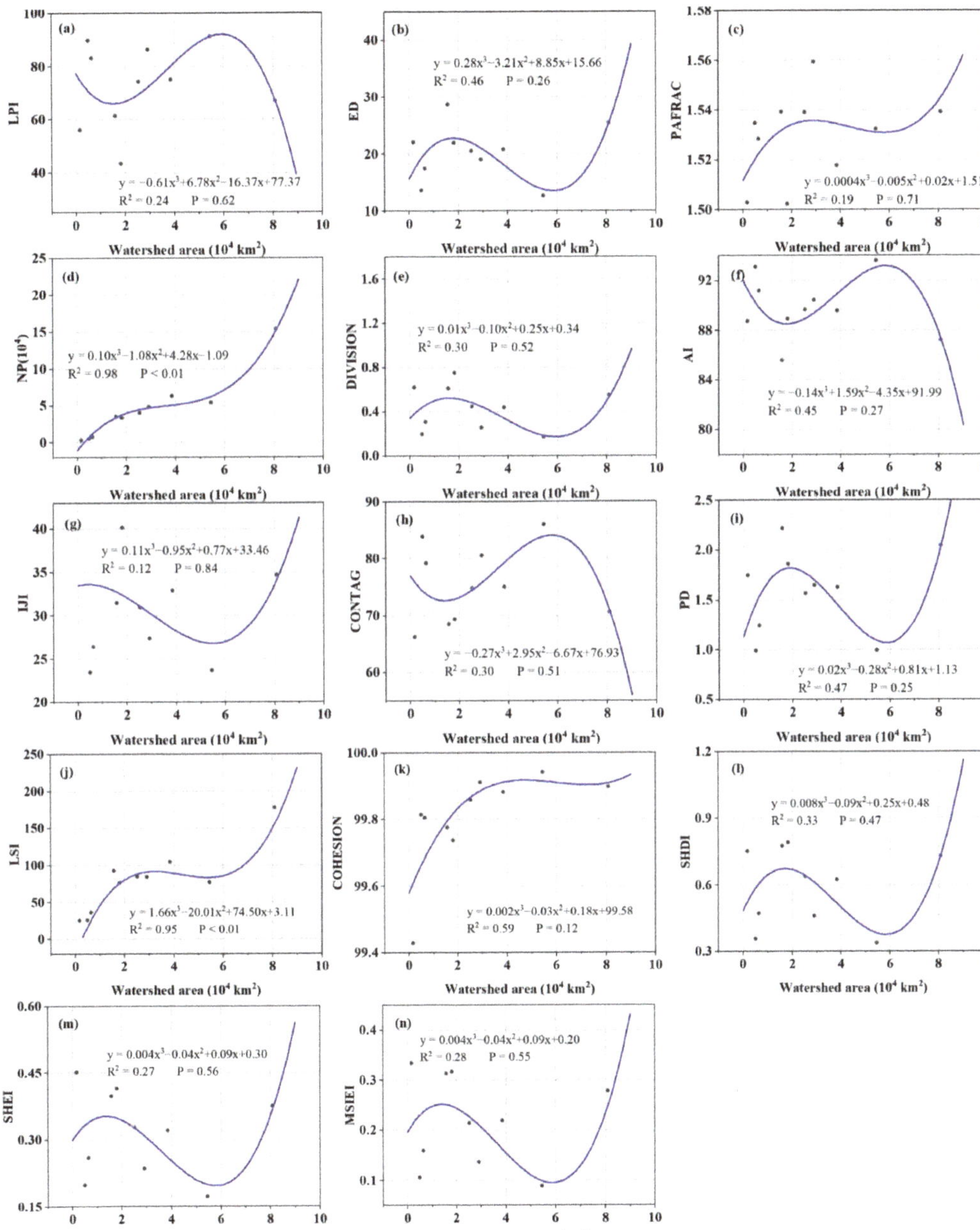

Figure 6. Statistics of the watershed area and average LMs for the chosen watersheds from 1990 to 2019. (**a–n**) represent LPI, ED, PAFRAC, NP, DIVISION, AI, IJI, CONTAG, PD, LSI, COHESION, SHDI, SHEI, and MSIEI, respectively.

Similar to previous studies, the relationships between the different hydrological indices and LMs in various watersheds had discrepancies [35]. In our study, some hydrological indices and LMs had the same correlations in most watersheds, and a few hydrological indices and some LMs even had the same correlation in all watersheds. For most watersheds ($\geq$7), WY was negatively correlated with LPI, AI, CONTAG, and COHESION and positively correlated with ED, PAFRAC, DIVISION, LSI, SHDI, SHEI, and MSIEI; RC was negatively correlated with LPI, CONTAG, and COHESION and positively correlated with PAFRAC, DIVISION, SHDI, SHEI, and MSIEI; SEM was negatively correlated with LPI and IJI; and SEM and SSC were positively correlated with PAFRAC, NP, PD, COHESION, and MSIEI. Interestingly, several LMs had the same relationships with WY and RC in all watersheds. As for SEM and SSC, there were no consistent correlations with any LMs for any watersheds. This discrepancy may have been caused by the soil erosion process being easier to disturb with human activity than the runoff process [68,69]. Hence, the relationships between LMs and SEM/SSC have discrepancies in various watersheds with differences in human activity intensity and mode.

LPI and ED were both related to the edge-area characteristic of LPs, and they had reverse effects on the runoff process. As mentioned before, LPI was negatively correlated with WY and RC, whereas ED was reversed. This result suggests that runoff increases in the subtropical monsoon climate zone when a watershed is dominated by a small patch of landscape. Similar results have also been reported in other regions [17], and a reverse result was also obtained [38]. In addition, SEM was negatively correlated with LPI in most watersheds, whereas it had no consistent correlation with ED. The dominant land use was forest and agricultural land, which may suggest that the dispersal of forest and agricultural land would increase soil erosion. The negative correlation between IJI and SEM also proved this result. A similar result was obtained by Zhang, Fan, Li and Yi [25]. In our study, seven aggregation LMs were chosen to analyze their correlation with four hydrological indices. For most watersheds, NP, PD, LSI, and DIVISION were positively correlated with WY and RC, whereas NP, PD, and COHESION were positively correlated with SEM and SSC. These results suggest that a proper increase in the fragmentation and physical connectivity of LPs was beneficial for water and soil conservation. A similar result was reported within the Upper Du River Basin in the middle of the Yangtze River [51], which is also within the subtropical monsoon climate zone. In addition, reverse results have been found in northern China, which belongs to the semi-arid continental monsoon climate zone [36,39] and may be caused by differences in rainfall patterns in different climate zones. The average annual precipitation in the study regions ranges from 1686 to 2162 mm (Table 1), whereas the average annual precipitation within the semi-arid continental monsoon climate zone ranges from 300 to 500 mm [70]. Furthermore, Ouyang, Skidmore, Hao, and Wang [33] pointed out that properly improving the fragile landscape status could prevent soil erosion, which may suggest that there is a threshold for landscape fragility. The changes in the fragmentation in the upper or lower fragmentation thresholds may be different, which may be another reason for the different correlations between relevant LMs and soil erosion. WY and RC were positively correlated with SHDI, SHEI, and MSIEI in most watersheds, which suggests that an increase in landscape diversity would increase runoff generation in the subtropical monsoon climate zone. As for SEM and SSC, the effect of landscape diversity varied in different regions [36,39,71], which may have resulted from discrepancies in land use type conversions. The main land use conversions from 1990 to 2019 within these watersheds were forest land–agricultural land, agricultural land–urban land, and agricultural land–forest land (Figure 4). Generally, small agricultural patches increased in a large forest patch, small urban patches increased in a large agricultural patch, and small forest patches increased in a large agricultural patch, which increased landscape diversity but may have affected soil erosion differently. A further investigation should be conducted to verify this hypothesis.

The correlations between LMs and RC were more obvious than those between LMs and WY (Table 4). There are no glaciers in the subtropical monsoon climate zone, and the main source of water yields was precipitation recharge [72]. Precipitation variation plays a more important role in interannual WY changes compared to LPs variations [73]. RC removed the effect of precipitation variations and watershed size to some extent compared to WY, and it was mainly affected by LULC, soil property, and slope [74]. These results suggest that relevant studies should choose RC to represent runoff generation character rather than WY. In addition, the relationships between LMs and SEM/SSC were not consistent in different watersheds. This may have been caused by the erosion process, which was also influenced by precipitation patterns, soil properties, slope, and topography [73,75–77]. WY/RC and SEM/SSC were both positively correlated with PAFRAC, NP, PD, and MSIEI in most watersheds. This shows that the soil erosion in these regions is mainly caused by rainfall erosion and surface runoff transportation [78]. In addition, the extensive construction projects brought about by rapid urban expansion also could have led to a lot of soil erosion [79], further influencing these results.

5.3. Recommendations for LM Selection in Future Relevant Studies

Our study selected 10 watersheds located in the subtropical monsoon climate zone in southeastern China. For the 10 chosen watersheds, the temporal variations and change trends of the 14 chosen LMs from 1990 to 2019 are shown in Figure A2 and Table 3. The results showed that the change trends of LPI and ED were opposite in all watersheds except for ZJ; the change trend of AI was opposite to LSI in all watersheds and similar to CONTAG and COHESION in all watersheds except for CA; and finally, SHDI and SHEI had similar change trends in all watersheds. In addition, the temporal variation in CONTAG was more obvious than those of AI and COHESION, but the change trend of COHESION was more significant compared to those of AI and CONTAG. SHDI was less disturbed by landscape changes compared to SHEI, as SHEI changed abruptly in some years, whereas similar changes did not appear for SHDI (Figure A2). In addition, the correlations with WY/RC for LPI and ED were opposite; the correlations with WY/RC for AI, CONTAG, and COHESION were similar, and COHESION had a positive correlation with SEM/SSC in most watersheds; the correlations with WY/RC for SHDI and SHEI were similar. Similar results were also reported in previous studies [17,35,37,80]. Hence, when analyzing the correlations between LPs and hydrological processes, it would be better to choose LPI, ED, PAFRAC, AI, COHESION, and SHDI as the representative LMs for fundamental LPs characteristics that include the edge area, shape, aggregation, and diversity aspects.

5.4. Implications, Limitations, and Prospects

Understanding the relationships between LPs and hydrological processes has a benefit for water and environmental management [27]. The results suggest that a proper decrease in landscape fragmentation and connectivity in the subtropical monsoon climate zone of southeastern China would benefit soil erosion prevention. In addition, meaningless land use conversions (forest land–agricultural land and agricultural land–forest land) should be decreased during local economic development, which would benefit decreasing landscape fragmentation and connectivity. In addition, the effects of LPs on SSC were also revealed in our study. Soil erosion and sediment transportation have a similar response to LPs changes, which indicates that it is possible to adjust LPs to mitigate river and reservoir siltation in some special locations. Overall, these results could benefit land use management for better soil and water conservation under the rapid development of southeastern China.

The effects of human activity on the results were not taken into consideration in our study. As mentioned before, construction engineering results in serious soil erosion events [79], and agricultural activity and industrial development consume many water resources [81,82]. A water use dataset should be collected to be used in relevant investigations in the future to reduce the influence of these human activities. In addition, it has been reported that the effects of various LPs on hydrological processes may have discrepancies in different seasons during a year [40,48]. The variations in hydrological processes characteristic in different seasons were not taken into consideration in our study due to the limitations of the hydrological data. A further study should be conducted to clarify these differences. Finally, though the watershed size in our study ranges from 1700 to 80,900 km^2, the number of watersheds used in this study was relatively few. More watersheds with various size should be taken into consideration to obtain more stable correlations between various LMs and hydrological processes.

6. Conclusions

In our study, the relationships between 14 LPs and 4 hydrological indices were investigated for 10 watersheds with various areas in the subtropical monsoon climate zone of southeastern China. The main conclusions are as follows:

(1) From 1990 to 2019, the change trends of WY and RC were not significant for any watersheds, and SEM and SSC decreased in all watersheds except for ZJ, HS, and LX. The main land use conversions were forest land–agricultural land, agricultural land–urban land, and agricultural land–forest land, and urban land expanded drastically in all watersheds. In addition, most LMs changed significantly ($p < 0.05$) for most watersheds, which demonstrates that LPs characteristics changed significantly.

(2) For most watersheds ($\geq$7), WY was negatively correlated with LPI, AI, CONTAG, and COHESION and positively correlated with ED, PAFRAC, DIVISION, LSI, SHDI, SHEI, and MSIEI; RC was negatively correlated with LPI, CONTAG, and COHESION and positively correlated with PAFRAC, DIVISION, SHDI, SHEI, and MSIEI; SEM was negatively correlated with LPI and IJI; SEM and SSC were positively correlated with PAFRAC, NP, PD, COHESION, and MSIEI. In addition, the effects of several LMs (IJI, SHDI, and SHEI) on WY, RC, and SEM had scale effects.

(3) In the subtropical monsoon climate zone, runoff increases when a watershed is dominated by a small patch of landscape. In addition, landscape fragmentation and diversity also increase runoff. Proper landscape fragmentation and physical connectivity would benefit soil erosion and river and reservoir siltation prevention.

More studies should be conducted in various climate zones to enhance the understanding of the relationships between LPs and various hydrological indices, and human activity and seasonal discrepancies should be taken into consideration.

Author Contributions: Conceptualization, C.W.; data curation, Y.M. and W.Z.; formal analysis, X.D., M.L., D.Y. and C.Z.; funding acquisition, X.D., Y.M. and D.Y.; investigation, C.W., W.Z. and M.L.; methodology, C.W.; software, D.Y.; supervision, X.D. and Y.M.; visualization, C.W.; writing—original draft, C.W.; writing—review and editing, X.D., Y.M. and B.S. All authors have read and agreed to the published version of the manuscript.

Funding: This research has been supported by the Second Tibetan Plateau Scientific Expedition and Research Program (STEP, grant number 2019QZKK0103), the European Space Agency (ESA) and National Remote Sensing Center of China (grant number 58516), the Power Construction Corporation of China (grant number DJ-ZDZX-2016-02-09), the Research Fund for Excellent Dissertation of China Three Gorges University (grant number No. 2021BSPY002), and the National Natural Science Foundation of China (grant number No. 52109058).

Data Availability Statement: Not applicable.

Conflicts of Interest: The authors declare no conflict of interest.

Glossary

Abbreviation	Full name
LP(s)	Landscape pattern(s)
WY	Water yields
SEM	Soil erosion modulus
SYL	Sediment yields load
NDCA	Number of disjunct core areas
LSI	Landscape shape index
DIVISION	Landscape division index
COHESION	Patch cohesion index
AI	Aggregation index
PLAND	Percentage of landscape
InVEST	Integrated Valuation of Ecosystem Services and Trade-offs
RUSLE	Revised Universal Soil Loss Equation
HRUs	Hydrological response units
DFK	Dufengkeng
LJD	Lijiadu
BL	Boluo
SJ	Shijiao
WZ	Waizhou
DA	Drainage area
PRE	Precipitation
NP	Number of patches
CONTAG	Contiguity index
R	Correlation coefficient
LMs	Landscape metrics
RC	Runoff coefficient
SSC	Suspended sediment concentration
p	Significance level
PD	Patch density
SHDI	Shannon's diversity index
LPI	Largest patch index
MSIEI	Modified Simpson's evenness index
ED	Edge density
SWAT	Soil and Water Assessment Tool
WaTEM/SEDEM	Water and Tillage Erosion Model and Sediment Delivery Model
IUH	Instantaneous Unit Hydrograph
ZJ	Zhuji
HS	Hushan
LX	Lanxi
CA	Chaoan
ZQ	Zhuqi
masl	Meters above the sea level
AE	Average elevation
PAFRAC	Perimeter area fractal dimension
IJI	Interspersion and Juxtaposition index
SHEI	Shannon's evenness index

Appendix A

Table A1. The descriptions of the selected landscape metrics and relevant references.

Categories	Metrics	Definition	Relevant Literature
Edge area	LPI	The ratio of the largest patch to the total landscape area. Unit (%)	[25,31,35,38,50,51]
	ED	The length of the edges per unit area. Unit (Meters/hectare)	[25,38,50,51]
Shape	PAFRAC	An index of patch shape complexity across a wide range of spatial scales.	[25,27,31,35,39,51]
Aggregation	NP	Extent of subdivision or fragmentation of the landscape pattern.	[31,35,38]
	DIVISION	Reflects the degree of fragmentation of the landscape. Unit (Proportion)	[17,35,36]
	AI	Connectivity between patches of landscape types. Unit (%)	[25,31,38,50]
	IJI	The observed interspersion over the maximum possible interspersion for the given number of patch types. Unit (%)	[27,31,39,51]
	CONTAG	An index measuring the extent to which patch types are aggregated or clumped. Unit (%)	[25,31,33,35,38,50,51]
	PD	The number of patches within 1 km^2. Unit (Number per 100 hectares)	[25,31,35,38,50,51]
	LSI	This index reflects the complexity of the boundaries of all patches within the region.	[31,35,50,51]
	COHESION	Measures the physical connectedness of the corresponding patch type.	[25,27,35,38,39,50,51]
Diversity	SHDI	The number of different patch types and the proportional area distribution among patch types.	[25,33,35,38,50,51]
	SHEI	The proportional abundance of each patch type.	[25,27,33,38,39]
	MSIEI	MSIEI equals minus the logarithm of the sum, across all patch types, of the proportional abundance of each patch type squared, which is then divided by the logarithm of the number of patch types.	[25,33]

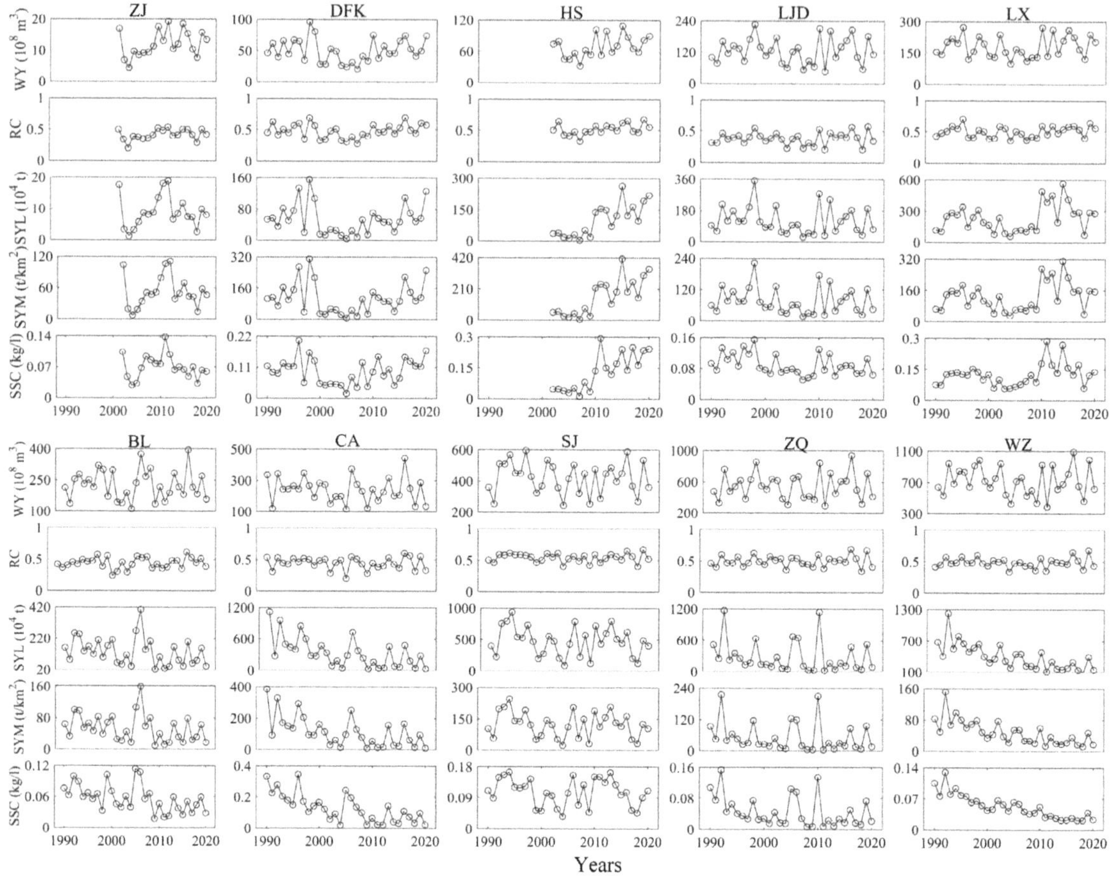

Figure A1. The temporal variations of WY, RC, SYL, SYM, and SSC in the ZJ, DFK, HS, LJD, LX, BL, CA, SJ, ZQ, and WZ watersheds.

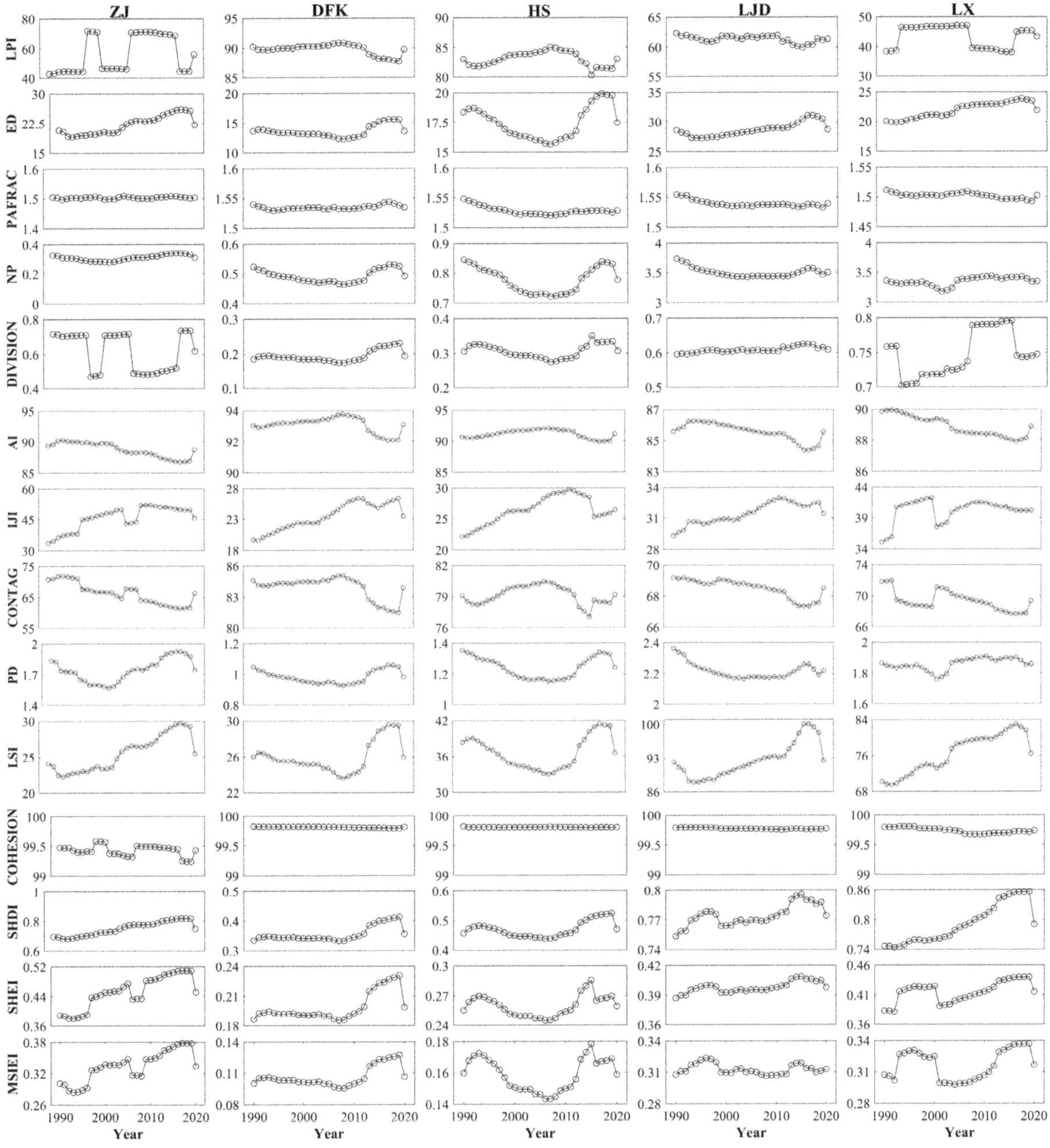

Figure A2. *Cont.*

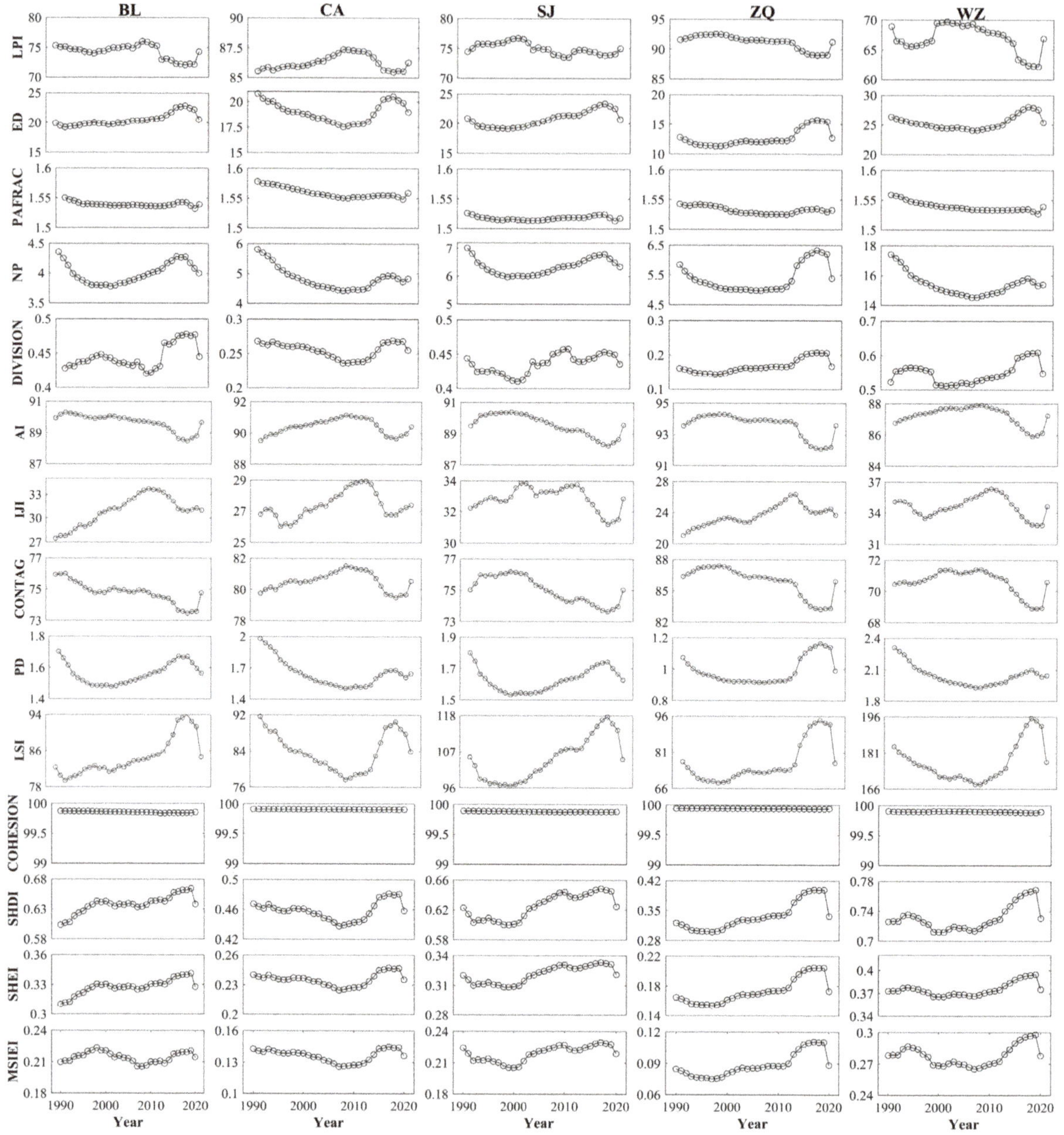

Figure A2. The temporal variations of LPI, ED, PAFRAC, NP, DIVISION, AI, IJI, CONTAG, PD, LSI, COHESION, SHDI, SHEI, and MSIEI in the ZJ, DFK, HS, LJD, LX, BL, CA, SJ, ZQ, and WZ watersheds.

References

1. Rasool, R.; Fayaz, A.; ul Shafiq, M.; Singh, H.; Ahmed, P. Land use land cover change in Kashmir Himalaya: Linking remote sensing with an indicator based DPSIR approach. *Ecol. Indic.* **2021**, *125*, 107447. [CrossRef]
2. Terêncio, D.; Varandas, S.; Fonseca, A.; Cortes, R.; Fernandes, L.; Pacheco, F.; Monteiro, S.; Martinho, J.; Cabral, J.; Santos, J. Integrating ecosystem services into sustainable landscape management: A collaborative approach. *Sci. Total Environ.* **2021**, *794*, 148538. [CrossRef] [PubMed]

3. Nafi'Shehab, Z.; Jamil, N.R.; Aris, A.Z.; Shafie, N.S. Spatial variation impact of landscape patterns and land use on water quality across an urbanized watershed in Bentong, Malaysia. *Ecol. Indic.* **2021**, *122*, 107254. [CrossRef]
4. Devátý, J.; Dostál, T.; Hösl, R.; Krása, J.; Strauss, P. Effects of historical land use and land pattern changes on soil erosion–Case studies from Lower Austria and Central Bohemia. *Land Use Policy* **2019**, *82*, 674–685. [CrossRef]
5. Sahin, V.; Hall, M.J. The effects of afforestation and deforestation on water yields. *J. Hydrol.* **1996**, *178*, 293–309. [CrossRef]
6. Costa, M.H.; Botta, A.; Cardille, J.A. Effects of large-scale changes in land cover on the discharge of the Tocantins River, Southeastern Amazonia. *J. Hydrol.* **2003**, *283*, 206–217. [CrossRef]
7. Kalantari, Z.; Lyon, S.W.; Folkeson, L.; French, H.K.; Stolte, J.; Jansson, P.E.; Sassner, M. Quantifying the hydrological impact of simulated changes in land use on peak discharge in a small catchment. *Sci. Total Environ.* **2014**, *466–467*, 741–754. [CrossRef]
8. Wang, Z.P.; Tian, J.C.; Feng, K.P. Response of runoff towards land use changes in the Yellow River Basin in Ningxia, China. *PLoS ONE* **2022**, *17*, e0265931. [CrossRef]
9. Wei, C.; Dong, X.; Yu, D.; Liu, J.; Reta, G.; Zhao, W.; Kuriqi, A.; Su, B. An alternative to the Grain for Green Program for soil and water conservation in the upper Huaihe River basin, China. *J. Hydrol. Reg. Stud.* **2022**, *43*, 101180. [CrossRef]
10. Molinero-Parejo, R.; Aguilera-Benavente, F.; Gómez-Delgado, M.; Shurupov, N. Combining a land parcel cellular automata (LP-CA) model with participatory approaches in the simulation of disruptive future scenarios of urban land use change. *Comput. Environ. Urban Syst.* **2023**, *99*, 101895. [CrossRef]
11. Liang, X.; Liu, X.; Li, X.; Chen, Y.; Tian, H.; Yao, Y. Delineating multi-scenario urban growth boundaries with a CA-based FLUS model and morphological method. *Landsc. Urban Plan.* **2018**, *177*, 47–63. [CrossRef]
12. Xu, G.; Cheng, Y.; Zhao, C.; Mao, J.; Li, Z.; Jia, L.; Zhang, Y.; Wang, B. Effects of driving factors at multi-spatial scales on seasonal runoff and sediment changes. *Catena* **2023**, *222*, 106867. [CrossRef]
13. Smetanova, A.; Follain, S.; David, M.; Ciampalini, R.; Raclot, D.; Crabit, A.; Le Bissonnais, Y. Landscaping compromises for land degradation neutrality: The case of soil erosion in a Mediterranean agricultural landscape. *J. Environ. Manag.* **2019**, *235*, 282–292. [CrossRef] [PubMed]
14. Lacher, I.L.; Ahmadisharaf, E.; Fergus, C.; Akre, T.; Mcshea, W.J.; Benham, B.L.; Kline, K.S. Scale-dependent impacts of urban and agricultural land use on nutrients, sediment, and runoff. *Sci. Total Environ.* **2019**, *652*, 611–622. [CrossRef] [PubMed]
15. Gonzales-Inca, C.A.; Kalliola, R.; Kirkkala, T.; Lepisto, A. Multiscale Landscape Pattern Affecting on Stream Water Quality in Agricultural Watershed, SW Finland. *Water Resour. Manag.* **2015**, *29*, 1669–1682. [CrossRef]
16. McGarigal, K. *FRAGSTATS: Spatial Pattern Analysis Program for Quantifying Landscape Structure*; US Department of Agriculture, Forest Service, Pacific Northwest Research Station: La Grande, OR, USA, 1995; Volume 351.
17. Bin, L.L.; Xu, K.; Xu, X.Y.; Lian, J.J.; Ma, C. Development of a landscape indicator to evaluate the effect of landscape pattern on surface runoff in the Haihe River Basin. *J. Hydrol.* **2018**, *566*, 546–557. [CrossRef]
18. Shi, P.; Qin, Y.-L.; Li, P.; Li, Z.-B.; Cui, L.-Z. Development of a landscape index to link landscape pattern to runoff and sediment. *J. Mt. Sci.* **2022**, *19*, 2905–2919. [CrossRef]
19. Wang, X.; Liu, X.; Wang, L.; Yang, J.; Wan, X.; Liang, T. A holistic assessment of spatiotemporal variation, driving factors, and risks influencing river water quality in the northeastern Qinghai-Tibet Plateau. *Sci. Total Environ.* **2022**, *851*, 157942. [CrossRef]
20. Stets, E.G.; Sprague, L.A.; Oelsner, G.P.; Johnson, H.M.; Murphy, J.C.; Ryberg, K.; Vecchia, A.V.; Zuellig, R.E.; Falcone, J.A.; Riskin, M.L. Landscape drivers of dynamic change in water quality of US rivers. *Environ. Sci Technol.* **2020**, *54*, 4336–4343. [CrossRef]
21. Liu, Y.; Shen, Y.; Cheng, C.; Yuan, W.; Gao, H.; Guo, P. Analysis of the influence paths of land use and landscape pattern on organic matter decomposition in river ecosystems: Focusing on microbial groups. *Sci. Total Environ.* **2022**, *817*, 153381. [CrossRef]
22. Kumar, G.; Baweja, P.; Gandhi, P.B. Impact of Anthropogenic Activities on Soil Patterns and Diversity. In *Structure and Functions of Pedosphere*; Springer: Berlin/Heidelberg, Germany, 2022; pp. 319–337.
23. Sadeghi, S.H.; Moradi Dashtpagerdi, M.; Moradi Rekabdarkoolai, H.; Schoorl, J.M. Sensitivity analysis of relationships between hydrograph components and landscapes metrics extracted from digital elevation models with different spatial resolutions. *Ecol. Indic.* **2021**, *121*, 107025. [CrossRef]
24. Zhao, X.; Huang, G. Exploring the impact of landscape changes on runoff under climate change and urban development: Implications for landscape ecological engineering in the Yangmei River Basin. *Ecol. Eng.* **2022**, *184*, 106794. [CrossRef]
25. Zhang, S.; Fan, W.; Li, Y.; Yi, Y. The influence of changes in land use and landscape patterns on soil erosion in a watershed. *Sci. Total Environ.* **2017**, *574*, 34–45. [CrossRef] [PubMed]
26. Brini, I.; Alexakis, D.D.; Kalaitzidis, C. Linking Soil Erosion Modeling to Landscape Patterns and Geomorphometry: An Application in Crete, Greece. *Appl. Sci.* **2021**, *11*, 5684. [CrossRef]
27. Zhou, Z.X.; Li, J. The correlation analysis on the landscape pattern index and hydrological processes in the Yanhe watershed, China. *J. Hydrol.* **2015**, *524*, 417–426. [CrossRef]
28. Li, J.; Zhou, Y.; Li, Q.; Yi, S.; Peng, L. Exploring the Effects of Land Use Changes on the Landscape Pattern and Soil Erosion of Western Hubei Province from 2000 to 2020. *Int. J. Environ. Res. Pub. Health* **2022**, *19*, 1571. [CrossRef]
29. Aghsaei, H.; Dinan, N.M.; Moridi, A.; Asadolahi, Z.; Delavar, M.; Fohrer, N.; Wagner, P.D. Effects of dynamic land use/land cover change on water resources and sediment yield in the Anzali wetland catchment, Gilan, Iran. *Sci. Total Environ.* **2020**, *712*, 136449. [CrossRef]
30. Yohannes, H.; Soromessa, T.; Argaw, M.; Dewan, A. Impact of landscape pattern changes on hydrological ecosystem services in the Beressa watershed of the Blue Nile Basin in Ethiopia. *Sci. Total Environ.* **2021**, *793*, 148559. [CrossRef]

31. Chen, C.; Zhao, G.; Zhang, Y.; Bai, Y.; Tian, P.; Mu, X.; Tian, X. Linkages between soil erosion and long-term changes of landscape pattern in a small watershed on the Chinese Loess Plateau. *Catena* **2023**, *220*, 106659. [CrossRef]
32. Wei, C.; Dong, X.; Ma, Y.; Gou, J.; Li, L.; Bo, H.; Yu, D.; Su, B. Applicability comparison of various precipitation products of long-term hydrological simulations and their impact on parameter sensitivity. *J. Hydrol.* **2023**, *618*, 129187. [CrossRef]
33. Ouyang, W.; Skidmore, A.K.; Hao, F.; Wang, T. Soil erosion dynamics response to landscape pattern. *Sci. Total Environ.* **2010**, *408*, 1358–1366. [CrossRef] [PubMed]
34. Xiao, R.; Cao, W.; Liu, Y.; Lu, B. The impacts of landscape patterns spatio-temporal changes on land surface temperature from a multi-scale perspective: A case study of the Yangtze River Delta. *Sci. Total Environ.* **2022**, *821*, 153381. [CrossRef] [PubMed]
35. Yang, Y.Y.; Li, Z.B.; Li, P.; Ren, Z.P.; Gao, H.D.; Wang, T.; Xu, G.C.; Yu, K.X.; Shi, P.; Tang, S.S. Variations in runoff and sediment in watersheds in loess regions with different geomorphologies and their response to landscape patterns. *Environ. Earth Sci.* **2017**, *76*, 517. [CrossRef]
36. Zhang, Y.; Bi, Z.L.; Zhang, X.; Yu, Y. Influence of Landscape Pattern Changes on Runoff and Sediment in the Dali River Watershed on the Loess Plateau of China. *Land* **2019**, *8*, 180. [CrossRef]
37. Gao, F.; He, B.; Xue, S.S.; Li, Y.Z. Impact of landscape pattern change on runoff processes in catchment area of the Ulungur River Basin. *Water Supply* **2020**, *20*, 1046–1058. [CrossRef]
38. Wei, C.; Zhang, Z.; Wang, Z.; Cao, L.; Wei, Y.; Zhang, X.; Zhao, R.; Xiao, L.; Wu, Q. Response of Variation of Water and Sediment to Landscape Pattern in the Dapoling Watershed. *Sustainability* **2022**, *14*, 678. [CrossRef]
39. Li, J.; Zhou, Z.X. Coupled analysis on landscape pattern and hydrological processes in Yanhe watershed of China. *Sci. Total Environ.* **2015**, *505*, 927–938. [CrossRef]
40. Chiang, L.-C.; Wang, Y.-C.; Chen, Y.-K.; Liao, C.-J. Quantification of land use/land cover impacts on stream water quality across Taiwan. *J. Clean Prod.* **2021**, *318*, 128443. [CrossRef]
41. Ke, Q.; Zhang, K. Patterns of runoff and erosion on bare slopes in different climate zones. *Catena* **2021**, *198*, 105069. [CrossRef]
42. Zhu, X.; Li, T.; Tian, Z.; Qu, L.; Liang, Y. Building pedotransfer functions for estimating soil erodibility in southeastern China. *Ecol. Indic.* **2022**, *145*, 109720. [CrossRef]
43. Wei, C.; Dong, X.; Yu, D.; Zhang, T.; Zhao, W.; Ma, Y.; Su, B. Spatio-temporal variations of rainfall erosivity, correlation of climatic indices and influence on human activities in the Huaihe River Basin, China. *Catena* **2022**, *217*, 106486. [CrossRef]
44. Yang, J.; Huang, X. The 30 m annual land cover dataset and its dynamics in China from 1990 to 2019. *Earth Syst. Sci. Data* **2021**, *13*, 3907–3925. [CrossRef]
45. Liang, L.; Yu, L.; Wang, Z. Identifying the dominant impact factors and their contributions to heatwave events over mainland China. *Sci. Total Environ.* **2022**, *848*, 157527. [CrossRef] [PubMed]
46. Wang, L.; Han, X.; Zhang, Y.; Zhang, Q.; Wan, X.; Liang, T.; Song, H.; Bolan, N.; Shaheen, S.M.; White, J.R. Impacts of land uses on spatio-temporal variations of seasonal water quality in a regulated river basin, Huai River, China. *Sci. Total Environ.* **2023**, *857*, 159584. [CrossRef]
47. Lisha, Q.; Qiuan, Z.; Chaofan, Z.; Jiang, Z. Monthly Precipitation Data Set with 1 km Resolution in China from 1960 to 2020[DS/OL]. Science Data Bank: Beijing, China, 2022; Available online: https://cstr.cn/31253.11.sciencedb.01607.CSTR:31253.11.sciencedb.01607 (accessed on 25 December 2022).
48. Xu, S.; Li, S.-L.; Zhong, J.; Li, C. Spatial scale effects of the variable relationships between landscape pattern and water quality: Example from an agricultural karst river basin, Southwestern China. *Agric. Ecosyst. Environ.* **2020**, *300*, 106999. [CrossRef]
49. Han, Y.; Wang, J.-l.; Li, P. Influences of landscape pattern evolution on regional crop water requirements in regions of large-scale agricultural operations. *J. Clean Prod.* **2021**, *327*, 129499. [CrossRef]
50. Jinying, X.; Yang, B.; Hailin, Y.; Xiaowei, W.; Zhifei, M.; Hongwei, Z. Water quality assessment and the influence of landscape metrics at multiple scales in Poyang Lake basin. *Ecol. Indic.* **2022**, *141*, 109096. [CrossRef]
51. Shi, Z.H.; Ai, L.; Li, X.; Huang, X.D.; Wu, G.L.; Liao, W. Partial least-squares regression for linking land-cover patterns to soil erosion and sediment yield in watersheds. *J. Hydrol.* **2013**, *498*, 165–176. [CrossRef]
52. Machado, R.E.; Cardoso, T.O.; Mortene, M.H. Determination of runoff coefficient (C) in catchments based on analysis of precipitation and flow events. *Int. Soil Water Conserv. Res.* **2022**, *10*, 208–216. [CrossRef]
53. Romano, G.; Abdelwahab, O.M.; Gentile, F. Modeling land use changes and their impact on sediment load in a Mediterranean watershed. *Catena* **2018**, *163*, 342–353. [CrossRef]
54. Wu, L.; Yao, W.W.; Ma, X.Y. Using the comprehensive governance degree to calibrate a piecewise sediment delivery ratio algorithm for dynamic sediment predictions: A case study in an ecological restoration watershed of northwest China. *J. Hydrol.* **2018**, *564*, 888–899. [CrossRef]
55. Wang, J.; Shi, B.; Zhao, E.; Yuan, Q.; Chen, X. The long-term spatial and temporal variations of sediment loads and their causes of the Yellow River Basin. *Catena* **2022**, *209*, 105850. [CrossRef]
56. Weisberg, S. *Applied Linear Regression*; John Wiley & Sons: Hoboken, NJ, USA, 2005; Volume 528.
57. Parhizkar, M.; Shabanpour, M.; Lucas-Borja, M.E.; Zema, D.A. Variability of rill detachment capacity with sediment size, water depth and soil slope in forest soils: A flume experiment. *J. Hydrol.* **2021**, *601*, 126625. [CrossRef]
58. Dall'Agnol, R.; Sahoo, P.K.; Salomao, G.N.; de Araujo, A.D.M.; da Silva, M.S.; Powell, M.A.; Ferreira, J.; Ramos, S.J.; Martins, G.C.; da Costa, M.F.; et al. Soil-sediment linkage and trace element contamination in forested/deforested areas of the Itacaiunas River Watershed, Brazil: To what extent land-use change plays a role? *Sci. Total Environ.* **2022**, *828*, 154327. [CrossRef] [PubMed]

59. Tiku, M. Tables of the power of the F-test. *J. Am. Stat. Assoc.* **1967**, *62*, 525–539. [CrossRef]
60. Meals, D.W.; Dressing, S.A.; Davenport, T.E. Lag Time in Water Quality Response to Best Management Practices: A Review. *J. Environ. Qual.* **2010**, *39*, 85–96. [CrossRef]
61. Li, Y.; Li, Y.R.; Fang, B.; Wang, Q.Y.; Chen, Z.F. Impacts of ecological programs on land use and ecosystem services since the 1980s: A case-study of a typical catchment on the Loess Plateau, China. *Land Degrad. Dev.* **2022**, *33*, 3271–3282. [CrossRef]
62. Fenicia, F.; McDonnell, J.J. Modeling streamflow variability at the regional scale: (1) perceptual model development through signature analysis. *J. Hydrol.* **2022**, *605*, 127287. [CrossRef]
63. Schilling, K.E.; Wolter, C.F. A GIS-based groundwater travel time model to evaluate stream nitrate concentration reductions from land use change. *Environ. Geol.* **2007**, *53*, 433–443. [CrossRef]
64. Barlow, K.M.; Weeks, A.; Christy, B. Modelling the response in streamflow to increased forestry plantations. In Proceedings of the 20th International Congress on Modelling and Simulation (MODSIM), Adelaide, Australia, 1–6 December 2013; pp. 538–544.
65. He, D.; Chen, Z.; Zhou, J.; Yang, T.; Lu, L. The Heterogeneous Impact of High-Speed Railway on Urban Expansion in China. *Remote Sens.* **2021**, *13*, 4914. [CrossRef]
66. Mostafa, E.; Li, X.; Sadek, M.; Dossou, J.F. Monitoring and Forecasting of Urban Expansion Using Machine Learning-Based Techniques and Remotely Sensed Data: A Case Study of Gharbia Governorate, Egypt. *Remote Sens.* **2021**, *13*, 4498. [CrossRef]
67. Sarkodie, S.A.; Owusu, P.A. Global land-use intensity and anthropogenic emissions exhibit symbiotic and explosive behavior. *Iscience* **2022**, *25*, 104741. [CrossRef] [PubMed]
68. Shi, P.; Zhang, Y.; Ren, Z.; Yu, Y.; Li, P.; Gong, J. Land-use changes and check dams reducing runoff and sediment yield on the Loess Plateau of China. *Sci. Total Environ.* **2019**, *664*, 984–994. [CrossRef] [PubMed]
69. Dong, H.; Song, Y.; Chen, L.; Liu, H.; Fu, X.; Xie, M. Soil erosion and human activities over the last 60 years revealed by magnetism, particle size and minerals of check dams sediments on the Chinese Loess Plateau. *Environ. Earth Sci.* **2022**, *81*, 162. [CrossRef]
70. Fan, J.; Xu, Y.; Ge, H.; Yang, W. Vegetation growth variation in relation to topography in Horqin Sandy Land. *Ecol. Indic.* **2020**, *113*, 106215. [CrossRef]
71. Zhang, Z.D.; Chen, S.J.; Wan, L.W.; Cao, J.; Zhang, Q.; Yang, C.X. The effects of landscape pattern evolution on runoff and sediment based on SWAT model. *Environ. Earth Sci.* **2021**, *80*, 2. [CrossRef]
72. Deng, Y.; Wu, S.; Ke, J.; Zhu, A. Effects of meteorological factors and groundwater depths on plant sap flow velocities in karst critical zone. *Sci. Total Environ.* **2021**, *781*, 146764. [CrossRef]
73. Yang, M.; Li, X.Z.; Hu, Y.M.; He, X.Y. Assessing effects of landscape pattern on sediment yield using sediment delivery distributed model and a landscape indicator. *Ecol. Indic.* **2012**, *22*, 38–52. [CrossRef]
74. Ma, B.; Wu, C.; Ding, F.; Zhou, Z. Predicting basin water quality using source-sink landscape distribution metrics in the Danjiangkou Reservoir of China. *Ecol. Indic.* **2021**, *127*, 107697. [CrossRef]
75. Ouyang, W.; Wu, Y.; Hao, Z.; Zhang, Q.; Bu, Q.; Gao, X. Combined impacts of land use and soil property changes on soil erosion in a mollisol area under long-term agricultural development. *Sci. Total Environ.* **2018**, *613*, 798–809. [CrossRef]
76. de Carvalho, D.F.; Macedo, P.M.S.; Pinto, M.F.; de Almeida, W.S.; Schultz, N. Soil loss and runoff obtained with customized precipitation patterns simulated by InfiAsper. *Int. Soil Water Conserv. Res.* **2022**, *10*, 407–413. [CrossRef]
77. Pijl, A.; Reuter, L.E.; Quarella, E.; Vogel, T.A.; Tarolli, P. GIS-based soil erosion modelling under various steep-slope vineyard practices. *Catena* **2020**, *193*, 104604. [CrossRef]
78. Zhu, X.; Liang, Y.; Tian, Z.; Wang, X. Analysis of scale-specific factors controlling soil erodibility in southeastern China using multivariate empirical mode decomposition. *Catena* **2021**, *199*, 105131. [CrossRef]
79. Wallace, M.; Mickovski, S.B.; Griffin, I. An innovative framework for selecting sustainable options to reduce the risk of soil erosion and environmental pollution incidents on road construction sites. In Proceedings of the XVII ECSMGE-2019: Geotechnical Engineering Foundation of the Future, Reykjavik, Iceland, 1–6 September 2019; The Icelandic Geotechnical Society: Reykjavik, Iceland, 2019.
80. Zhang, Y.F.; Wang, N.; Tang, C.J.; Zhang, S.Q.; Song, Y.J.; Liao, K.T.; Nie, X.F. A New Indicator to Better Represent the Impact of Landscape Pattern Change on Basin Soil Erosion and Sediment Yield in the Upper Reach of Ganjiang, China. *Land* **2021**, *10*, 990. [CrossRef]
81. Rufí-Salís, M.; Petit-Boix, A.; Villalba, G.; Sanjuan-Delmás, D.; Parada, F.; Ercilla-Montserrat, M.; Arcas-Pilz, V.; Munoz-Liesa, J.; Rieradevall, J.; Gabarrell, X. Recirculating water and nutrients in urban agriculture: An opportunity towards environmental sustainability and water use efficiency? *J. Clean Prod.* **2020**, *261*, 121213. [CrossRef]
82. Trang, T.T.; Bush, S.R.; van Leeuwen, J. Enhancing institutional capacity in a centralized state: The case of industrial water use efficiency in Vietnam. *J. Ind. Ecol.* **2022**, *27*, 210–222. [CrossRef]

remote sensing

MDPI

Article

Dynamic Effects of Atmosphere over and around the Tibetan Plateau on the Sustained Drought in Southwest China from 2009 to 2014

Yiwei Ye [1,2], Rongxiang Tian [1,*] and Zhan Jin [1]

1 Key Laboratory of Geoscience Big Data and Deep Resource of Zhejiang Province, School of Earth Sciences, Zhejiang University, Hangzhou 310027, China; yeyiwei21@mails.ucas.ac.cn (Y.Y.); 22038003@zju.edu.cn (Z.J.)
2 Key Laboratory of Cloud-Precipitation Physics and Severe Storms, Institute of Atmospheric Physics, Chinese Academy of Sciences, Beijing 100029, China
* Correspondence: trx@zju.edu.cn; Tel.: +86-13515817319

Abstract: The two westerly branches have a significant impact on the climate of the area on the eastern side of the Tibetan Plateau when flowing around it. A continuous drought event in Southwest China from the winter of 2009 to the spring of 2014 caused huge economic losses. This research focuses on the dynamic field anomalies over the Tibetan Plateau during this event using statistical analysis, attempts to decipher its mechanism on drought in Southwest China, and provides a regression model. We established that the anticyclone and downdraft over the Tibetan Plateau were weaker than usual during the drought, which would reduce the southward cold airflow on the northeast of the Tibetan Plateau and strengthen the west wind from dry central Asia on the south of the plateau. As a result, a larger area of the southwest region in China was controlled by the warm and dry air mass, which was acting against precipitation. The results will be of reference value to the drought forecast for Southwest China, and also encourage further research about how the Tibetan Plateau influence the climate on its eastern side.

Keywords: Tibetan Plateau; drought in Southwest China; dynamic effects; anticyclone on Tibetan Plateau

Citation: Ye, Y.; Tian, R.; Jin, Z. Dynamic Effects of Atmosphere over and around the Tibetan Plateau on the Sustained Drought in Southwest China from 2009 to 2014. *Remote Sens.* **2023**, *15*, 2198. https://doi.org/10.3390/rs15082198

Academic Editor: Xianjun Hao

Received: 28 February 2023
Revised: 11 April 2023
Accepted: 19 April 2023
Published: 21 April 2023

1. Introduction

The Tibetan Plateau, located in the interior of the Asian continent, is the highest plateau in the world. It has a total area of about 2.5×10^6 km^2 and an average elevation >4000 m (about 600 hPa) with a variety of landforms. The climate of the Tibetan Plateau is controlled by cold high-pressure systems (anticyclones) in winter, whereas in summer, the main controls are thermal depressions (cyclones) in the lower atmosphere and the South Asian high at upper levels [1]. The thermodynamic and dynamic effects of the atmosphere over and around the Tibetan Plateau affect the local atmospheric circulation and also propagate outward via Rossby waves [2]. This region, therefore, has a significant impact on the climate of Asia and the wider world [3,4].

There has been much research over a long period of time on the dynamic fields of the atmosphere over the Tibetan Plateau. The huge Tibetan Plateau has a significant blocking effect [5]. Mid-latitude westerly winds encounter the plateau while moving eastward. Part of the airflow is forced to lift over the plateau, whereas the other part is deflected into two branches to the north and south, both of which continue to move eastward along the edges of the plateau. The flows around the plateau have more significant effects and change with the season [6]. The northern branch is generally stronger than the southern branch [7].

The two airflows pass through different areas and lead to different types of weather. The northern branch generates an anticyclonic circulation north of the plateau, leading to cold advection on the northeast side [8]. Variations in the intensity of the flow affect temperatures in its eastern and southern regions—for example, in Northeast China and the Jianghuai region (downstream along the Yangtze River) to the east of the Tibetan Plateau [9,10]. The northern branch also influences the intensity of precipitation on the southeastern side of the plateau in South China [11]. By contrast, the southern branch bends cyclonically on the south side of the plateau, forming a semi-permanent low-pressure trough (the south trough), which splits the subtropical high in this region [8] and forms southwesterly warm advection to the southeast of the plateau. This branch affects the outbreak and persistence of the South Asian summer monsoon, in addition to the spatial and temporal distribution of precipitation in Eastern China [12–16]. There has not been enough attention paid to how the dynamic effects of the atmosphere over and around the Tibetan Plateau influence the strength of advection in the northern and southern branches, further affecting the climate on the eastern side.

The cold northern branch and the warm southern branch converge on the eastern side of the plateau, resulting in different climatic phenomena at different times of the year. In spring and summer, the northern and southern branches flowing around the plateau converge in the downstream areas of the Yangtze River (in Jianghuai region), affecting the local distribution of temperature and rainfall [7,17–19]. In winter, guided by the anticyclone circulation formed by thermal action over the Tibetan Plateau, the northern airflow merges with cold air from the north and moves southward. Where it is blocked by the terrain of the Yunnan–Guizhou Plateau, and it meets the southwesterly warm air brought by the southern branch to form a quasi-stationary front (the Kunming quasi-stationary front) between Kunming and Guiyang. The weather varies on the two sides of the front. It is relatively wet with rain or sometimes snow on the cold northern side (e.g., in Guizhou, Sichuan and Chongqing). The warm southern side in Yunnan and other regions is controlled by a warm air mass, with sunshine and little rainfall. If the cold air flow is stronger, then the quasi-stationary front moves southward, leading to more rain in Yunnan [20,21]. There have been few studies of the mechanisms by which the dynamic effects of the atmosphere over and around the Tibetan Plateau influence the position of the intersection between warm and cold advection.

Winter rainfall in Southwest China is primarily generated by the quasi-stationary front in this region and should therefore be affected by the westerly flow around the Tibetan Plateau. Southwest China suffered severe droughts in winter and spring for five consecutive years from 2009 to 2014 (compared to the average of the same seasons in 1960–2019) (Figure 1). There has been a variety of research about drought using different metrics or indexes. For example, Aksoy et al. [22] raised a methodology related to the standardized precipitation index (SPI) to develop critical drought intensity–duration–frequency (IDF) curves; Huang et al. [23] used SPI and Standardized Runoff Index (SRI), as well as Copula functions, to study the temporal and spatial evolution of meteorological drought and hydrological drought in the Jinghe River Basin by driving SWAT model. Here, we adopted the Percentage of Precipitation Anomaly, complying with the Chinese official document GBT20481-2017 [24], to characterize this drought event in Southwest China by its spatial and temporal distributions, as well as its severity. In autumn 2009, precipitation was more than 50% lower than the average in central and eastern Yunnan, western Guizhou, northwestern Guangxi and the southern margin of Chongqing. Some areas in central Yunnan had a >80% reduction in precipitation, reaching the level of extreme drought (regions in the darkest red). The drought worsened during the winter of 2009. Almost the whole of Yunnan, southwest Guizhou and a major part of Sichuan experienced moderate-to-severe drought, and large areas of Yunnan and southwest Sichuan were affected by severe drought. This drought caused water shortages for 25 million people in the winter of 2009 and the following spring, resulting in 660 × 104 ha of crop failure in Yunnan, Guizhou, Guangxi and Sichuan provinces and >40 billion yuan of direct economic losses [25–27]. In 2011, a severe drought in this region affected 39.52 million people, with 11.882 million people suffering shortages of freshwater and total economic losses of 21.8 billion yuan. The drought in Southwest China in 2013–2014 was rated as one of the top ten weather and climate events in China [25–27].

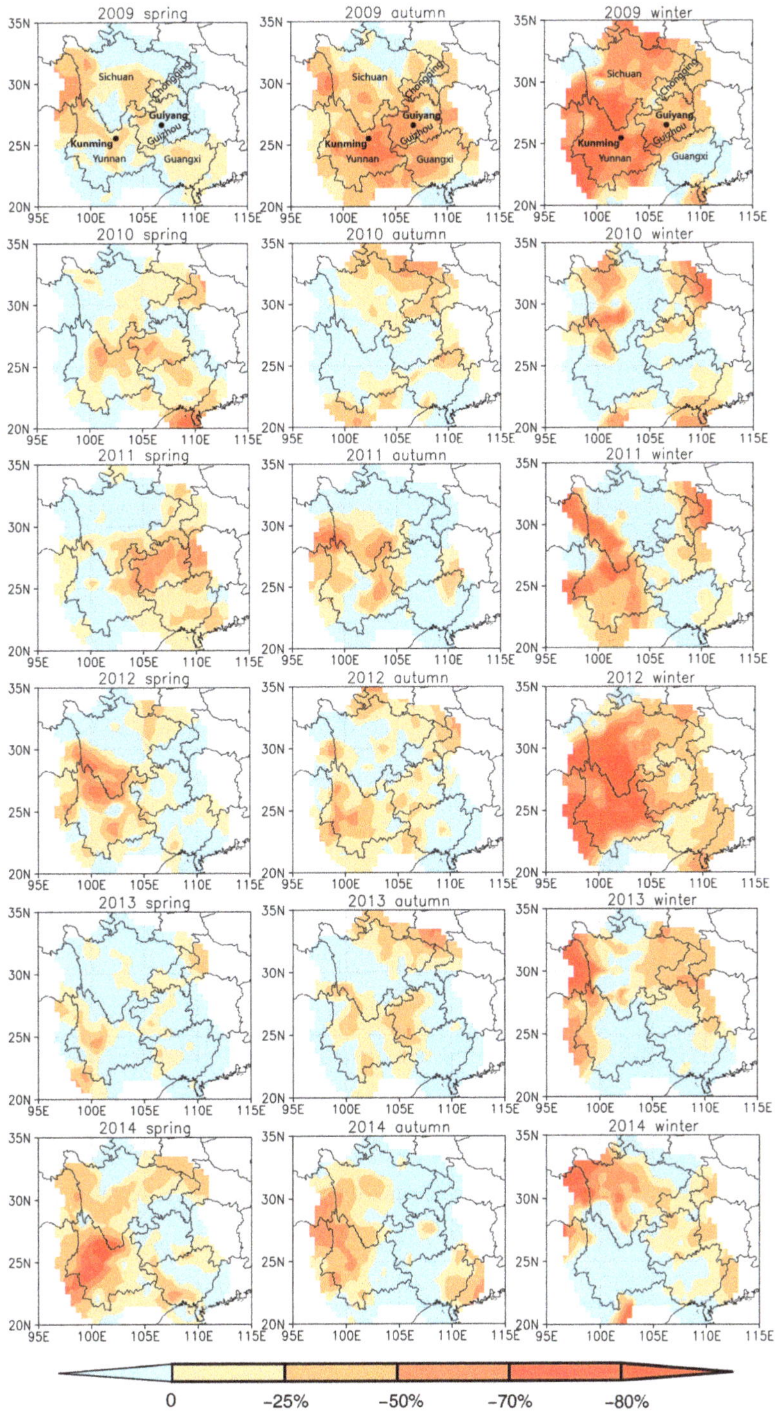

Figure 1. Percentage of Precipitation Anomaly in southwest China during spring, autumn and winter periods from 2009 to 2014 (comparing to the average of the same seasons in 1960–2019). The rating in the color bar indicates different drought levels given by the Chinese official document GBT20481-2017 [24]: mild drought from −50% to −25%, moderate drought from −70% to −50%, severe drought from −80% to −70% and extreme drought under −80%.

Anomalous precipitation in Southwest China is closely related to the position of the confluence between warm and cold advection. However, it is unclear whether the circulation over the Tibetan Plateau during this period of extreme drought was different from the average field, so there would be unusual changes in the westerly flow around the Tibetan Plateau, leading to abnormal rainfall in Southwest China.

The work reported here is based on a statistical analysis of the dynamic fields over the Tibetan Plateau and precipitation in Southwest China from 1960 to 2019. It focuses on the continuous drought event in Southwest China from 2009 to 2014 and explores the mechanisms by which the dynamic fields over the Tibetan Plateau affected the precipitation anomaly in Southwest China. A regression model is proposed for the dynamic fields over the plateau and precipitation in Southwest China. The results of this study will contribute to the prediction of precipitation in Southwest China and encourage further research about how the flow around the Tibetan Plateau affects the climate on its eastern side.

2. Data and Methods

The data and methods used in the paper and the results are shown in Figure 2. Two kinds of data were used in the paper. The first was the reanalysis data (including dynamic elements such as relative divergence, relative vorticity and horizontal and vertical wind velocity) from the Japanese 55-year Reanalysis (JRA-55), which was used to analyze the dynamical characteristics of the atmosphere over and around the Tibetan Plateau. The other was a precipitation observation from meteorological stations in Southwest China, which was used to introduce the drought event that happened in 2009–2014. In Sections 3 and 4, to study the relation between the Tibetan Plateau and the severe drought in Southwest China, these two kinds of data were connected using statistical methods, such as empirical orthogonal function (EOF) and regression.

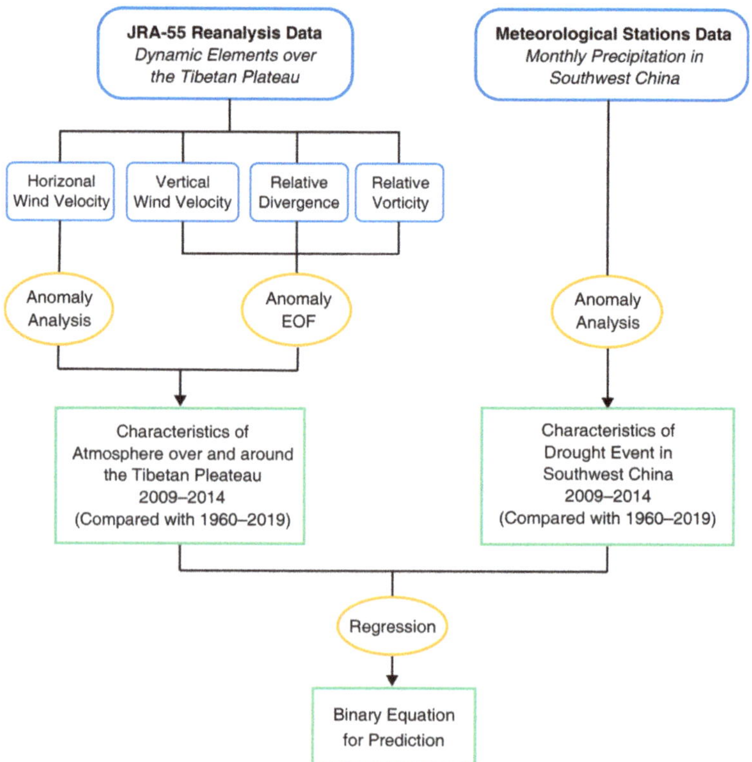

Figure 2. Flowchart including data, methods and results.

2.1. Data

It was found that the Japanese Reanalysis and the European Centre for Medium-Range Weather Forecasts (ECMWF) ReAnalysis (–ERA) datasets can better describe surface air temperatures and vertical wind velocity in China. In addition, the Japanese 55-year Reanalysis (JRA-55) is the most stable dataset over China [28,29]. Therefore, the dynamic field data of the Tibetan Plateau used in this paper was taken from the JRA-55 reanalysis data set, and the time range was 1960.1–2020.2. The latitude and longitude range of the Tibetan Plateau is 25°N–46°N, 75°E–105°E (see Figure 3), the spatial resolution of which was $1.25° \times 1.25°$. The meteorological elements we used included relative divergence, relative vorticity, vertical velocity and wind speed. For the isobaric surface, 600 hPa, 500 hPa, 400 hPa, 300 hPa, 200 hPa and 100 hPa were selected. As the mean altitude of the Tibetan Plateau is over 3000 m (about 700 hPa), selections begin at 600 hPa to avoid any effects from the boundary layer.

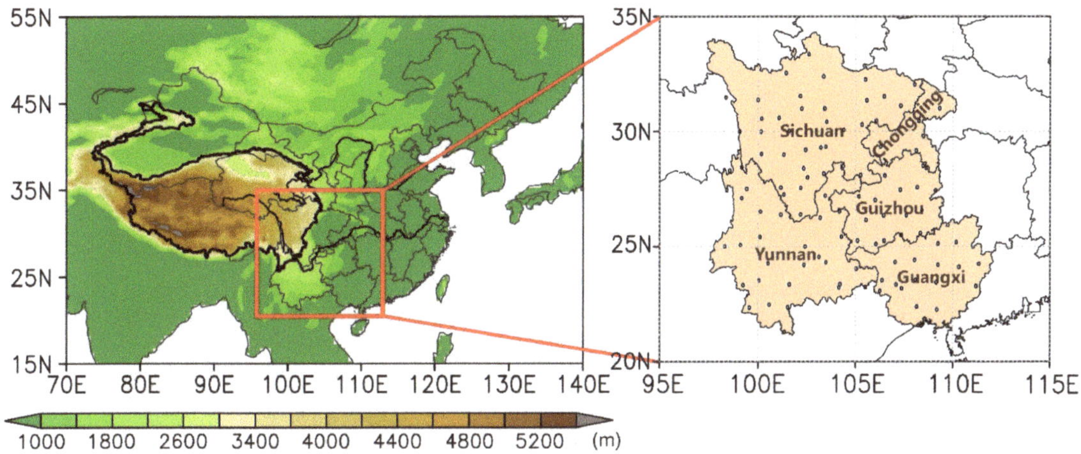

Figure 3. Terrain of the Tibetan Plateau (the color bar referring to the altitude measured in meters) and Meteorological National Stations (dots) in Southwest China.

In order to ensure the reliability of the data, the ERA-5 dataset with higher accuracy $0.25° \times 0.25°$ was also analyzed with the same method and compared with the JRA-55 data, and it was found that the distribution of elements was similar. Therefore, the JRA-55 dataset was considered reliable.

The precipitation data of Southwest China used in this was from 'Dataset of monthly climate data from Chinese surface stations' (http://101.200.76.197:93/data/cdcdetail/dataCode/SURF_CLI_CHN_MUL_MON.html, accessed on 1 February 2021). The data is quality controlled as the true rate of each meteorological element item data is generally more than 99%, and the accuracy rate of data is close to 100%. The observation data in the files of the monthly surface report from the national station from 1951 till now were repeatedly tested and controlled for quality, during which the incorrect records were corrected, and missing data were supplemented, which significantly improved the data quality. In addition, the quality controls and verifications for meteorological indices from 1951 till now were carried out during the production of the dataset, and the suspicious and incorrect values were generally checked and corrected manually. Finally, all meteorological indices were marked with quality control codes.

In this paper, Southwest China refers to Yunnan, Sichuan, Guangxi, Guizhou provinces and Chongqing Municipality (see Figure 3). Only national stations established before 1960 with data records were counted, which yielded a total of 104 stations.

2.2. Methods

Empirical Orthogonal Function (EOF) is a method of analyzing the structural features of matrices and extracting the characteristic quantities of major data. It was first introduced into meteorological and climate research in 1956 by Lorenz [30]. This method has been widely extended and applied in geosciences, including meteorology, climate and other subjects.

The basic principle of EOF decomposition is to decompose the meteorological element field with m observation space points and n time series points. The anomalous values x_{ij} corresponding to space point i and time point j in the field can be regarded as a linear combination of m spatial function v_{ik} and temporal function y_{ki} (k = 1, 2, ..., p), expressed as:

$$x_{ij} = \sum_{k=1}^{m} v_{ik} y_{kj} = v_{i1} y_{1j} + v_{i2} y_{2j} + \ldots + v_{im} y_{mj} \tag{1}$$

It can also be written as matrix form: X = VY.

X is m × n spatial–temporal data matrix. The element X_{ij} is the anomaly value. V is the spatial function matrix. Y is the temporal function matrix. V and Y are orthogonal.

Decomposition method: $XX^T = VYY^TV^T = V\Lambda V^T$. Λ is the diagonal matrix composed of the eigenvalues of XX^T matrix. V is the spatial matrix composed of eigenvectors, where each column corresponds to an eigenvector of an eigenvalue.

It can be inferred that $YY^T = \Lambda$.

Therefore, the spatial function matrix V can be obtained by finding the eigenvectors of XX^T matrix, while the temporal function Y can be obtained by $Y = V^TX$.

The spatial functions can generalize the spatial distribution characteristics of variables to a certain extent, and the time functions consist of linear combinations of variables, which means rotating the original coordinate system to create a new coordinate system, in order to make the variance as large as possible. The time coefficient is also called the principal component. If the variance contribution of several principal components accounts for the main part of the total variance of all variables in the original field, these principal components can be used to approximately show the main information of the original factor field. Therefore, the analysis of the temporal variation of the meteorological element anomaly field in this paper can be simplified to the analysis of the characteristics of the principal component of the meteorological element anomaly field changing with time, so as to analyze and explain the physical characteristics of the element field.

In this work, the eigenvectors show the spatial distribution characteristics of the elements in the modes, and the time coefficients represent the change in the eigenvectors with time. To consider this fact, the product of an eigenvector and its corresponding time coefficient is generally taken [31]. To determine whether the first spatial eigenvector has a physical meaning, we used the rule suggested by North et al. [32] to test the results.

$$e_j = \lambda_j \left(\frac{2}{N}\right)^{\frac{1}{2}} \tag{2}$$

where e_j is the error range of the eigenvalue λ_j and $N = 60$ is the sample size. When the adjacent eigenvalues satisfied $\lambda_j - \lambda_j + 1 \geq e_j$, we considered that the EOFs corresponding to these two eigenvalues were significant.

Regression analysis is a common meteorological statistical analysis method, which is based on the existing elements of the meteorological data for statistical analysis and model fitting, aimed at looking for a statistical relationship between variables and the development of a regression model. By using this model, the future value of the elements can be evaluated [31]. Only linear regression is considered in this paper.

The unary regression analysis includes a predicting factor and a predictor variable, and mainly constructs the statistical relationship between them.

The general unary linear regression equation can be written as: $\hat{y}_i = b_0 + bx_i$, where $\hat{y}_i$ is the estimated value of the predictor variable. It is easy to deduce that, for all x_i, if the total deviation between $\hat{y}_i$ and y_i (the observed values) is the minimum, this equation can be considered to best represent the dispersion pattern of all measured points. To ensure that the deviation will not affect the results, the variance is generally taken to represent the deviation level between the observed value and the forecast value. The sum of the variances between all observed values y_i and the regression forecast value $\hat{y}_i$ is denoted as:

$$Q(b_0, b) = \sum_{i=1}^{n} (y_i - \hat{y}_i)^2 \tag{3}$$

According to the extremum principle, it is required that:

$$\frac{\partial Q}{\partial b_0} = 0 \; \frac{\partial Q}{\partial b} = 0$$

The regression coefficient can be determined:

$$\begin{cases} b_0 = \overline{y} - b\overline{x} \\ b = \frac{\sum_{i=1}^{n} x_i y_i - \frac{1}{n}\left(\sum_{i=1}^{n} x_i\right)\left(\sum_{i=1}^{n} y_i\right)}{\sum_{i=1}^{n} {x_i}^2 - \frac{1}{n}\left(\sum_{i=1}^{n} x_i\right)^2} \end{cases}$$

Multiple linear regression is similar to the unary.

3. Results

3.1. Dynamic Anomaly on the Tibetan Plateau in 2009–2014

Figure 4 shows the EOFs of the anomaly field and the mean field of the relative divergence at 500 hPa over the Tibetan Plateau from November to next April 2009–2014.

It can be found that there was little difference in the explained variances of the first two modes, which can both be regarded as relatively significant modes. Compared with the mean field (Figure 4e), it can be seen that the divergence anomaly in the first mode was almost completely opposite to the average divergence, that is, there were negative anomalies in the Himalayan Mountains (on the south) and the central plateau, and positive anomalies in the southern Tibetan Valley (on the north of the Himalayan Mountains), the north and the mid-eastern part of the plateau (Figure 4a). The time coefficients indicated that this feature was more obvious from the winter of 2009 to the spring of 2010, and the opposite feature was found for the period of the 2011 winter to 2012 spring (Figure 4b). In the second mode, there was a positive anomaly in the Himalayan mountains, a negative anomaly in the main region of the plateau, and a positive anomaly in the northern part (Figure 4c). With the time coefficients also taken into consideration, this feature was obvious in the spring and winter of 2010, and in the winter of 2011 to the winter of 2012 (Figure 4d). In general, in 2009 and 2012, when the drought was severe, divergence anomalies of the main part of the plateau (including the central plateau and the northern mountains) were obviously opposite to the mean field (white circles in Figure 4). In other words, in the period of severe drought in Southwest China, the relative divergence field of 500 hPa over the Tibetan Plateau tended to weaken, no matter whether it was convergence or divergence.

The EOF of the anomaly field and the mean field of the relative vorticity at 500 hPa over the Tibetan Plateau from November to next April 2009–2014 were shown in Figure 5. The first two modes have passed the North Test [32].

It can be seen that the plateau vorticity anomaly at 500 hPa was mainly positive, with the positive vorticity strengthening and the negative vorticity weakening, which was contrary to the large-area negative vorticity in the mean field (Figure 5c). In other words, the relative vorticity field near the plateau surface also tended to weaken in the winter season.

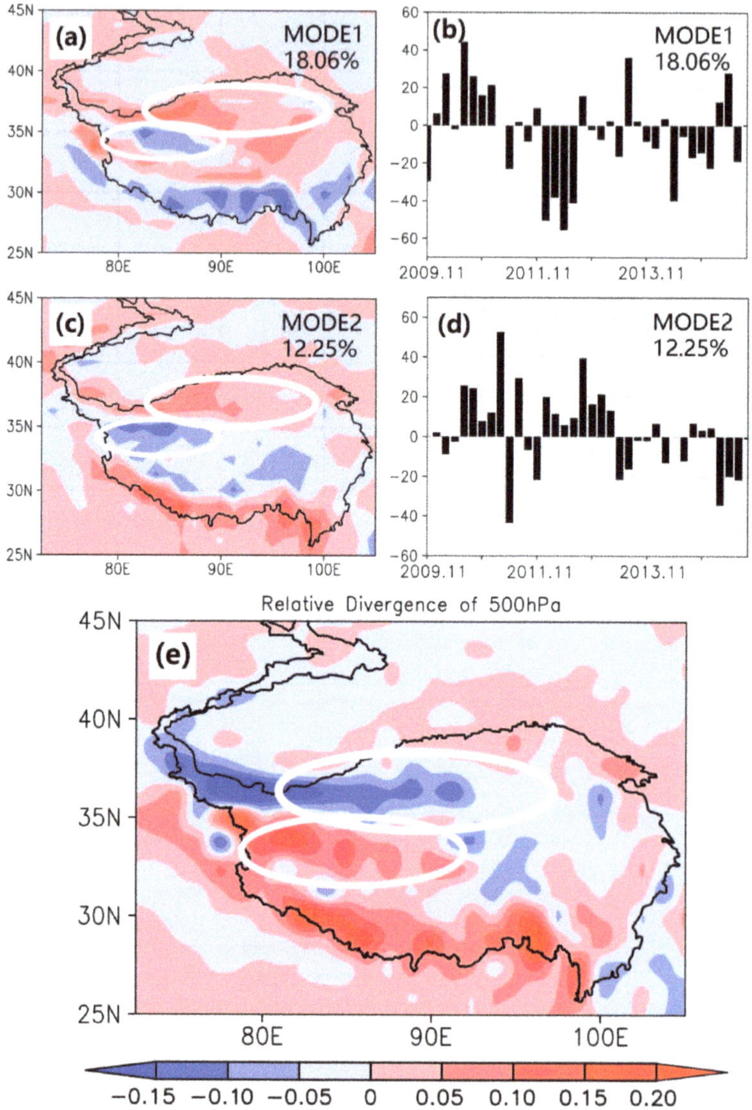

Figure 4. EOF of the relative divergence anomaly field and mean field of the relative divergence at 500 hPa over the Tibetan Plateau from November to next April 2009–2014 (10^{-4} s^{-1}). (**a**) Eigenvectors of the first mode (with explained variance of 18.06%); (**b**) Time coefficients of the first mode; (**c**) Eigenvectors of the second mode (with explained variance of 12.25%); (**d**) Time coefficients of the second mode; (**e**) Mean field. The first two modes have passed the North Test [32]. The white ovals highlight the opposite part in the EOF modes and the mean field.

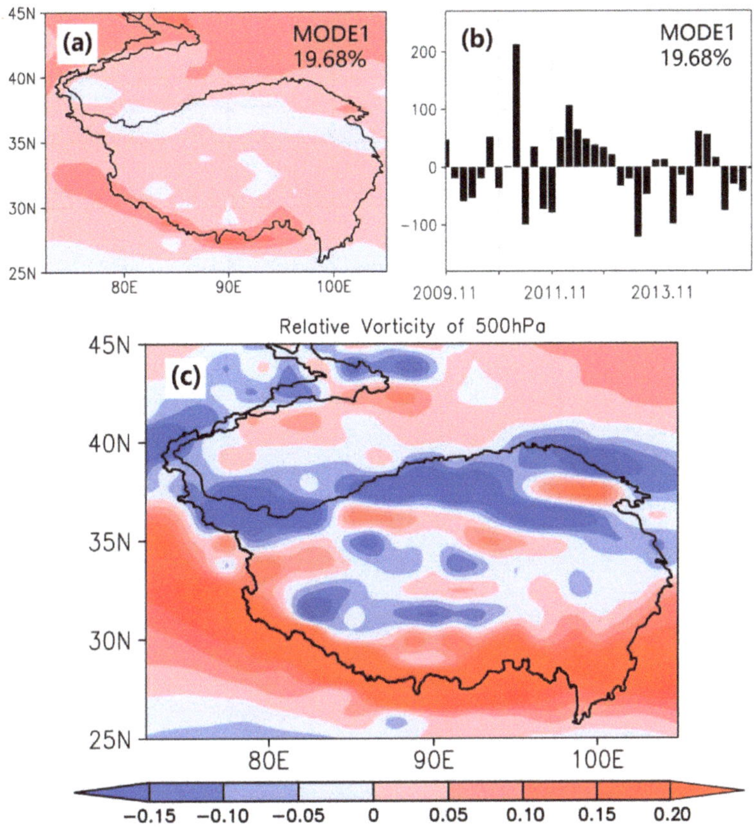

Figure 5. EOF of the relative vorticity anomaly field and mean field of the relative vorticity at 500 hPa over the Tibetan Plateau from November to next April 2009–2014 (10^{-4} s^{-1}). (**a**) Eigenvectors of the first mode (with explained variance of 19.68%); (**b**) Time coefficients of the first mode; (**c**) Mean field.

Finally, the EOF of the anomaly field and the mean field of the vertical wind component velocity at 500 hPa over the Tibetan Plateau from November to next April 2009–2014 was shown in Figure 6.

The main vertical movement over the Tibetan Plateau in winter was the downdraft caused by both thermal and dynamic action, which can also be reflected in the mean field (Figure 6e). The first and the second EOF mode, both with significant explained variances (22.68% and 17.58%, respectively), showed that the main region of the Tibetan Plateau was negative anomalous with dominant negative time coefficients, that is, during the drought in Southwest China, the downdraft over the Tibetan Plateau had an abnormal tendency to be weaker than the average state.

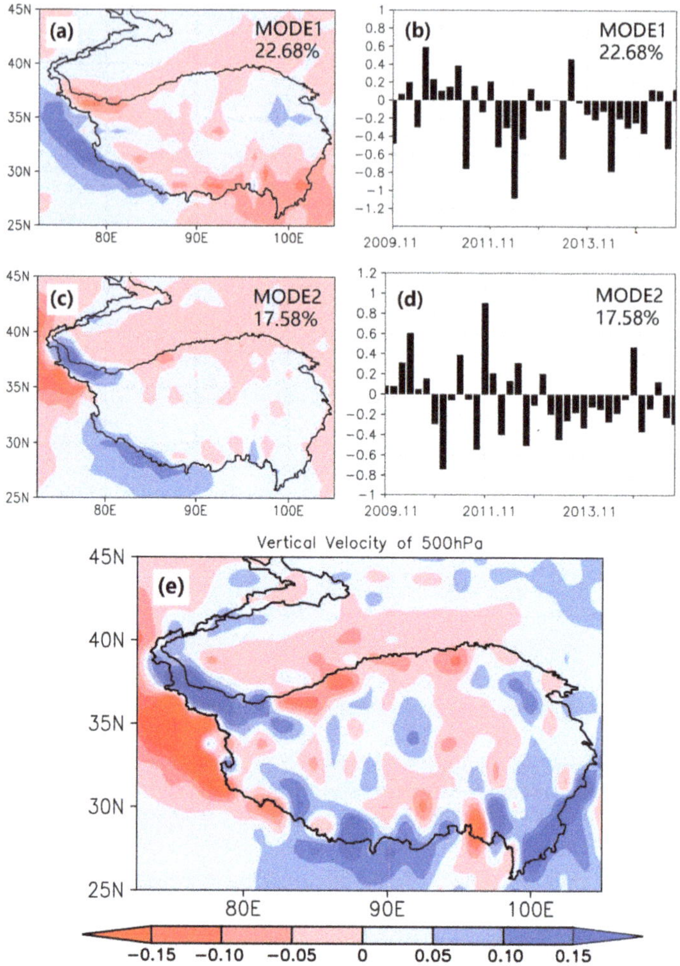

Figure 6. EOF of the vertical velocity anomaly field and mean field of the vertical velocity at 500 hPa over the Tibetan Plateau from November to next April 2009–2014 (Pa/s). (**a**) Eigenvectors of the first mode (with explained variance of 22.68%); (**b**) Time coefficients of the first mode; (**c**) Eigenvectors of the second mode (with explained variance of 17.58%), (**d**) Time coefficients of the second mode; (**e**) Mean field. The first two modes have passed the North Test [32].

3.2. The Flow of Westerly Wind and the Intersection of Cold and Warm Air

As mentioned above, Southwest China is alternatively controlled by winter and summer monsoons, and the distribution of precipitation is uneven among the seasons. The rainy season is formed under the influence of the summer monsoon system, and the dry season is formed under the influence of the winter monsoon system [33]. Until now, it has been generally believed that the main precipitation system in winter in Southwest China is the Kunming quasi-stationary front (in the southwest) and the South China quasi-stationary front (in the southeast). The South China quasi-stationary front is located in the east of Southwest China and only affects part of Guangxi. In winter, the westerlies are blocked by the Tibetan Plateau, which they bypass. The northern branch forms a ridge of high pressure, which merges with the cold air from the north and moves southward under the influence of the anticyclone formed by the thermal effect over the Tibetan Plateau. Then, it is blocked again by the Yunnan–Guizhou Plateau. The southern branch bypasses the

Tibetan Plateau to form the south trough near the Bay of Bengal, and merges with the warm current towards the north. The cold and dry continental air flowing along the Northern Slope of Yunnan–Guizhou Plateau intersects with the warm air between Kunming and Guiyang, forming the Kunming quasi-stationary front. The weather on two sides of the front is quite different: on the cold air (northeast) side of the Kunming quasi-stationary front, Guizhou, Sichuan, Chongqing and other places are wet with rain (or snow) from time to time, while Yunnan and other areas on the warm (southwest) side are controlled by the warm air mass, hence are sunny with less rain. Once the cold air is strong and the quasi-stationary front moves westward and southward crossing over the mountains, there may be more rainy days in winter in Yunnan [20,21].

During the drought events in Southwest China from 2009 to 2014, the convergence of the dynamic field in the upper level and the divergence in the lower level of the Tibetan Plateau both weakened, the absolute value of relative vorticity in the lower level was smaller, and the downdraft was also weaker. It can be concluded that during the drought in Southwest China, the high-pressure anticyclone formed by the thermal effect over the Tibetan Plateau tended to diminish.

When the clockwise circulation of the anticyclone slowed down, the intensity of the northern and southern air flows formed by the westerlies meeting the plateau terrain would also change. The flow on the northeast towards the south weakened, while the flow on the southeast towards the north strengthened. By comparing the mean wind field of the dry season of 2009–2014 with that of 1960–2019 (Figure 7), it can be found that at about 106°E at 700 hPa (red dotted lines), at the mean field of 1960–2019 (a) and (c), the latitude where the northward and southward branch airflow met on the east side of the Tibetan Plateau was lower than 35°N. However, under the mean state during the drought event in Southwest China (b) and (d), the intersection of northward and southward branch airflows exceeded 35°N, which was about 1° north of that in the average state.

At 850 hPa where warm and cold air often intersect (see Figure 8), there were two air currents controlling the Southwest China climate: one was the southeast air flow coming from the Pacific Ocean, and the other was the southwest current. Part of the latter was from the dry continent in central Asia, and the other part was the current turned from the southeast Pacific trade winds. According to the 1960–2019 average, the east and west wind converged near 101°E at around 20°N, while during the 2009–2014 drought, the convergence line moved eastward to near 104°E, with a gap of 2–3°. As can be seen from Figure 8e, on the west side of the convergence line, there was a strong positive abnormal center of southwest wind speed, while on the east side, there was also a southwest wind anomaly where the easterly (weaker) air wind flow used to be (which means a weaker east wind flow), indicating that the strong dry west wind pushed the southeast wind from the Pacific to the east, resulting in the reduction in water vapor content in Southwest China. This was also confirmed by the abnormal increase in the dew point depression (see Figure 9). During the drought event, the dew point depression in the region at 700 hPa and 850 hPa was 0.5–1 °C higher than the 1960–2019 dry season average. It could be demonstrated that the difference between the dew point temperature and the actual temperature during this period was larger, and the relative humidity was lower.

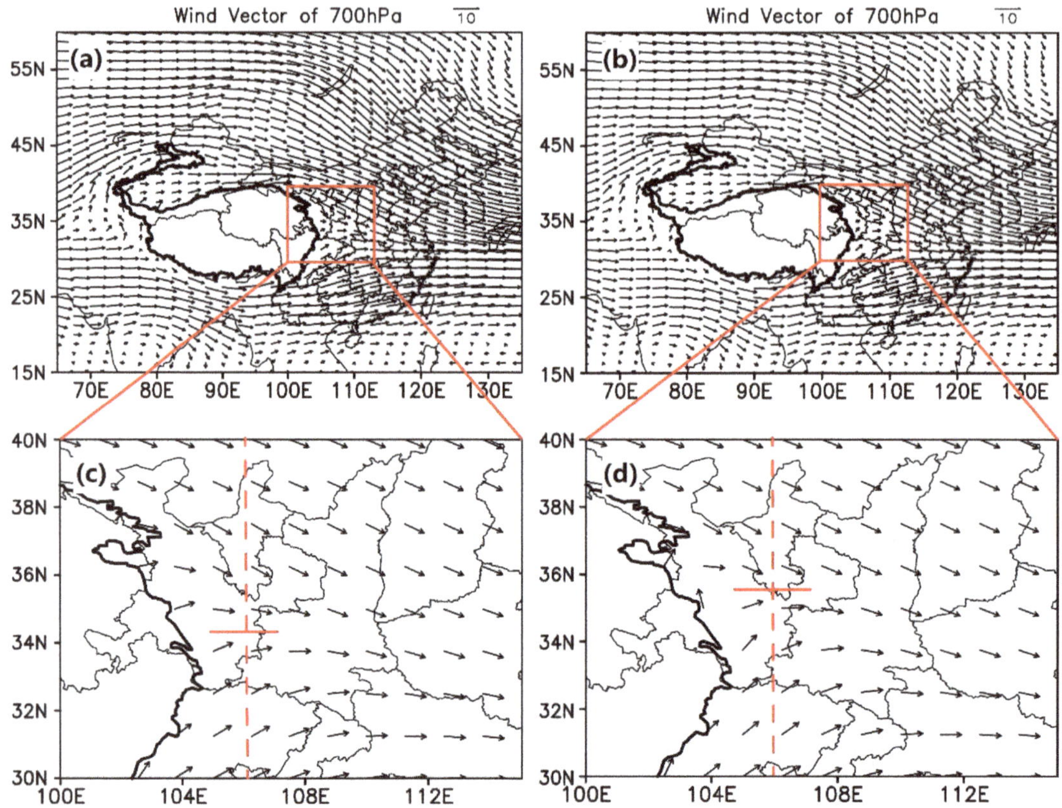

Figure 7. Average wind field at 700 hPa over east Asia (m/s). (**a**) Mean field from November to next April during 1960–2019; (**b**) Mean field from November to next April during 2009–2014. (**c**,**d**) enlarge the regions of red rectangles in (**a**,**b**). The arrow marks with numbers near (**a**,**b**) represent the size of the vectors in the graphs. The red solid lines in (**c**,**d**) refer to the boundaries between south and north wind. And the red dotted lines are referential lines at 106°E.

3.3. The Regression Analysis of Dynamic Anomaly of the Tibetan Plateau and Precipitation in Southwest China

As long as we confirmed that the dynamic field anomalies on the Tibetan Plateau were inherently related to the precipitation in Southwest China, it was reasonable to try a quantitative method to measure this relationship. Here, we chose regression. It was found that each dynamic field element had similar distribution characteristics at 600 hPa to 400 hPa (low level) and 300 hPa to 100 hPa (high level), respectively. Therefore, a total of six dynamic field elements at 500 hPa and 200 hPa levels were selected in the regression analysis. They were relative divergence, relative vorticity and vertical velocity at 500 hPa and 200 hPa, respectively. Each factor was averaged from November to April (the months affected by winter monsoon), and then EOF analysis with 59 years of time series from 1960 to 2019 was conducted. At last, unary and multiple (up to four variables) linear regression analyses were conducted between the time coefficients of EOF and the anomaly percentage of precipitation average of all stations in the same period in Southwest China. All the below-listed regression equations have passed the F-test.

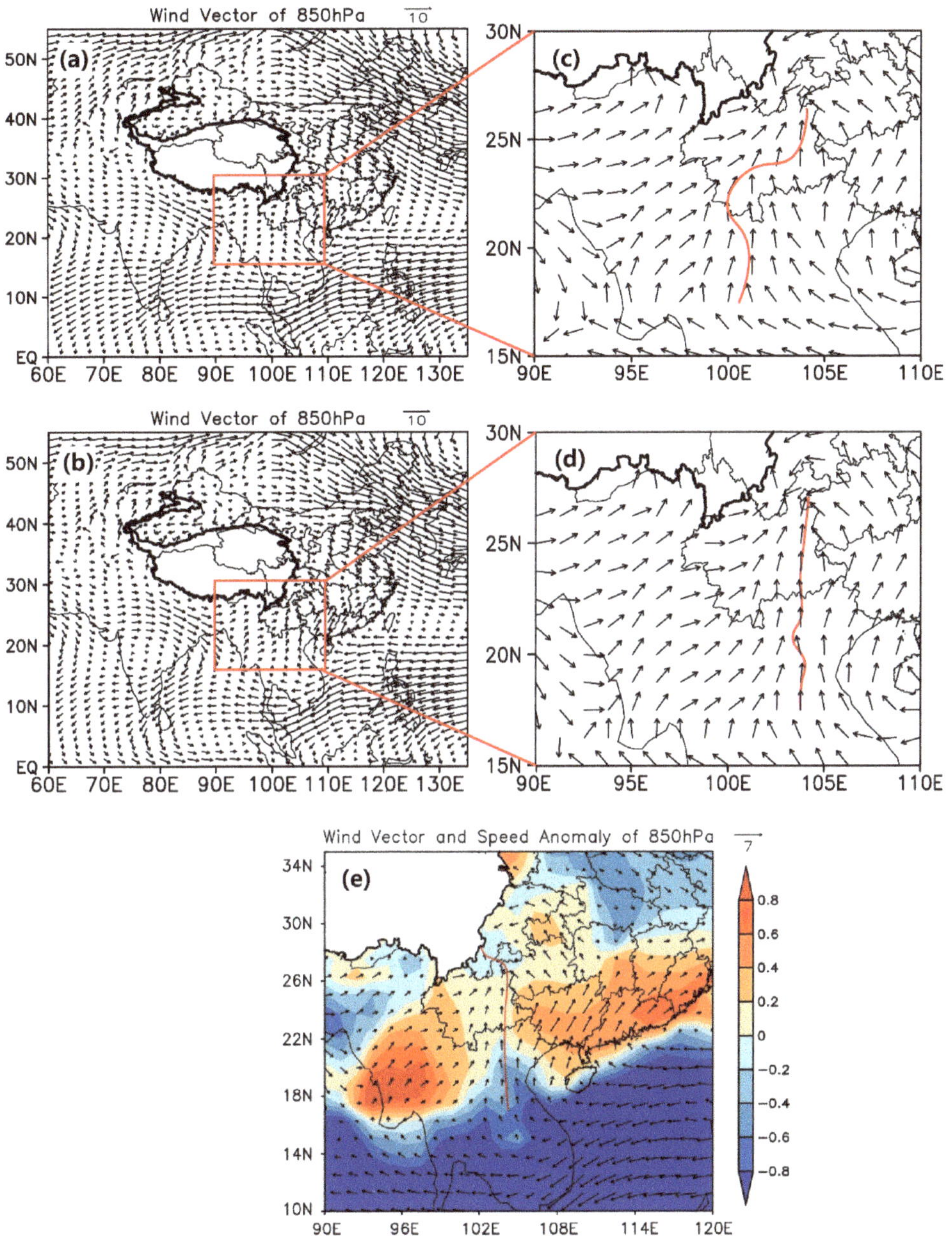

Figure 8. Average wind field and wind speed anomaly at 850 hPa over east Asia (m/s). (**a**) Mean field from November to next April during 1960–2019, (**b**) Mean field from November to next April during 2009–2014, (**e**) Mean field of wind (vector) and wind speed anomaly (shaded) at 850 hPa over east Asia from 2009 to 2014. (**c**,**d**) enlarge the regions of red rectangles in (**a**,**b**). The arrow marks with numbers near (**a**,**b**,**e**) represent the size of the vectors in the graphs. The red solid lines refer to the boundaries between west and east wind.

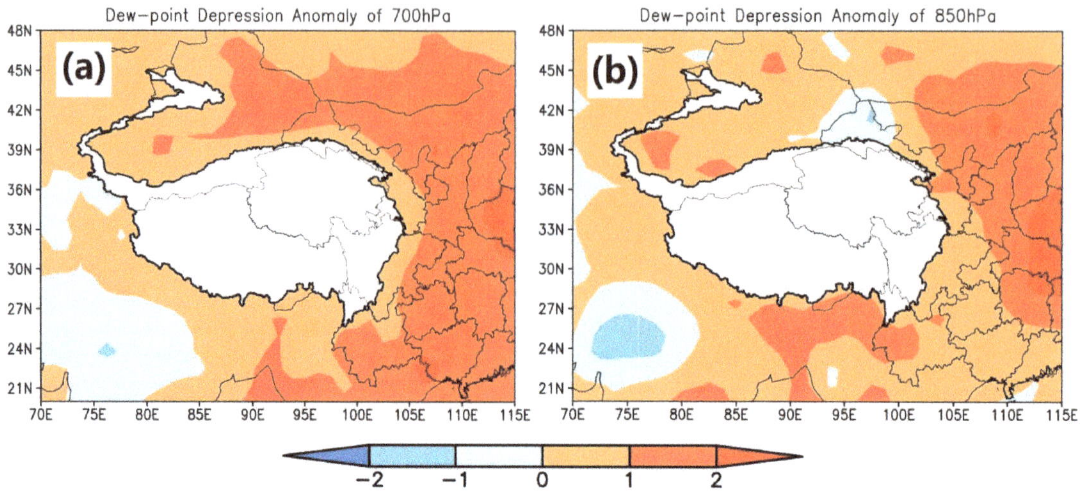

Figure 9. Dew point depression anomaly over east Asia from 2009 to 2014 (K). (**a**) 700 hPa, (**b**) 850 hPa.

Among all the tested linear regression equations, the one that worked best was the following binary regression equation:

$$y = 0.063 + 0.138x_1 + 0.297x_2$$

In the formula, y denotes the forecast value of precipitation anomaly percentage, x_1 denotes the EOF time coefficient of 500 hPa relative vorticity anomaly in 59 years, and x_2 denotes the EOF time coefficient of 500 hPa relative divergence anomaly in 59 years.

Next, the time coefficients were substituted into the regression equation to obtain the forecast values; the comparison between the forecast value and the observed value is described below. The Mean Absolute Error (MAE) of the equation was 0.106, and the Root Mean Square Error (RMSE) was 0.138, both of which were acceptable (when the difference between the maximum and the minimum of the observation was about 0.79). The Pearson correlation coefficient between the forecast value and the observed value was 0.408, which exceeded the critical value of correlation coefficient 0.336 (degree of freedom: $N - M - 1 = 56$) at the significance level of 0.01, which means passing the reliability test with 99%.

It can be seen from Figure 10 that the regression equation could predict the overall trend and most peak and valley situations, while the absolute value prediction of some peak and valley values was not accurate.

We also made further attempts. In unitary linear regression analysis, the best result was the equation with relative vorticity at 500 hPa, whose correlation coefficient was 0.368 between the forecast value and the observed value, and it was not as accurate as the binary equation in the prediction of peak and valley values. For the equation with four unknowns, when the value was close to 0, the difference of the quaternionic regression equation was larger than that of the binary equation, and it was more prone to qualitative forecast errors, which had a great influence on the accuracy of the forecast.

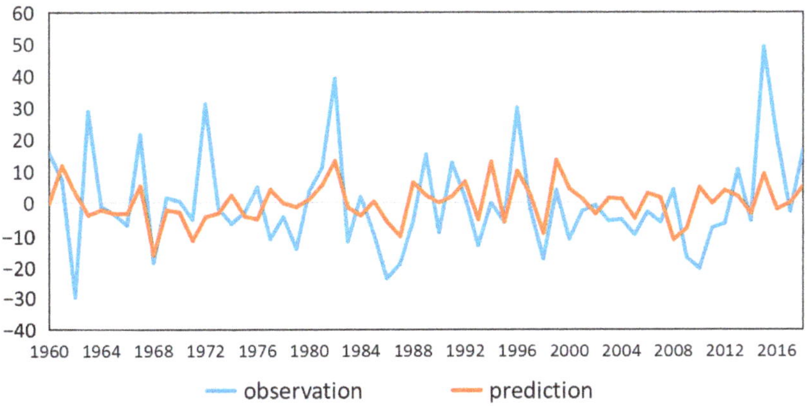

Figure 10. The forecast results of the regression equations, variables including relative divergence and relative vorticity at 500 hPa. The blue line represents the observed percentage of precipitation anomaly, and the yellow line represents the prediction.

4. Discussion

When comparing the surface temperature field (see Figure 11a) with precipitation anomalies (see Figure 11b) during drought events in Southwest China, it was noticed that the areas with positive temperature anomalies coincided greatly with the areas with drought, and the boundary of the drought was close to the boundary of positive and negative temperature anomalies. In 2009 and 2012, when the winter droughts were quite severe, almost the entire area of Southwest China experienced warmer temperatures than usual. There were also quite high Pearson correlation coefficients between two groups of grid data, such as 0.222 for 2010 and 0.427 for 2013 (with 818 grids) at the significance level of 0.001. All these observations indicate that the location changes of cold and warm air masses did have a strong relationship with precipitation in Southwest China, which is consistent with our results.

To summarize the above-listed results, we raised a conceptual model of the dynamic effect influencing drought in Southwest China (see Figure 12). It could be inferred that during the drought events, the momentum of cold air moving southward from the north weakened, while the westerly wind from the dry region of Central Asia was strong in the south. The convergence area of cold and warm air was more to the north and east than usual, and the area controlled by warm air mass in the southwest was larger, which was not conducive to precipitation.

To some extent, the above result is also in line with some previous research. A few studies have attributed this drought event to the negative phase of the Arctic Oscillation [34–39], which caused the cold air in East Asia to move through an eastern route, so that less air flows southwards along the plateau, consistent with that illustrated in this model.

Through the EOF analysis and regression analysis above, it has been found that during the same period of drought in Southwest China, the anticyclone was abnormally weak over the Tibetan Plateau, and its influence mechanism on drought in Southwest China was discussed. However, there are still other problems worthy of additional research.

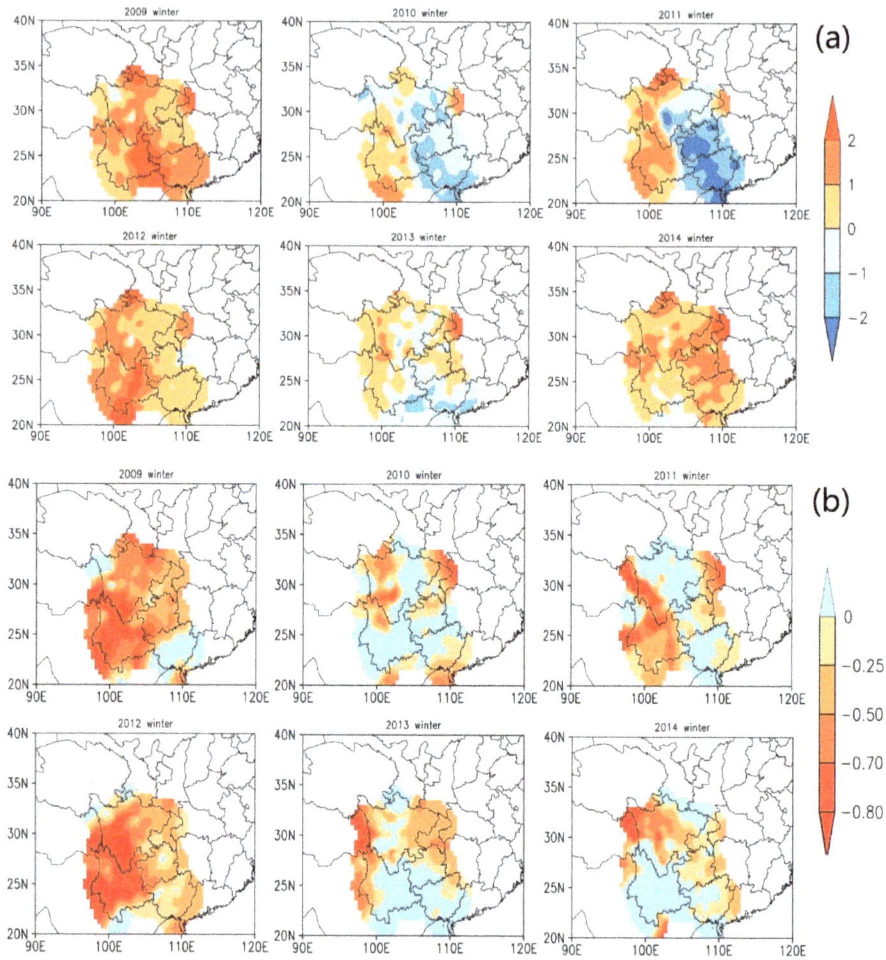

Figure 11. Temperature anomalies and percentage of precipitation anomalies in southwest China in winter of 2009–2014. (**a**) Temperature anomalies (°C), (**b**) percentage of precipitation anomalies.

Firstly, the reasons for the weakening of the anticyclone over the Tibetan Plateau in winter deserve further investigation. The reasons may include abnormal circulation caused by global warming, that is, abnormal weakening of the sinking branch of Hadley circulation or Ferrer circulation in the subtropical zone [40]; the thermal anomaly of the Tibetan Plateau itself, that is, the higher temperature at a low level, which may be caused by global warming and the change of snow-covered area [41].

In addition, it is expected that more ground and upper air observation stations will be built on the Tibetan Plateau and its adjacent areas along with the regular collection of more complete data, so that we can obtain a more accurate prediction of drought events in Southwest China.

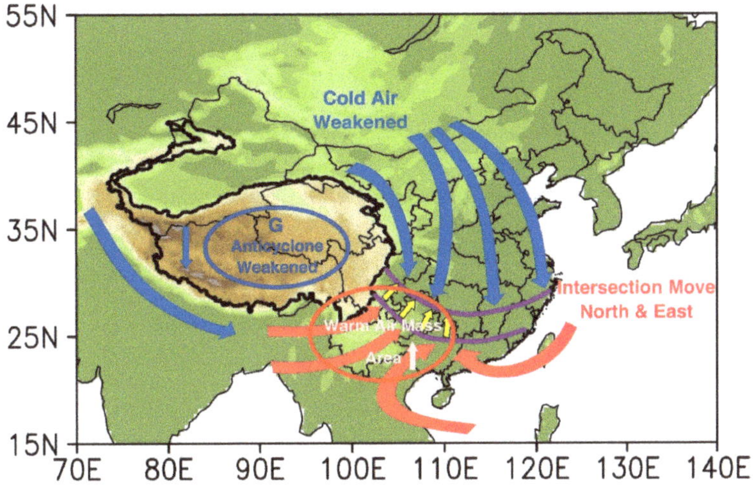

Figure 12. Conceptual Model of dynamic effect influencing drought in southwest China. The blue arrows represent cold airflows, while the red ones represent warm airflows. Note: The South China Sea has not been marked due to layout reasons.

5. Conclusions

The westerlies are separated into two branches, i.e., the northern branch and southern branch, when encountering the Tibetan Plateau. These two branches have great impacts on the weather and climate of the area where they pass, and then converge on the eastern side of the plateau, affecting the local distribution of temperature and rainfall. However, there have been few studies on how the Tibetan Plateau influence these two branches of the westerlies. It has not been clear whether there was a relationship between the Tibetan Plateau, the branches of the westerlies and the severe drought event that happened in Southwest China in 2009–2014. In addition, most previous research did not regard this continuous drought as a whole, as there was more specific work about the drought in one single year. This paper tried to deal with the problems above using some statistical methods such as EOF analysis and regression analysis. The following conclusions have been drawn:

(1) Most areas in Southwest China experienced different degrees of continuous abnormal drought in spring, autumn, and especially winter, between 2009 and 2014.

(2) The relative divergence, relative vorticity and vertical velocity at a low level over the Tibetan Plateau all showed the abnormal characteristics of a smaller absolute value of elements from November to April in the period of 2009–2014. Namely, during the drought in Southwest China, the downdraft over the Tibetan Plateau and the anticyclone both weakened.

(3) When the anticyclone weakened over the Tibetan Plateau in the winter season, the cold air from the northern branch of the westerly weakened, such that the cold air flowing southward on the northeast of the plateau weakened dwindled. In the south, the westerly wind from the arid region of Central Asia strengthened, and the intersection of cold and warm air moved eastward and northward, so that a larger area of Southwest China was controlled by a dry and warm air mass, which was not conducive to precipitation.

(4) A binary linear regression forecast model for precipitation in Southwest China was established: $y = 0.063 + 0.138x_1 + 0.297x_2$, where y denotes the forecast value of precipitation anomaly percentage, x_1 denotes the EOF time coefficient of 500 hPa relative vorticity anomaly from 1960 to 2019, and x_2 denotes the EOF time coefficient of 500 hPa relative divergence anomaly from 1960 to 2019. The correlation coefficient between the forecast value and the observed value was 0.408, passing the reliability test with 99%.

Author Contributions: Conceptualization, Y.Y. and R.T.; methodology, R.T.; formal analysis, Y.Y. and Z.J.; investigation, Y.Y. and Z.J.; writing—original draft preparation, Y.Y.; writing—review and editing, Y.Y., R.T. and Z.J.; supervision, R.T.; project administration, R.T. All authors have read and agreed to the published version of the manuscript.

Funding: This research was funded by the Second Tibetan Plateau Scientific Expedition and Research Program (STEP), grant number 2019QZKK0103.

Data Availability Statement: The dynamic field data of the Tibetan Plateau used in this paper is from JRA-55 reanalysis data set, which can be found at: https://jra.kishou.go.jp/JRA-55/index_en.html, accessed on 1 April 2020. The precipitation data of southwest China used in this paper is the data of national stations from the Monthly Data set of China Surface Climatic Data from China Meteorological Data Network, which can be found at: http://101.200.76.197:93/data/cdcdetail/dataCode/SURF_CLI_CHN_MUL_MON.html, accessed on 1 February 2021.

Acknowledgments: This work was supported by the Second Tibetan Plateau Scientific Expedition and Research Program (STEP), Grant No. 2019QZKK0103 and the National Natural Science Foundation of China (Grant Nos 41775142). We acknowledge the use of meteorological data collected from Japan Meteorological Agency and China Meteorological Administration. All data in the research can be obtained by contacting the corresponding author, Rongxiang Tian (trx@zju.edu.cn). We also appreciate all the help and advice given to this research, as well as constructive comments from all the reviewers.

Conflicts of Interest: The authors declare no conflict of interest.

References

1. Qian, Z.; Jiao, Y. Advances and Problems on Qinghai-Xizang plateau meteorology research. *Adv. Earth Sci.* **1997**, *12*, 207–216.
2. Wu, G.; Li, W.; Guo, H. *Sensible Heat Driven Air-Pump over the Tibetan Plateau and Its Impacts on the Asian Summer Monsoon*; Science Press: Beijing, China, 1997.
3. Ye, D.; Gao, Y. *Meteorology of Tibetan Plateau*; Science Press: Beijing, China, 1979.
4. Huang, R. *Introduction to Atmospheric Science*; China Meteorological Press: Beijing, China, 2005.
5. Queney, P. The Problem of Air Flow Over Mountains: A Summary of Theoretical Studies. *Bull. Am. Meteorol. Soc.* **1948**, *29*, 16–26. [CrossRef]
6. Bolin, B. On the Influence of the Earth's Orography on the General Character of the Westerlies. *Tellus* **1950**, *2*, 184–195. [CrossRef]
7. Qiao, Y.; Zhou, S.; Ma, Y.; Wang, C.; Li, Q. Dynamic Effect of Tibetan Plateau and Its Impact on Weather and Climate in China. *Meteorol. Sci. Technol.* **2014**, *42*, 1039–1046. [CrossRef]
8. Suo, M.; Ding, Y. The Structures and Evolutions of the Wintertime Southern Branch Trough in the Subtropical Westerlies. *Chin. J. Atmos. Sci.* **2009**, *33*, 425–442.
9. Jiang, Y.; He, J.; Wen, M.; Qi, L. Characteristic of a Couple of Vortexes on the East Side of Tibetan Plateau from Winter to Spring and Their Impact on the Weather and Climate in China. *Plateau Meteorol.* **2009**, *28*, 945–954.
10. Xue, D.; Zhang, Y.; Wang, P.; Sun, F.; Yin, J. Distinct influences of cold vortex over Northeast China on local precipitation in early summer and midsummer. *Clim. Dyn.* **2022**, *59*, 3701–3716. [CrossRef]
11. Liang, X.; Liu, Y.; Wu, G. The Impact of Qinghai-Xizang Plateau Uplift on Asian General Circulation in Spring and Summer. *Plateau Meteorol.* **2005**, *24*, 837–845.
12. Wan, R.; Wu, G. Study on the Climatic Mechanism of Spring Rain in Jiangnan. *Sci. China Ser. D Earth Sci.* **2006**, *36*, 936–950. [CrossRef]
13. Li, Y.; Li, D.; Yang, S.; Liu, C.; Zhong, A.; Li, Y. Characteristics of the precipitation over the eastern edge of the Tibetan Plateau. *Meteorol. Atmos. Phys.* **2009**, *106*, 49–56. [CrossRef]
14. Park, H.-S.; Chiang, J.C.H.; Bordoni, S. The Mechanical Impact of the Tibetan Plateau on the Seasonal Evolution of the South Asian Monsoon. *J. Clim.* **2012**, *25*, 2394–2407. [CrossRef]
15. Wei, J.; He, J.; Su, Z.; Yao, L. Characteristics of South Wind in East Southern Tibetan Plateau and Its Relationship with Eastern China Spring Rain. *Meteorol. Mon.* **2013**, *39*, 129–136. [CrossRef]
16. Son, J.; Seo, K.; Wang, B. How Does the Tibetan Plateau Dynamically Affect Downstream Monsoon Precipitation? *Geophys. Res. Lett.* **2020**, *47*, e2020GL090543. [CrossRef]
17. Yeh, T.C.; Dao, S.Y.; Li, M.T. *The Abrupt Change of Circulation over the Northern Hemisphere during June and October*; Rockefeller Institute Press: New York, NY, USA, 1959.
18. Kong, W.; Chiang, J.C.H. Interaction of the Westerlies with the Tibetan Plateau in Determining the Mei-Yu Termination. *J. Clim.* **2020**, *33*, 339–363. [CrossRef]
19. Wang, M.; Wang, J.; Duan, A.; Yang, J.; Liu, Y. Quasi-biweekly impact of the atmospheric heat source over the Tibetan Plateau on summer rainfall in Eastern China. *Clim. Dyn.* **2019**, *53*, 4489–4504. [CrossRef]
20. Zhang, B. The Intersection of Air mass and Weather in China. *Acta Meteorol. Sin.* **1949**, *20*, 32–37.

21. Suo, M.; Ding, Y. The Discovery and Study of Kunming Quasi-Stationary Front. *Adv. Meteorol. Sci. Technol.* **2016**, *6*, 6–16. [CrossRef]
22. Aksoy, H.; Cetin, M.; Eris, E.; Burgan, H.I.; Cavus, Y.; Yildirim, I.; Sivapalan, M. Critical drought intensity-duration-frequency curves based on total probability theorem-coupled frequency analysis. *Hydrol. Sci. J.* **2021**, *66*, 1337–1358. [CrossRef]
23. Huang, T.; Wu, Z.; Xiao, P.; Sun, Z.; Liu, Y.; Wang, J.; Wang, Z. Possible Future Climate Change Impacts on the Meteorological and Hydrological Drought Characteristics in the Jinghe River Basin, China. *Remote Sens.* **2023**, *15*, 1297. [CrossRef]
24. *GB/T 20481-2017*; Grades of Meteorological Drought. Standardization Administration of the People's Republic of China: Beijing, China, 2017.
25. China Meteorological Administration. *Yearbook of Meteorological Disasters in China in 2010*; China Meteorological Press: Beijing, China, 2010.
26. China Meteorological Administration. *Yearbook of Meteorological Disasters in China in 2012*; China Meteorological Press: Beijing, China, 2012.
27. China Meteorological Administration. *Yearbook of Meteorological Disasters in China in 2015*; China Meteorological Press: Beijing, China, 2015.
28. Xu, J.; Tian, R.; Feng, S. Comparison of Atmospheric Vertical Motion over China in ERA-Interim, JRA-55, and NCEP/NCAR Reanalysis Datasets. *Asia-Pacific J. Atmos. Sci.* **2021**, *57*, 773–786. [CrossRef]
29. Zhao, T.; Fu, C. Applicability Evaluation of Surface Air Temperature from Several Reanalysis Datasets in China. *Plateau Meteorol.* **2009**, *28*, 594–606. [CrossRef]
30. Lorenz, E. Empirical Orthogonal Functions and Statistical Weather Prediction. *Stat. Forecast.* **1956**, *1*, 52.
31. Huang, J. *Meteorological Statistical Analysis and Forecasting Methods*; China Meteorological Press: Beijing, China, 1990.
32. North, G.R.; Bell, T.L.; Cahalan, R.F.; Moeng, F.J. Sampling Errors in the Estimation of Empirical Orthogonal Functions. *Mon. Weather. Rev.* **1982**, *110*, 699–706. [CrossRef]
33. Yan, H.; Li, Q.; Sun, C.; Yuan, Y.; Li, D. Criterion for Determining the Onset and End of the Rainy Season in Southwest China. *Chin. J. Atmos. Sci.* **2013**, *37*, 1111–1128. [CrossRef]
34. Cheng, Q.P.; Gao, L.; Zhong, F.L.; Zuo, X.A.; Ma, M.M. Spatiotemporal variations of drought in the Yunnan-Guizhou Plateau, southwest China, during 1960–2013 and their association with large-scale circulations and historical records. *Ecol. Indic.* **2020**, *112*, 106041. [CrossRef]
35. Hu, X.; Xu, P.; Ning, G.; Wang, S.; Shang, K. Causes of Continuous Drought in Southwest China from Autumn of 2012 to Spring of 2013. *J. Desert Res.* **2015**, *35*, 763–773. [CrossRef]
36. Huang, R.; Liu, Y.; Wang, L.; Wang, L. Analyses of the causes of severe drought occurring in Southwest China from the fall of 2009 to the spring of 2010. *Chin. J. Atmos. Sci.* **2012**, *36*, 443–457. [CrossRef]
37. Yang, H.; Song, J.; Yan, H.; Li, C. Cause of the Severe Drought in Yunnan Province during Winter of 2009 to 2010. *Clim. Environ. Res.* **2012**, *17*, 315–326. [CrossRef]
38. Zhong, Y. Circulation Anomalies Related to the Autumn-to-Winter Drought in Southwest China and Possible Impacts of the North Atlantic SSTs. Master's Thesis, Nanjing University of Information Science and Technology, Nanjing, China, 2020.
39. Zhou, L.; Cai, R.; Lan, M.; Deng, C.; Wen, P.; Zhang, J.; Liu, L. Temporal and Spatial Distribution Characteristics of Drought in Southwest China during 2011–2012 and Its Diagnostic Analysis of Dynamic Causes. *Hubei Agric. Sci.* **2018**, *57*, 28–33+45. [CrossRef]
40. Kjellsson, J. Weakening of the global atmospheric circulation with global warming. *Clim. Dyn.* **2014**, *45*, 975–988. [CrossRef]
41. Dawei. *Thermodynamic Effects of the Tibetan Plateau on the Sustained Drought in Southwest China*; Zhejiang University: Zhejiang, China, 2021.

remote sensing

Article

Evaluation of Spatial and Temporal Variations in the Difference between Soil and Air Temperatures on the Qinghai–Tibetan Plateau Using Reanalysis Data Products

Xiqiang Wang and Rensheng Chen *

Qilian Alpine Ecology and Hydrology Research Station, Key Laboratory of Ecological Safety and Sustainable Development in Arid Lands, Northwest Institute of Eco-Environment and Resources, Chinese Academy of Sciences, Lanzhou 730000, China; wangxq@lzb.ac.cn
* Correspondence: crs2008@lzb.ac.cn

Abstract: Many extreme meteorological events are closely related to the strength of land–atmosphere interactions. In this study, the heat exchange regime between the shallow soil layer and the atmosphere over the Qinghai–Tibetan Plateau (QTP) was investigated using a reanalysis dataset. The analysis was conducted using a simple metric ΔT, defined as the difference between the temperatures of the shallow soil and the air. First, the performance of 4 widely used reanalysis data products (GLDAS-Noah, NCEP-R2, ERA5 and ERA5-land) in estimating ΔT on the QTP at soil depths of 0~7 or 0~10 cm was evaluated during the baseline period (1981–2010); the ERA5-land product was selected for subsequent analysis, because it yielded a better performance in estimating the annual and seasonal ΔT and finer spatial resolution than the other datasets. Using the soil temperature at depths of 0~7 cm and the air temperature at 2 m above the ground, as provided by the ERA5-Land reanalysis dataset, the entire QTP was found to be dominated by a positive ΔT both annually and seasonally during the baseline period, with large differences in the spatial distribution of the seasonal values of ΔT. From 1950 to 2021, the QTP experienced a significant decreasing trend in the annual ΔT at a rate of -0.07 °C/decade, and obvious decreases have also been detected at the seasonal level (except in spring). In the southern and northeastern parts of the QTP, rapid rates of decrease in the annual ΔT were detected, and the areas with significantly decreasing trends in ΔT were found to increase in size gradually from summer, through autumn, to winter. This study provides a holistic view of the spatiotemporal variations in ΔT on the QTP, and the findings can improve our understanding of the land–atmosphere thermal interactions in this region and provide important information pertaining to regional ecological diversity, hydrology, agricultural activity and infrastructural stability.

Keywords: land–atmosphere interaction; soil temperature; air temperature; ERA5-land; Qinghai–Tibetan Plateau

Citation: Wang, X.; Chen, R. Evaluation of Spatial and Temporal Variations in the Difference between Soil and Air Temperatures on the Qinghai–Tibetan Plateau Using Reanalysis Data Products. *Remote Sens.* **2023**, *15*, 1894. https://doi.org/10.3390/rs15071894

Academic Editors: Massimo Menenti, Yaoming Ma, Li Jia and Lei Zhong

Received: 31 January 2023
Revised: 25 March 2023
Accepted: 30 March 2023
Published: 31 March 2023

1. Introduction

Land–atmosphere interactions play a major role in shaping and projecting regional climates [1], and studies have identified a close relationship between the strength of such interactions and extreme meteorological events such as heat waves, droughts, and heavy precipitation [2–7]. Variations in atmospheric conditions can directly alter the soil hydrothermal status by modulating meteorological conditions, such as air temperature and precipitation, while changes in the soil temperature and moisture level can trigger changes in the surface energy distribution and water balance and, ultimately, affect atmospheric processes [8–11]. Climate warming, which is caused primarily by greenhouse gas emissions, has intensified in high-latitude and mountainous regions [12,13]; inevitably, the land–atmosphere interactions in these regions can be altered and, subsequently, exert a profound impact on energy and moisture exchanges, carbon release, agricultural activity, ecosystem diversity, engineering construction and hydrological processes [14–19].

Some variables, such as the surface net radiation, sensible heat flux, and latent heat flux and ground heat flux, can directly reflect the exchange of heat between the land and atmosphere; however, long-term series of measurements of these variables are rare and difficult to obtain, especially in high-latitude and mountainous regions [20]. In contrast, variations in air and soil temperatures, which are the result of energy and moisture partitioning at the surface, have long been observed and recorded, and the difference between these two temperatures, ΔT, is commonly used to assess land–atmosphere heat exchange regimes over long periods [10]. ΔT has been identified as having a crucial influence on climates and environments at the regional and even global level [21]. Over the past few decades, considerable research has been performed to investigate the spatiotemporal variations in ΔT, as well as the relationships of this variable with environmental factors on the regional, national and hemispheric scales, using station observations, reanalysis data or satellite remote-sensing data [9,10,22–36].

Regarded as the "Third Pole of the World", the Qinghai–Tibetan Plateau (QTP) has unique topographic and thermodynamic effects that have exerted an obvious influence on the weather and climate systems of China and East Asia and have even been shown to have global effects [20,37–41]. Over the past few decades, the QTP has experienced rapid surface air warming and humidification, leading to changes in atmospheric and hydrological cycles and profound effects on regional land–atmosphere heat exchange [42,43]. Under such circumstances, increased knowledge of the spatiotemporal variations in ΔT on the QTP can inform analyses of the regional climate, ecology and other parameters. Studies have used ground observations to investigate the relationship of soil temperature at different depths with air temperature on the QTP [44–46]. However, the results of such analyses may have been subject to low-elevation bias, because the study sites were sparsely distributed, with most being less than 4000 m above sea level due to the harsh natural environment and high cost of observation. By combining in situ observations with remote sensing data, the release of reanalysis datasets provide substantial data for large-scale studies of land–atmosphere interactions with relatively high spatial and temporal resolutions [47–49], and such datasets have been used in a series of scientific studies on the QTP [11,47–54]. Using multi-source reanalysis datasets, Wang et al. (2020) investigated spatiotemporal variations in the difference between skin (0 cm of the land surface) and air temperatures from 1979 to 2018 [11]. As ecological, hydrological and biological activities mainly occur within the near-surface soil layer, understanding the exchange of heat between the shallow soil layer and the atmosphere is crucial for understanding temperature-dependent processes in fields such as regional ecology and agriculture. However, there has been little research on ΔT between shallow soil and air on the QTP.

Following the identification of a reanalysis data product with a relatively high simulation accuracy of ΔT, this study investigated the spatiotemporal variations in ΔT between shallow soil and air over the QTP. The dataset used in this study was selected from four widely used reanalysis data products on the QTP [11,50,54,55]: the Global Land Data Assimilation System (GLDAS) Version 2.0, National Centers for Environmental Prediction (NCEP) and the Department of Energy (DOE) Reanalysis 2, the fifth-generation European Centre for Medium-Range Weather Forecasts (ECMWF) atmospheric reanalysis (ERA5) and an enhanced-resolution version of the ERA5 (ERA5-land).

2. Materials and Methods

2.1. Study Area

The QTP stretches more than 2945 km from east to west and more than 1532 km from south to north, within the approximate geographic boundaries of 25°N–40°N and 73°E–105°E (Figure 1). Known as the "Roof of the World," the QTP has an average elevation exceeding 4000 m, which has promoted the development of an alpine climate characterized by strong radiation, low temperatures and a large diurnal temperature range. The QTP has an annual mean temperature of approximately −2.5 °C and annual mean precipitation of approximately 380 mm.

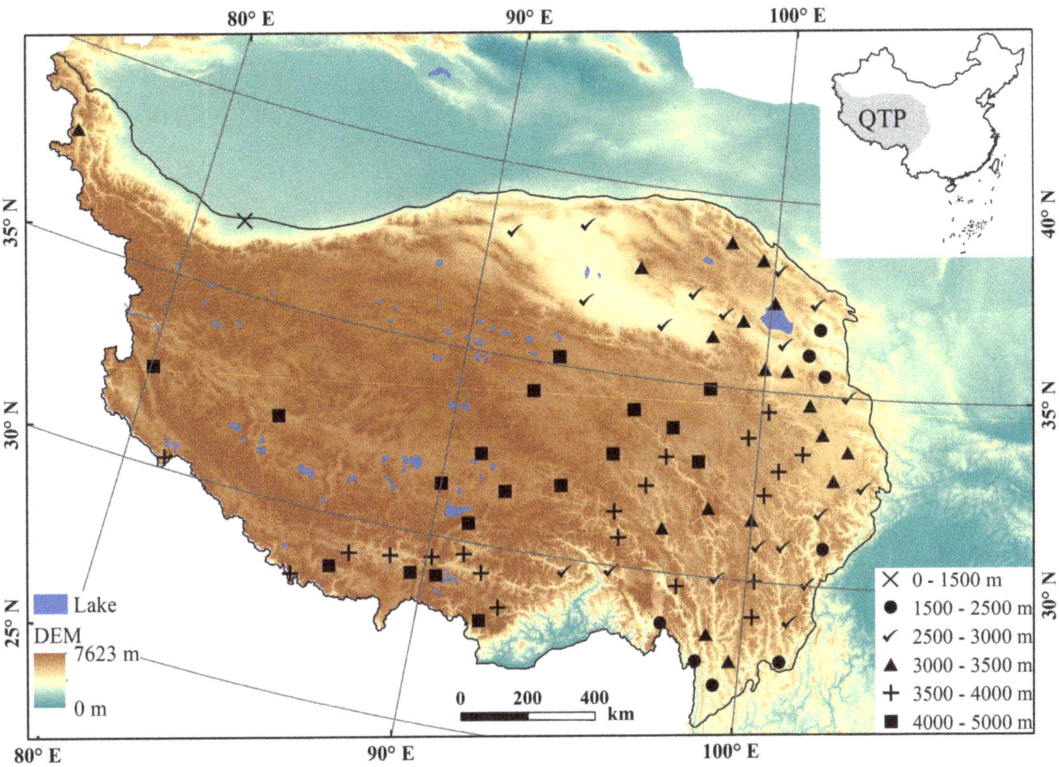

Figure 1. Location of the Qinghai–Tibetan Plateau (QTP) and the spatial distribution of 84 meteorological stations. The boundary of the QTP is provided by the National Tibetan Plateau Data Center (https://data.tpdc.ac.cn/zh-hans/data/0231c972-8460-4691-a187-70e4cc356f60 (accessed on 1 October 2022)). The background coloration reflects the altitude. DEM: digital elevation model.

2.2. Data

2.2.1. Reanalysis Data

Four reanalysis data products that contain long-term series data on soil and air temperatures were evaluated in this study (Table 1). GLDAS is a global, high-resolution, offline terrestrial modeling system that can produce optimal fields of land surface states and fluxes in near-real time [56]. GLDAS-Noah is the Noah Land Surface Model driven by GLDAS V2.0; it has a spatial resolution of 0.25° × 0.25° and provides time series from 1948 to 2015 [56,57]. NCEP-DOE Reanalysis 2 (hereafter, NCEP-R2) is an improved version of the NCEP Reanalysis 1 model; in NCEP-R2, errors have been corrected, and updated parameterizations of physical processes are provided [58] along with data at a spatial resolution of 1.875° (longitude) × 1.889° (latitude) and a time range from 1979 to the present.

Table 1. Summary of the four reanalysis datasets evaluated in this study.

Dataset	Institutes	Time Resolution (h)	Spatial Resolution	Time Period	Soil Temperature (cm)
GLDAS-Noah	NASA/NCEP	3 h	0.25° × 0.25°	1948~2015	0~10, 10~40, 40~100, 100~200
NCEP-R2	NCEP	6 h	1.875° × 1.889°	1979~present	0~10, 10~200
ERA5	ECMWF	1 h	0.25° × 0.25°	1959~present	0~7, 7~28, 28~100, 100~289
ERA5-Land	ECMWF	1 h	0.1° × 0.1°	1950~present	0~7, 7~28, 28~100, 100~289

NASA: National Aeronautics and Space Administration, USA. NCEP: National Centers for Environmental Prediction, USA. ECMWF: the European Centre for Medium-range Weather Forecasts, European.

As the fifth-generation reanalysis data product of the ECMWF, ERA5 uses data assimilation to combine model data with global observations and provides hourly values for a large number of atmospheric, ocean–wave and land–surface parameters at a spatial resolution of $0.25° \times 0.25°$ and a time range from 1959 to the present [59]. Compared with ERA5, ERA5-land provides a consistent view of the land variables over several decades at an enhanced resolution (i.e., $0.1° \times 0.1°$) from 1950 to the present and is forced by ERA5 atmospheric parameters with lapse rate correction [60–62].

2.2.2. Observational Data

In accordance with the ground meteorological observation standard of the China Meteorological Administration (CMA), meteorological stations can provide soil temperatures at depths of 0, 5, 10, 15, 20, 40, 80, 160 and 320 cm [63]. To better match the soil layer of the reanalysis products (Table 1) with the observation data, this study focused on the shallow soil layer corresponding to depths of 0~10 cm for GLDAS-Noah and NCEP-R2 products and 0~7 cm for ERA5-land and ERA5 products. A set of observed soil temperature (at depths of 0, 5 and 10 cm) collected at 84 meteorological stations within and around the QTP were used to calculate the mean shallow soil temperature at depths of 0~10 cm (Figure 1). For all 84 stations, the average elevation was 2866 m, ranging from 1422 m to 4800 m. Time series of daily air and soil temperature observations within and around the QTP were obtained from the CMA; these were measured at 02:00, 08:00, 14:00 and 20:00 Beijing time (18:00, 00:00, 06:00 and 12:00 UTC) at a height of 1.5 m above the ground and depths of 0, 5 and 10 cm, respectively. Time series of daily air temperature from the 4 reanalysis datasets corresponded to a height of 2.0 m above the ground.

2.3. *Methods*

We used ΔT as a metric for evaluating land–atmosphere interactions. To evaluate the performances of the four reanalysis datasets in estimating ΔT, the annual and seasonal ΔT values calculated from in situ observations within and around the QTP were compared with those calculated from the four reanalysis datasets using three evaluation indexes: the root mean square error (RMSE), mean absolute error (MAE) and mean relative error (MRE). The soil and air temperature data were extracted from the 4 products using ArcGIS software for the grid cells, corresponding to the longitude and latitude of each of the considered 84 stations.

Inevitably, the high altitude and harsh natural environment of the QTP have led to a lack of soil temperature observation data. In this study, the mean annual and seasonal soil temperatures at depths of 0, 5 and 10 cm were calculated for each station only when all monthly values were available in a year, and the values for each month were computed when daily values were available for more than three quarters of a month. The data were also divided into four seasons as follows: spring, March to May; summer, June to August; autumn, September to November; and winter, December to February of the next year. The mean shallow soil temperature was the arithmetic means of soil temperatures at depths of 0, 5 and 10 cm.

It is noteworthy that the soil temperature data provided by the ERA5-land and ERA5 products were obtained at depths of 0~7 cm, which was used to correspond with observed soil temperatures collected at depths of 0~10 cm [50]. During the baseline period (i.e., 1981–2010), the comparative analysis on the performance of the four reanalysis datasets in estimating annual and seasonal ΔT were performed. After comparison, we selected the reanalysis data product with relatively better performance in estimating ΔT and used this dataset in further in-depth investigations of the spatiotemporal variations in annual and seasonal ΔT over the QTP. Long-term trends of the variations in ΔT at the grid levels were assessed using the modified Mann–Kendall test and the Sen's slope estimator method [64,65], and the trend of ΔT across the entire QTP was estimated using a time series of anomalies (with respect to the mean ΔT from 1981 to 2010).

3. Results

3.1. Performance of the Four Reanalysis Datasets in Estimating ΔT

For the estimation of the annual ΔT, the ERA5-land and ERA5 products yielded the lowest RMSE value of 1.57 °C, followed by the GLDAS-Noah product (3.30 °C), while the NCEP-R2 product yielded the largest RMSE value of 9.62 °C. Similarly, the ERA5-land and ERA5 products yielded the smallest MAE and MRE values, whereas the NCEP-R2 product yielded the largest MAE and MRE values (Table 2).

Table 2. Error statistics of the ΔT values estimated using the 4 reanalysis data products against those calculated from the observation data of 84 stations across the QTP from 1981 to 2010.

	ERA5			ERA5-Land			GLDAS-Noah			NCEP-R2		
	RMSE	MAE	MRE	RMSE	MAE	MRE	RMSE	MAE	MRE	RMSE	MAE	MRE
Spring	1.85	1.76	0.39	1.94	1.86	0.41	2.29	2.22	0.49	12.31	11.61	3.27
Summer	1.33	1.22	0.25	1.32	1.22	0.24	1.74	1.64	0.30	6.30	6.16	1.51
Autumn	2.30	2.22	0.62	2.15	2.07	0.58	1.70	1.62	0.38	10.56	9.94	3.40
Winter	2.59	2.44	1.87	2.38	2.23	1.66	1.49	1.35	1.52	12.21	11.64	18.53
Annual	1.57	1.47	0.41	1.57	1.47	0.41	3.30	2.85	0.58	9.62	8.94	2.96

The ΔT values estimated using the ERA5-land and ERA5 products represent the difference between the soil temperature at depths of 0~7 cm and the air temperature at a height of 2 m above the ground. The ΔT values estimated using the GLDAS-Noah and NCEP-R2 products represent the difference between the soil temperature at depths of 0~10 cm and the air temperature at a height of 2 m.

For the estimation of the seasonal ΔT, the ERA5-land and ERA5 products performed best in spring and summer ΔT, with slight differences in the RMSE, MAE and MRE. For the autumn and winter ΔT, the GLDAS-Noah product yielded the smallest RMSE, MAE and MRE values, followed by the ERA5-land product, while NCEP-R2 product yielded the largest RMSE, MAE and MRE.

Overall, the ERA5-land and ERA5 products were superior to the other products in terms of estimating the annual, spring and summer ΔT. The GLDAS-Noah product performed best in terms of estimating the autumn and winter ΔT, followed by the ERA5-land product. Given the high spatial and time resolution of the ERA5-land product (Table 1), the ERA5-land reanalysis data product was selected for further investigation of the detailed spatiotemporal variations in the annual and seasonal ΔT over the QTP.

3.2. Spatial Distribution of ΔT

The annual and seasonal spatial distributions of the mean ΔT varied widely over the QTP during the 30-year baseline period (1981–2010) (Figure 2). The mean annual ΔT across the QTP was 4.13 °C, with a range from −8.08 °C to 11.15 °C. Positive values indicated a relatively high annual shallow soil temperature compared with the corresponding air temperature. Areas with positive ΔT values accounted for the majority of the QTP, whereas areas with negative values were found only in the southeastern and southern margins of the QTP. The areas with a mean annual ΔT exceeding 6 °C were mainly concentrated in the southeastern QTP, whereas the areas with a mean annual ΔT of 0–3 °C were mainly concentrated in the northern edge of the plateau and around the Qaidam Basin, with sporadic distribution in the western QTP.

The mean ΔT of the entire QTP was largest in winter (5.66 °C), followed by autumn (4.46 °C) and spring (3.25 °C), and smallest in summer (3.13 °C). Compared with the other three seasons, the total area with a negative value of ΔT was larger in spring and distributed mainly along the edge of the QTP, especially in the northwestern and southeastern margins. The eastern part of the plateau had a low mean ΔT in summer, which increased gradually in autumn and reached a maximum in winter (Figure 2).

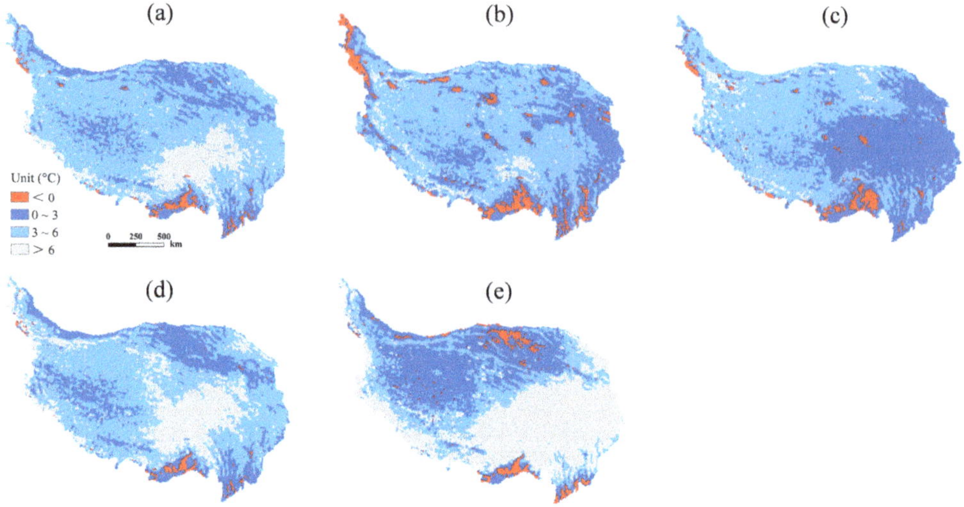

Figure 2. Geographic distribution of the mean ΔT on the QTP during the baseline period (i.e., 1981–2010); annual (**a**), spring (**b**), summer (**c**), autumn (**d**) and winter (**e**).

3.3. Changes in ΔT

Across the QTP, a significant decreasing trend in anomalies of the annual ΔT ($p < 0.01$) was observed from 1950 to 2021 at a rate of -0.07 °C/decade (Figure 3). Similarly, significant decreasing trends in ΔT ($p < 0.01$) were detected seasonally, with the largest rate of decrease in winter (-0.14 °C/decade), followed by those in autumn (-0.11 °C/decade) and summer (-0.03 °C/decade). Notably, the spring ΔT exhibited a nonsignificant increasing trend ($p = 0.43$) but no evident decreasing trend.

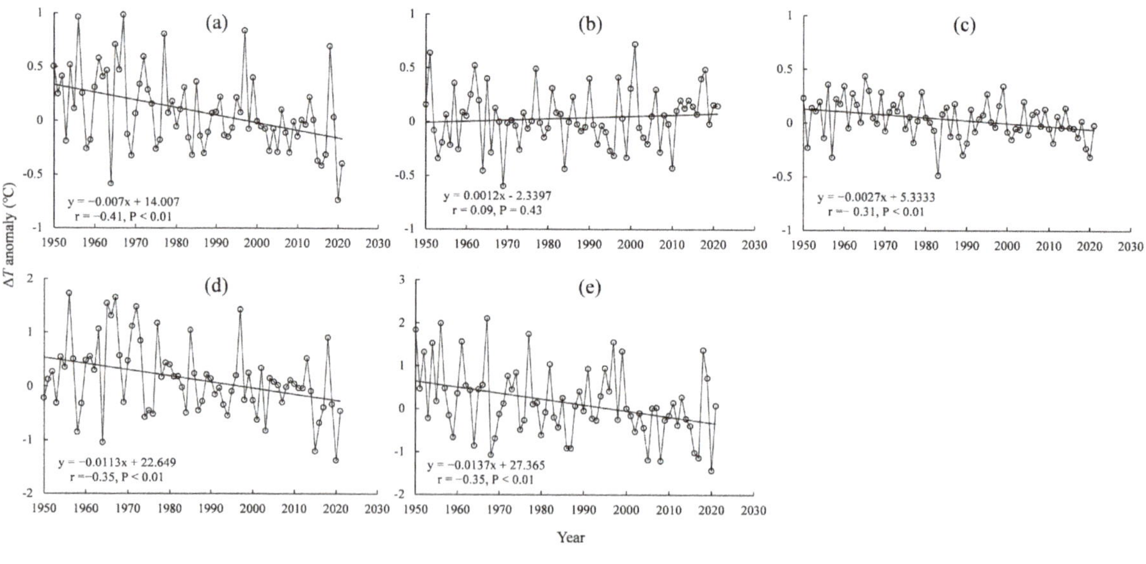

Figure 3. Changes in the mean ΔT on the QTP based on a time series of anomalies (with respect to the mean of the 30-year baseline period, i.e., 1981–2010) from 1950 to 2021; annual (**a**), spring (**b**), summer (**c**), autumn (**d**) and winter (**e**).

Compared with the mean ΔT from 1950 to 1985, the mean annual and autumn ΔT from 1986 to 2021 has decreased widely across the QTP (Figure 4). The areas with increased ΔT exceeded 50% of the entire QTP in spring, but less than 30% in summer. In winter, the magnitude of the decrease in ΔT (<-1 °C) was large in the southeastern part of the plateau. From 1950 to 2021, a large area with decreasing trends in the annual ΔT was detected, with the largest rates of decrease (e.g., -0.2 to -0.1 °C/decade, and -0.3 to -0.2 °C/decade) mainly concentrated in the southern and northeastern parts of the QTP. Spatial trends in ΔT were observed by season, with most areas showing an increasing trend in spring, especially in the central and northern parts of the plateau. During summer, autumn and winter, the proportion of the QTP with a decreasing trend in ΔT was larger than the proportion with an increasing trend. From summer, through autumn, to winter, the area with a significant decreasing trend in ΔT gradually increased in size. The largest decrease in ΔT was detected in winter, when the decreased rate of ΔT was more than 0.3 °C/decade in the southeastern part of the QTP (Figure 5).

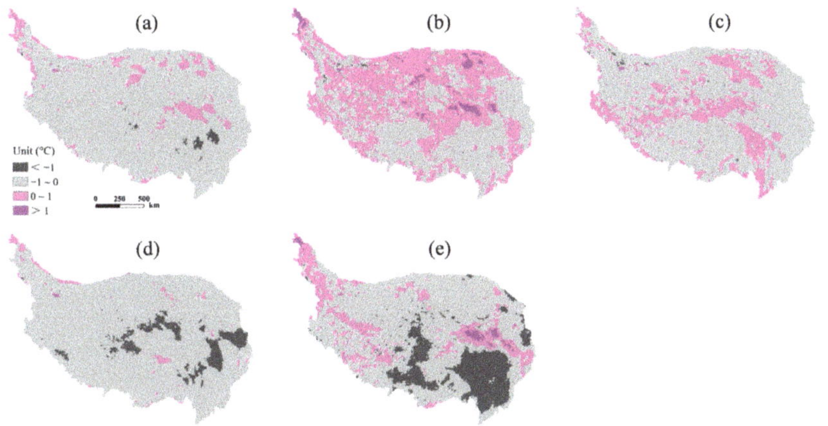

Figure 4. Spatial distributions of the difference in mean ΔT between 1986–2021 and 1950–1985 on the QTP; annual (**a**), spring (**b**), summer (**c**), autumn (**d**) and winter (**e**).

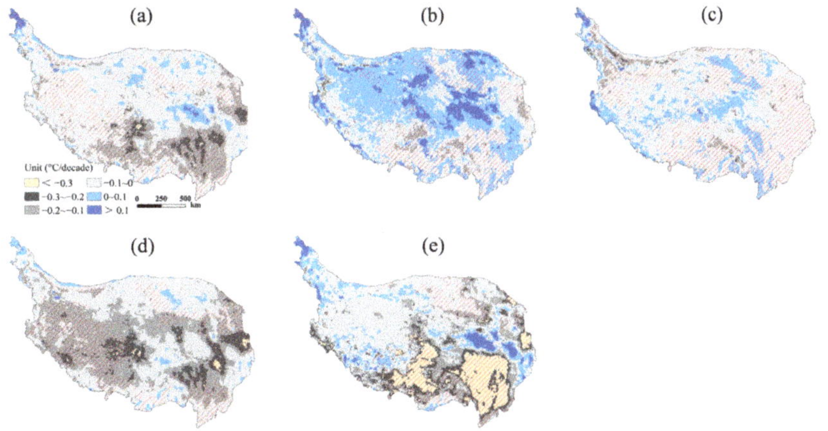

Figure 5. Geographic distribution of the rates of change in ΔT annually (**a**) and in spring (**b**), summer (**c**), autumn (**d**) and winter (**e**) on the QTP from 1950 to 2021. The oblique lines indicate significant changes in the trends at the 95% confidence level.

4. Discussion

Employing a simple metric, ΔT, this study aimed to assess the land–atmosphere heat exchange regime over the QTP using reanalysis data products. Similar studies have been conducted to investigate spatiotemporal variations in ΔT at the regional, national and hemispheric scales over the past few decades [9,10,22–36]. As ecological, hydrological and biological activities mainly occur in the near-surface soil layer, this study focused on the annual and seasonal spatiotemporal variations in ΔT between the shallow soil and air over the QTP.

From 1950 to 2021, both air and soil temperatures have exhibited warming trends on the QTP, especially in winter (Table 3). The asynchronism of air temperature and soil temperature variations could lead to the differences in the changes of the mean ΔT on the QTP both annually and seasonally (Figure 3). For example, a relatively large difference of increase in soil temperature and air temperature in winter could result in a relatively large decrease in winter ΔT during the study period (Table 3). Here, we investigated the possible influence of snow cover and soil moisture on ΔT during the study period, and precipitation is not considered, because it mainly affects the soil thermal state via soil moisture feedback in the warm season [22,28] and in the form of snow cover in the cold season. Results showed that snow depth was closely associated with the ΔT dynamics during the study period, except in summer (Table 3). Although snow only covers the ground during cold seasons, snow cover could modulate the land–atmosphere relationship, owing to its high albedo, low thermal conductivity and latent heat of phase changes [66]. Wang et al. (2017) reported a close association of ΔT with winter snow depth across China and also pointed out that such relationship might be complex and nonlinear [32]. Soil moisture conditions control the energy–water balance between the land surface and atmosphere, which can affect the surface albedo and heat capacity and then regulate the local net radiation flux and heat exchange between the land and atmosphere [3]. In this study, soil moisture was significantly and negatively correlated with summer ΔT during the study period (Table 3), which may be due to the soil moisture feedback. Relatively wet soil during summer may increase the energy consumption for evaporation and eventually cool the soil. During the soil freeze phase, relatively wet soil will have more liquid water freeze into ice compared with the soil with relatively low water content. This phase change leads to a greater release of latent heat and then slows down the cooling of the soil. This can partially explain why soil moisture was significantly and positively correlated with ΔT during autumn and winter, especially in winter (Table 3). In reality, snow cover and soil moisture are just two possible elements that influence the variation of ΔT. Some other regional variables, such as vegetation, air pollution and land albedo, could also affect the ΔT variation. Future research should, therefore, pay more attention to the investigations on the links of multiple environmental variables with ΔT variation.

Table 3. Changes in the mean shallow soil temperature and air temperature and correlation between the ΔT with snow depth and soil moisture across the QTP from 1950 to 2021.

	Change Rate (°C/Decade)		Correlation Coefficient	
	Soil Temperature	**Air Temperature**	**Snow Depth**	**Soil Moisture**
Spring	0.07	0.05	**0.25**	0.12
Summer	0.05	**0.08**	-	**−0.32**
Autumn	**0.08**	**0.18**	**0.84**	**0.37**
Winter	**0.14**	**0.26**	**0.86**	**0.46**
Annual	**0.09**	**0.17**	**0.61**	**0.25**

The soil temperature (0~7 cm), air temperature (2 m above ground), snow depth and soil moisture (0~7 cm) data were obtained from the ERA5-land products. Bold values indicate statistical significance of $p < 0.05$.

Great attention had been paid to investigate the potential causes of errors for the reanalysis products, and soil properties, input parameters, underlying surface (such as vegetation, snow cover), model structures and geographical conditions (such as altitude, aspect, slope) were thought to partially affect the simulation accuracy of air and soil temperatures [50,53,54,67–72]. In this study, the accuracy of the 4 tested reanalysis data products was compared with the observation data of 84 meteorological stations in terms of estimating ΔT and was supported by metrics, namely the MAE, RMSE and MRE. The distribution of these meteorological stations is very heterogeneous, with almost no stations in the central and western parts of the QTP (Figure 1), and this undoubtedly affected the accuracy of the validation of the estimated ΔT values when using the four reanalysis products. Therefore, more station observations and field survey data should be included in future studies.

In a previous study, the ERA5 product was found to yield the most accurate simulation of soil temperature at depths of 0~7 cm over the QTP among four tested reanalysis products (GLDAS-Noah, ERA5-land, the Climate Forecast System Reanalysis version 2 and the ECMWF interim reanalysis) [50]. Another study recommended the use of ERA5 and GLDAS-2.1 to represent air temperatures over the QTP [54]. As the ERA5-land product yielded better annual and seasonal ΔT values in our study, and its finer spatial resolution could provide more detailed spatial information about extreme events, we selected the ERA5-land reanalysis data product for our investigation of spatial and temporal changes in the annual and seasonal values of ΔT over the QTP. It is important to note, however, that the ERA5-land product was not superior to the GLDAS-Noah product in estimating the autumn and winter ΔT (Table 2). To obtain more reasonable estimations of the annual and seasonal ΔT on the QTP, future studies should consider including multiple, downscaled and remote sensing datasets with high spatial and temporal resolutions.

5. Conclusions

This study evaluated the performance of four widely used reanalysis data products, namely GLDAS-Noah, NCEP-R2, ERA5 and ERA5-land, in estimating ΔT over the QTP. The ERA5-land product provided superior estimations of ΔT both annually and seasonally, and its fine spatial resolution enabled it to provide more detailed spatial information on extreme events. Using the soil temperature at depths of 0~7 cm and air temperature at a height of 2 m above the ground, which were provided by the ERA5-Land reanalysis data product, the spatiotemporal variations in the annual and seasonal ΔT over the QTP were assessed in detail. Positive values of ΔT dominated the entire QTP both annually and seasonally during the baseline period (1981–2010), with negative annual values of ΔT only in the southeastern and southern margins of the QTP. The spatial distribution of the seasonal ΔT varied greatly, and a large area of the QTP had a negative value of ΔT in spring. In the eastern QTP, the mean ΔT was relatively small in summer, increased basically from spring to autumn and reached a maximum in winter. From 1950 to 2021, the QTP experienced significant decreasing trends in both annual and seasonal ΔT, except in spring when a nonsignificant increasing trend was observed. Spatially, the areas with higher rates of decrease in the annual ΔT were concentrated mainly in the southern and northeastern parts of the QTP, and the areas with a significant decreasing trend in ΔT gradually increased in size from summer, through autumn, to winter.

This study has some limitations. The heterogeneous distribution of the observation network on the QTP, with sparse coverage of some areas, may have reduced the accuracy of validation of the ΔT values estimated using the four reanalysis data products, and future studies should include additional station observations and field survey data. To obtain more accurate estimations of the annual and seasonal ΔT on the QTP, future studies should include multiple, high-resolution datasets that combine more station observations and field survey data.

Author Contributions: Conceptualization, X.W. and R.C.; methodology, X.W. and R.C.; validation, R.C.; formal analysis, X.W. and R.C.; data curation, X.W.; writing—original draft preparation, X.W.; funding acquisition, X.W. and R.C. All authors have read and agreed to the published version of the manuscript.

Funding: This work was supported by the National Natural Science Foundation of China (42171145 and 41901084), the National Key Research and Development Program of China (2019YFC1510505), Joint Research Project of Three-River Headwaters National Park, Chinese Academy of Sciences and The People's Government of Qinghai Province (LHZX-2020-11), and the CAS "Light of West China" Program.

Data Availability Statement: The ERA5-Land reanalysis dataset used in this study can be accessed online (https://cds.climate.copernicus.eu/cdsapp#!/dataset/reanalysis-era5-land?tab=overview (accessed on 10 October 2022)). The ERA5 reanalysis dataset used in this study can be accessed online (https://cds.climate.copernicus.eu/cdsapp#!/dataset/reanalysis-era5-single-levels?tab=overview (accessed on 10 October 2022)). The soil temperature and air temperature data for GLDAS Noah Land Surface Model L4 monthly 0.25 × 0.25 degree V2.0 can be accessed online (https://disc.gsfc.nasa.gov/datasets/GLDAS_NOAH025_3H_2.0/summary (accessed on 15 October 2022)). The soil temperature and air temperature data for NCEP-DOE Reanalysis 2 can be accessed online (https://psl.noaa.gov/data/gridded/data.ncep.reanalysis2.html (accessed on 19 October 2022)). The Daily Meteorological Dataset of basic meteorological elements of China National Surface Weather Stations provided by the China Meteorological Administration can be accessed online (http://data.cma.cn (accessed on 15 March 2022)).

Acknowledgments: The authors would like to thank the European Centre for Medium-range Weather Forecasts (ECMWF) (https://cds.climate.copernicus.eu/, (accessed on 10 October 2022)) for providing the ERA5-Land and ERA5 reanalysis datasets, National Aeronautics and Space Administration (NASA) and National Centers for Environmental Prediction (NCEP) (https://disc.gsfc.nasa.gov/, (accessed on 15 October 2022)) for providing the GLDAS Noah Land Surface Model L4 monthly 0.25 × 0.25 degree V2.0 dataset, NCEP (https://psl.noaa.gov/, (accessed on 19 October 2022)) for providing the NCEP-DOE Reanalysis 2 dataset, and China Meteorological Administration (http://data.cma.cn, (accessed on 15 March 2022)) for providing the Daily Meteorological Dataset of basic meteorological elements of China National Surface Weather Stations. The authors also would like to thank all team members at the Qilian Alpine Ecology and Hydrology Research Station, Northwest Institute of Eco-Environment and Resources, Chinese Academy of Sciences.

Conflicts of Interest: The authors declare no conflict of interest.

References

1. Council, N.R. *Assessment of Intraseasonal to Interannual Climate Prediction and Predictability*; The National Academies Press: Washington, DC, USA, 2010; p. 192.
2. Fischer, E.M.; Seneviratne, S.I.; Lüthi, D.; Schär, C. Contribution of land-atmosphere coupling to recent European summer heat waves. *Geophys. Res. Lett.* **2007**, *34*. [CrossRef]
3. Seneviratne, S.I.; Corti, T.; Davin, E.L.; Hirschi, M.; Jaeger, E.B.; Lehner, I.; Orlowsky, B.; Teuling, A.J. Investigating soil moisture–climate interactions in a changing climate: A review. *Earth-Sci. Rev.* **2010**, *99*, 125–161. [CrossRef]
4. Zhang, J.; Wang, W.-C.; Leung, L.R. Contribution of land-atmosphere coupling to summer climate variability over the contiguous United States. *J. Geophys. Res. Atmos.* **2008**, *113*, 11. [CrossRef]
5. Lorenz, R.; Argüeso, D.; Donat, M.G.; Pitman, A.J.; van den Hurk, B.; Berg, A.; Lawrence, D.M.; Chéruy, F.; Ducharne, A.; Hagemann, S.; et al. Influence of land-atmosphere feedbacks on temperature and precipitation extremes in the GLACE-CMIP5 ensemble. *J. Geophys. Res. Atmos.* **2016**, *121*, 607–623. [CrossRef]
6. Vogel, M.M.; Orth, R.; Cheruy, F.; Hagemann, S.; Lorenz, R.; van den Hurk, B.J.J.M.; Seneviratne, S.I. Regional amplification of projected changes in extreme temperatures strongly controlled by soil moisture-temperature feedbacks. *Geophys. Res. Lett.* **2017**, *44*, 1511–1519. [CrossRef]
7. García-García, A.; Cuesta-Valero, F.J.; Beltrami, H.; González-Rouco, F.; García-Bustamante, E.; Finnis, J. Land surface model influence on the simulated climatologies of temperature and precipitation extremes in the WRF v3.9 model over North America. *Geosci. Model Dev.* **2020**, *13*, 5345–5366. [CrossRef]
8. Koster, R.D.; Dirmeyer, P.A.; Guo, Z.; Bonan, G.; Chan, E.; Cox, P.; Gordon, C.T.; Kanae, S.; Kowalczyk, E.; Lawrence, D.; et al. Regions of strong coupling between soil moisture and precipitation. *Science* **2004**, *305*, 1138–1140. [CrossRef] [PubMed]
9. Feng, H.; Zou, B. A greening world enhances the surface-air temperature difference. *Sci. Total Environ.* **2019**, *658*, 385–394. [CrossRef] [PubMed]
10. García-García, A.; Cuesta-Valero, F.J.; Beltrami, H.; Smerdon, J.E. Characterization of air and ground temperature relationships within the CMIP5 historical and future climate simulations. *J. Geophys. Res. Atmos.* **2019**, *124*, 3903–3929. [CrossRef]
11. Wang, X.; Chen, D.; Pang, G.; Ou, T.; Yang, M.; Wang, M. A climatology of surface–air temperature difference over the Tibetan Plateau: Results from multi-source reanalyses. *Int. J. Climatol.* **2020**, *40*, 6080–6094. [CrossRef]

12. McBean, G. Arctic Climate: Past and Present (Chapter 2). In *Arctic Climate Impact Assessment*; Symon, C., Arris, L., Heal, B., Eds.; ACIA Scientific Report; Cambridge University Press: Cambridge, UK, 2005; pp. 21–60.
13. Pepin, N.; Bradley, R.S.; Diaz, H.F.; Baraer, M.; Caceres, E.B.; Forsythe, N.; Fowler, H.; Greenwood, G.; Hashmi, M.Z.; Liu, X.D.; et al. Elevation-dependent warming in mountain regions of the world. *Nat. Clim. Chang.* **2015**, *5*, 424–430. [CrossRef]
14. Nelson, F.E.; Anisimov, O.A.; Shiklomanov, N.I. Subsidence risk from thawing permafrost. *Nature* **2001**, *410*, 889–890. [CrossRef] [PubMed]
15. Walvoord, M.A.; Kurylyk, B.L. Hydrologic impacts of thawing permafrost—A review. *Vadose Zone J.* **2016**, *15*. [CrossRef]
16. Turetsky, M.R.; Abbott, B.W.; Jones, M.C.; Anthony, K.W.; Olefeldt, D.; Schuur, E.A.G.; Grosse, G.; Kuhry, P.; Hugelius, G.; Koven, C.; et al. Carbon release through abrupt permafrost thaw. *Nat. Geosci.* **2020**, *13*, 138–143. [CrossRef]
17. Ganjurjav, H.; Gao, Q.; Gornish, E.S.; Schwartz, M.W.; Liang, Y.; Cao, X.; Zhang, W.; Zhang, Y.; Li, W.; Wan, Y. Differential response of alpine steppe and alpine meadow to climate warming in the central Qinghai–Tibetan Plateau. *Agric. For. Meteorol.* **2016**, *223*, 233–240. [CrossRef]
18. Aalto, J.; Scherrer, D.; Lenoir, J.; Guisan, A.; Luoto, M. Biogeophysical controls on soil-atmosphere thermal differences: Implications on warming Arctic ecosystems. *Environ. Res. Lett.* **2018**, *13*, 074003. [CrossRef]
19. Wu, Q.; Zhang, Z.; Gao, S.; Ma, W. Thermal impacts of engineering activities and vegetation layer on permafrost in different alpine ecosystems of the Qinghai-Tibet Plateau, China. *Cryosphere* **2016**, *10*, 1695–1706. [CrossRef]
20. Ma, Y.; Yao, T.; Zhong, L.; Wang, B.; Xu, X.; Hu, Z.; Ma, W.; Sun, F.; Han, C.; Li, M.; et al. Comprehensive study of energy and water exchange over the Tibetan Plateau: A review and perspective: From GAME/Tibet and CAMP/Tibet to TORP, TPEORP, and TPEITORP. *Earth-Sci. Rev.* **2023**, *237*, 104312. [CrossRef]
21. Carruthers, D.J.; Stull, R.B. *An Introduction to Boundary Layer Meteorology*; Atmospheric Sciences Library: Dordrecht, The Netherlands; Kluwer: Alphen aan den Rijn, The Netherlands, 1988; Volume 13, p. 670.
22. Zhang, T.; Barry, R.G.; Gilichinsky, D.; Bykhovets, S.S.; Sorokovikov, V.A.; Ye, J. An amplified signal of climatic change in soil temperatures during the last century at Irkutsk, Russia. *Clim. Chang.* **2001**, *49*, 41–76. [CrossRef]
23. Beltrami, H.; Kellman, L. An examination of short- and long-term air–ground temperature coupling. *Glob. Planet. Chang.* **2003**, *38*, 291–303. [CrossRef]
24. Isard, S.A.; Schaetzl, R.J.; Andresen, J.A. Soils cool as climate warms in the Great Lakes region: 1951–2000. *Ann. Assoc. Am. Geogr.* **2007**, *97*, 467–476. [CrossRef]
25. Romanovsky, V.E.; Sazonova, T.S.; Balobaev, V.T.; Shender, N.I.; Sergueev, D.O. Past and recent changes in air and permafrost temperatures in eastern Siberia. *Glob. Planet. Chang.* **2007**, *56*, 399–413. [CrossRef]
26. Woodbury, A.D.; Bhuiyan, A.K.M.H.; Hanesiak, J.; Akinremi, O.O. Observations of northern latitude ground-surface and surface-air temperatures. *Geophys. Res. Lett.* **2009**, *36*. [CrossRef]
27. Lawrence, D.M.; Slater, A.G. The contribution of snow condition trends to future ground climate. *Clim. Dyn.* **2010**, *34*, 969–981. [CrossRef]
28. Qian, B.; Gregorich, E.G.; Gameda, S.; Hopkins, D.W.; Wang, X.L. Observed soil temperature trends associated with climate change in Canada. *J. Geophys. Res. Atmos.* **2011**, *116*. [CrossRef]
29. Park, H.; Sherstiukov, A.B.; Fedorov, A.N.; Polyakov, I.V.; Walsh, J.E. An observation-based assessment of the influences of air temperature and snow depth on soil temperature in Russia. *Environ. Res. Lett.* **2014**, *9*, 064026. [CrossRef]
30. Streletskiy, D.A.; Sherstiukov, A.B.; Frauenfeld, O.W.; Nelson, F.E. Changes in the 1963–2013 shallow ground thermal regime in Russian permafrost regions. *Environ. Res. Lett.* **2015**, *10*, 125005. [CrossRef]
31. Wang, W.; Rinke, A.; Moore, J.C.; Ji, D.; Cui, X.; Peng, S.; Lawrence, D.M.; McGuire, A.D.; Burke, E.J.; Chen, X.; et al. Evaluation of air–soil temperature relationships simulated by land surface models during winter across the permafrost region. *Cryosphere* **2016**, *10*, 1721–1737. [CrossRef]
32. Wang, Y.; Hu, Z.-Z.; Yan, F. Spatiotemporal variations of differences between surface air and ground temperatures in China. *J. Geophys. Res. Atmos.* **2017**, *122*, 7990–7999. [CrossRef]
33. Luo, D.; Jin, H.; Marchenko, S.S.; Romanovsky, V.E. Difference between near-surface air, land surface and ground surface temperatures and their influences on the frozen ground on the Qinghai-Tibet Plateau. *Geoderma* **2018**, *312*, 74–85. [CrossRef]
34. Shati, F.; Prakash, S.; Norouzi, H.; Blake, R. Assessment of differences between near-surface air and soil temperatures for reliable detection of high-latitude freeze and thaw states. *Cold Reg. Sci. Technol.* **2018**, *145*, 86–92. [CrossRef]
35. Zhang, Y.; Sherstiukov, A.B.; Qian, B.; Kokelj, S.V.; Lantz, T.C. Impacts of snow on soil temperature observed across the circumpolar north. *Environ. Res. Lett.* **2018**, *13*, 044012. [CrossRef]
36. Chen, L.; Aalto, J.; Luoto, M. Decadal changes in soil and atmosphere temperature differences linked with environment shifts over northern Eurasia. *J. Geophys. Res. Earth Surf.* **2021**, *126*. [CrossRef]
37. Wu, G.; Duan, A.; Liu, Y.; Mao, J.; Ren, R.; Bao, Q.; He, B.; Liu, B.; Hu, W. Tibetan Plateau climate dynamics: Recent research progress and outlook. *Natl. Sci. Rev.* **2015**, *2*, 100–116. [CrossRef]
38. Fu, Y.; Ma, Y.; Zhong, L.; Yang, Y.; Guo, X.; Wang, C.; Xu, X.; Yang, K.; Xu, X.; Liu, L.; et al. Land-surface processes and summer-cloud-precipitation characteristics in the Tibetan Plateau and their effects on downstream weather: A review and perspective. *Natl. Sci. Rev.* **2020**, *7*, 500–515. [CrossRef]
39. Duan, A.; Li, F.; Wang, M.; Wu, G. Persistent weakening trend in the spring sensible heat source over the Tibetan Plateau and its impact on the Asian summer monsoon. *J. Clim.* **2011**, *24*, 5671–5682. [CrossRef]

40. Ma, Y.M.; Hu, Z.Y.; Tian, L.D.; Zhang, F.; Yang, Y.P. Study process of the Tibet Plateau climate system change and mechanism of its impact on East Asia. *Adv. Earth Sci.* **2014**, *29*, 207–215.
41. Liu, Y.; Lu, M.; Yang, H.; Duan, A.; He, B.; Yang, S.; Wu, G. Land–atmosphere–ocean coupling associated with the Tibetan Plateau and its climate impacts. *Natl. Sci. Rev.* **2020**, *7*, 534–552. [CrossRef]
42. Yang, K.; Wu, H.; Qin, J.; Lin, C.; Tang, W.; Chen, Y. Recent climate changes over the Tibetan Plateau and their impacts on energy and water cycle: A review. *Glob. Planet. Chang.* **2014**, *112*, 79–91. [CrossRef]
43. You, Q.; Chen, D.; Wu, F.; Pepin, N.; Cai, Z.; Ahrens, B.; Jiang, Z.; Wu, Z.; Kang, S.; AghaKouchak, A. Elevation dependent warming over the Tibetan Plateau: Patterns, mechanisms and perspectives. *Earth-Sci. Rev.* **2020**, *210*, 103349. [CrossRef]
44. Fang, X.; Luo, S.; Lyu, S. Observed soil temperature trends associated with climate change in the Tibetan Plateau, 1960–2014. *Theor. Appl. Climatol.* **2019**, *135*, 169–181. [CrossRef]
45. Zhu, F.; Lan, C.; Zhang, Y.; Luo, J.J.; Lettenmaier, D.P.; Lin, Y.; Zhe, L. Spatiotemporal variations of annual shallow soil temperature on the Tibetan Plateau during 1983–2013. *Clim. Dyn.* **2018**, *51*, 2209–2227. [CrossRef]
46. Wang, X.; Chen, R.; Han, C.; Yang, Y.; Liu, J.; Liu, Z.; Guo, S.; Song, Y. Response of shallow soil temperature to climate change on the Qinghai–Tibetan Plateau. *Int. J. Climatol.* **2021**, *41*, 1–16. [CrossRef]
47. Gao, K.; Duan, A.; Chen, D.; Wu, G. Surface energy budget diagnosis reveals possible mechanism for the different warming rate among Earth's three poles in recent decades. *Sci. Bull.* **2019**, *64*, 1140–1143. [CrossRef] [PubMed]
48. Hinkelman, L.M. The global radiative energy budget in MERRA and MERRA-2: Evaluation with respect to CERES EBAF Data. *J. Clim.* **2019**, *32*, 1973–1994. [CrossRef]
49. Yang, J.; Huang, M.; Zhai, P. Performance of the CRA-40/Land, CMFD, and ERA-Interim datasets in reflecting changes in surface air temperature over the Tibetan Plateau. *J. Meteorol. Res.* **2021**, *35*, 663–672. [CrossRef]
50. Yang, S.; Li, R.; Wu, T.; Hu, G.; Xiao, Y.; Du, Y.; Zhu, X.; Ni, J.; Ma, J.; Zhang, Y.; et al. Evaluation of reanalysis soil temperature and soil moisture products in permafrost regions on the Qinghai-Tibetan Plateau. *Geoderma* **2020**, *377*, 114583. [CrossRef]
51. Qin, Y.; Zhang, P.; Liu, W.; Guo, Z.; Xue, S. The application of elevation corrected MERRA2 reanalysis ground surface temperature in a permafrost model on the Qinghai-Tibet Plateau. *Cold Reg. Sci. Technol.* **2020**, *175*, 103067. [CrossRef]
52. Xu, D.; Xin, L.A.I.; Guangzhou, F.A.N.; Jun, W.E.N.; Yuan, Y.; Xin, W.; Zuoliang, W.; Lihua, Z.H.U.; Yongli, Z.; Bingyun, W. Analysis on the applicability of reanalysis soil temperature and moisture datasets over Qinghai-Tibetan Plateau. *Plateau Meteorol.* **2018**, *37*, 626–641.
53. Hu, G.; Zhao, L.; Li, R.; Wu, X.; Wu, T.; Xie, C.; Zhu, X.; Su, Y. Variations in soil temperature from 1980 to 2015 in permafrost regions on the Qinghai-Tibetan Plateau based on observed and reanalysis products. *Geoderma* **2019**, *337*, 893–905. [CrossRef]
54. Liu, L.; Gu, H.; Xie, J.; Xu, Y.-P. How well do the ERA-Interim, ERA-5, GLDAS-2.1 and NCEP-R2 reanalysis datasets represent daily air temperature over the Tibetan Plateau? *Int. J. Climatol.* **2021**, *41*, 1484–1505. [CrossRef]
55. Zou, H.; Zhu, J.; Zhou, L.; Li, P.; Ma, S. Validation and application of reanalysis temperature data over the Tibetan Plateau. *J. Meteorol. Res.* **2014**, *28*, 139–149. [CrossRef]
56. Rodell, M.; Houser, P.R.; Jambor, U.; Gottschalck, J.; Mitchell, K.; Meng, C.J.; Arsenault, K.; Cosgrove, B.; Radakovich, J.; Bosilovich, M.; et al. The global land data assimilation system. *Bull. Am. Meteorol. Soc.* **2004**, *85*, 381–394. [CrossRef]
57. Rodell, M.; Beaudoing, H.K. *NASA/GSFC/HSL, GLDAS Noah Land Surface Model L4 Monthly 0.25 x 0.25 Degree V2.0*; Goddard Earth Sciences Data and Information Services Center (GES DISC): Greenbelt, MD, USA, 2019. [CrossRef]
58. Kanamitsu, M.; Ebisuzaki, W.; Woollen, J.; Yang, S.-K.; Hnilo, J.J.; Fiorino, M.; Potter, G.L. NCEP–DOE AMIP-II Reanalysis (R-2). *Bull. Am. Meteorol. Soc.* **2002**, *83*, 1631–1644. [CrossRef]
59. Hersbach, H.; Bell, B.; Berrisford, P.; Biavati, G.; Horányi, A.; Muñoz Sabater, J.; Nicolas, J.; Peubey, C.; Radu, R.; Rozum, I.; et al. ERA5 Hourly Data on Single Levels from 1959 to Present. Copernicus Climate Change Service (C3S) Climate Data Store (CDS). 2018. Available online: https://cds.climate.copernicus.eu/cdsapp#!/dataset/reanalysis-era5-single-levels?tab=overview (accessed on 10 October 2022).
60. Muñoz-Sabater, J.; Dutra, E.; Agustí-Panareda, A.; Albergel, C.; Arduini, G.; Balsamo, G.; Boussetta, S.; Choulga, M.; Harrigan, S.; Hersbach, H.; et al. ERA5-Land: A state-of-the-art global reanalysis dataset for land applications. *Earth Syst. Sci. Data* **2021**, *13*, 4349–4383. [CrossRef]
61. Muñoz Sabater, J. ERA5-Land Hourly Data from 1981 to Present. Copernicus Climate Change Service (C3S) Climate Data Store (CDS). 2018. Available online: https://cds.climate.copernicus.eu/cdsapp#!/dataset/reanalysis-era5-land?tab=overview (accessed on 10 October 2022).
62. Muñoz Sabater, J. ERA5-Land Hourly Data from 1950 to 1980. Copernicus Climate Change Service (C3S) Climate Data Store (CDS). 2021. Available online: https://cds.climate.copernicus.eu/cdsapp#!/dataset/reanalysis-era5-land?tab=overview (accessed on 10 October 2022).
63. China Meteorological Administration. *The Norm of Surface Meteorological Observation*; China Meteorological Press: Beijing, China, 2003; p. 151.
64. Sen, P.K. Estimates of the regression coefficient based on Kendall's Tau. *J. Am. Statal Assoc.* **1968**, *63*, 1379–1389. [CrossRef]
65. Hamed, K.H.; Rao, A.R. A modified Mann-Kendall trend test for autocorrelated data. *J. Hydrol.* **1998**, *204*, 182–196. [CrossRef]
66. Zhang, T. Influence of the seasonal snow cover on the ground thermal regime: An overview. *Rev. Geophys.* **2005**, *43*, RG4002. [CrossRef]

67. You, Q.; Kang, S.; Pepin, N.; Flügel, W.-A.; Yan, Y.; Behrawan, H.; Huang, J. Relationship between temperature trend magnitude, elevation and mean temperature in the Tibetan Plateau from homogenized surface stations and reanalysis data. *Glob. Planet. Chang.* **2010**, *71*, 124–133. [CrossRef]
68. Ding, L.; Zhou, J.; Zhang, X.; Liu, S.; Cao, R. Downscaling of surface air temperature over the Tibetan Plateau based on DEM. *Int. J. Appl. Earth Obs. Geoinf.* **2018**, *73*, 136–147. [CrossRef]
69. Hu, G.; Zhao, L.; Wu, X.; Li, R.; Wu, T.; Su, Y.; Hao, J. Evaluation of reanalysis air temperature products in permafrost regions on the Qinghai-Tibetan Plateau. *Theor. Appl. Climatol.* **2019**, *138*, 1457–1470. [CrossRef]
70. Peng, X.; Frauenfeld, O.W.; Jin, H.; Du, R.; Qiao, L.; Zhao, Y.; Mu, C.; Zhang, T. Assessment of Temperature Changes on the Tibetan Plateau During 1980–2018. *Earth Space Sci.* **2021**, *8*, e2020EA001609. [CrossRef]
71. Bi, H.; Ma, J.; Zheng, W.; Zeng, J. Comparison of soil moisture in GLDAS model simulations and in situ observations over the Tibetan Plateau. *J. Geophys. Res. Atmos.* **2016**, *121*, 2658–2678. [CrossRef]
72. Chen, Y.; Yang, K.; Qin, J.; Zhao, L.; Tang, W.; Han, M. Evaluation of AMSR-E retrievals and GLDAS simulations against observations of a soil moisture network on the central Tibetan Plateau. *J. Geophys. Res. Atmos.* **2013**, *118*, 4466–4475. [CrossRef]

remote sensing

Communication

Simulated Trends in Land Surface Sensible Heat Flux on the Tibetan Plateau in Recent Decades

Shuzhou Wang [1,2,*], Yaoming Ma [3,4,5,6,7,8] and Yuxin Liu [1,2]

1 Collaborative Innovation Center on Forecast and Evaluation of Meteorological Disasters/Nanjing University of Information Science and Technology, Nanjing 210044, China
2 School of Atmospheric Sciences, Nanjing University of Information Science and Technology, Nanjing 210044, China
3 Land-Atmosphere Interaction and Its Climatic Effects Group, State Key Laboratory of Tibetan Plateau Earth System, Resources and Environment (TPESRE), Institute of Tibetan Plateau Research, Chinese Academy of Sciences, Beijing 100101, China
4 College of Earth and Planetary Sciences, University of Chinese Academy of Sciences, Beijing 100049, China
5 College of Atmospheric Sciences, Lanzhou University, Lanzhou 730000, China
6 National Observation and Research Station for Qomolangma Special Atmospheric Processes and Environmental Changes, Dingri 858200, China
7 Kathmandu Center of Research and Education, Chinese Academy of Sciences, Beijing 100101, China
8 China-Pakistan Joint Research Center on Earth Sciences, Chinese Academy of Sciences, Islamabad 45320, Pakistan
* Correspondence: wangsz@nuist.edu.cn

Abstract: The spatial distribution and temporal variation of land surface sensible heat (SH) flux on the Tibetan Plateau (TP) for the period of 1981–2018 were studied using the simulation results from the Noah-MP land surface model. The simulated SH fluxes were also compared with the simulation results from the SEBS model and the results derived from 80 meteorological stations. It is found that, much larger annual mean SH fluxes occurred on the western and central TP compared with the eastern TP. Meanwhile, the inter-annual variations of SH fluxes on the central and western TP were larger than that on the eastern TP. The SEBS simulation showed much larger inter-annual variations than did the Noah-MP simulation across most of the TP. There was a trend of decrease in SH flux from the mid-1980s to the beginning of the 21st century in the Noah-MP simulations. Both Noah-MP and SEBS showed an increasing SH flux trend after this period of decrease. The increasing trend appeared on the eastern TP later than on the western and central TP. In the Noah-MP simulation, the western and central TP showed larger values of temperature difference between the ground surface and air (*Ts–Ta*) than did the eastern TP. Both mean *Ts–Ta* and wind speed decreased from the mid-1980s to approximately 2000, and then increased slightly. However, the *Ts–Ta* transition occurred later than that of wind speed. Changes in mean ground surface temperature (*Ts*) were the main cause of the decreasing and increasing trends in SH flux on the TP. Meanwhile, changes in wind speed contributed substantially to the decreasing trend in SH flux before 1998.

Keywords: Tibetan Plateau; sensible heat flux; Noah-MP; SEBS

Citation: Wang, S.; Ma, Y.; Liu, Y. Simulated Trends in Land Surface Sensible Heat Flux on the Tibetan Plateau in Recent Decades. *Remote Sens.* **2023**, *15*, 714. https://doi.org/10.3390/rs15030714

Academic Editor: Constantinos Cartalis

Received: 15 December 2022
Revised: 13 January 2023
Accepted: 18 January 2023
Published: 25 January 2023

1. Introduction

As the Earth's third pole, the Tibetan Plateau (TP) plays a critical role in influencing regional and global climate. Over the past few decades, the TP has experienced evident climate change, which has contributed to amplifying environmental changes on the global scale. The large heat source in the mid-troposphere provided by the TP is perceived as an important factor contributing to the formation and variation of the Asian summer monsoon [1–3]. Land surface sensible heat (SH) flux is a major component of this heat source, and plays a considerable role in modulating large scale atmospheric circulation and the summer monsoon precipitation patterns [4–9]. Hence, many studies have attempted to estimate the spatiotemporal changes in SH flux over the TP using meteorological data [10–13], reanalysis data [14–16], remote sensing products [17–19] or model simulations [20,21].

Duan and Wu [10] calculated the SH flux on the TP for the period 1980–2003 using data from meteorological stations and satellites. They found that the SH flux on the TP exhibited a large diurnal range, but much smaller annual range. The SH flux also showed a significant decreasing trend from the mid-1980s to the beginning of the 21st century, with a subdued surface wind speed contributing highly to the decreasing trend. Further, Duan et al. [11] reexamined the SH flux trend during 1980–2008, and confirmed the weakening trend. They found that the trend was induced mainly by a reduction in surface wind speed, despite a sharp increase in the ground-air temperature difference in 2004–2008. They also considered the trend to be a primary response to the spatial nonuniformity of large scale warming over the East Asian continent. Yang et al. [13] investigated the differences between different methods in estimating SH flux trends on the TP using meteorological station data for the period 1984–2006. Their results showed that different schemes produced different trends, and they claimed that the SH flux on the TP weakened by 2% per decade using their own newly developed method. They suggested that a decrease in wind speed and increase in ground–air temperature difference may moderate the trend of the heat transfer coefficient, which in turn may influence the SH flux trend. Wang and Ma [21] performed several simulation experiments on the TP using the Noah-MP land surface model. Their results showed that the SH flux is very sensitive to the thermal roughness length parameterization scheme, and also supported the SH flux weakening trend from the mid-1980s onwards.

However, the SH flux weakening trend did not continue during the last decade [22–25]. Zhang et al. [23] calculated the SH flux over the TP from 1970 to 2015 using meteorological station data, and analyzed the temporal characteristics of SH flux. They pointed out that there was an increasing trend in SH flux during the period 2001–2015. Chen et al. [25] investigated the spatiotemporal variability in SH over the TP from 1980 to 2015 using data from meteorological station and reanalysis products. They also confirmed that an increasing trend followed the earlier decreasing trend, and stated that the declines in SH prior to 2000 resulted from changes in wind speed, while the subsequent recovery can be attributed to increases in both wind speed and air–surface temperature gradient.

Most previous studies estimated the SH flux on the TP using reanalysis datasets or meteorological station data from the China Meteorological Administration (CMA), and bulk transfer algorithms. There are many uncertainties in estimating the SH flux on the TP using data from unevenly distributed meteorological stations and/or diverse reanalysis datasets. Hence, the decadal SH flux trends, especially when and how the trends changed, remain controversial. This study presents high resolution simulations of SH flux using the Noah-MP land surface model and SEBS (surface energy balance system) model, and compares these with the results from CMA meteorological stations. The climatic features of SH flux on the TP and possible causes of the annual variation trends are then analyzed. In the next section, observations and the experimental design will be described.

2. Materials and Methods

To evaluate the simulation results from the Noah-MP model, SH flux datasets collected from the eddy covariance (EC) systems at three observation stations were used. The QOMS station is situated at the bottom of the lower Rongbuk Valley, to the north of Mt. Everest; here, the EC system was installed at a height of 3.25 m above ground level. The Nam Co station is located on the southeast shore of the Nam Co Lake on the central TP; here, the EC system was installed at a height of 3.06 m above ground level. The Ali station is located within a flat and open mountain valley on the northwestern TP, where the Indian monsoon and westerly wind interact intensively; here, the EC system was installed at a height of 2.75 m above ground level. Figure 1 shows the locations of the stations (yellow dots), and Table 1 gives the observation periods and EC system information. The details of the stations and data have been introduced in previous studies [26,27]. All the datasets were quality-controlled using the TK3 software [28].

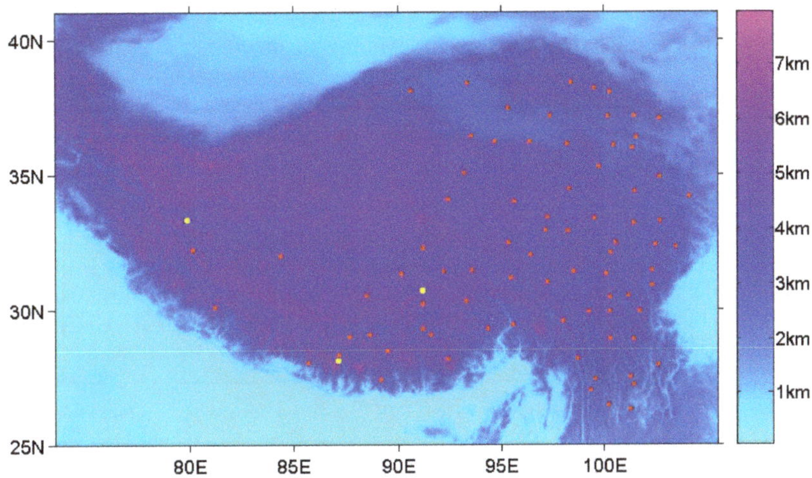

Figure 1. Locations of the three EC stations (yellow dots) and 80 CMA stations (red dots) on the TP.

Table 1. Details of the three EC systems and observation periods.

Station	Location	Land Use	Sonic Anemometer	Period (yr)
QOMS	28.21°N, 86.95°E	Bare soil	CSAT3	2011–2015
Nam Co	30.77°N, 90.98°E	Alpine steppe	CSAT3	2011–2015
Ali	33.39°N, 79.70°E	Desert	CSAT3	2011–2015

The Noah-MP land surface model was developed from the Noah land surface model (version 3.0) with multiple parameterization options [29,30]. This model is suitable for land surface process simulation over the TP according to previous studies [31,32]. In our simulation, the SIMGM runoff scheme, Noah β-factor scheme, and BATS snow surface albedo scheme were selected. Table 2 gives the schemes for the key processes.

Table 2. Parameterization schemes for the key processes in the Noah-MP model.

Process	Scheme
Runoff	SIMGM
β-factor	Noah
Snow surface albedo	BATS
Stomatal resistance	Ball-Berry
Frozen soil permeability	Koren99

The thermal roughness length scheme proposed by Zeng and Dickinson [33] was selected for our simulation experiment, because this scheme provided the best estimation of monthly mean SH flux in terms of squared correlation coefficients [21]. The simulation experiment was conducted at a spatial resolution of 0.1° and temporal resolution of 3 h using the CMFD (China meteorological forcing dataset) data [34] as the atmospheric forcing dataset for the TP area. The Noah-MP model was run continuously for the period from 1 January 1981 to 31 December 2018. The 38-yr simulation results were calculated and analyzed in this study.

We also used the 18-yr dataset produced by the SEBS model, which was developed by Su [35] and updated by Chen et al. [36] and Han et al. [37,38]. In the process of calculating the turbulent flux, the sub-grid scale topography drag parameterization scheme was introduced to improve the simulation of SH flux [39]. The simulation was conducted for the period 2001–2018 based on satellite remote sensing data (MODIS) and meteorological data (CMFD). The MODIS monthly land surface products, including land surface temperature and emissivity, land surface albedo, and vegetation index, provides the land surface conditions for the SEBS model. The SH flux was computed using the Monin-Obukhov similarity theory. The values of land surface variables in the MODIS monthly products were derived by compositing and averaging the values from the corresponding month of MODIS daily files. Detailed information on the MODIS land surface variables is listed in Table 3.

Table 3. Input datasets for the SEBS model. MOD11C3 is the short name of the MODIS/Terra land surface temperature and emissivity monthly L3 global products, MOD09CMG is the short name of the MODIS/Terra surface reflectance daily L3 global products, MOD13C2 is the short name of the MODIS/Terra vegetation indices' monthly L3 global products, GLAS stands for the Geoscience Laser Altimeter System, and SPOT stands for the Systeme pour l'Observation de la Terre sensor.

Variables	Data Source	Data Period	Temporal Resolution	Spatial Resolution
Downward shortwave	CMFD	2001–2018	3 h	0.1°
Downward longwave	CMFD	2001–2018	3 h	0.1°
Air temperature	CMFD	2001–2018	3 h	0.1°
Specific humidity	CMFD	2001–2018	3 h	0.1°
Wind speed	CMFD	2001–2018	3 h	0.1°
Land surface temperature	MOD11C3	2001–2018	Monthly	0.05°
Land surface emissivity	MOD11C3	2001–2018	Monthly	0.05°
Height of canopy	GLAS and SPOT	2001–2018	Monthly	0.01°
Albedo	MOD09CMG	2001–2018	Daily	0.05°
NDVI	MOD13C2	2001–2018	Monthly	0.05°

The simulated SH fluxes from the two models were also compared with the SH flux dataset estimated using routine meteorological data from CMA stations. This CMA station-based dataset was provided by Duan et al. [40], and was derived from 80 CMA stations (Figure 1) using a physical method developed by Yang et al. [41]. The daily mean heat flux is calculated by

$$\overline{SH} = \overline{\rho c_p C_H u(T_s - T_a)} \tag{1}$$

where ρ is air density, c_p is the specific heat at a constant pressure, u is the wind speed at a reference level, Ts is the ground surface temperature, Ta is the air temperature at a reference level, and C_H is the heat transfer coefficient. C_H is not a given constant value, and its value can be obtained by different parameterization schemes [42].

In this method, C_H is determined from micro-meteorological theory and experimental analysis. It exhibits clear diurnal variations, which significantly affect the estimation of SH flux [41]. This method (hereafter, the Yang method) produces higher SH fluxes than conventional empirical methods that are widely used in climatological studies [25,43,44].

To assess the simulations, we used statistical methods including the correlation coefficient, squared correlation coefficient (R^2), linear least-squares regression, and Mann–Kendall trend test.

3. Results

3.1. Assessment of the Simulation

Before analyzing the trends in simulated SH flux, a validation was performed using the in-situ observation data at the 3 stations. The linear regression method was applied here. Figure 2a shows the comparison between simulated and observed monthly mean SH fluxes at the QOMS station for 5 yrs (2011–2015). The blue dots represent the monthly mean values, and the blue line is the best-fit line. The squared correlation coefficient (R^2)

was 0.81, with a root-mean-square error (RMSE) of 8.27 Wm^{-2}. Overall, the simulated SH fluxes were higher than the corresponding observations. At the Nam Co station, there were many missing observations in the dataset during the observation period (2011–2015). Hence, the mean values of SH flux in 5 months were not obtained. The R^2 was 0.68, with a RMSE of 9.65 Wm^{-2}. The simulated mean SH fluxes fit well with observations during the equivalent period (Figure 2b). Figure 2c show the comparison between simulated and observed monthly mean SH fluxes at the Ali station for 5 yrs (2011–2015). As there were many missing values in the dataset, only 36 monthly mean values were obtained here. The R^2 was 0.72, and the RMSE was larger those at the other two stations.

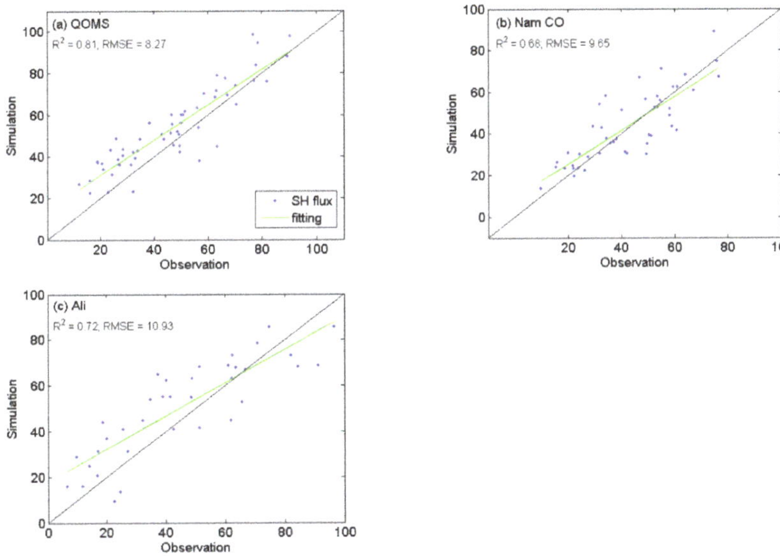

Figure 2. Comparison of simulated and observed monthly mean SH fluxes (Wm^{-2}) at the QOMS station (**a**), Nam CO station (**b**), and Ali station (**c**). The green line is the best-fit line.

Han et al. [38,39] previously evaluated the dataset produced by SEBS during the period 2007–2012. The SH flux was underestimated, with a mean bias of 4.7 Wm^{-2} at the QOMS station, and 7.8 Wm^{-2} at the Nam Co station. The correlation coefficients between the SEBS simulation and EC observations were 0.41 at the QOMS station, and 0.63 at the Nam Co station, respectively.

3.2. Simulated SH Flux and Its Trends

From the analysis performed in the section above, the simulations of SH flux were deemed to be suitable overall. The spatial distribution of annual mean SH flux (2001–2018) based on Noah-MP is shown in Figure 3a, and the corresponding SEBS-based annual mean SH flux is shown in Figure 3b. In Figure 3a, it can be seen that the eastern TP received much weaker mean SH fluxes than the central and western TP during the period 2001–2018. The areas with mean SH fluxes greater than 50 Wm^{-2} were mainly in the southwestern TP region. In the SEBS simulation (Figure 3b), most areas of the western TP received more than 60 Wm^{-2} SH fluxes. Overall, SEBS gave larger SH flux values in the northwestern TP region, but smaller SH flux values in the southeastern TP region than did Noah-MP.

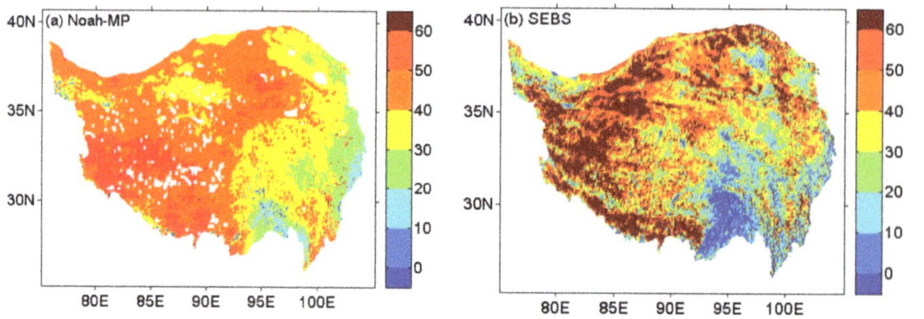

Figure 3. Annual mean SH fluxes (Wm^{-2}) simulated by (**a**) Noah-MP and (**b**) SEBS on the TP during the period 2001–2018.

Figure 4 shows the variations in the annual mean SH flux during the period 2001–2018. A large standard deviation (STD) value indicates large inter-annual variation in the SH flux, and vice versa. In the Noah-MP simulation (Figure 4a), large STD values were mainly observed over the central and western TP, while small STD values were mainly observed over areas to the east of 92°E. This means that the central and western TP recorded larger inter-annual variations in SH flux than did the eastern TP. The SEBS simulation (Figure 4b) showed much larger inter-annual variation in SH flux than did the Noah-MP simulation (Figure 4a) in most areas of the TP.

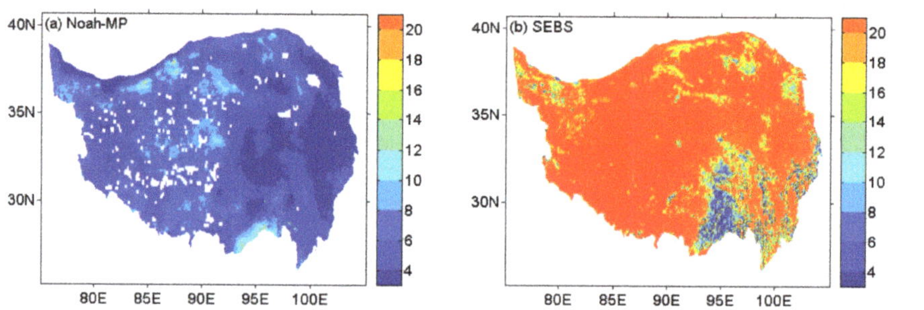

Figure 4. Standard deviation (STD) of annual mean SH fluxes (Wm^{-2}) simulated by (**a**) Noah-MP and (**b**) SEBS on the TP during the period 2001–2018.

We also analyzed the annual variation trends in the SH flux over the whole TP. We divided the TP into 3 climate zones: dry western TP (west of 85°E), transitional central TP (85°E–95°E), and wet eastern TP (east of 95°E). The weather stations established by the CMA are unevenly distributed on the TP, with most located in the eastern and central regions of the TP. Figure 5 shows the annual variations in SH flux from Noah-MP, SEBS, and CMA station data for the western, central, eastern and entire TP. In the Noah-MP simulation, the SH fluxes among the 3 sub-regions and over the entire TP all exhibited significant decreasing trends beginning in the mid-1980s, as was also reported in previous studies [11,13]. These decreasing trends were more obvious on the western and central TP than on the eastern TP. However, all regions showed an increasing trend from the beginning of the 21st century, with the eastern TP experiencing the increasing trend later than the western and central TP. Overall, the decreasing trend was approximately 0.31 W m^{-2} per yr, while the increasing trend was 0.64 W m^{-2} per yr. Both the decreasing and increasing trends were more significant than those obtained from CAM station-based results. In the SEBS simulation on the western TP (Figure 5a), the annual mean SH fluxes were much stronger than those in the Noah-MP simulation, and the trend was not significant during the period 2001–2018. There was a big difference between the two simulation results during the period 2001–2004. The SH flux from SEBS showed an upward trend beginning in 2001, while Noah-MP gave an increasing trend only after a decrease in 2002. On the central TP

(Figure 5b), both Noah-MP and SEBS gave an increasing trend in the annual mean SH flux. The CAM station-based SH fluxes were larger than those from the Noah-MP and SEBS simulations. On the eastern TP (Figure 5c), trends obtained from the two models were relatively consistent. Meanwhile, the annual mean SH fluxes from SEBS became weaker than those from Noah-MP from 2008 onwards. However, both Noah-MP and SEBS gave smaller values of SH flux compared with the CAM station-based results. Moreover, the CAM station-based SH fluxes began to increase earlier than those of Noah-MP and SEBS. Figure 5d shows the variations in annual mean TP-averaged SH fluxes from Noah-MP, SEBS, and CMA station data. The changes in CAM station-based SH flux were similar to the simulated results on the eastern TP (Figure 5c), because 54 of the 80 stations were installed on the eastern TP. Overall, simulated TP-averaged SH fluxes began to increase later than the CAM station-based results. Chen et al. [25] also reported similar results, showing that the reversal of trends derived from reanalysis products (ER-Interim and NCEP) happened later than for those derived from CAM station data.

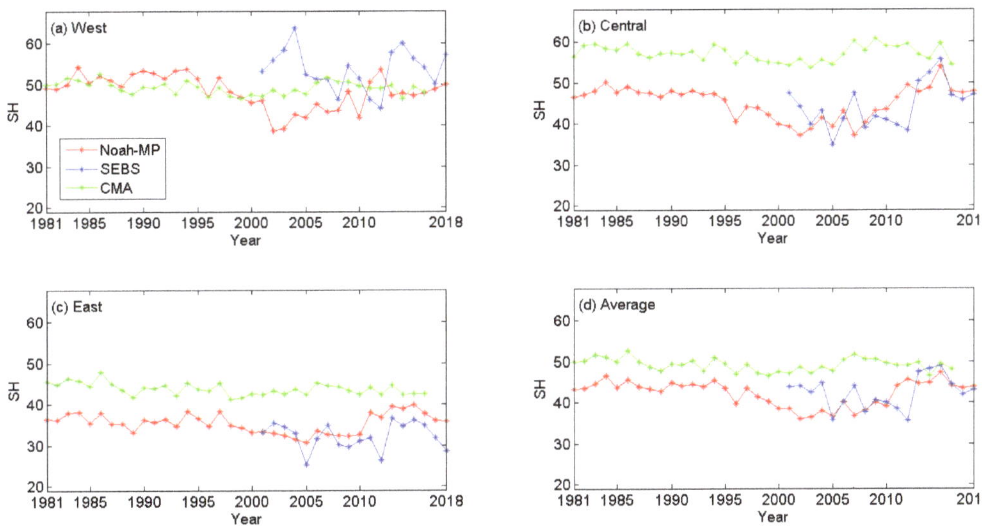

Figure 5. Variations in annual mean SH flux (Wm^{-2}) from Noah-MP, SEBS, and CMA data for the western (**a**), central (**b**), eastern (**c**), and entire TP (**d**) during the period 1981–2018.

3.3. Causes of the Trends

The bulk transfer coefficient method is typically used to calculate the surface SH flux. The surface wind speed, temperature difference between the ground surface and air (*Ts–Ta*), and heat transfer coefficient (C_H) are key determinants of SH flux. Hence, *Ts–Ta*, wind speed, and C_H were analyzed to determine the main causes of the increasing and decreasing trends in the Noah-MP simulations.

Figure 6 shows the variations in annual mean *Ts–Ta* in the Noah-MP simulations on the TP during the period 1981–2018. There was an obvious decreasing trend from the mid-1980s to the beginning of the 21st century on the western TP (Figure 6a). Then, the trendline turned sharply upwards. This phenomenon also existed on the central TP (Figure 6b). Figure 6c shows the annual mean values of *Ts–Ta* and the corresponding trends on the eastern TP. The values of *Ts–Ta* were smaller than those of the western and central TP. Meanwhile, both the decreasing and increasing trends were not as significant as those of the western and central TP. The TP-average trends are shown in Figure 6d. Both the decreasing trend and increasing trend were significant, similar to the trends based on Noah-MP shown in Figure 5d. The correlation coefficient between annual mean SH flux and *Ts–Ta* was 0.74. This conflicts with the results of Yang et al. [9], who reported an increase in *Ts–Ta* from approximately 1970 to 2006. However, the *Ts–Ta* trends reported herein agree with the results based on reanalysis products (ERA-Interim and NCEP) reported by Chen et al. [25].

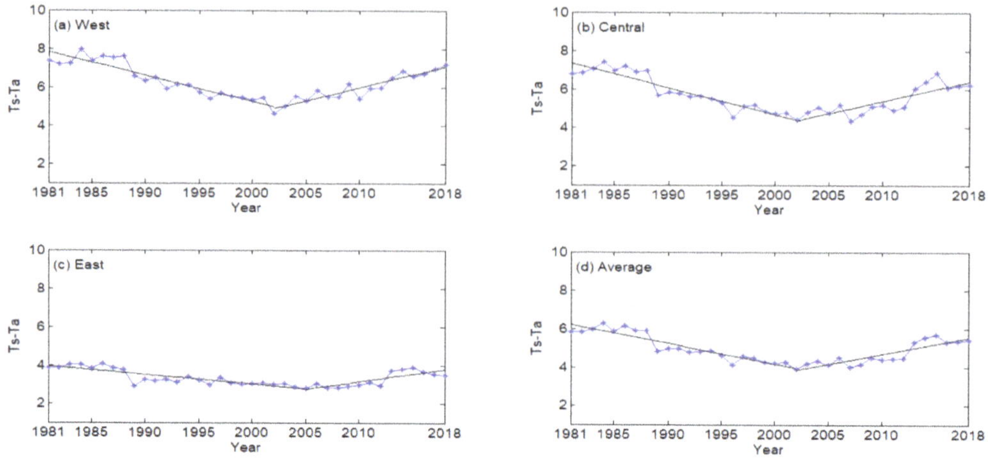

Figure 6. Variations in annual *Ts–Ta* (°C) in the Noah-MP simulation for the western (**a**), central (**b**), eastern (**c**), and entire TP (**d**) during the period 1981–2018. All of the trends are derived from linear least–squares regressions and are significant at the 95% confidence level using the Mann–Kendall trend test.

For the whole TP, the variations in *Ts* and *Ta* in the Noah-MP simulation were also calculated, and are shown in Figure 7. We found that, *Ta* showed an increase during the period 1981–2018. This increasing trend was also reported in previous studies [9,11]. However, *Ts* experienced a period of decrease before 2002; from then on, *Ts* also showed an increase. This shift resulted in the transition of *Ts–Ta* in approximately 2002. In the Noah-MP and SEBS simulations, SH fluxes began to increase later than in the CAM station-based results. The annual *Ts* variation trends of correlate well with those of *Ts–Ta* and SH flux on the TP.

Figure 8 shows the annual variation in wind speed (10 m above the ground), which was derived from the atmospheric forcing data (CMFD). The central and western TP experienced greater wind speeds than did the eastern TP during the 38-yr study period. The values of annual mean wind speed in all the 3 sub-regions declined from the mid-1980s, and then increased from approximately 1998. However, these changes were not very clear, especially on the eastern TP. Overall, the wind speed over the TP decreased from 1981 to 1998, and then increased after 1998 (Figure 8d). The changes in wind speed contributed substantially to the SH flux before 1998. Duan et al. [11] also considered this reduction in surface wind speed was also considered to be the cause of the decrease in SH flux. Yang et al. [9] attributed the changes in surface wind speed to the downward momentum transport of upper-air wind modulated by the upper-air pressure gradient.

Nevertheless, the heat transfer coefficient C_H in the Noah-MP simulations was also analyzed to determine its annual variations. However, this showed were no clear trends across the whole TP. The changes in surface wind speed and ground-air temperature gradient were the two major factors that contributed to the changes in the SH flux, as reported in previous studies [9,11]. From the analysis above, we found that surface wind speed began to increase from approximately 1998, while *Ts–Ta* increased sharply after 2002. During the period 1998–2002, *Ts–Ta* retained a decreasing trend, and this contributed to the decreasing trend in SH flux during this period. This was caused by *Ts* retaining a decreasing trend during this period. Overall, *Ts–Ta* trends were similar to those of annual mean SH flux in both Noah-MP and SEBS. Both *Ts–Ta* and SH flux showed a transition at the beginning of the 21st century in the Noah-MP and SEBS simulations and in the reanalysis products (ERA-Interim and NCEP) reported by Chen et al. [25].

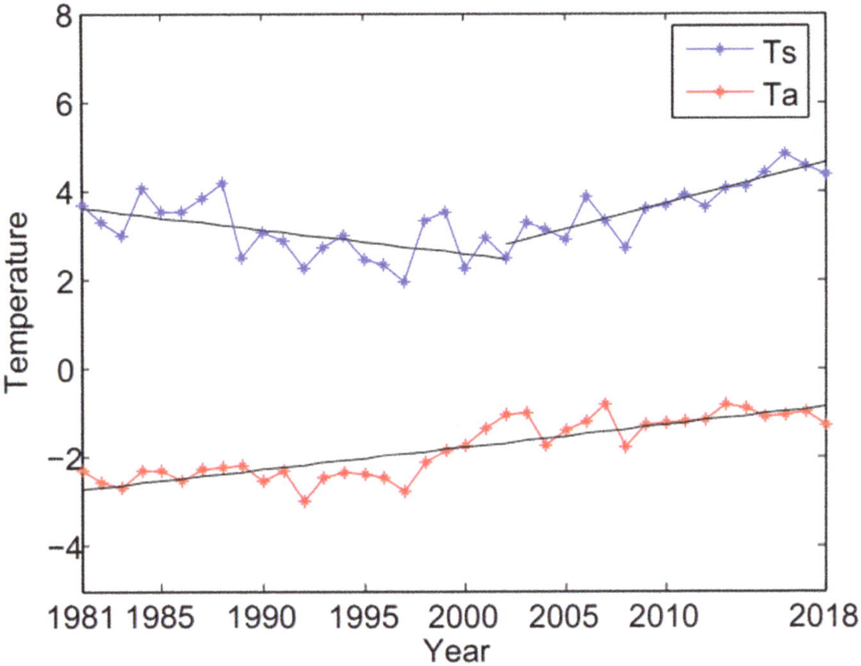

Figure 7. Variations in annual *Ts* (°C) and *Ta* (°C) in the Noah-MP simulation for the entire TP during the period 1981–2018. All of the trends are derived from linear least–squares regressions and are significant at the 95% confidence level using the Mann–Kendall trend test.

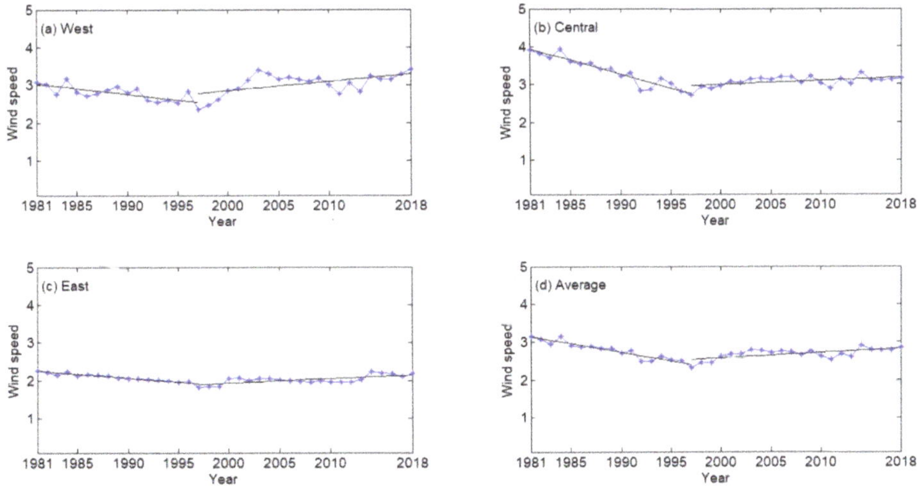

Figure 8. Variations in annual mean wind speed (m/s) derived from CMFD reanalysis data for the western (**a**), central (**b**), eastern (**c**), and entire TP (**d**) during the period 1981–2018. All of the trends are derived from linear least−squares regressions and are significant at the 95% confidence level using the Mann–Kendall trend test.

4. Conclusions and Discussion

Decadal trends in SH flux on the TP, especially when and how the trends changed, remain controversial. Here, we analyzed the climatic features of the SH flux on the TP using the high-resolution Noah-MP and SEBS simulations, and compared these results with the CMA station-based results.

The western and central TP witnessed much larger mean SH fluxes than did the eastern TP in both the Noah-MP and SEBS simulations. Meanwhile, the inter-annual variations in SH fluxes on the central and western TP were larger than that on the eastern TP. The SEBS simulation showed much larger inter-annual variations than did the Noah-MP simulation in most areas of the TP. The Yang method [41] applied to CMA station data here produced higher SH fluxes compared with the Noah-MP and SEBS simulations. We confirmed the decreasing trend in SH flux on the TP from the mid-1980s to the beginning of the 21st century, as also reported in previous studies [11,13]. Overall, the subsequent increasing trend in SH flux began in approximately 2002. The SEBS simulation showed much stronger SH fluxes on the western TP, but weaker SH fluxes on the eastern TP during 2001–2018 compared with the Noah-MP simulation. Both the Noah-MP and SEBS simulations showed an increasing trend in the TP-averaged SH flux from the beginning of the 21st century. The CAM station-based results showed an increasing trend before 2000, earlier than that recorded by both Noah-MP and SEBS. The model-based results showed consistent trends with those obtained from reanalysis products (ERA-Interim and NCEP) reported by Chen et al. [25].

Overall, there was a close relationship between changes in mean *Ts–Ta* and trends in the SH flux. Annual mean *Ts* experienced a period of decrease period before 2002, and then increased significantly. This shift in *Ts* resulted in the *Ts–Ta* transition in approximately 2002. Hence, the changes in *Ts* were the main cause of the changing trends in SH flux on the TP. The central and western TP experienced greater wind speeds than did the eastern TP during the 38-yr study period. Overall, the wind speed over the TP decreased from 1981 to 1998, and then increased after 1998. This explains why SH fluxes derived from CMA station data using conventional methods showed an early transition before 2000 in previous studies [13,25,43]. The changes in wind speed after 1998 were not as important as *Ts–Ta* in terms of modulating the SH flux trends. During the period 1998–2002, *Ts–Ta* continued to decrease, and this contributed to the decreasing trend in SH flux during this period.

It should be pointed out that the wind speed derived from reanalysis products was recorded at a height of 10 m above the ground. Wind speed at 10 m has been widely used to calculate C_H and SH flux based on the bulk transfer coefficient method. This may result in larger SH fluxes than compared with using the EC system. From the analysis of the Noah-MP simulation, the SH flux on the TP showed a decreasing trend from the mid-1980s, but an increasing trend since approximately 2002. The changes in mean *Ts–Ta* are here credited as the main cause of the decreasing and increasing trends in SH flux on the TP. The nonconformity of changes in surface wind speed and ground-air temperature gradient may result in disagreements as to the cause of SH flux trends on the TP, as highlighted by some previous studies. Additional in situ and remote sensing-based datasets are needed to clarify this issue.

Author Contributions: Conceptualization, S.W. and Y.M.; analysis, S.W. and Y.M.; investigation, S.W. and Y.L.; writing-original draft preparation, S.W.; writing-review and editing, S.W. and Y.M.; visualization, S.W. and Y.L.; supervision, Y.M. All authors have read and agreed to the published version of the manuscript.

Funding: This work was funded by the Second Tibetan Plateau Scientific Expedition and Research (STEP) program (2019QZKK0103), and the National Natural Science Foundation of China (42075156 and 91837208).

Data Availability Statement: The datasets generated and analyzed during the current study are not publicly available, but are available from the corresponding author on reasonable request.

Conflicts of Interest: The authors declare no conflict of interest.

References

1. Chen, L.; Reiter, E.; Feng, Z. The atmospheric heat source over the Tibetan Plateau: May–August 1979. *Mon. Weather Rev.* **1985**, *113*, 1771–1790. [CrossRef]
2. Ye, D.; Wu, G. The role of the heat source of the Tibetan Plateau in the general circulation. *Meteorol. Atmos. Phys.* **1998**, *67*, 181–198. [CrossRef]
3. Wu, G.; Liu, Y.; He, B.; Bao, Q.; Duan, A.; Jin, F.-F. Thermal controls on the Asian summer monsoon. *Sci. Rep.* **2012**, *2*, 404. [CrossRef] [PubMed]
4. Zhao, P.; Chen, L. Study on climatic features of surface turbulent heat exchange coefficients and surface thermal sources over the Qinghai-Xizang (Tibetan) Plateau. *J. Meteorol. Res.* **2000**, *14*, 13–29.
5. Duan, A.; Wang, M.; Lei, Y.; Cui, Y. Trends in summer rainfall over China associated with the Tibetan Plateau sensible heat source during 1980–2008. *J. Clim.* **2013**, *26*, 261–275. [CrossRef]
6. Duan, A.; Sun, R.; He, J. Impact of surface sensible heating over the Tibetan Plateau on the western Pacific subtropical high: A land-air-sea interaction perspective. *Adv. Atmos. Sci.* **2017**, *34*, 157–168. [CrossRef]
7. Wang, Z.; Duan, A.; Wu, G. Time-lagged impact of spring sensible heat over the Tibetan Plateau on the summer rainfall anomaly in East China: Case studies using the WRF model. *Clim. Dyn.* **2014**, *42*, 2885–2898. [CrossRef]
8. Yanai, M.; Li, C.; Song, Z. Seasonal heating of the Tibetan Plateau and its effects on the evolution of the Asian summer monsoon. *J. Meteorol. Soc. Jpn.* **1992**, *70*, 319–351. [CrossRef]
9. Yang, K.; Wu, H.; Qin, J.; Lin, C.; Tang, W.; Chen, Y. Recent climate changes over the Tibetan Plateau and their impacts on energy and water cycle: A review. *Glob. Planet. Chang.* **2014**, *112*, 79–91. [CrossRef]
10. Duan, A.; Wu, G. Weakening trend in the atmospheric heat source over the Tibetan Plateau during recent decades. Part I: Observations. *J. Clim.* **2008**, *21*, 3149–3164. [CrossRef]
11. Duan, A.; Li, F.; Wang, M.; Wu, G. Persistent weakening trend in the spring sensible heat source over the Tibetan Plateau and its impact on the Asian summer monsoon. *J. Clim.* **2011**, *24*, 5671–5682. [CrossRef]
12. Guo, X.; Yang, K.; Chen, Y. Weakening sensible heat source over the Tibetan Plateau revisited: Effects of the land-atmosphere thermal coupling. *Theor. Appl. Climatol.* **2011**, *104*, 1–12. [CrossRef]
13. Yang, K.; Guo, X.; Wu, B. Recent trends in surface sensible heat flux on the Tibetan Plateau. *Sci. China Ser. D* **2011**, *54*, 19–28. [CrossRef]
14. Zhu, X.; Liu, Y.; Wu, G. An assessment of summer sensible heat flux on the Tibetan Plateau from eight data sets. *Sci. China Earth Sci.* **2012**, *55*, 779–786. [CrossRef]
15. Shi, Q.; Liang, S. Surface-sensible and latent heat fuxes over the Tibetan Plateau from ground measurements, reanalysis, and satellite data. *Atmos. Chem. Phys.* **2014**, *14*, 5659–5677. [CrossRef]
16. Xie, J.; Yu, Y.; Li, J.; Ge, J.; Liu, C. Comparison of surface sensible and latent heat fluxes over the Tibetan Plateau from reanalysis and observations. *Meteorol. Atmos. Phys.* **2019**, *131*, 567–584. [CrossRef]
17. Ma, Y.; Zhong, L.; Wang, B.; Ma, W.; Chen, X.; Li, M. Determination of land surface heat fluxes over heterogeneous landscape of the Tibetan Plateau by using the MODIS and in situ data. *Atmos. Chem. Phys.* **2011**, *11*, 10461–10469. [CrossRef]
18. Ma, Y.; Wang, B.; Zhong, L.; Ma, W. The regional surface heating field over the heterogeneous landscape of the Tibetan Plateau using MODIS and in-situ data. *Adv. Atmos. Sci.* **2012**, *29*, 47–53. [CrossRef]
19. Ge, N.; Zhong, L.; Ma, Y.; Cheng, M.; Wang, X.; Zou, M.; Huang, Z. Estimation of land surface heat fluxes based on Landsat 7 ETM+ data and field measurements over the northern Tibetan Plateau. *Remote Sens.* **2019**, *11*, 2899. [CrossRef]
20. Ma, W.; Ma, Y. Modeling the influence of land surface flux on the regional climate of the Tibetan Plateau. *Theor. Appl. Climatol.* **2016**, *125*, 45–52. [CrossRef]
21. Wang, S.; Ma, Y. On the simulation of sensible heat flux over the Tibetan Plateau using different thermal roughness length parameterization schemes. *Theor. Appl. Climatol.* **2019**, *137*, 1883–1893. [CrossRef]
22. Zhu, L.; Huang, G.; Fan, G.; Qu, X.; Zhao, G.; Hua, W. Evolution of surface sensible heat over the Tibetan Plateau under the recent global warming hiatus. *Adv. Atmos. Sci.* **2017**, *34*, 1249–1262. [CrossRef]
23. Zhang, C.; Tian, R.; Mao, H.; Zhang, Z.-F.; Shen, Z.-W. Temporal and spatial variation of surface sensitive heat fluxes over the middle and eastern Tibetan Plateau. *Adv. Clim. Chang. Res.* **2018**, *14*, 127–136.
24. Wang, M.; Wang, J.; Chen, D.; Duan, A.; Liu, Y.; Zhou, S.; Guo, D.; Wang, H.; Ju, W. Recent recovery of the boreal spring sensible heating over the Tibetan Plateau will continue in CMIP6 future projections. *Environ. Res. Lett.* **2019**, *14*, 124066. [CrossRef]
25. Chen, L.; Pryor, S.C.; Wang, H.; Zhang, R. Distribution and variation of the surface sensible heat flux over the central and eastern Tibetan Plateau: Comparison on of station observations and multireanalysis products. *J. Geophys. Res. Atmos.* **2019**, *124*, 6191–6206. [CrossRef]
26. Li, M.; Babel, W.; Chen, X.; Zhang, L.; Sun, F.; Wang, B.; Ma, Y.; Hu, Z.; Foken, T. A 3-year dataset of sensible and latent heat fluxes from the Tibetan Plateau derived using eddy covariance measurements. *Theor. Appl. Climatol.* **2015**, *122*, 457–469. [CrossRef]
27. Ma, Y.; Hu, Z.; Xie, Z.; Ma, W.; Wang, B.; Chen, X.; Li, M.; Zhong, L.; Sun, F.; Gu, L.; et al. A long-term (2005–2016) dataset of hourly integrated land-atmosphere interaction observations on the Tibetan Plateau. *Earth Syst. Sci. Data* **2020**, *12*, 2937–2957. [CrossRef]

28. Mauder, M.; Foken, T. *Documentation and Instruction Manual of the Eddy Covariance Software Package TK3*; Work Report; University of Bayreuth: Bayreuth, Germany, 2011.
29. Niu, G.; Yang, Z.; Mitchell, K.E.; Chen, F.; Ek, M.B.; Barlage, M.; Kumar, A.; Manning, K.; Niyogi, D.; Rosero, E.; et al. The community Noah land surface model with multiple parameterization options (Noah-MP): 1. Model description and evaluation with local-scale measurements. *J. Geophys. Res.* **2011**, *16*, D12109. [CrossRef]
30. Yang, Z.; Niu, G.; Mitchell, K.E.; Chen, F.; Ek, M.B.; Barlage, M.; Longuevergne, L.; Manning, K.; Niyogi, D.; Tewari, M.; et al. The community Noah land surface model with multiple parameterization options (Noah-MP): 2. Evaluation over global river basins. *J. Geophys. Res.* **2011**, *116*, D12110. [CrossRef]
31. Gao, Y.; Li, K.; Chen, F.; Jiang, Y.; Lu, C. Assessing and improving Noah-MP land model simulations for the central Tibetan Plateau. *J. Geophys. Res. Atmos.* **2015**, *120*, 9258–9278. [CrossRef]
32. Zhang, G.; Chen, F.; Chen, Y.; Li, J.; Peng, X. Evaluation of Noah-MP Land-Model Uncertainties over Sparsely Vegetated Sites on the Tibet Plateau. *Atmosphere* **2020**, *11*, 458. [CrossRef]
33. Zeng, X.; Dickinson, R.E. Effect of surface sublayer on surface skin temperature and fluxes. *J. Clim.* **1998**, *11*, 537–550. [CrossRef]
34. He, J.; Yang, K.; Tang, W.; Lu, H.; Qin, J.; Chen, Y.; Li, X. The first high-resolution meteorological forcing dataset for land process studies over China. *Sci. Data* **2020**, *7*, 25. [CrossRef] [PubMed]
35. Su, Z. The surface energy balance system (SEBS) for estimation of turbulent heat fluxes. *Hydrol. Earth Syst. Sci.* **2002**, *6*, 85–100. [CrossRef]
36. Chen, X.; Su, Z.; Ma, Y.; Liu, S.; Yu, Q.; Xu, Z. Development of a 10-year (2001–2010) 0.1° data set of land-surface energy balance for mainland China. *Atmos. Chem. Phys.* **2014**, *14*, 13097–13117. [CrossRef]
37. Han, C.; Ma, Y.; Su, Z.; Chen, X.; Zhang, L.; Li, M.; Sun, F. Estimates of effective aerodynamic roughness length over mountainous areas of the Tibetan Plateau. *Q. J. R. Meteorol. Soc.* **2015**, *141*, 1457–1465. [CrossRef]
38. Han, C.; Ma, Y.; Chen, X.; Su, Z. Trends of land surface heat fluxes on the Tibetan Plateau from 2001 to 2012. *Int. J. Climatol.* **2017**, *37*, 4757–4767. [CrossRef]
39. Han, C.; Ma, Y.; Wang, B.; Zhong, L.; Ma, W.; Chen, X.; Su, Z. Long-term variations in actual evapotranspiration over the Tibetan Plateau. *Earth Syst. Sci. Data* **2021**, *13*, 3513–3524. [CrossRef]
40. Duan, A.; Liu, S.; Zhao, Y.; Gao, K.; Hu, W. Atmospheric heat source/sink dataset over the Tibetan Plateau based on satellite and routine meteorological observations. *Big Earth Data* **2018**, *2*, 11. [CrossRef]
41. Yang, K.; Qin, J.; Guo, X.; Zhou, D.; Ma, Y. Method development for estimating sensible heat flux over the Tibetan Plateau from CMA data. *J. Appl. Meteorol. Clim.* **2009**, *48*, 2474–2486. [CrossRef]
42. Chen, F.; Zhang, Y. On the coupling strength between the land surface and the atmosphere: From viewpoint of surface exchange coefficients. *Geophys. Res. Lett.* **2009**, *36*, L10404. [CrossRef]
43. Wang, H.; Hu, Z.; Li, D.; Dai, Y. Estimation of the surface heat transfer coefficient over the east-central Tibetan Plateau using satellite remote sensing and field observation data. *Theor. Appl. Climatol.* **2019**, *108*, 169–183. [CrossRef]
44. Wang, H.; Zhang, J.; Chen, L.; Li, D. Relationship between summer extreme precipitation anomaly in Central Asia and surface sensible heat variation on the Central-Eastern Tibetan Plateau. *Clim. Dyn.* **2022**, *59*, 685–700. [CrossRef]

remote sensing

MDPI

Article

A New Drought Monitoring Index on the Tibetan Plateau Based on Multisource Data and Machine Learning Methods

Meilin Cheng [1], Lei Zhong [1,2,3,4,*], Yaoming Ma [5,6,7,8,9,10], Xian Wang [1], Peizhen Li [1], Zixin Wang [1] and Yuting Qi [1]

1 School of Earth and Space Sciences, University of Science and Technology of China, Hefei 230026, China
2 CAS Center for Excellence in Comparative Planetology, Hefei 230026, China
3 Jiangsu Collaborative Innovation Center for Climate Change, Nanjing 210023, China
4 Frontiers Science Center for Planetary Exploration and Emerging Technologies, University of Science and Technology of China, Hefei 230026, China
5 Land-Atmosphere Interaction and Its Climatic Effects Group, State Key Laboratory of Tibetan Plateau Earth System, Resources and Environment (TPESRE), Institute of Tibetan Plateau Research, Chinese Academy of Sciences, Beijing 100101, China
6 College of Earth and Planetary Sciences, University of Chinese Academy of Sciences, Beijing 100049, China
7 College of Atmospheric Science, Lanzhou University, Lanzhou 730000, China
8 National Observation and Research Station for Qomolongma Special Atmospheric Processes and Environmental Changes, Dingri 858200, China
9 Kathmandu Center of Research and Education, Chinese Academy of Sciences, Beijing 100101, China
10 China-Pakistan Joint Research Center on Earth Sciences, Chinese Academy of Sciences, Islamabad 45320, Pakistan
* Correspondence: zhonglei@ustc.edu.cn

Abstract: Drought is a major disaster over the Tibetan Plateau (TP) that exerts great impacts on natural ecosystems and agricultural production. Furthermore, most drought indices are only useful for assessing drought conditions on a coarse temporal scale. Drought indices that describe drought evolution at a fine temporal scale are still scarce. In this study, four machine learning methods, including random forest regression (RFR), k-nearest neighbor regression (KNNR), support vector regression (SVR), and extreme gradient boosting regression (XGBR), were used to construct daily drought indices based on multisource remote sensing and reanalysis data. Through comparison with in situ soil moisture (SM) over the TP, our results indicate that the drought index based on the XGBR model outperforms other models (R^2 = 0.76, RMSE = 0.11, MAE = 0.08), followed by RFR (R^2 = 0.74, RMSE = 0.11, MAE = 0.08), KNNR (R^2 = 0.73, RMSE = 0.11, MAE = 0.08) and SVR (R^2 = 0.66, RMSE = 0.12, MAE = 0.1). A new daily drought index, the standardized integrated drought index (SIDI), was developed by the XGBR model for monitoring agricultural drought. A comparison with ERA5-Land SM and widely used indices such as SPI-6 and SPEI-6 indicated that the SIDI depicted the dry and wet change characteristics of the plateau well. Furthermore, the SIDI was applied to analyze a typical drought event and reasonably characterize the spatiotemporal patterns of drought evolution, demonstrating its capability and superiority for drought monitoring over the TP. In addition, soil properties accounted for 59.5% of the model output, followed by meteorological conditions (35.8%) and topographic environment (4.7%).

Keywords: drought monitoring; machine learning method; Tibetan Plateau

Citation: Cheng, M.; Zhong, L.; Ma, Y.; Wang, X.; Li, P.; Wang, Z.; Qi, Y. A New Drought Monitoring Index on the Tibetan Plateau Based on Multisource Data and Machine Learning Methods. *Remote Sens.* **2023**, *15*, 512. https://doi.org/10.3390/rs15020512

Academic Editors: Gabriel Senay and Xianjun Hao

Received: 6 December 2022
Revised: 1 January 2023
Accepted: 17 January 2023
Published: 15 January 2023

1. Introduction

Drought is one of the most widespread and costly natural disasters that not only affects agricultural and livestock production but also leads to a series of ecological and socioeconomic problems [1,2]. Globally, dry areas are increasing at a rate of approximately 1.74% per decade from 1950 to 2008 [3]. The Tibetan Plateau (TP) is known as the "Asian Water Tower". Nevertheless, the areas of the arid and semiarid regions of the TP account for 23% and 44%, respectively [4], and approximately 62% of the plateau area is covered by alpine meadows and grasslands [5]. The ecosystems over the TP are fragile due to their high elevation and unique geographical location. Meanwhile, the TP is highly susceptible to global climate change. Zhong et al. [6] noted that the air temperature over the TP has been 1.27 °C higher than normal since 2014. The average increase was 2.2 times the global average (0.57 °C). Overall, it easily suffers from drought under the combined effects of climate change and fragile ecosystems. Gao et al. [7] calculated the ratio of precipitation to potential evapotranspiration (P/PET) at 83 stations in the TP between 1979 and 2011 and found that the eastern TP was becoming drier. Wang et al. [8] investigated the plateau drought variation based on the self-calibrating Palmer drought severity index (scPDSI) from 1961 to 2009 and revealed that the southern TP experienced a wetting trend even though the northern TP became significantly drier, particularly in spring and autumn. Feng et al. [9] calculated the standardized precipitation evapotranspiration index (SPEI) using data from 274 meteorological stations over the TP during 1970–2017, indicating that severe drought frequency in winter and drought risk in summer showed an increasing trend. According to the statistics of the China Meteorological Administration, drought is the most dominant meteorological disaster among all meteorological disasters over the TP [10]. Therefore, a comprehensive understanding of drought characteristics on the TP is of great importance for drought early warning, prevention, and mitigation.

In general, a universal objective definition of drought is impractical and does not exist without knowledge of the climatologically expected values for the availability of stored water for a given need [11]. Wilhite and Glantz in 1985 [12] classified droughts into four categories: meteorological droughts, agricultural droughts, hydrological droughts, and socioeconomic droughts, which have been widely recognized by the scientific community [13]. Recently, some new drought types were proposed, such as ground water droughts [14], ecological droughts [15], agroecological droughts [16], and environmental droughts [17]. The aridity index, defined as the ratio between precipitation and evapotranspiration, is usually used to classify climatic zones and monitor drought [18]. The aim of this study is to develop a comprehensive index characterizing the complex drought conditions affected by climatic, hydrometeorological, and environmental factors. Accordingly, the drought index was adopted instead of the aridity index. To objectively quantify the onset, intensity, and spatial extent of drought, more than one hundred drought indices have been developed thus far. Among these indices, the Palmer drought severity index (PDSI) [19], the standardized precipitation index (SPI) [20], and the standardized precipitation evapotranspiration index (SPEI) [21] are the most popular and widely used drought indices. However, most of these indices are developed and evaluated on a monthly time scale. Although some daily drought indices, such as the standardized drought and flood potential index (SDFPI) [22] and drought potential index (DPI) [23], have been developed recently, drought indices with high resolution are generally scarce and need to be further investigated. With an average elevation of approximately 4000 m, the TP has the largest frozen soil zone in the mid-latitudes. In the freeze-thaw process, especially for the seasonal transitional period, the soil water phase and energy budget have dramatic changes at the daily temporal scale, which can affect the soil-vegetation-atmosphere interaction. Drought indices with high resolution have been expected to reflect this variation [24]. On the other hand, drought indices with high resolution can provide valuable drought information, such as the onset, end, and duration of drought, which are capable of guiding vulnerable agricultural and livestock production over the TP. In addition, some of these indices are based on a single variable, such as the vegetation condition index (VCI) [25] and temperature condition index (TCI) [26], which mainly reflect one specific aspect of drought. In terms of agricultural drought related to meteorology, soil, and vegetation systems, they cannot adequately capture the complex features of drought evolution. Integrating multiple drought-related variables and indices is an effective method for addressing this issue [27]. For example, Huang et al. [28] constructed an integrated drought index (IDI) based on precipitation, runoff, and soil moisture using the entropy weight method. It is an objective method for weight determination and gives soil moisture a low weight, causing the insensitivity of IDI to agricultural drought. Lu et al. [29] developed the integrated scaled drought index (ISDI) based on precipitation, the normalized difference vegetation index (NDVI), soil moisture, and land surface temperature. ISDI is a linear combination of four drought factors. It cannot reflect the nonlinear relationships between hydrometeorological factors and drought. Previous studies have demonstrated that these integrated drought indices improved the capacity of drought monitoring. To date, there are three types of fusion methods: linear combination [30], copula-based methods [31,32], and machine learning (ML). Hao and Singh [33] noted that the former two approaches may suffer from the linearity assumption. In comparison, the ML approach has a strong capability of extracting target information from a large amount of random, noisy data and capturing the nonlinear characteristics of physical processes. It has therefore recently been favored by many scholars for the construction of drought indices. For example, Liu et al. [34] proposed an integrated agricultural drought index (IDI) based on remote sensing data and the backpropagation (BP) neural network, and it can effectively monitor drought events on the North China Plain.

Therefore, the objectives of this study are to (1) compare daily integrated drought indices developed by the four ML methods, namely, random forest regression (RFR), k-nearest neighbor regression (KNNR), support vector regression (SVR), and extreme gradient boosting regression (XGBR), based on multisource remote sensing and reanalysis data with the in situ soil moisture and obtain the optimal drought index as the new standardized integrated drought index (SIDI); (2) evaluate the SIDI performance in dry and wet changes against SPI-6, SPEI-6, and European Centre for Medium-Range Weather Forecasts Reanalysis 5 Land (ERA5-Land) soil moisture; and (3) assess the spatiotemporal applicability of SIDI for a typical drought event. The SIDI is expected to monitor plateau droughts with more detail and accuracy for agricultural water resource management.

2. Materials and Methods

2.1. Materials

In situ soil moisture data are obtained from the time-lapse observation dataset of soil temperature and humidity on the Tibetan Plateau from 2015 to 2020 [35–38] and a long-term dataset of integrated land–atmosphere interaction observations on the Tibetan Plateau from 2008 to 2016 [39]. The data were downloaded through the National Tibetan Plateau Data Center (https://data.tpdc.ac.cn/ (accessed on 20 April 2021)). The locations of the field observation stations can be found in Figure 1. These stations over different climates and underlying surface conditions are representative. Soil moisture data in units of percent with an hourly temporal resolution were converted to m^3/m^3 and further averaged to daily means. Soil moisture has been one of the most direct indicators of agricultural drought. Approximately 62% of the plateau area is covered by alpine meadows and grasslands with shallow root systems [5]. Therefore, the depth of soil moisture data used in this study was 10 cm. Moreover, drought during the critical stage of vegetation growth is more destructive to agricultural production. Hence, the soil moisture data collected at a 10 cm depth during the growing season (from May to October) were selected as the "ground truth" for assessing the performance of drought indices [34,40].

To develop the integrated drought index, we comprehensively consider the overall effects of the meteorological conditions, vegetation information, soil properties, and topographic environment. Thirteen variables in total were selected as predictor variables and summarized as follows [34,41]: four near-surface meteorological elements, including 2 m air temperature (TEMP), specific humidity (SHUM), 10 m wind speed (WIND), and precipitation rate (PREC), are provided by the China meteorological forcing dataset (CMFD) with a temporal resolution of three hours and a spatial resolution of 10 km (http://poles.tpdc.ac.cn (accessed on 20 April 2021)). Land surface temperature (LST) was acquired by a daily 1 km all-weather land surface temperature dataset for western China from 2000 to 2021 (http://data.tpdc.ac.cn (accessed on 15 July 2022)). Evaporation (EVAP) is provided by the ERA5-Land hourly reanalysis dataset at a spatial resolution of 10 km from 1950 to the present (https://cds.climate.copernicus.eu (accessed on 12

July 2019)). The fraction of absorbed photosynthetically active radiation (FAPAR) was obtained from the daily global QA4ECV FAPAR product at 5 km × 5 km during 1982–2016 (http://www.qa4ecv-land.eu (accessed on 16 February 2018)). Five soil characteristic data are provided by the SoilGrids 250 m 2.0 product, which includes bulk density (BDOD), clay, silt, sand, and soil organic carbon (SOC) at six depths: 0–5 cm, 5–15 cm, 15–30 cm, 30–60 cm, 60–100 cm, and 100–200 cm (https://soilgrids.org (accessed on 4 May 2020)). The 5–15 cm soil depth characteristic data were chosen as the input variables. Digital elevation model (DEM) data at 1 km are derived from the Resource and Environment Science and Data Center (https://www.resdc.cn (accessed on 1 September 2008)). The final resolution of all input data is 10 km × 10 km by using bilinear interpolation.

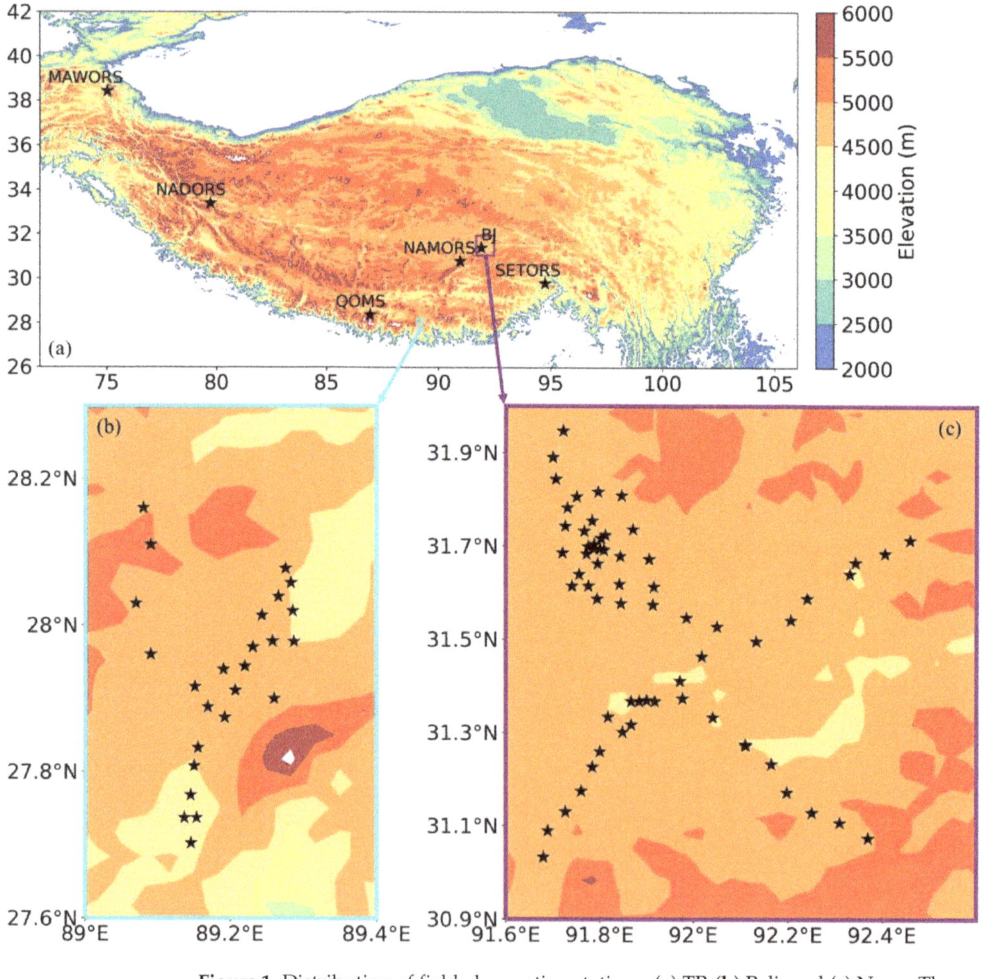

Figure 1. Distribution of field observation stations: (**a**) TP, (**b**) Pali, and (**c**) Naqu. The contour color represents different elevations. Pentagrams represent the station locations.

SPI-6 and SPEI-6 represent six months of rainfall and evaporation anomalies. Zhao et al. investigated the correlations of SPI and SPEI at different timescales (1, 3, 6, and 9 months) with NDVI at 33 stations around the Gannan region in the eastern TP and found that SPI-6 and SPEI-6 have good correlations with NDVI. SPI-6 and SPEI-6 are suitable for monitoring the drought conditions of alpine meadows [42]. Therefore, referring to McKee et al. [20] and Vicente-Serrano et al. [21], the SPI-6 and SPEI-6 were calculated. SPI-6, SPEI-6, and the

ERA5-Land hourly volume of water (m^3/m^3) at the 7–28 cm soil layer (ERA5-Land SM) (https://cds.climate.copernicus.eu/ (accessed on 12 July 2019)) were chosen as contrasts to evaluate the performance of the SIDI over the entire TP. The datasets used in this study are shown in Table 1.

Table 1. Datasets used in this study.

Variables	Temporal Interval	Spatial Resolution	Data Source
In situ soil moisture (SM)	Hour	-	https://data.tpdc.ac.cn (accessed on 20 April 2021)
2 m air temperature (TEMP), Specific humidity (SHUM), 10 m wind speed (WIND), and Precipitation rate (PREC)	3 h	10 km	http://poles.tpdc.ac.cn (accessed on 20 April 2021)
Land surface temperature (LST)	Day	1 km	http://data.tpdc.ac.cn (accessed on 15 July 2022)
Evaporation (EVAP), ERA5-Land SM	Hour	10 km	https://cds.climate.copernicus.eu (accessed on 12 July 2019)
Fraction of absorbed photosynthetically active radiation (FAPAR)	Day	5 km	http://www.qa4ecv-land.eu (accessed on 16 February 2018)
Bulk density (BDOD), Clay, Silt, Sand, and Soil organic carbon (SOC)	-	250 m	https://soilgrids.org (accessed on 4 May 2020)
Digital elevation model (DEM)	-	1 km	https://www.resdc.cn (accessed on 1 September 2008)

2.2. Methods

2.2.1. Machine Learning Models

RFR is an ensemble model that consists of multiple decision trees [43]. "Random" means that the input of each tree is randomly extracted from the training dataset, and a subset of features at each tree node is randomly selected from the available features to expand the tree. The model's final output is calculated as the average of predictions created by all individual trees. Consequently, RFR decreases the overall variance and avoids overfitting. For both small sample sizes and high-dimensional data, RFR captures nonlinear relationships between features and target variables and hence provides reliable results [34,44].

SVR is a supervised machine learning algorithm. SVR employs the kernel function where the input feature is projected into a high-dimensional feature space for building the optimal hyperplane to regress the training dataset with the minimum loss [45]. The performance of SVR depends on the proper selection of the kernel function. Many applications have demonstrated that the Gaussian radial basis function (RBF) is an excellent kernel function in SVR. In this study, we used SVR with the RBF kernel function.

KNNR is a nonparametric model. Its major advantage is its simplicity and efficiency. Given a data point, KNNR searches for the closest K data points based on the distance between that point and the remaining points in the training dataset. The model finally outputs the average of the target predictions for these K neighbors [46]. The target predictions of K neighbors are equally weighted in the KNNR we used.

XGBR is an advanced ensemble model designed by Chen et al. [47] on the basis of a gradient boosting machine (GBM). Compared with the traditional GBM, XGBR implementation adopts a regularized boosting technique and parallel processing, which help to reduce overfitting and speed up. Therefore, XGBR is a powerful machine learning model, especially when speed and accuracy are taken into consideration.

The detailed technical framework for the development of SIDI can be found in Figure 2.

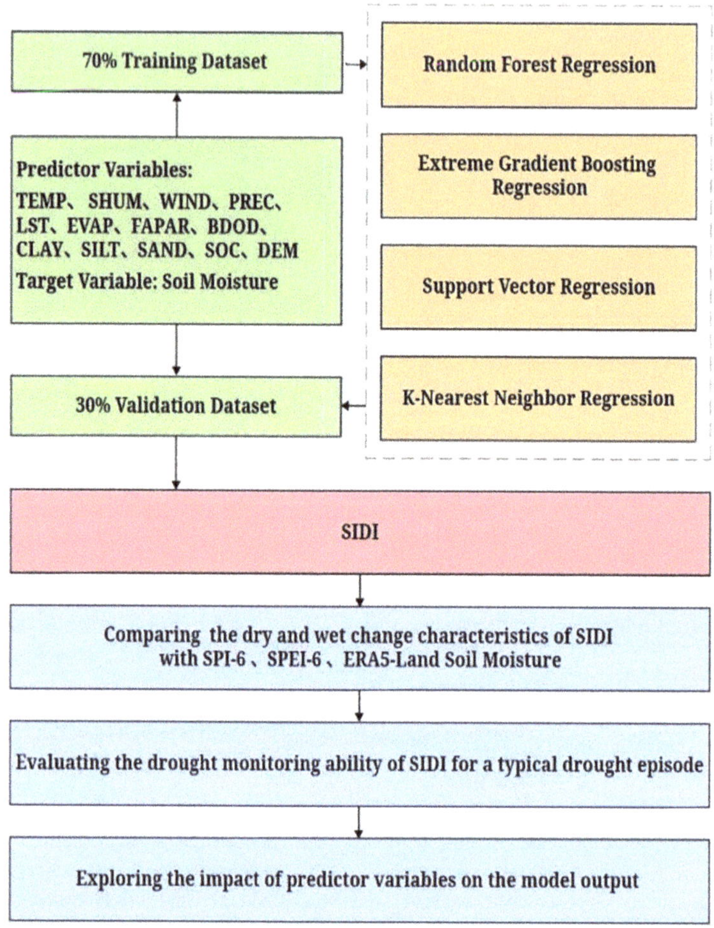

Figure 2. The technical framework for the development of SIDI. Green boxes represent the input data; yellow boxes represent the machine learning models; red boxes represent the model output; blue boxes represent the validation processes.

2.2.2. Statistical Indicators

The root mean square error (RMSE), mean absolute error (MAE), and coefficient of determination (R^2) were calculated to evaluate the performance of the ML models, as defined below [48,49].

$$\text{RMSE} = \sqrt{\frac{\sum_{i=1}^{N}(P_i - O_i)^2}{N}} \quad (1)$$

$$\text{MAE} = \frac{\sum_{i=1}^{N}(|P_i - O_i|)}{N} \quad (2)$$

$$R^2 = 1 - \frac{\sum_{i=1}^{N}(P_i - O_i)^2}{\sum_{i=1}^{N}(P_i - \overline{P})^2} \quad (3)$$

where N is the number of observations. P_i and O_i are the predictions and observations, respectively. $\overline{P}$ represents the average of the predictions.

2.2.3. Shapley Additive Explanation

SHAP was first proposed by Shapley in 1953 to calculate the contribution of each player and allocate the value created by them in a collaborative game [50]. Lundberg and Lee in 2017 first introduced the SHAP to explain the output of machine learning models regarded as black boxes [51]. The SHAP value can be used to interpret individual machine learning predictions. The main idea of the SHAP value is to obtain the marginal contribution across all the possible permutations of the features and then take the average. The expression of the SHAP value is as follows [52]:

$$\varnothing_i = \sum_{s \in N\{i\}} \frac{|S|!(N - |S| - 1)!}{n!}[v(S \cup \{i\}) - v(S)] \tag{4}$$

where, $\varnothing_i$ is the contribution of feature i, N is the set of features, n is the number of features in N, S is the subset of N that contains feature i, and $v(N)$ is the base value meaning the predicted outcome for each feature in N without knowing the feature values.

The sum of the SHAP value of each feature for each observation is considered the model outcome for each observation. Therefore, the explanation model g is formulated as follows:

$$g(z') = \varnothing_0 + \sum_{i=1}^{M} \varnothing_i z'_i \tag{5}$$

where, $z \in \{0, 1\}^M$ and M is the number of features.

3. Results

3.1. Construction and Comparison of Drought Monitoring Index with In Situ Soil Moisture Measurements, SPI-6 and SPEI-6

3.1.1. Construction of Drought Monitoring Index

To achieve an independent assessment of the performance of ML models, we used 70% of the input data (thirteen predictor variables and in situ SM) as the training dataset to construct the drought index, and the remaining 30% was employed as the validation dataset to assess the performance of the drought index. Moreover, min-max standardization was performed on the input data to eliminate the effects of dimensionality and accelerate the convergence speed. Regarding the in situ SM during the growing season as the ground truth, the drought indices were outputted by four ML regression models at the station scale. Figure 3 shows the comparisons of drought indices using four machine learning models with the in situ soil moisture in the training process. All 32,509 samples in the training dataset were used. The drought index based on the SVR model had a poor performance with a low R^2 of 0.66 and a high RMSE of 0.12. This may be because the RBF kernel function is not suitable for these complicated data sets. Another reason may be that SVR hyperparameters such as C and gamma are not good, although they have been optimized by the grid search. In contrast, the drought indices based on the RFR, XGBR, and KNNR models performed well, with R^2 values of 0.86, 0.82, and 0.78; RMSE values of 0.08, 0.09, and 0.1; and MAE values of 0.06, 0.07, and 0.08, respectively. The results indicate that the three ML models have good application potential in the construction of the drought index in our study.

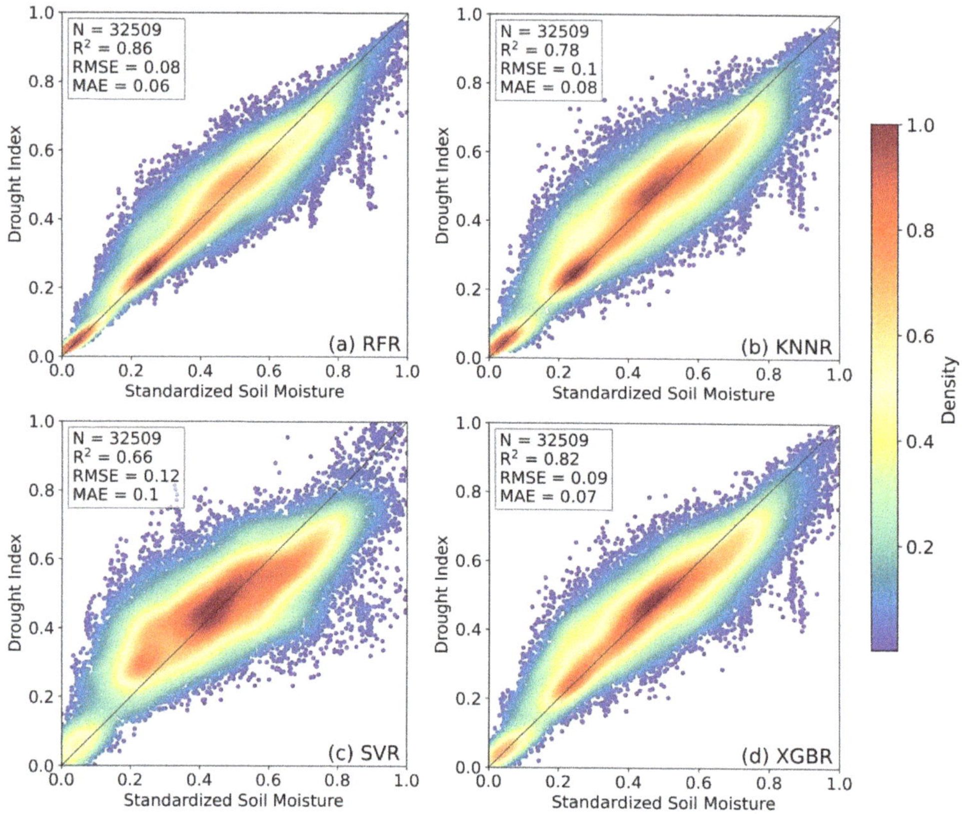

Figure 3. Comparisons of drought indices using four machine learning models with the in situ soil moisture in the training process: (**a**) RFR, (**b**) KNNR, (**c**) SVR, and (**d**) XGBR. Colors represent data density: the redder the color, the larger the data density is.

3.1.2. Comparison of Drought Monitoring Index with In Situ Soil Moisture Measurements, SPI-6 and SPEI-6

The prediction ability of the model is determined by the validation dataset because it was not involved in its construction. Hence, comparisons of drought indices using four machine learning models with in situ soil moisture were performed in the validation dataset (Figure 4). There are a total of 13,933 data points in the validation dataset. The result in the validation dataset for each model was slightly worse than that of the training dataset, indicating a high generalization level for each model due to the independence between the training and validation datasets [53]. In the validation process, the drought index based on the XGBR model outperformed that based on other models (R^2 = 0.76, RMSE = 0.11, MAE = 0.08), followed by RFR (R^2 = 0.74, RMSE = 0.11, MAE = 0.08), KNNR (R^2 = 0.73, RMSE = 0.11, MAE = 0.08), and SVR (R^2 = 0.66, RMSE = 0.12, MAE = 0.1) [54,55]. The two ensemble models of XGBR and RFR achieve better accuracy and higher correlation compared with other models, suggesting the superiority of ensemble learning. The primary advantage of XGBR lies in the fact that a regularization term is added to the cost function to control the complexity of the model (regularization boosting technique). This technique reduces the variance and overfitting and makes the model simpler and faster [56]. The result confirms that the optimal drought index is the XGBR-based drought index, and thus, the XGBR model is used to construct the SIDI.

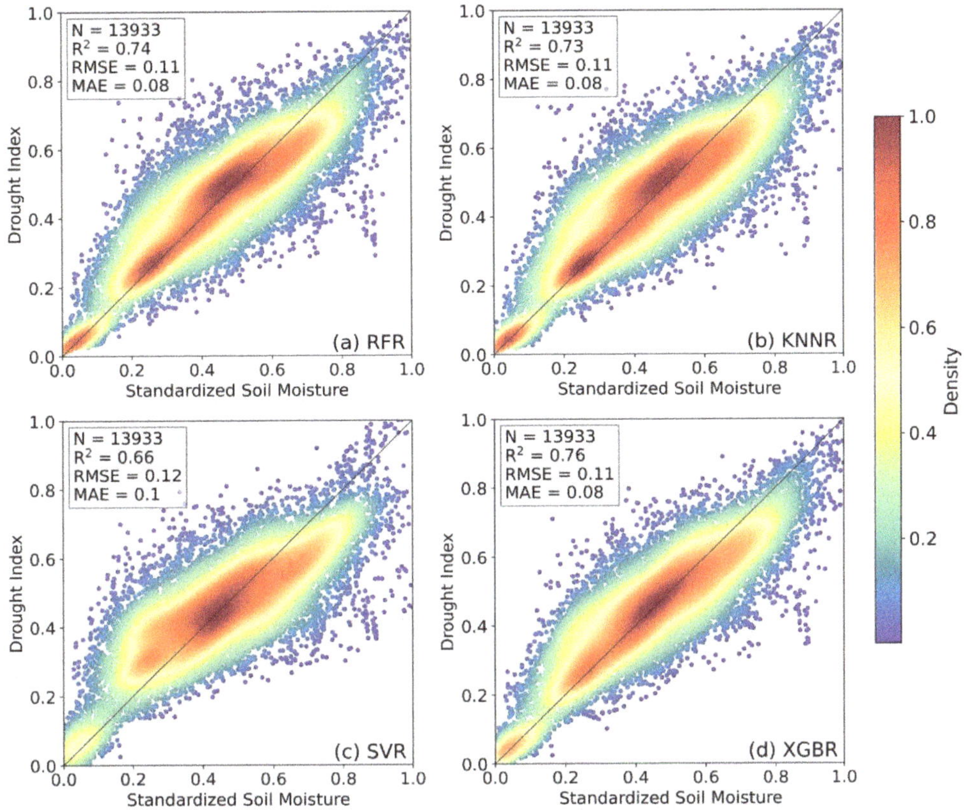

Figure 4. Comparisons of drought indices using four machine learning models with the in situ soil moisture in the validation process: (**a**) RFR, (**b**) KNNR, (**c**) SVR, and (**d**) XGBR. Colors represent data density: the redder the color, the larger the data density is.

Furthermore, the SIDI from 2000 to 2016 was obtained by inputting the spatial maps of thirteen predictor variables into the XGBR model. SIDI was switched to the monthly average, and then the spatial and temporal distributions of the monthly averaged SIDI were compared against ERA5-Land SM, SPI-6, and SPEI-6 in 2012, as shown in Figure 5. In terms of four variables, a smaller value corresponds to a drier area, and vice versa. In Figure 5a–d, the SM decreases gradually from the southeast to the northwest and exhibits obvious seasonal variability with a small (large) value in winter (summer). The SIDI has a similar spatial pattern and seasonal variability to the SM (Figure 5e–h). The spatial distributions of the SPI-6 and SPEI-6 are generally consistent, and there are some differences compared with the SM. For example, the SPI-6 and SPEI-6 display arid characteristics in the relatively humid southeastern region. This result suggests that they are not able to capture the dry and wet characteristics of the plateau well (Figure 5i–p). SPI-6 is formulated based on precipitation, ignoring the impact of temperature on drought. Considering this, the SPEI-6 is calculated based on precipitation and potential evaporation. However, previous studies found that the Thornth-waite algorithm failed to calculate the potential evapotranspiration when the average monthly temperature was below 0 °C, resulting in the poor applicability of SPEI-6 in arid and alpine regions such as the TP [57].

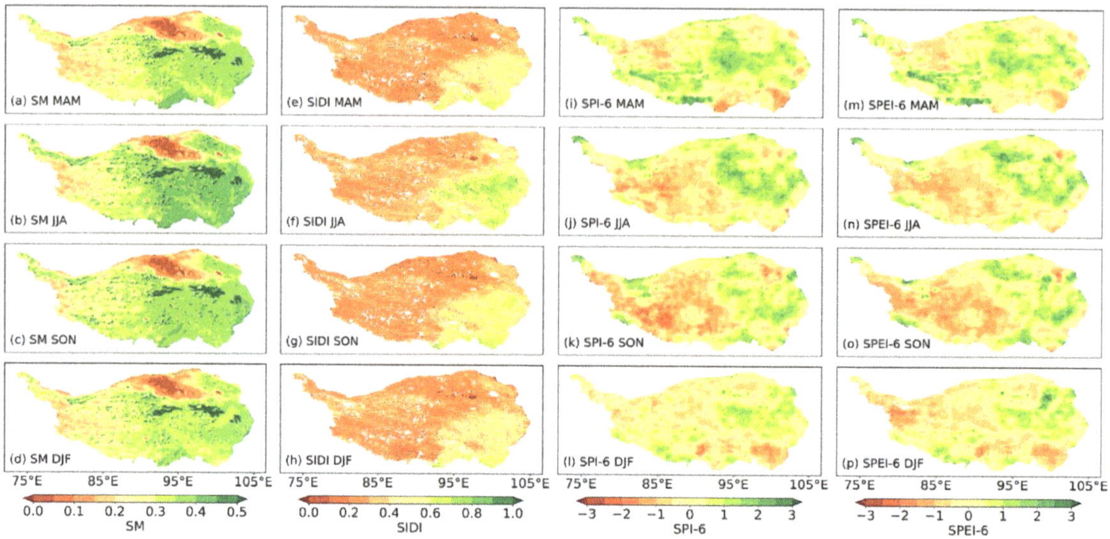

Figure 5. Comparisons of spatial and seasonal distributions of SIDI (**e–h**) against ERA5-Land SM at 7–28 cm ((**a–d**), units: m^3/m^3), SPI-6 (**i–l**), and SPEI-6 (**m–p**) in 2012. Colors represent drought degree: the redder the color, the stronger the drought degree is.

3.2. Drought Monitoring Performance for Typical Drought Events

The drought monitoring ability of the SIDI was evaluated for a typical drought event. This event occurred in the Xizang Autonomous Region in May and June of 2009 [58,59]. SIDI classifications are divided into 9 levels using the percentile threshold method (Table 2) [57]. Figure 6 illustrates the spatial evolution characteristics of the drought process captured by the SIDI at an interval of nine days from 22 April to 1 July 2009. As shown in Figure 6a,b, an abnormal drought occurred over most regions of Xizang on 22 April, and then the drought intensity rapidly increased. In particular, the central part of Xizang exhibited extreme drought. Subsequently, the drought eased, whether in drought range or intensity, to a large extent on 10 May (Figure 6c). The drought area constantly expanded from the northwestern to southeastern regions of Xizang, with a higher drought magnitude on 19 May (Figure 6d). Thereafter, the drought continued to weaken (Figure 6e–h).

Table 2. Dry and wet classifications for the SIDI.

Classification	Percentile Chance k (%)	SIDI
Extreme drought (Edry)	$k \leq 2$	0.04~0.14
Severe drought (Sdry)	$2 < k \leq 10$	0.14~0.18
Moderate drought (Mdry)	$10 < k \leq 20$	0.18~0.20
Abnormal drought (Adry)	$20 < k \leq 30$	0.20~0.22
Normal	$30 < k \leq 70$	0.22~0.33
Abnormal wet (Awet)	$70 < k \leq 80$	0.33~0.38
Moderate wet (Mwet)	$80 < k \leq 90$	0.38~0.48
Severe wet (Swet)	$90 < k < 98$	0.48~0.62
Extreme wet (Ewet)	≥ 98	0.62~0.84

In addition, the temporal evolution characteristics of this drought event are depicted in Figure 7. There are two valleys (1 May and 18 May) in the SIDI. A valley indicates drought aggravation. Compared with the SM variation, the SIDI is more sensitive and detailed in capturing the key points of the drought process. In addition, with the increase in precipitation, the SIDI increased gradually, revealing that the drought had been relieved. Overall, the SIDI is capable of accurately describing the evolution process of drought events in space and on a daily temporal scale, owing to its combination with the meteorological, vegetation, soil, and topographic environmental factors. Therefore, the SIDI is a reliable and comprehensive indicator for drought assessment.

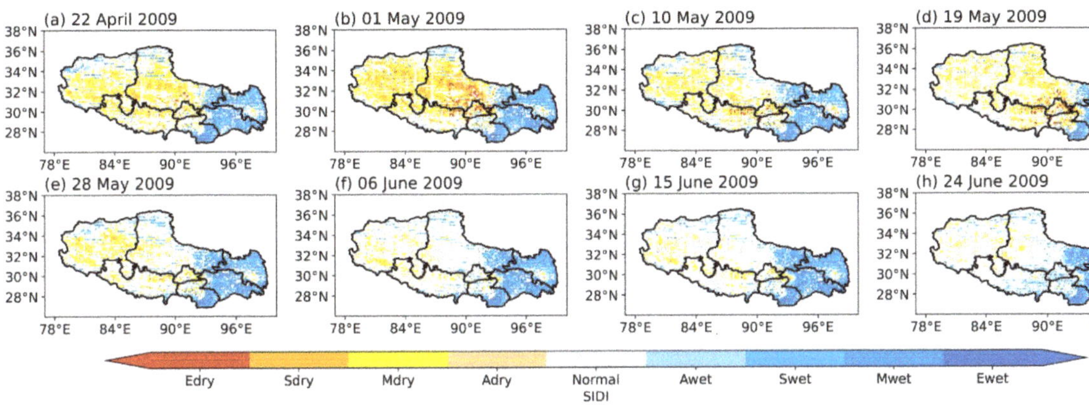

Figure 6. Spatial evolution characteristics of the drought process captured by the SIDI in the Xizang Autonomous Region from 22 April to 1 July 2009 at an interval of nine days: (**a**) 22 April 2009, (**b**) 1 May 2009, (**c**) 10 May 2009, (**d**) 19 May 2009, (**e**) 28 May 2009, (**f**) 06 June 2009, (**g**) 15 June 2009, and (**h**) 24 June 2009. Colors represent drought degree: the redder the color, the stronger the drought degree is.

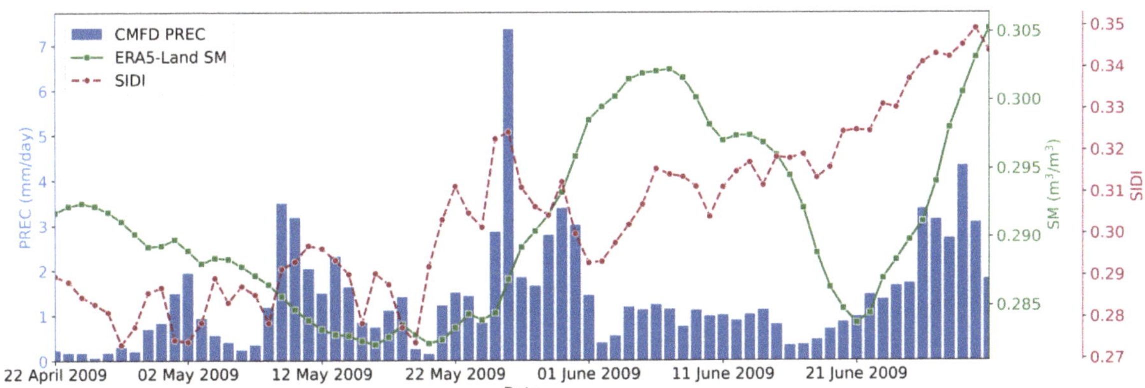

Figure 7. Temporal evolution characteristics of drought processes captured by SIDI in the Xizang Autonomous Region from 22 April to 01 July 2009, against ERA5-Land SM and CMFD PREC.

Furthermore, the SIDI and daily SPEI were adopted to identify drought characteristics at the BJ station, a representative station covering the alpine meadow on the central TP based on run theory [60] (Figure 8). A short-term drought event lasting 19 days was identified by the SIDI. However, the SPEI cannot identify it. It began on 26 March 2001, and lasted until 13 April 2001 (Figure 8a). The SIDI detected that the drought developed quickly and reached the intensity of extreme drought in several days. Thereafter, the drought eased slowly. Referring to the dry and wet classifications in Table 2, this drought event belonged to a moderate drought (I = 0.19). However, the SPEI shows wet conditions during this period (Figure 8b). The SIDI has a superior ability to identify drought information compared with other traditional indices, such as the SPEI.

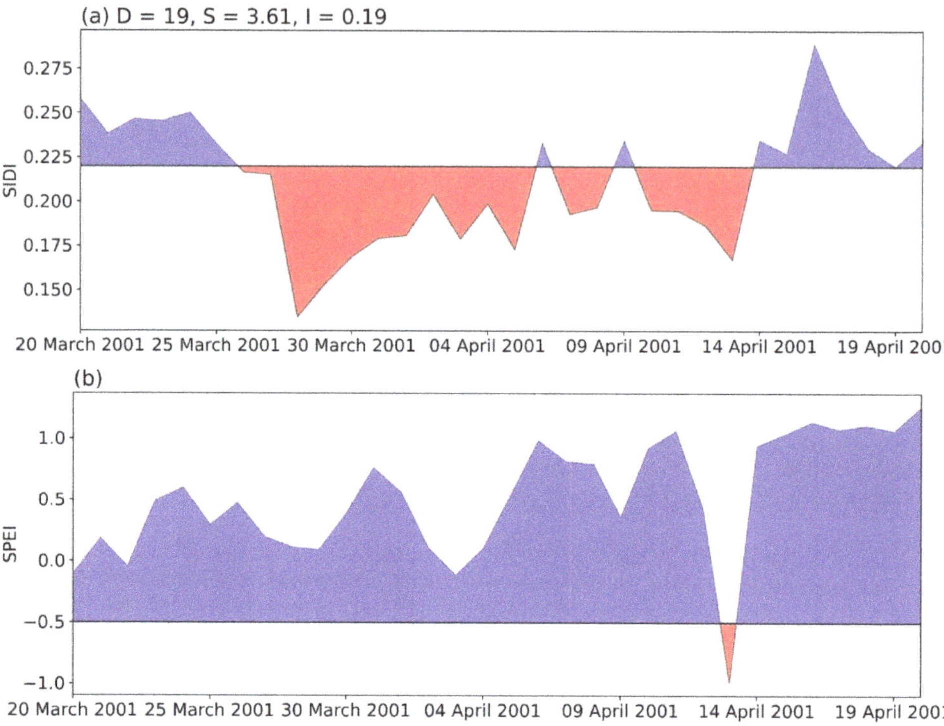

Figure 8. Daily drought evolution of a drought event identified by the SIDI (**a**) and SPEI (**b**) at the BJ station based on run theory. D: drought duration; S: drought severity, which is the cumulative sum of drought conditions on D days; I: drought intensity, which can be calculated by dividing S by D. Red patterns represent dry conditions, and blue patterns represent wet conditions.

3.3. Importance of the Predictor Variables

The performance of ML models has significant advantages for large data volumes with multiple predictor variables. Relationships between these predictor variables and model output are complicated and poorly identified due to the multiple predictor variables involved in the models and the black-box nature of ML models. How much did each predictor variable (feature value) contribute to the model output? To explain this, the SHAP was introduced into the XGBR model [61]. The average SHAP value of every feature and the SHAP value of every feature for every sample in the training dataset are presented in Figure 9. The color bar represents the feature value (red high, blue low). According to the averaged SHAP value, soil properties contributed 59.5% to the model output, followed by the meteorological (35.8%) and topographic environmental conditions (4.7%). The top three features that influence the model output are bulk density (BDOD), soil organic carbon (SOC), and silt, which are soil properties (Figure 9a). Moreover, the other two features, in addition to bulk density, have positive impacts on the model output (Figure 9b).

To interpret the interactions among predictor variables toward the model output, the BJ station covered by the representative alpine meadow on the central TP was chosen to apply the SHAP for different drought conditions (Figure 10). Figure 10 is the individual SHAP force plot, which includes three important characteristics: model output f(x), base value (the average of model outputs), and colors. The red color pushes the model output higher, whereas the blue color pushes the model output lower. For extreme and severe drought conditions, bulk density, land surface temperature, and soil organic carbon were the three major contributors to decreasing the model output (Figure 10a,b). In the case of moderate drought, abnormal drought, and nondrought conditions (Figure 10c–e), bulk density and soil organic carbon decreased the model output, while specific humidity slightly increased it. Overall, bulk density (bdod) and soil organic carbon (soc) were two major contributors influencing the model output for all drought conditions, which is consistent with the result in Figure 9b.

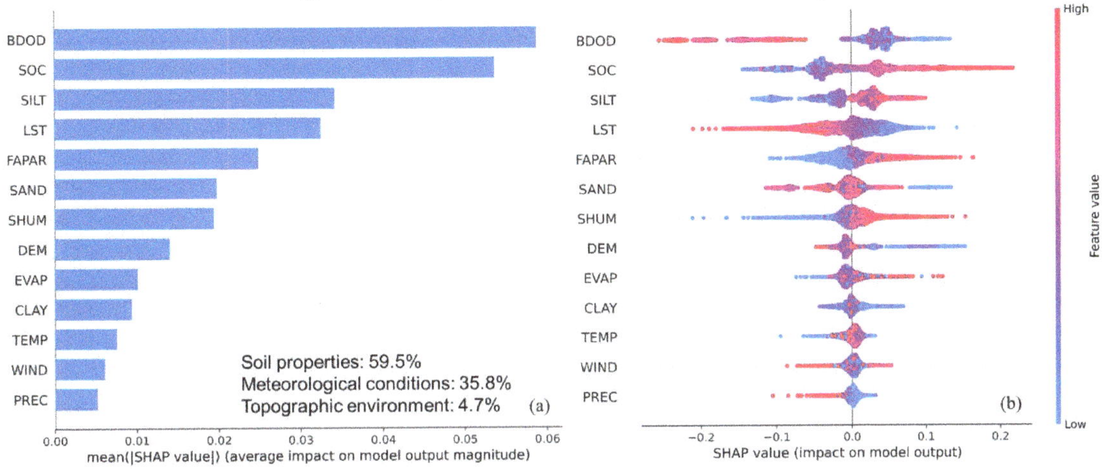

Figure 9. Average SHAP value of every feature (**a**) and SHAP value of every feature for every sample (**b**) in the XGBR model. Soil properties include BDOD, SOC, SLIT, SAND and CLAY; meteorological conditions include LST, FAPAR, SHUM, EVAP, TEMP, WIND and PREC; and topographic environment includes DEM.

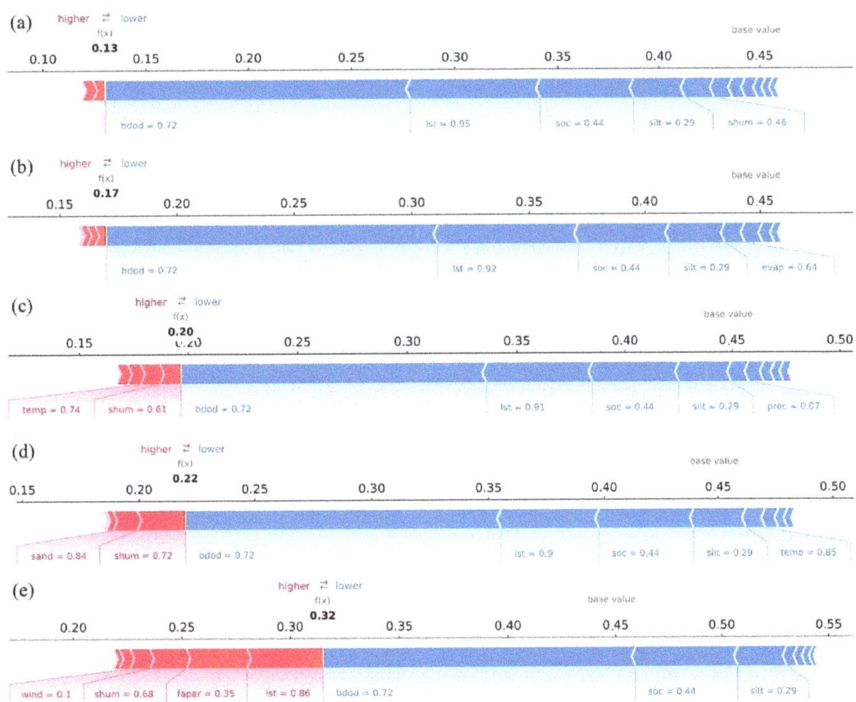

Figure 10. The individual force plots at the BJ station for different drought conditions: (**a**) extreme drought on 6 August 2006, (**b**) severe drought on 28 May 2013, (**c**) moderate drought on 11 June 2015, (**d**) abnormal drought on 27 June 2012 and (**e**) nondrought on 19 September 2012.

4. Discussion

The aforementioned results showed that the XGBR-based SIDI can serve as an efficient drought index for TP drought monitoring. Traditional vegetation-based remote sensing indices at 8 days or 16 days are capable of monitoring agricultural drought. However, these vegetation-based indices usually identify drought characteristics through vegetation conditions, which reflect one specific aspect of drought. In fact, agricultural drought is related to meteorology, soil, and vegetation systems. In comparison, the new drought index, SIDI, comprehensively considered multiple factors. Nevertheless, detailed drought information, such as the onset, end, and intensity of drought, can be identified by the SIDI, which is capable of guiding vulnerable agricultural and livestock production, particularly for the growing season over the TP. The time span of the SIDI is from 2000 to 2016 and, thus, can be used to analyze drought change characteristics over a long time scale. However, there remain some issues with the construction of the SIDI. First, in the process of building the ML model, the model input is only from several stations in the central and western parts of the plateau. Moreover, some special underlying surfaces, such as deserts and glaciers, were not included in the model input. In the future, it is expected that more newly collected in situ data will be added to optimize the model, especially data at new stations where no data have been collected before. On the other hand, this new data can be used for model precision evaluation. Second, despite the development of SIDI, which considers multivariate factors, the mutual response relationship between factors is ignored. For example, the memory of soil moisture leads to a time lag effect between soil moisture and meteorological factors. Liu et al. [34] considered the lagging effect of NDVI on LST and precipitation changes in the newly developed integrated agricultural drought index (IDI). Qing et al. [62] constructed a comprehensive agricultural drought index (CADI) that comprehensively integrated the lagging times of soil moisture with precipitation and evapotranspiration. It is also worthwhile to further investigate whether the time lag effect will improve the prediction accuracy. In addition, many drought indices were developed under the assumption of a statistically stationary distribution of meteorological variables. In fact, the meteorological variables are not stationary due to the influences of climate change and human activities. Considering this, some nonstationary drought indices, such as the standardized nonstationary precipitation index (SnsPI) [63] and the nonstationary standardized runoff index (SRINS) [64] have been developed. Therefore, the nonstationarity of meteorological variables should be taken into account when optimizing the drought index. Third, the definition of the drought threshold level is a crucial step for drought severity categorization. However, drought studies currently focus on drought identification rather than categorization, resulting in various drought categorizations [65]. Fixed threshold levels and the percentile method are the two commonly used drought categorizations, which are not applicable to every region. Standardization of drought categorization is still an issue.

It is worth discussing the interesting finding revealed by the SHAP results: soil characteristics are more important than some meteorological variables in modeling drought. The climatological point is striking due to the negligence of soil in previous studies and deserves more attention. In addition, it remains unclear whether the inclusion of soil bulk density and soil organic content is what makes this model an improvement over others, which needs to be validated in future work.

5. Conclusions

In this study, based on multisource remote sensing and reanalysis data, daily drought indices developed by four machine learning methods, including RFR, KNNR, SVR, and XGBR, were compared. The optimal drought index was selected as the SIDI. Furthermore, the drought monitoring ability of the SIDI was investigated based on the SPI-6, SPEI-6, ERA5-Land SM, and a typical drought event. In addition, the impact of predictor variables on the model output was also explored. The main conclusions are as follows.

(1) By comparing drought indices from four ML models with in situ SM data during the growing season, the drought index based on the XGBR model outperformed that based on other models (R^2 = 0.76, RMSE = 0.11, MAE = 0.08), followed by RFR (R^2 = 0.74, RMSE = 0.11, MAE = 0.08), KNNR (R^2 = 0.73, RMSE = 0.11, MAE = 0.08), and SVR (R^2 = 0.66, RMSE = 0.12, MAE = 0.1). The result proves the superiority of the XGBR model, and this model is used to develop the SIDI.

(2) Compared with the spatial and seasonal distributions of SPI-6, SPEI-6, and ERA5-Land SM, the SIDI reflects the spatial characteristics of the plateau, which is dry in the northwest and humid in the southeast. It also depicts obvious seasonal variability, with large values in winter and small values in summer. For a typical drought event that occurred in the Xizang Autonomous Region in May and June of 2009, the SIDI accurately describes the evolution process of drought spatial evolution on a daily timescale, demonstrating its application potential in drought detection.

(3) Of the thirteen prediction variables, the contribution of 59.5% to model output was from soil properties, 35.8% was from meteorological conditions, and 4.7% was from the topographic environment. The top three variables that influence the model output are bulk density, soil organic carbon, and silt. Moreover, except for bulk density, the other two features have positive impacts on the model output. This suggests that soil information is an important factor affecting drought evolution, which should be taken into account in the construction of the drought index in the future.

Author Contributions: Conceptualization: M.C., L.Z. and Y.M.; methodology: M.C.; software: M.C. and L.Z.; validation: M.C., L.Z. and X.W.; formal analysis: M.C. and P.L.; investigation: X.W., Z.W. and Y.Q.; resources: L.Z.; data curation: M.C. and P.L.; writing—original draft preparation: M.C.; writing—review and editing: L.Z. and Y.M.; visualization: M.C. and Z.W.; supervision: M.C. and Y.Q.; project administration: L.Z.; funding acquisition: L.Z. All authors have read and agreed to the published version of the manuscript.

Funding: This study was supported by National Natural Science Foundation of China (Grant No. 41875031); the Second Tibetan Plateau Scientific Expedition and Research (STEP) Program, Ministry of Science and Technology of the People's Republic of China (Grant No. 2019QZKK0103); National Natural Science Foundation of China (Grant No. 41522501, 41275028, 42230610); CLIMATE-Pan-TPE in the framework of the ESA-MOST Dragon 5 Programme (Grant No. 58516).

Data Availability Statement: Datasets used in this study can be obtained according to the links in Table 1. The newly created SIDI data are available from the authors upon reasonable request as the data need further use.

Conflicts of Interest: The authors declare no conflict of interest.

References

1. Yao, N.; Li, Y.; Lei, T.; Peng, L. Drought evolution, severity and trends in mainland China over 1961–2013. *Sci. Total Environ.* **2018**, *616–617*, 73–89. [CrossRef] [PubMed]
2. Zhang, X.; Su, Z.; Lv, J.; Liu, W.; Ma, M.; Peng, J.; Leng, G. A set of satellite-based near real-time meteorological drought monitoring data over China. *Remote Sens.* **2019**, *11*, 453. [CrossRef]
3. Dai, A. Characteristics and trends in various forms of the Palmer Drought Severity Index during 1900–2008. *J. Geophys. Res. Atmos.* **2011**, *116*, D12115. [CrossRef]
4. Wang, M.; Zhou, C.; Wu, L.; Xu, X.; Ouyang, H. Aridity pattern of Tibetan Plateau and its influential factors in 2001–2010. *Progress. Inquisitiones Mutat. Clim.* **2012**, *8*, 11–17.
5. Zhong, L.; Ma, Y.; Salama, M.S.; Su, Z. Assessment of vegetation dynamics and their response to variations in precipitation and temperature in the Tibetan Plateau. *Clim. Chang.* **2010**, *103*, 519–535. [CrossRef]
6. Zhong, L.; Ma, Y.; Xue, Y.; Piao, S. Climate change trends and impacts on vegetation greening over the Tibetan Plateau. *J. Geophys. Res. Atmos.* **2019**, *124*, 7540–7552. [CrossRef]
7. Gao, Y.; Li, X.; Ruby Leung, L.; Chen, D.; Xu, J. Aridity changes in the Tibetan Plateau in a warming climate. *Environ. Res. Lett.* **2015**, *10*, 034013. [CrossRef]
8. Wang, Z.; Li, J.; Lai, C.; Zeng, Z.; Zhong, R.; Chen, X.; Zhou, X.; Wang, M. Does drought in China show a significant decreasing trend from 1961 to 2009? *Sci. Total Environ.* **2017**, *579*, 314–324. [CrossRef]

9. Feng, W.; Lu, H.; Yao, T.; Yu, Q. Drought characteristics and its elevation dependence in the Qinghai-Tibet plateau during the last half-century. *Sci. Rep.* **2020**, *10*, 14323. [CrossRef]
10. Liu, G. *Encyclopedia of Meteorological Disasters in China: Tibet Volume*; China Meteorological Press: Beijing, China, 2008; pp. 157–229.
11. Lloyd-Hughes, B. The impracticality of a universal drought definition. *Theor. Appl. Climatol.* **2014**, *117*, 607–611. [CrossRef]
12. Dracup, J.A.; Glantz, M.H. Understanding the drought phenomenon: The role of definitions. *Water Int.* **1985**, *10*, 111–120.
13. Society, A.M. Meteorological drought-policy statement. *Bull. Am. Meteorol. Soc.* **1997**, *78*, 847–849.
14. Mishra, A.K.; Singh, V.P. A review of drought concepts. *J. Hydrol.* **2010**, *391*, 202–216. [CrossRef]
15. Crausbay, S.D.; Ramirez, A.R.; Carter, S.L.; Cross, M.S.; Hall, K.R.; Bathke, D.J.; Betancourt, J.L.; Colt, S.; Cravens, A.E.; Dalton, M.S.; et al. Defining Ecological Drought for the Twenty-First Century. *Bull. Am. Meteorol. Soc.* **2017**, *98*, 2543–2550. [CrossRef]
16. Tramblay, Y.; Koutroulis, A.; Samaniego, L.; Vicente-Serrano, S.M.; Volaire, F.; Boone, A.; Le Page, M.; Llasat, M.C.; Albergel, C.; Burak, S.; et al. Challenges for drought assessment in the Mediterranean region under future climate scenarios. *Earth-Sci. Rev.* **2020**, *210*, 103348. [CrossRef]
17. Vicente-Serrano, S.M.; Quiring, S.M.; Pena-Gallardo, M.; Yuan, S.; Dominguez-Castro, F. A review of environmental droughts; increased risk under global warming? *Earth-Sci. Rev.* **2020**, *201*, 102953. [CrossRef]
18. Maliva, R.; Thomas, M. Aridity and drought. In *Arid Lands Water Evaluation and Management*; Springer: Berlin/Heidelberg, Germany, 2012; pp. 21–39.
19. Alley, W.M. The Palmer drought severity index: Limitations and assumptions. *J. Appl. Meteorol. Climatol.* **1984**, *23*, 1100–1109. [CrossRef]
20. Mckee, T.B.; Doesken, N.J.; Kleist, J. The Relationship of Drought Frequency and Duration to Time Scales. In Proceedings of the 8th Conference on Applied Climatology, Anaheim, CA, USA, 17–22 January 1993.
21. Vicente-Serrano, S.M.; Beguería, S.; López-Moreno, J.I. A Multiscalar Drought Index Sensitive to Global Warming: The Standardized Precipitation Evapotranspiration Index. *J. Clim.* **2010**, *23*, 1696–1718. [CrossRef]
22. Xiong, J.; Guo, S.; Abhishek; Li, J.; Yin, J. A novel standardized drought and flood potential index based on reconstructed daily GRACE data. *J. Hydrometeorol.* **2022**, *23*, 1419–1438. [CrossRef]
23. Kinouchi, T.; Sayama, T. A comprehensive assessment of water storage dynamics and hydroclimatic extremes in the Chao Phraya River Basin during 2002–2020. *J. Hydrol.* **2021**, *603*, 126868.
24. Yang, M.C.; Yao, T.D.; Gou, X.H.; Koike, T.; He, Y. The soil moisture distribution, thawing-freezing processes and their effects on the seasonal transition on the Qinghai-Xizang (Tibetan) Plateau. *J. Asian Earth Sci.* **2003**, *21*, 457–465. [CrossRef]
25. Kogan, F.; Sullivan, J. Development of global drought-watch system using NOAA/AVHRR data. *Adv. Space Res.* **1993**, *13*, 219–222. [CrossRef]
26. Kogan, F.N. Application of vegetation index and brightness temperature for drought detection. *Adv. Space Res.* **1995**, *15*, 91–100. [CrossRef]
27. Wu, B.; Ma, Z.; Yan, N. Agricultural drought mitigating indices derived from the changes in drought characteristics. *Remote Sens. Environ.* **2020**, *244*, 111813. [CrossRef]
28. Huang, S.; Chang, J.; Leng, G.; Huang, Q. Integrated index for drought assessment based on variable fuzzy set theory: A case study in the Yellow River basin, China. *J. Hydrol.* **2015**, *527*, 608–618. [CrossRef]
29. Lu, J.; Carbone, G.J.; Gao, P. Mapping the agricultural drought based on the long-term AVHRR NDVI and North American Regional Reanalysis (NARR) in the United States, 1981–2013. *Appl. Geogr.* **2019**, *104*, 10–20. [CrossRef]
30. Rhee, J.; Im, J.; Carbone, G.J. Monitoring agricultural drought for arid and humid regions using multi-sensor remote sensing data. *Remote Sens. Environ.* **2010**, *114*, 2875–2887. [CrossRef]
31. Hao, Z.; AghaKouchak, A. A nonparametric multivariate multi-index drought monitoring framework. *J. Hydrometeorol.* **2014**, *15*, 89–101. [CrossRef]
32. Won, J.; Choi, J.; Lee, O.; Kim, S. Copula-based Joint Drought Index using SPI and EDDI and its application to climate change. *Sci. Total Environ.* **2020**, *744*, 140701. [CrossRef]
33. Hao, Z.; Singh, V.P. Drought characterization from a multivariate perspective: A review. *J. Hydrol.* **2015**, *527*, 668–678. [CrossRef]
34. Liu, X.; Zhu, X.; Zhang, Q.; Yang, T.; Pan, Y.; Sun, P. A remote sensing and artificial neural network-based integrated agricultural drought index: Index development and applications. *Catena* **2020**, *186*, 104394. [CrossRef]
35. Su, Z.; Wen, J.; Dente, L.; van der Velde, R.; Wang, L.; Ma, Y.; Yang, K.; Hu, Z. The Tibetan Plateau observatory of plateau scale soil moisture and soil temperature (Tibet-Obs) for quantifying uncertainties in coarse resolution satellite and model products. *Hydrol. Earth Syst. Sci.* **2011**, *15*, 2303–2316. [CrossRef]
36. van der Velde, R.; Su, Z.; van Oevelen, P.; Wen, J.; Ma, Y.; Salama, M.S. Soil moisture mapping over the central part of the Tibetan Plateau using a series of ASAR WS images. *Remote Sens. Environ.* **2012**, *120*, 175–187. [CrossRef]
37. Dente, L.; Vekerdy, Z.; Wen, J.; Su, Z. Maqu network for validation of satellite-derived soil moisture products. *Int. J. Appl. Earth Obs.* **2012**, *17*, 55–65. [CrossRef]
38. Su, Z.; de Rosnay, P.; Wen, J.; Wang, L.; Zeng, Y. Evaluation of ECMWF's soil moisture analyses using observations on the Tibetan Plateau. *J. Geophys. Res. Atmos.* **2013**, *118*, 5304–5318. [CrossRef]
39. Ma, Y.; Hu, Z.; Xie, Z.; Ma, W.; Wang, B.; Chen, X.; Li, M.; Zhong, L.; Sun, F.; Gu, L.; et al. A long-term (2005–2016) dataset of hourly integrated land–atmosphere interaction observations on the Tibetan Plateau. *Earth Syst. Sci. Data* **2020**, *12*, 2937–2957. [CrossRef]

40. Zhang, H.; Ding, M.; Li, L.; Liu, L. Continuous wetting on the Tibetan Plateau during 1970–2017. *Water* **2019**, *11*, 2605. [CrossRef]
41. Rahmati, O.; Falah, F.; Dayal, K.S.; Deo, R.C.; Mohammadi, F.; Biggs, T.; Moghaddam, D.D.; Naghibi, S.A.; Bui, D.T. Machine learning approaches for spatial modeling of agricultural droughts in the south-east region of Queensland Australia. *Sci. Total Environ.* **2020**, *699*, 134230. [CrossRef]
42. Zhao, X.L.; Li, W.L.; Guo, X.L.; Yu, C.; Zhao, Y.T.; Xu, J. The responses of Pa, SPI, SPEI to dry climate in alpine meadows of eastern Qing-Tibet Plateau. *Pratacultural Sci.* **2017**, *34*, 273–282.
43. Breiman, L. Random Forests. *Mach. Learn.* **2001**, *45*, 5–32. [CrossRef]
44. Park, S.; Im, J.; Jang, E.; Rhee, J. Drought assessment and monitoring through blending of multi-sensor indices using machine learning approaches for different climate regions. *Agric. For. Meteorol.* **2016**, *216*, 157–169. [CrossRef]
45. Gholami, R.; Nikoo, F. Support vector machine: Principles, parameters, and applications. In *Handbook of Neural Computation*; Academic Press: Cambridge, MA, USA, 2017; pp. 515–535.
46. Kramer, O. Dimensionality reduction by unsupervised k-nearest neighbor regression. In Proceedings of the 2011 10th International Conference on Machine Learning and Applications and Workshops, Honolulu, HI, USA, 18–21 December 2011; pp. 275–278.
47. Chen, T.; Guestrin, C. Xgboost. In Proceedings of the 22nd ACM SIGKDD International Conference on Knowledge Discovery and Data Mining, San Francisco, CA, USA, 13–17 August 2016; pp. 785–794.
48. Cheng, M.; Zhong, L.; Ma, Y.; Zou, M.; Ge, N.; Wang, X.; Hu, Y. A study on the assessment of multi-source satellite soil moisture products and reanalysis data for the Tibetan Plateau. *Remote Sens.* **2019**, *11*, 1196. [CrossRef]
49. Ge, N.; Zhong, L.; Ma, Y.; Cheng, M.; Wang, X.; Zou, M.; Huang, Z. Estimation of land surface heat fluxes based on Landsat 7 ETM+ data and field measurements over the Northern Tibetan Plateau. *Remote Sens.* **2019**, *11*, 2899. [CrossRef]
50. Shapley, L.S. A value for n-person games. In *Contributions to the Theory of Games*; Princeton University Press: Princeton, NJ, USA, 1953; pp. 307–317.
51. Lundberg, S.; Lee, S.-I. A unified approach to interpreting model predictions. *arXiv* **2017**, arXiv:1705.07874.
52. Dikshit, A.; Pradhan, B. Interpretable and explainable AI (XAI) model for spatial drought prediction. *Sci. Total Environ.* **2021**, *801*, 149797. [CrossRef]
53. Zhu, F.; Li, X.; Qin, J.; Yang, K.; Cuo, L.; Tang, W.; Shen, C. Integration of multisource data to estimate downward longwave radiation based on deep neural networks. *IEEE Trans. Geosci. Remote Sens.* **2022**, *60*, 1–15. [CrossRef]
54. Dikshit, A.; Pradhan, B.; Alamri, A.M. Temporal hydrological drought index forecasting for New South Wales, Australia using machine learning approaches. *Atmosphere* **2020**, *11*, 585. [CrossRef]
55. Wang, X.Y.; Li, J.; Xing, L.T. Comparative agricultural drought monitoring based on three machine learning methods. *Arid. Zone Res.* **2022**, *39*, 322–332.
56. Nguyen-Sy, T.; Wakim, J.; To, Q.; Vu, M.; Nguyen, T.; Nguyen, T. Predicting the compressive strength of concrete from its compositions and age using the extreme gradient boosting method. *Constr. Build. Mater.* **2020**, *260*, 119757. [CrossRef]
57. Zhang, X.; Duan, Y.; Duan, J.; Jian, D.; Ma, Z. A daily drought index based on evapotranspiration and its application in regional drought analyses. *Sci. China Earth Sci.* **2022**, *65*, 317–336. [CrossRef]
58. Liu, Y.; Feng, W.; Zha, X.; Liang, L.; Li, T. Remote sensing monitoring of drought level in North Tibet based on MODIS TVDI and fuzzy mathematics. *Arid. Zone Res.* **2020**, *37*, 86–96.
59. Liu, J.; He, Z.; Duan, Y. Monitoring the droughts in Tibet based on remote sensing using MODIS data. *J. Arid. Land Resour. Environ.* **2013**, *27*, 134–139.
60. Dracup, J.A.; Lee, K.S.; Paulson, E.G. On the definition of droughts. *Water Resour.* **1980**, *16*, 297–302. [CrossRef]
61. Lundberg, S.M.; Erion, G.G.; Lee, S.I. Consistent individualized feature attribution for tree ensembles. *arXiv* **2018**, arXiv:1802.03888.
62. Tian, Q.; Lu, J.; Chen, X. A novel comprehensive agricultural drought index reflecting time lag of soil moisture to meteorology: A case study in the Yangtze River basin, China. *Catena* **2022**, *209*, 105804. [CrossRef]
63. Russo, S.; Dosio, A.; Sterl, A.; Barbosa, P.; Vogt, J. Projection of occurrence of extreme dry-wet years and seasons in Europe with stationary and nonstationary Standardized Precipitation Indices. *J. Geophys. Res. Atmos.* **2013**, *118*, 7628–7639. [CrossRef]
64. Jehanzaib, M.; Shah, S.A.; Yoo, J.; Kim, T. Investigating the impacts of climate change and human activities on hydrological drought using non-stationary approaches. *J. Hydrol.* **2020**, *588*, 125052. [CrossRef]
65. Bazrkar, M.H.; Zhang, J.; Chu, X. Hydroclimatic aggregate drought index (HADI): A new approach for identification and categorization of drought in cold climate regions. *Stoch. Environ. Res. Risk Assess.* **2020**, *34*, 1847–1870. [CrossRef]

remote sensing

MDPI

Article

Responses of Soil Freeze–Thaw Processes to Climate on the Tibetan Plateau from 1980 to 2016

Chunwei Fu [1,2,3], Zeyong Hu [1,3,*], Yaoxian Yang [1,3], Mingshan Deng [1], Haipeng Yu [1,3], Shan Lu [4], Di Wu [1,2,3] and Weiwei Fan [1,2,3]

1 Key Laboratory of Land Surface Process and Climate Change in Cold and Arid Regions, Northwest Institute of Eco-Environment and Resources, Chinese Academy of Sciences, Lanzhou 730000, China
2 University of Chinese Academy of Sciences, Beijing 100049, China
3 Nagqu Station of Plateau Climate and Environment, Northwest Institute of Eco-Environment and Resources, Chinese Academy of Sciences, Nagqu 852000, China
4 Shaanxi Meteorological Service Center, Xi'an 710014, China
* Correspondence: zyhu@lzb.ac.cn

Abstract: Soil freeze–thaw processes are remarkable features of the land surface across the Tibetan Plateau (TP). Soil moisture and temperature fluctuate during the freeze–thaw cycle, affecting the soil water and energy exchange between the land and atmosphere. This study investigates variations in the soil temperature, humidity, and freeze–thaw state and their responses to air temperature and precipitation on the TP from 1981 to 2016. Regional simulations of the TP using Community Land Model version 4.5 demonstrate that the climate of the TP has become warmer and wetter over the past 37 years, with increases in both regional average temperature and precipitation. Using empirical orthogonal function analysis and the Mann–Kendall trend test of air temperature, we show that 1980–1998 was relatively cold, and 1999–2016 was relatively warm. Soil temperature and moisture in most areas of the TP were affected by air temperature and precipitation, and both showed an upward trend during the past 37 years. Overall, from 1981 to 2016, the freezing date of the TP has become delayed, the thawing date has been hastened, and the duration of the freeze–thaw state has shortened. The surface soil freezes and thaws first, and these processes pervade deeper soil with the passage of time; freeze–thaw processes have an obvious hysteresis. Precipitation and air temperature had marked effects on the freeze–thaw processes. Higher air temperatures delay the freezing date, hasten the thawing date, and shorten the freeze–thaw period. Areas with the highest precipitation saw late soil freeze, early thaw, and the shortest freeze–thaw duration. Areas with less vegetation froze earlier and thawed later. The freeze–thaw duration increased in the northwest of the plateau and decreased on the rest of the plateau. This article informs research on frozen soil change in the context of global warming.

Keywords: freezing–thawing processes; climate change; air temperature and precipitation; Community Land Model; Tibetan Plateau

Citation: Fu, C.; Hu, Z.; Yang, Y.; Deng, M.; Yu, H.; Lu, S.; Wu, D.; Fan, W. Responses of Soil Freeze–Thaw Processes to Climate on the Tibetan Plateau from 1980 to 2016. *Remote Sens.* **2022**, *14*, 5907. https://doi.org/10.3390/rs14235907

Academic Editor: Marouane Temimi

Received: 31 August 2022
Accepted: 16 November 2022
Published: 22 November 2022

1. Introduction

With an average elevation of more than 4000 m, the Tibetan Plateau (TP), also known as the "Roof of the World", lies adjacent to the subtropical zone in the south and approaches mid-latitudes in the north [1,2]. It is the highest plateau on Earth and has the most complicated topology. Owing to its unique geographical position and altitude and the complex underlying surface, the TP creates a unique regional atmospheric circulation; this has a considerable influence on the weather and climate of China, the atmospheric circulation in East Asia, and even the global climate [3–5]. Land surface processes are closely related to atmospheric movement and constitute momentum, energy, and water exchange activities between the land and atmosphere. Land surface processes and climate change have a reciprocal relationship. While simultaneously being affected by climate change, the land surface also further influences atmospheric turbulence and ocean movement by participating in land–atmosphere and land–ocean energy and water interactions; these, in turn, have a notable impact on climate change, providing further feedback to influence the land surface [6–9]. The realistic simulation of land surface processes, particularly the process of hydrothermal transport, is critical for climate change research.

Frozen soil is a major forcing factor on land surface processes and is a highly sensitive indicator of climate change [10]. The substantial warming of the climate over the past few decades has caused a significant increase in surface temperature and a sharp degradation of frozen soil [11–13]. Moreover, the accelerated climate change has driven a widespread area expansion of lakes across the Changtang Plateau (CP) and desertification in the northern TP [14,15]. Frozen soil is widely distributed across the planet. According to the duration of freezing, frozen soil can be divided into two types: permafrost and seasonally frozen soil. Currently, permafrost covers 23–25% of the Earth's land surface, and a further 25% is covered by seasonally frozen soil. The TP contains approximately 2.7×10^6 km^2 of frozen soil, with permafrost and seasonally frozen soil covering 1.5×10^6 and 1.2×10^6 km^2, respectively [16,17]. Frozen soil has unique hydrothermal characteristics, which play an essential role in land surface processes. The presence of ice in frozen soil renders the process of soil water and heat transfer more complicated; it also affects the infiltration of precipitation, ice melt, and snow melt, which can regulate energy and water circulation between the land and atmosphere. The processes of soil freezing and thawing not only play an important role in the transition between dry and wet seasons on the TP but also greatly affect the exchange of water and energy between the land and atmosphere. Features of the spatiotemporal distribution of temperature and moisture, as well as continual changes in the processes of soil moisture and heat transport, are critical to the alteration of the plateau ecosystem [18–21]. Furthermore, there is a good correlation between summer precipitation in China and atmospheric circulation on the TP and East Asia [22–24].

Soil moisture and temperature are the two core elements of land surface processes. Soil moisture controls the energy–water balance between the land surface and atmosphere; it can affect the surface albedo and heat capacity, with the former regulating the local net radiation flux and the latter affecting heat exchange between the land and atmosphere. Soil moisture also affects surface evaporation and vegetation transpiration, which together determine water vapor and latent heat transport to the atmosphere. Soil temperature directly controls sensible heat transport from the land to the atmosphere and the presence and phase of water in the soil; this, in turn, affects soil moisture and the series of physical and biochemical processes that it controls. Soil moisture and temperature are also the most direct reflections of the freeze–thaw processes on the plateau. Studies have shown that freeze–thaw processes can directly change the hydrothermal properties of the soil, thereby affecting the surface energy budget and hydrological cycle and, further, affecting the exchange of energy and water between the land and atmosphere.

Soil moisture plays an important role in the climate system, second only to sea surface temperature (SST), and its role even exceeds that of SST on land [25]. Evaporation from the land surface accounts for 65% of precipitation on land [26], and soil moisture is one of the key elements that influences evaporation. Changes in surface albedo, heat capacity, and sensible and latent heat delivered to the atmosphere all influence climate change [26,27]. The freeze–thaw processes of soil and the spatiotemporal distribution of soil temperature are notably affected by changes in the soil moisture on the TP [28]. Frozen soil contains nearly twice as much organic carbon as the atmosphere [29]; organic carbon initially trapped in frozen soil will be released into the atmosphere as a result of climate change, enhancing the greenhouse effect. Therefore, the study of the soil freeze–thaw state is of great significance for global climate change.

2. Study Area and Data

2.1. Study Area

The study area of this paper is the Tibetan Plateau (25°N~40°N, 74°E~104°E). This part of China starts from the Pamir Plateau in the west and the Hengduan Mountains in the east. It spans 31 degrees of longitude and is about 2945 km long from east to west. The width is about 1532 km. The average altitude of the TP is about 4000 m. China is the third-largest country with frozen soil in the world, and its frozen soil is mainly distributed on the TP, which is known as the "third pole in the world"; the spatial distribution of the TP's permafrost is shown in Figure 1a [30]. In addition, the map of vegetation types is given in Figure 1b.

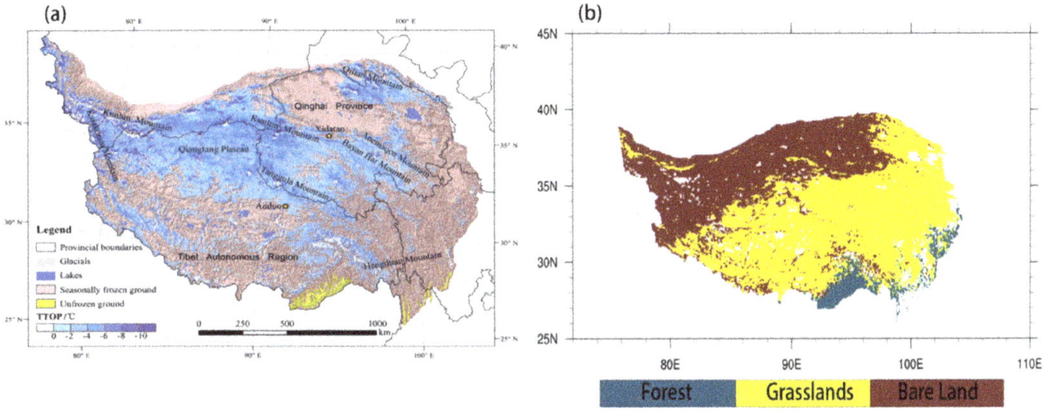

Figure 1. (**a**) Spatial distribution of permafrost on the TP [30]. (**b**) Spatial patterns of different vegetation types on the TP. The vegetation type map is derived from the MODIS land–cover product.

2.2. Data

Land–cover data. A collection of the 5.1 MODIS land-cover product (MCD12C1) was used in this study for statistical analysis. There were five land–cover classification schemes in this dataset, and the International Geosphere and Biosphere Programme (IGBP), at a spatial resolution of 0.05° from 2001–2008, was developed. This dataset consists of 17 general land–cover types, which include 11 natural vegetation classes, three developed and mosaicked land classes, and three non-vegetated land classes [31,32]. We selected three vegetation types: forest, grassland, and bare land.

Atmospheric forcing data for our study, comprising shortwave radiation, precipitation, pressure, specific humidity, surface air temperature, and wind, were obtained from the China Meteorological Forcing Dataset (CMFD), which was created by the Chinese Academy of Sciences' Institute of TP Research. This dataset, which was produced by merging a variety of data sources (i.e., remote sensing data, reanalysis data, station data), spans the years 1979–2012 and has a 0.1° spatial resolution and a 3 h temporal resolution. The CMFD has been utilized in many previous studies [33], and its precision has been confirmed to be sufficient for modeling. The data sources used to produce the forcing data include:

1. Wind, relative humidity, sunshine duration, air temperature, precipitation, and surface pressure observations from China Meteorological Administration (CMA) weather stations for the years 1980 to 2016; true values of the meteorological parameters calculated using the observed data and radiation data estimated from the observed sunshine duration.
2. Tropical Rainfall Measuring Mission (TRMM) satellite precipitation analysis data (3B42) for the years 1998 to 2016 and Global Land Data Assimilation System (GLDAS) precipitation for the years 1980 to 2016.

3. GLDAS downward shortwave radiation data for the years January 1980–June 1983 and January 2008–December 2016 and Global Energy and Water Exchanges-Surface Radiation Budget (GEWEX-SRB) downward shortwave radiation data for the years July 1983–December 2007.
4. The modern era-retrospective analysis for research and applications (MERRA) surface pressure for the years 1979 to 2015. GLDAS surface pressure data after 2015.
5. GLDAS air temperature, wind, and relative humidity data for the years 1979 to 2018.

To investigate the responses of soil freeze–thaw processes to climate change on the entire TP, the atmospheric forcing data from the CMFD on the TP (25°–40°N, 75°–105°E) was used to drive the regional simulation of CLM4.5. We ran CLM4.5 for 37 years from 1980 to 2016, and the spatial resolution of CLM4.5 is 10 km × 10 km.

The observed data. The relevant research is based on field weather station data from the BJ site, Amdo site, and NewD66 site of the Nagqu Plateau Climate and Environment (NPCE) station on the TP. The location of the BJ site is 31.37°N, 91.90°E, the Amdo site is 32.24°N, 91.62°E, and the NewD66 site is 35.43°N, 93.59°E.

3. Methods

3.1. Model Description

The land surface process model utilized in this study was the Community Land Model (CLM) version 4.5 [34], which is the land surface module of the Community Earth System Model (CESM) developed by the National Center for Atmospheric Research (NCAR). The CLM is currently one of the most widely used land surface process models; it includes biogeophysical processes, hydrological cycle processes, surface heterogeneity, dynamic processes of the biological system, and biochemical processes, among others. Compared with the previous version, CLM4.5 has expanded the performance of the model, updated the atmospheric and surface forcing datasets, added the concept of surface water storage, replaced the original wetland unit of the model, adjusted the photosynthetic parameters, and improved some parameterization schemes. Improvements to the frozen soil water conduction parameterization scheme [35–37], as well as changes to the plant canopy radiation scheme, mean that CLM4.5 can better simulate water and heat transfer processes within the soil.

The simulation of soil temperature is related to the surface energy balance equation; the equation describing the surface energy balance in CLM4.5 is:

$$h = S_{soil} - L_{soil} - H_{soil} - \lambda E_{soil}, \tag{1}$$

where h is the heat flux entering the soil (W/m^2); S_{soil} is the solar short-wave radiation absorbed by the surface (W/m^2); L_{soil} is the long-wave radiation absorbed by the surface (W/m^2, defined with upwards being the positive direction); H_{soil} is the surface sensible heat flux (W/m^2); and λE_{soil} is the surface latent heat flux. The heat that enters the soil from the atmosphere is determined by the net surface radiation and surface sensible and latent heat; in turn, the surface sensible and latent heat are determined by the difference in temperature and humidity between the land and atmosphere. Precipitation affects changes in surface temperature. The simulation of soil temperature is, thus, related to both air temperature and precipitation.

In CLM4.5, the change in soil moisture with time is calculated by the surface water balance equation:

$$\frac{\mathrm{dW}}{\mathrm{dt}} = Pr - ET - \mathrm{Rof}, \tag{2}$$

where w is soil moisture (mm); t is time (s); Pr is precipitation rate (mm/s); ET is evapotranspiration (mm/s); and Rof is surface runoff (mm/s). The strength of evaporation depends on the temperature. Runoff is affected by precipitation and the water-holding capacity of the soil. Hence, changes in soil moisture are also related to temperature and precipitation [38,39].

To better analyze the characteristics of freeze–thaw processes based on soil temperature, the soil temperature simulated by the CLM4.5 was validated by the observation data of three field sites. The correlation between the simulated soil temperature values and the observed values can reach more than 90% (Figure 2); in other words, the reliability of the results by CLM4.5 simulation was validated, so the CLM4.5 model simulation results was used to analyze freeze–thaw processes in this paper.

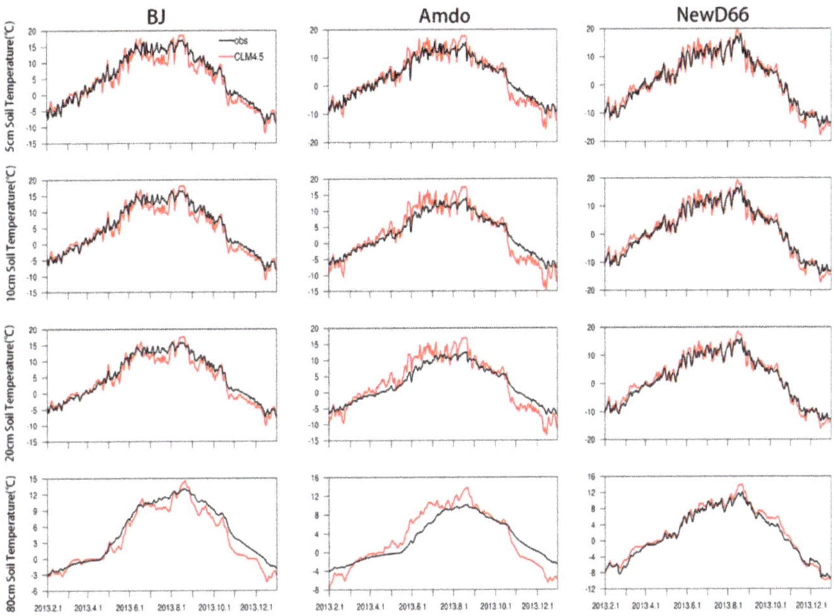

Figure 2. Comparison of simulated and observed soil temperature in 4 soil layers at BJ site, Amdo site, and NewD66 site.

3.2. Empirical Orthogonal Function (EOF) Analysis

Empirical orthogonal function analysis is a common method in meteorology, which has fast convergence and a good reflection of the basic structural characteristics of meteorological elements. In this paper, the EOF analysis method was used to study the spatiotemporal distribution characteristics of temperature on the TP. The original meteorological data were projected on an orthogonal basis in the EOF analysis. In addition, the orthogonal basis was calculated using the eigenvector of the spatially weighted anomaly covariance matrix, with the associated eigenvalues reflecting the percentage variance explained by each pattern. As a result, the EOFs of spatiotemporal physical processes can reflect mutually orthogonal spatial patterns in the data change set, with the first pattern accounting for the majority of the variance, the second pattern accounting for the majority of the residual variance, and so on. We selected the matching principal component (PC) of the dominant mode as the reference time series for air temperature, because the PC of an EOF mode illustrates how the spatial pattern of this mode oscillates over time.

In this paper, the CMFD data was used to force CLM4.5 to simulate soil temperature and moisture on the Tibetan Plateau from 1980 to 2016, empirical orthogonal function analysis and the Mann–Kendall trend test of air temperature were used to divide the 37 years of TP into different climate situations, and the characteristics of freeze–thaw processes based on soil temperature were analyzed.

4. Results

4.1. Regional Climate Change on the TP

The CMFD provided temperature and precipitation data from 1980 to 2016, which were used to calculate the climatology and the trends of temperature and precipitation (Figure 3). The climatology of regional average temperature and precipitation is −0.15 °C and 465.14 mm, respectively, and their regional average trend is 0.41 °C/decade and 6.44 mm/decade, respectively. The regional distributions of precipitation and temperature are similar, as the monsoon advances, with values roughly increasing gradually from the northwest to the southeast of the TP. The southeast region of the TP has the highest annual average temperature and the most annual average precipitation; moreover, high temperature and low precipitation occur in the Qaidam Basin. The trends of temperature and precipitation show that the temperature and precipitation in most parts of TP have increased significantly in the past 37 years as a result of global warming, though the temperature in the Karakoram area has a decreased trend, and, at the same time, precipitation has increased in this area. The Karakoram Mountains have always been the focus of scientists due to their unusually stable glaciers. According to a study published in *Nature Geoscience*, the glaciers in the Karakoram Mountains have risen rather than reduced during the last 10 years, which contradicts the global trend of glacial melting [40]; the combination of temperature and precipitation trends may be the cause of this phenomenon.

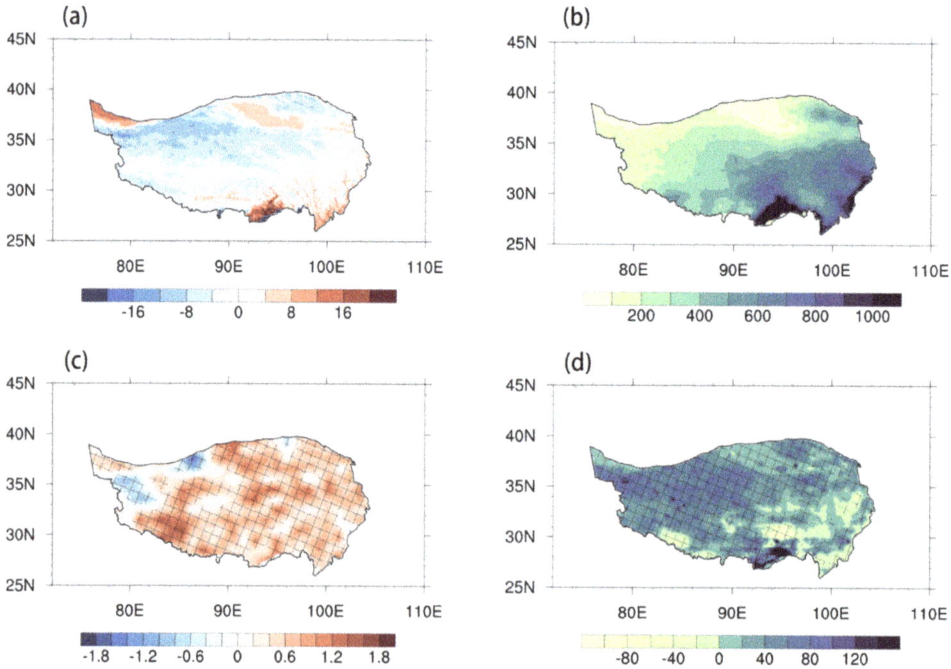

Figure 3. Climatology and trends in temperature and precipitation on the TP from 1980 to 2016. (**a**) Climatology of temperature (unit: °C). (**b**) Climatology of precipitation (unit: mm). (**c**) Trend in temperature (unit: °C/decade). (**d**) Trend in precipitation (unit: mm/decade). Grid points with statistically significant anomalies at the 90% confidence level are denoted by an oblique line.

To identify the change characteristics of the annual average temperature, EOF analysis was performed using the air temperature data and was carried out on the covariance matrix. The spatial pattern of the first dominant mode (EOF1) is depicted in Figure 4a, which was created using the air temperature data. The spatial distribution of EOF1 is marked by consistent variations in temperature in the entire TP, with an explained variance of 50.7%.

The corresponding normalized PC1 shows that the temperature has changed with time, with a significant increase since 1980. The Mann–Kendall trend test was conducted on the average temperature of the TP from 1980 to 2016 and showed that the obvious warming of the TP after the 1990s was an abrupt phenomenon, and the average temperature mutational change occurred in 1998 (Figure 4c). Therefore, according to the time when this mutational change occurred (1998), the average temperature of the TP was divided into two sections for discussion; the years from 1980 to 1998 were relatively cold (hereafter RC), and the years from 1999 to 2016 were relatively warm (hereafter RW).

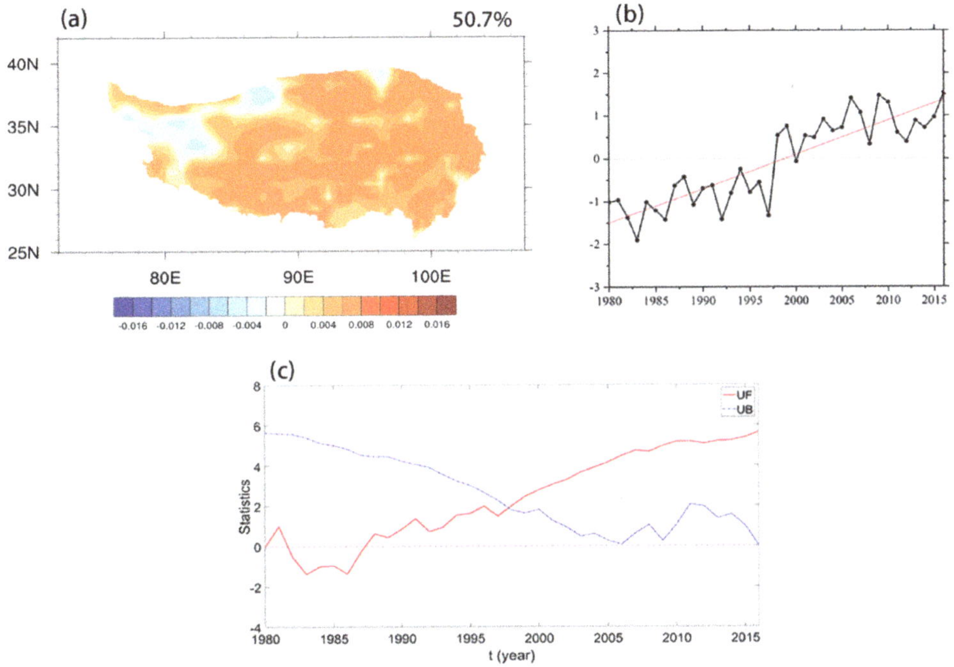

Figure 4. Temperature patterns on the TP from 1980 to 2016. (**a**) EOF1 spatial pattern. (**b**) Time coefficient. (**c**) Mann–Kendall trend test.

4.2. Responses of Soil Temperature and Moisture

As shown in Figures 5 and 6, on average, both the soil temperature and soil moisture across most parts of the TP have increased between 1980 and 2016. The trend in soil temperature is closely related to that in air temperature. The trends in soil temperature and moisture at different soil depths across the TP are essentially the same. In areas where the air temperature has risen, the soil temperature has also risen, indicating that soil temperature is directly affected by the air temperature. The average soil temperature trends at soil depths of 10, 20, 40, and 60 cm were 0.294, 0.291, 0.287, and 0.282 °C/decade, respectively. The average soil moisture trends at soil depths of 10, 20, 40, and 60 cm were 0.00241, 0.00242, 0.00250, and 0.00265 $m^3 \cdot m^{-3}$/decade, respectively. With an increase in soil depth, soil temperature increased at a lower rate, while soil moisture increased at a slightly higher rate, indicating that the shallow soil temperature is more susceptible to air temperature. The trend seen in soil moisture is roughly the same as that seen in precipitation. In areas where precipitation has increased, soil moisture has also tended to increase; where precipitation has decreased, soil moisture has also decreased, indicating that changes in soil moisture are affected by precipitation. In the northern part of the TP, the soil has become noticeably wetter, while in some parts of the southern TP, soil moisture has noticeably decreased.

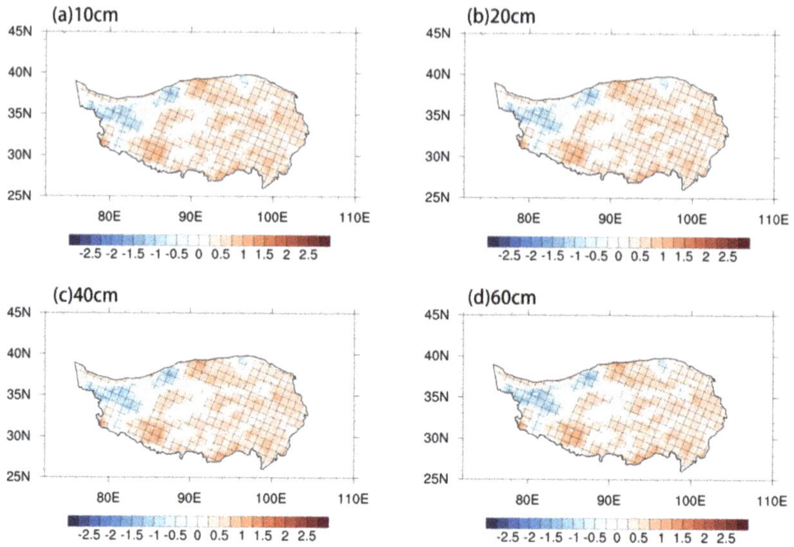

Figure 5. Trends in soil temperature at four different soil depths across the TP from 1980 to 2016 (unit: °C/decade). (**a**) 10 cm; (**b**) 20 cm; (**c**) 40 cm; (**d**) 60 cm.

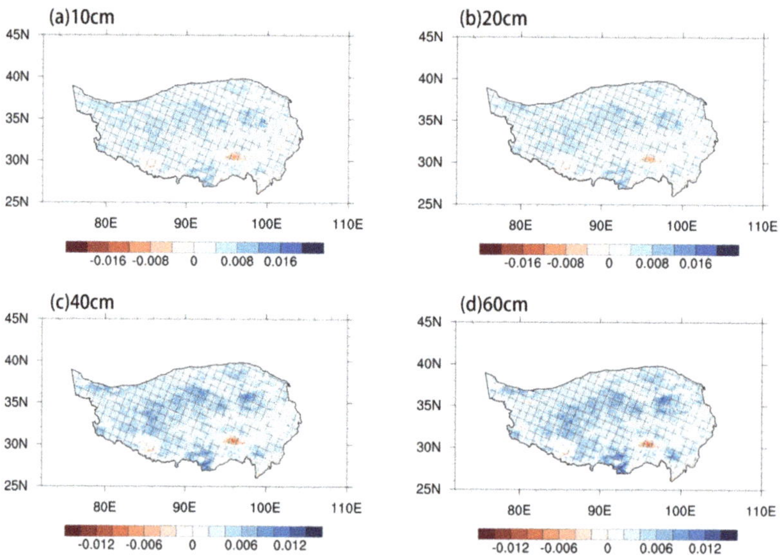

Figure 6. Same as Figure 5, but for soil moisture (unit: $m^3 \cdot m^{-3}$/decade).

To better study the influence of air temperature on soil temperature, soil moisture, and freeze–thaw processes, Figure 7 shows comparisons of the annual average soil temperature between RW (from 1999 to 2016) and RC (from 1980 to 1998); the average soil temperature differences, at depths of 10, 20, 40, and 60 cm between RW and RC, were 0.568, 0.562, 0.555, and 0.546 °C, respectively. The soil temperature differences across most of the TP were positive (and above 0.5 °C) at all four depths, but the differences in the northwestern TP were negative. In addition, the maximum difference was found in the southwest, indicating that the soil temperature in this region is most sensitive to changes in air temperature. As the soil depth increased, the difference in soil temperature decreased slightly. The spatial distribution of the difference in soil temperature is consistent with the temperature trend.

On the one hand, this shows that shallow soil will be affected by the air temperature, to a certain extent; on the other hand, it shows that the vertical gradient in soil temperature is not obvious, and temperature changes are not sensitive to changes in depth. Figure 8 depicts the differences in annual average soil moisture between RW and RC; at depths of 10, 20, 40, and 60 cm, these differences were 0.00586, 0.00584, 0.00591, and 0.00611 m^3/m^3, respectively. As the soil depth increased, the vertical change in soil moisture is not obvious, and the difference value increased slightly. In terms of the type of land cover, the soil in alpine steppe areas was considerably wetter than that in alpine meadow and alpine desert areas.

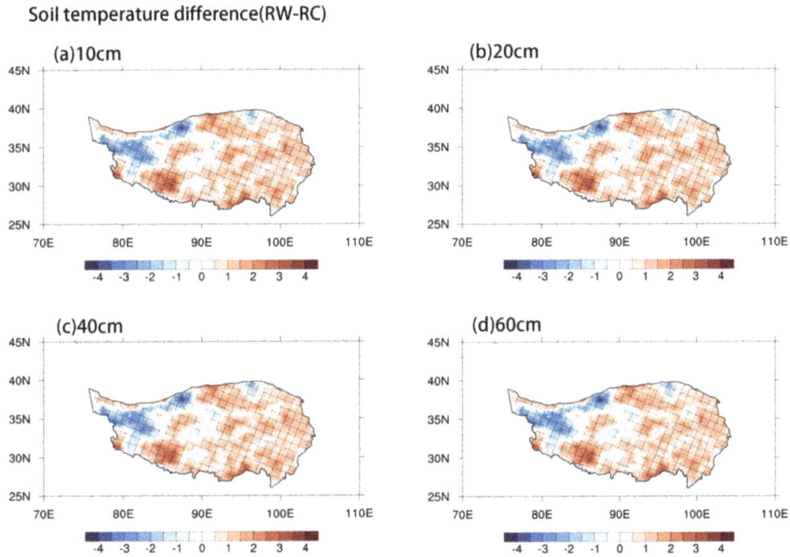

Figure 7. Same as Figure 5, but for differences in average soil temperature between RW and RC.

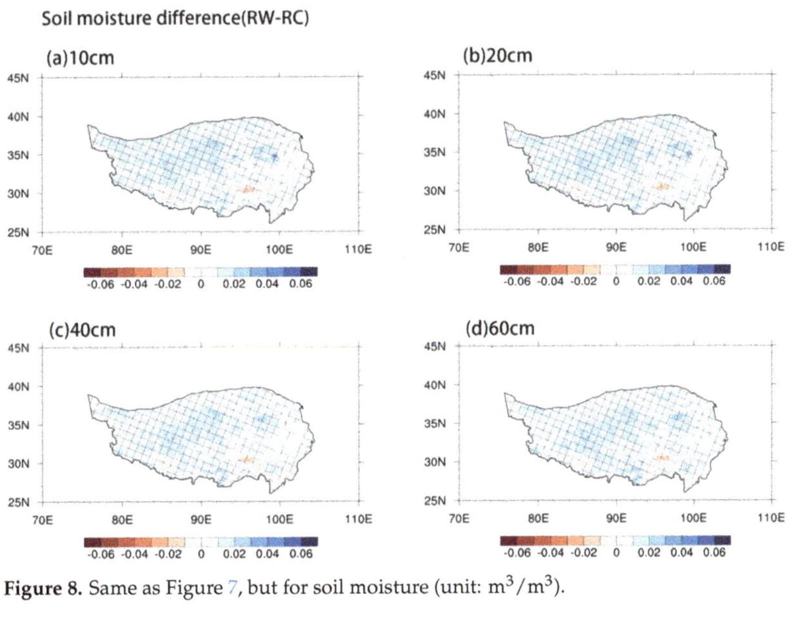

Figure 8. Same as Figure 7, but for soil moisture (unit: m^3/m^3).

4.3. Responses of Soil Freeze–Thaw Processes to Climate

To analyze the characteristics of freeze–thaw processes, this study divided the soil state into two periods: frozen and unfrozen. To consider a complete freeze–thaw cycle, we adjusted the data to start on 1 September. The criteria for judging whether a freeze–thaw occurred were as follows: (a) if the soil temperature was below 0 °C for five consecutive days, the current time was regarded as the beginning of the freezing period; (b) if the soil temperature was above 0 °C for five consecutive days, the current time was regarded as the end of the freezing period. The freezing date means the start date of the freezing season, and the thawing date is the end date of the freezing season.

As can be seen from Figure 9 and Table 1, over the past 37 years, the multi-year means of the freezing and thawing dates at the four soil depths were delayed with increasing soil depth; the shallow soil of the TP began to freeze from early November, while the soil at a 60 cm depth was frozen, on average, for 84.16 days after 1 September. The soil at 10 cm depth started to thaw in early April, on average, while the soil at s 60 cm depth began to thaw approximately 20 days later. On the entire TP, the surface soil freezes and thaws first, and these processes pervade deeper soil layers as time goes by; these freezing and thawing processes have an obvious hysteresis. Since the average freezing and thawing times of each layer are both delayed with increasing soil depth, the average period over which the soil remains frozen is similar, ranging from 144 to 146 days. The spatial distributions of the soil freeze–thaw processes in each layer also reflect the characteristics of air temperature distribution. Where temperatures were high, each layer of the soil froze later and thawed earlier. We can see in the northeast of the TP—the Qaidam Basin, the freezing date was later than that of the surrounding area, the thawing date was earlier than that of the surrounding area, and the freeze–thaw duration was shorter than that of the surrounding area. The most likely reason for this phenomenon is mainly due to the drought and little rainfall in this area, as the soil is dominated by sandy soil, the vegetation is relatively sparse, and the coverage rate is low.

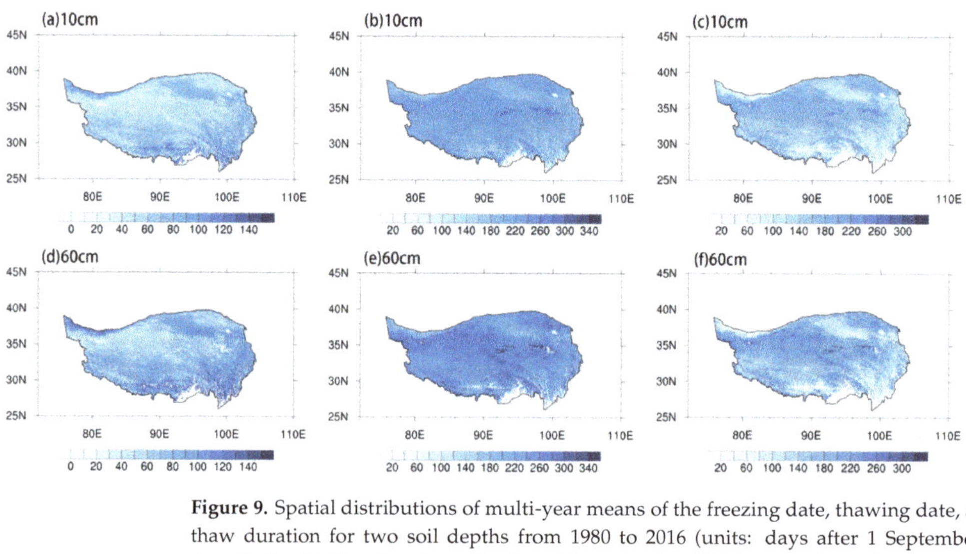

Figure 9. Spatial distributions of multi-year means of the freezing date, thawing date, and freeze–thaw duration for two soil depths from 1980 to 2016 (units: days after 1 September (**a**,**b**,**d**,**e**); days (**c**,**f**)). (**a**) Freezing date of soil at 10 cm depth. (**b**) Thawing date of soil at 10 cm depth. (**c**) Freeze–thaw duration of soil at 10 cm depth. (**d**) Freezing date of soil at 60 cm depth. (**e**) Thawing date of soil at 60 cm depth. (**f**) Freeze–thaw duration of soil at 60 cm depth.

Table 1. Multi-year means of soil freeze–thaw processes at different soil depths.

Soil Depth	Freezing Date (Days after 1 September)	Thawing Date (Days after 1 September)	Freeze–Thaw Duration (Days)
10 cm	67.42	212.12	144.70
20 cm	70.62	215.42	144.78
40 cm	76.12	220.70	144.59
60 cm	84.16	230.45	146.59

As shown in Figure 10 and Table 2, across most of the TP, all soil layers exhibited that the freezing date moved later from 1980 to 2016. On average, the decreasing trend was >2 days decade^{-1}, and the trend was more pronounced in deeper soil layers. Meanwhile, there were also delaying trends in thawing date and freeze–thaw duration, with these change trends being less pronounced in deeper soil layers. As the global climate changed, the freeze–thaw processes also changed dramatically, especially in shallow soils. The thawing date varied more significantly than the freezing date, which also explains changes in the duration of frozen soil. In the Karakoram Mountains and parts of the Qiangtang Plateau, the freezing date has tended to advance (i.e., the soil freezes earlier in the year), while the rest of the plateau essentially shows a delayed trend; the thawing date in the two aforementioned regions has tended to be delayed, while the thawing date of the rest of the plateau has advanced. Therefore, the freeze–thaw duration of each layer of soil in the northwestern part of the TP has increased, while the permafrost in the rest of the region has declined.

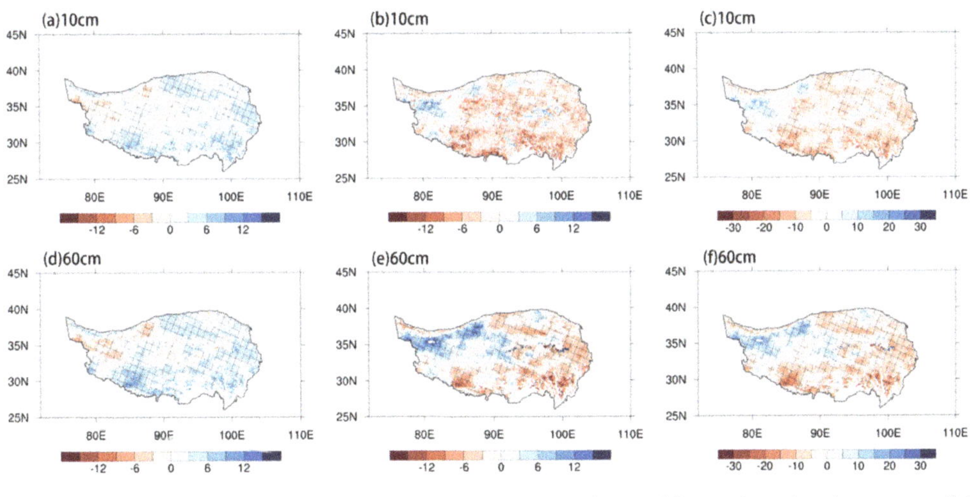

Figure 10. Trends in freezing date, thawing date, and freeze–thaw duration at two soil depths from 1980 to 2016 (unit: day/decade). Grid points with statistically significant anomalies at the 90% confidence level are denoted by an oblique line.

Table 2. Trends in soil freeze–thaw processes at different soil depths (unit: day/decade).

Soil Depth	Freezing Date	Thawing Date	Freeze–Thaw Duration
10 cm	2.15	−2.17	−4.32
20 cm	2.15	−2.23	−4.38
40 cm	2.23	−1.62	−3.85
60 cm	2.30	−1.49	−3.79

Figure 11 and Table 3 illustrate the differences in annual average soil freeze–thaw processes between the RW and RC periods. Air temperature was one of the most important factors affecting the freeze–thaw processes on the TP. In recent decades, a clear signal of elevation-dependent warming (EDW) of the air temperature has been observed on the TP, based on observational and satellite data [41–43]. The decrease in maximum freezing depth with elevation is a reaction to the EDW of the air temperature. The freezing dates of soil at depths of 10, 20, 40, and 60 cm were 3.87, 3.70, 4.28, and 4.41 days, respectively, while their thawing dates were −5.76, −5.22, −3.76, and −3.39 days, respectively; hence, their freeze–thaw durations were −9.54, −8.92, −8.05, and −7.80 days, respectively. The effect of air temperature on freeze–thaw processes can, thus, clearly be seen. The spatial distribution of freeze–thaw processes between the RW and RC periods is similar to that shown in Figure 10; shallow soils are more susceptible to climate change, and their changes are more dramatic.

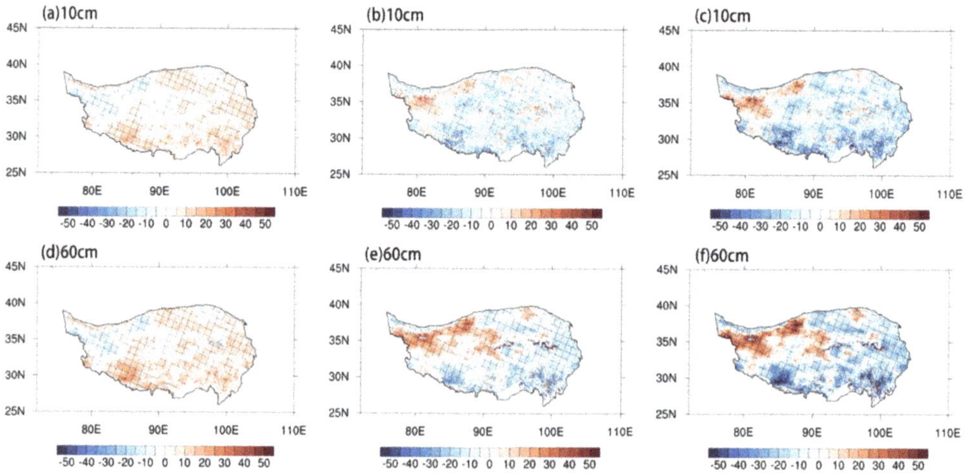

Figure 11. Differences in freezing date, thawing date, and freeze–thaw duration at two soil depths between the RW and RC periods (unit: day).

Table 3. Differences in soil freeze–thaw processes between the RW and RC periods (unit: day).

Soil Depth	Freezing Date	Thawing Date	Freeze–Thaw Duration
10 cm	3.87	−5.67	−9.54
20 cm	3.70	−5.22	−8.92
40 cm	4.28	−3.76	−8.05
60 cm	4.41	−3.39	−7.80

Satellite data (satellite-based 2 m air temperature (SBAT), Moderate Resolution Imaging Spectroradiometer (MODIS)-based LST, snow cover, and daytime/nighttime cloud extent) indicate that the annual mean 2 m air temperature above 4500 m on the TP rapidly declined between 2001 and 2015 [44]. As a result, long-term freeze–thaw trends are complicated, because other local parameters play important roles; in addition to changes in air temperature, precipitation is another important meteorological element. To better study the effect of precipitation on freeze–thaw processes, we divided the TP into climate zones according to average annual precipitation. Areas with annual precipitation <200 mm were classified as arid areas, those between 200 and 400 mm as semiarid, those between 400 and 800 mm as subhumid, and those >800 mm as humid. Figure 12 shows the climate zones on the TP. The average annual precipitation decreases sequentially from southeast to northwest, while its spatial distribution of values is similar to that of soil temperature and soil moisture.

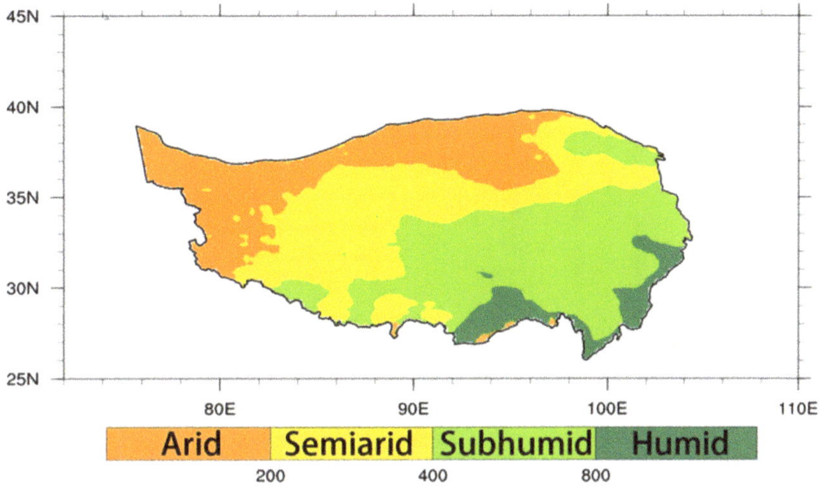

Figure 12. Climate zones on the Tibetan Plateau.

As shown in Figure 13, each layer of soil has specific freeze–thaw characteristics; that is, with an increase in soil depth, there are changes in the freezing date, thawing date, and freeze–thaw duration, which can reflect the hysteresis of freeze–thaw processes. There are obvious differences between the climate zones, notably in humid areas. Areas with the least precipitation freeze first, and other areas freeze sequentially, mirroring increases in the average annual precipitation. On average, the soil at a 10 cm depth in arid areas began to freeze at the end of October, while soil at a 60 cm depth in humid areas began to freeze in mid-December; freezing started much later in humid regions than in other regions. The soil in humid areas began to thaw in mid-March, while soil at a 60 cm depth in the rest of the regions began to melt at the end of April. Humid regions also have a much shorter freeze–thaw duration than other regions, with the freeze–thaw duration of every soil layer being over 40 days less than in other areas. Hence, the effect of precipitation on freeze–thaw processes is very clear.

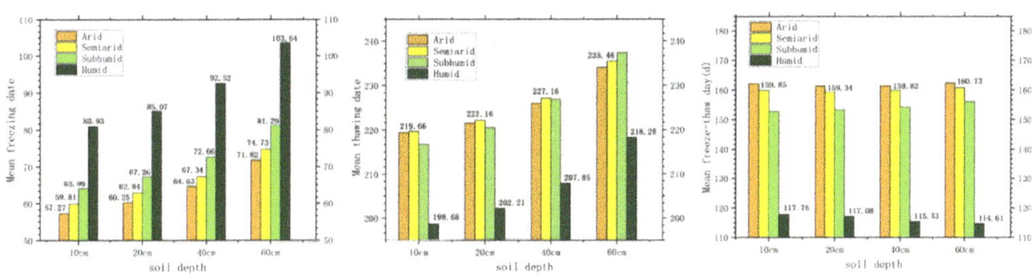

(a) (b) (c)

Figure 13. Multi-year means of the (**a**) freezing date, (**b**) thawing date, and (**c**) freeze–thaw duration at four different soil depths from 1980 to 2016, plotted with respect to climate zone (unit: day).

As shown in Figure 14, the soil freeze–thaw processes of the underlying surface of different vegetation types are very different, so we can also see the hysteresis of soil freeze–thaw processes with increasing soil depth. Compared with the underlying surface of other types of vegetation, the soil underlying the surface of the forest froze the latest, began to thaw the earliest, and had the shortest freeze–thaw duration. The freeze–thaw duration of the bare land underlying the surface of each layer of soil was more than one month longer than that of the underlying surface of the forest. Different vegetation types have a great influence on soil freeze–thaw processes. The soil freeze–thaw processes of the underlying surface of the bare land were more sensitive to climate change.

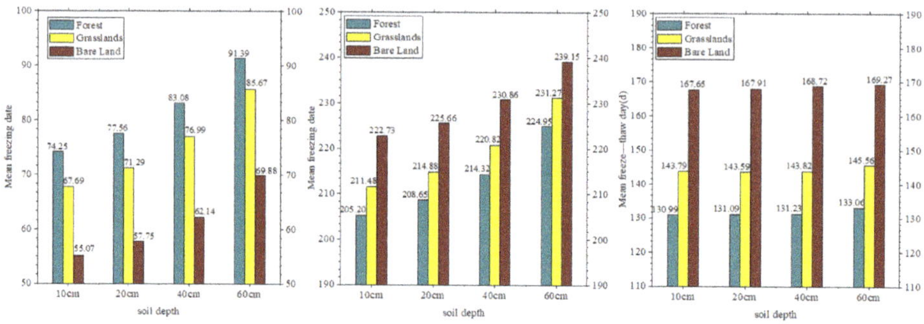

Figure 14. Same as Figure 13, but plotted with respect to vegetation types (unit: day).

5. Discussion

Freeze–thaw processes on the TP have a considerable influence on both regional and Asia-wide climate. Therefore, it is critical to understand and forecast such freeze–thaw processes. This study first investigated the characteristics of the temperature-dominant mode using data from 1980 to 2016. According to EOF analysis, a pattern of interdecadal increase in air temperature has emerged over almost all of the TP between 1980 and 2016. According to the latest publication from the Intergovernmental Panel on Climate Change's Sixth Assessment Report Working Group I, the global climate has warmed significantly over the last century (since 1850–1900); the average global surface temperature has risen by approximately 1 °C, and the average temperature rise over the next 20 years is expected to approach or surpass 1.5 °C [45]. Changes in the environment, such as permafrost, glacier, and ice cap melting, are expected to be exacerbated as a result of global warming, and this will have an impact on human existence. There are also substantial differences in the responses of different underlying substrates to global warming. Arctic sea ice and permafrost are particularly sensitive to a warming climate. The soil at various depths also has a clear warming tendency under a background of rising air temperature. Previous research has demonstrated that the warming trend of soil temperature in each layer was more significant in western China from 1980 to 2017 than before 1980 [46]. Changes in the features and spatial distribution of permafrost have been caused by the influence of global warming and human activities in recent years, manifesting as the elevation of permafrost's lower limits, rising ground temperature, and thickening of the active layer. Permafrost has been diminished in certain areas, resulting in seasonally frozen soil. Nearly half of the permafrost on the TP will be reduced to seasonally frozen soil by the end of the century under the Representative Concentration Pathway 4.5 (RCP 4.5) emissions scenario [47]. The maximum freezing depth and freeze–thaw duration of seasonally frozen soil in western China have shown downward trends in recent decades as a result of climate change, whereby there have been delayed freezing start dates and early melting end dates [48]. Before soil freezes, soil moisture exists in the form of liquid water, which is conducive to water transmission, and surface soil moisture is transported upwards in the form of liquid or vapor; precipitation will directly change the value of soil moisture, causing surface sensible and latent heat by affecting surface albedo; energy and water exchange between the earth and atmosphere is relatively frequent at this time. When soil freezes, moisture in the soil freezes into ice, and the amount of liquid water in the soil reduces dramatically, making soil moisture transport difficult. As shown herein, freeze–thaw processes vary with air temperature and precipitation. Indeed, the two most important factors impacting freeze–thaw processes on the TP are air temperature and precipitation. Freeze–thaw process trends are complicated, because other local elements or parameters play important roles, in addition to changes in air temperature and precipitation. Hence, further research into how freeze–thaw processes respond to climate change is required.

Frozen soil is a kind of soil that is very sensitive to temperature and is extremely unstable [49]. As the global temperature increases, the exchange process between the surface and the atmosphere of energy, water, soil temperature, etc., rises, and the ice melts, the soil moisture increases, the freezing date is delayed, the thawing date is advanced, and the freeze–thaw duration is shortened; the freezing–thawing process significantly affects the energy balance and water cycle of the soil surface through water phase transformation [50,51]. In addition, the abnormal soil moisture caused by snowmelt and the thawing process has an impact on summer precipitation in eastern China [52]. The multi-process interaction and complex influence mechanism of hydrothermal elements and vegetation activities have formed obvious regional differentiation characteristics. Moreover, areas with lush vegetation generally have more precipitation and higher temperatures.

6. Conclusions

On the basis of the remote sensing data and CLM4.5 model simulation results, this study investigated the main modes of air temperature on the TP, as well as the spatiotemporal distribution and changing trends in air temperature and precipitation. In addition, the characteristics of freeze–thaw processes and their response to climate change were also studied. The main conclusions are as follows:

1. The climate of the TP has become warmer and wetter over the past 37 years; the rates of the increase in regional average temperature and precipitation were 0.41 °C/decade and 6.44 mm/decade, respectively. As the monsoon moved forward, the regional distributions of precipitation and temperature were similar, with values steadily rising from the northwest to the southeast of the TP. We depicted the spatial pattern of the first dominant mode (EOF1), which was created using air temperature data. The spatial distribution of EOF1 was marked by consistent variations in temperature on the whole TP; the years from 1980 to 1998 were relatively cold, and the years from 1999 to 2016 were relatively warm.
2. Soil temperature and moisture across most parts of the TP showed an increasing trend. Soil temperature and moisture were shown to be affected by air temperature and precipitation.
3. Surface soil was first to freeze and thaw on the TP; freezing and thawing then pervaded deeper soil as time passed, with an obvious hysteresis in the freeze–thaw cycle. On average, the four analyzed layers of soil on the TP began to freeze in November and began to thaw in April. The mean freeze–thaw duration of these four layers of soil was 144–146 days. Between 1980 and 2016, the freezing date of each soil layer in most regions of the TP has moved later in the year, with an average rate of >2 days decade^{-1}. Meanwhile, the thawing date has moved earlier in the year, and the freeze–thaw duration has declined.
4. Areas with the least amount of precipitation were the first to freeze, with other areas freezing sequentially, in line with increasing average annual precipitation. Soil thawing occurred sooner in areas with more precipitation. Hence, precipitation appears to have a substantial impact on freeze–thaw processes.
5. The areas under the bare land were the first to freeze, and the areas under the forest were the first to thaw. Different vegetation types had a major impact on the freeze–thaw process.

Author Contributions: Conceptualization, C.F. and Z.H.; methodology, C.F. and Y.Y.; software, C.F. and D.W.; investigation, C.F. and Z.H.; writing—original draft preparation, C.F., M.D., Y.Y. and H.Y.; writing—review and editing, C.F., Z.H., Y.Y., M.D., H.Y., S.L., D.W. and W.F. All authors have read and agreed to the published version of the manuscript.

Funding: This research was supported by the Second Tibetan Plateau Scientific Expedition and Research (STEP) program (grant no. 2019QZKK0103), the National Natural Science Foundation of China (grant no. 91837208), the Strategic Priority Research Program of the Chinese Academy of Sciences (grant no. XDA20060101), the Systematic Major Project of the China Railway (grant no. P2021G047), and the National Key Research and Development Program of China (grant no. 2018YFC1505701).

Data Availability Statement: The MODIS land-cover product (MCD12C1) is available online at: https://lpdaac.usgs.gov/resources/data-action/ (accessed on 29 June 2022); the China Meteorological Forcing Dataset (CMFD) can be obtained at: http://data.tpdc.ac.cn/en/data/8028b944-daaa-4511-8769-965612652c49/ (accessed on 7 November 2019).

Conflicts of Interest: The authors declare no conflict of interest.

References

1. Zhang, Y.; Li, B.; Zheng, D. A discussion on the boundary and area of the Tibetan Plateau in China. *Geogr. Res.* **2002**, *21*, 1–8. [CrossRef]
2. Ma, Y.; Yao, T.; Hu, Z.; Wang, J. The cooperative study on energy and water cycle over the Tibetan Plateau. *Adv. Earth Sci.* **2009**, *24*, 1280.
3. Liu, X.; Hui, X.; Chen, B. Influence of heat source anormal of underlying surface over Tibet Plateau and western tropical Pacific on short-term climate in China. *Plateau Meteorol.* **1991**, *10*, 305–316. (In Chinese)
4. Wang, L.; Zheng, Q.; Song, Q. Numerical simulation of the influence of the underlying surface of the western Qinghai-Tibet Plateau on the seasonal transition of the atmospheric circulation in East Asia. *Plateau Meteorol.* **2003**, *22*, 179–184. (In Chinese)
5. Wu, G.X.; Zhu, B.Z.; Gao, D.Y. The Impact of the Tibetan Plateau on Local and Regional Climate. In *Theoretical Research Progress of the Second Qinghai-Tibet Plateau Atmospheric Science Experiment (1)*; Meteorological Press: Beijing, China, 1999; pp. 257–273.
6. Duan, A.; Liu, Y.; Wu, G. Heating status of the Tibetan Plateau from April to June and rainfall and atmospheric circulation anomaly over East Asia in midsummer. *Sci. China Ser. D Earth Sci.* **2005**, *48*, 250–257. [CrossRef]
7. Duan, A.; Li, F.; Wang, M.; Wu, G. Persistent weakening trend in the spring sensible heat source over the Tibetan Plateau and its impact on the Asian summer monsoon. *J. Clim.* **2011**, *24*, 5671–5682. [CrossRef]
8. Davidson, E.A.; Janssens, I.A. Temperature sensitivity of soil carbon decomposition and feedbacks to climate change. *Nature* **2006**, *440*, 165–173. [CrossRef] [PubMed]
9. Hu, Q.; Feng, S. How have soil temperatures been affected by the surface temperature and precipitation in the Eurasian continent? *Geophys. Res. Lett.* **2005**, *32*, L14711. [CrossRef]
10. Lawrence, D.M.; Slater, A.G.; Swenson, S.C. Simulation of present-day and future permafrost and seasonally frozen ground conditions in CCSM4. *J. Clim.* **2012**, *25*, 2207–2225. [CrossRef]
11. Zhang, T.; Barry, R.; Gilichinsky, D.; Bykhovets, S.; Sorokovikov, V.A.; Ye, J. An amplified signal of climatic change in soil temperatures during the last century at Irkutsk, Russia. *Clim. Change* **2001**, *49*, 41–76. [CrossRef]
12. Yeşilırmak, E. Soil temperature trends in Büyük Menderes Basin, Turkey. *Meteorol. Appl.* **2014**, *21*, 859–866. [CrossRef]
13. Woodbury, A.D.; Bhuiyan, A.K.M.H.; Hanesiak, J.; Akinremi, O.O. Observations of northern latitude ground-surface and surface-air temperatures. *Geophys. Res. Lett.* **2009**, *36*, L07703. [CrossRef]
14. Gao, Y.; Li, X.; Leung, L.R.; Chen, D.; Xu, J. Aridity changes in the Tibetan Plateau in a warming climate. *Environ. Res. Lett.* **2015**, *10*, 034013. [CrossRef]
15. Yao, F.; Wang, J.; Yang, K.; Wang, C.; Walter, B.A.; Crétaux, J.F. Lake storage variation on the endorheic Tibetan Plateau and its attribution to climate change since the new millennium. *Environ. Res. Lett.* **2018**, *13*, 064011. [CrossRef]
16. Sun, S.F. *Parameterization Study of Physical and Biochemical Mechanism in Land Surface Process*; Meterology Press: Beijing, China, 2005.
17. Zhou, Y.; Guo, D.; Qiu, G.; Cheng, G.; Li, S. *China Permafrost*; Science Press: Beijing, China, 2000; pp. 145–151.
18. Yang, M.X.; Yao, T.D. A review of the study on the impact of snow cover in the Tibetan Plateau on Asian monsoon. *J. Glaciol. Geocryol.* **1998**, *20*, 186–192.
19. Li, S.X.; Nan, Z.T.; Zhao, L. Impact of freezing and thawing on energy exchange between the system and environment. *J. Glaciol. Geocryol.* **2002**, *24*, 109–115.
20. Qingbai, W.; Yongping, S.; Bin, S. Relationship between frozen soil together with its water-heat process and ecological environment in the Tibetan Plateau. *J. Glaciol. Geocryol.* **2003**, *25*, 250–255.
21. Gao, R.; Zhong, H.L.; Dong, W.J.; Wei, Z.G. Impact of snow cover and frozen soil in the Tibetan Plateau on summer precipitation in China. *J. Glaciol. Geocryol.* **2011**, *33*, 254–260.
22. Chenghai, W.; Wenjie, D.; Zhigang, W. Study on relationship between the frozenthaw process in Qinghai Xizang Plat eau and circulation in East Asia. *Chin. J. Geo Phys.* **2003**, *46*, 310–312.
23. Wang, C.; Shang, D. Effect of the variation of the soil temperature and moisture in the transition from dry-season to wet-season over northern Tibet Plateau. *Plateau Meteorol.* **2007**, *26*, 677–685.
24. Yang, M.; Yao, T.; Gou, X.; Hirose, N.; Fujii, H.Y.; Hao, L.; Levia, D.F. Diurnal freeze/thaw cycles of the ground surface on the Tibetan Plateau. *Chin. Sci. Bull.* **2007**, *52*, 136–139. [CrossRef]

25. Guo, W.D.; Ma, Z.G.; Wang, H.J. Soil Moisture—An Important Factor of Seasonal Precipitation Prediction and Its Application. *Clim. Environ. Res.* **2007**, *12*, 20–28. (In Chinese)
26. Chahine, M.T. The hydrological cycle and its influence on climate. *Nature* **1992**, *359*, 373–380. [CrossRef]
27. Trenberth, K.E. Atmospheric moisture recycling: Role of advection and local evaporation. *J. Clim.* **1999**, *12*, 1368–1381. [CrossRef]
28. Yang, M.; Yao, T.; Gou, X.; Koike, T.; He, Y. The soil moisture distribution, thawing–freezing processes and their effects on the seasonal transition on the Qinghai–Xizang (Tibetan) plateau. *J. Asian Earth Sci.* **2003**, *21*, 457–465. [CrossRef]
29. Tarnocai, C.; Canadell, J.G.; Schuur, E.A.G.; Kuhry, P.; Mazhitova, G.; Zimov, S. Soil organic carbon pools in the northern circumpolar permafrost region. *Glob. Biogeochem. Cycles* **2009**, *23*, GB2023. [CrossRef]
30. Zou, D.; Zhao, L.; Sheng, Y.; Chen, J.; Hu, G.; Wu, T.; Wu, J.; Xie, C.; Wu, X.; Pang, Q.; et al. A new map of permafrost distribution on the Tibetan Plateau. *Cryosphere* **2017**, *11*, 2527–2542. [CrossRef]
31. Friedl, M.A.; McIver, D.K.; Hodges, J.C.F.; Zhang, X.Y.; Muchoney, D.; Strahler, A.H.; Woodcock, C.E.; Gopal, S.; Schneider, A.; Cooper, A.; et al. Global land cover mapping from MODIS: Algorithms and early results. *Remote Sens. Environ.* **2002**, *83*, 287–302. [CrossRef]
32. Friedl, M.A.; Sulla-Menashe, D.; Tan, B.; Schneider, A.; Ramankutty, N.; Sibley, A.; Huang, X. MODIS Collection 5 global land cover: Algorithm refinements and characterization of new datasets. *Remote Sens. Environ.* **2010**, *114*, 168–182. [CrossRef]
33. Chen, Y.; Yang, K.; He, J.; Qin, J.; Shi, J.; Du, J.; He, Q. Improving land surface temperature modeling for dry land of China. *J. Geophys. Res. Atmos.* **2011**, *116*, D20104. [CrossRef]
34. Oleson, K.W.; Lawrence, D.M.; Bonan, G.B.; Fisher, R.A.; Koven, C.D.; Swenson, S.C.; Collier, N.; Ghimire, B.; van Kampenhout, L.; Kennedy, D.; et al. *Technical Description of Version 4.5 of the Community Land Model (CLM)*; (NCAR Tech. Note NCAR/TN-5031STR); National Center for Atmospheric Research: Boulder, CO, USA, 2013; 420p.
35. Niu, G.Y.; Yang, Z.L. Effects of frozen soil on snowmelt runoff and soil water storage at a continental scale. *J. Hydrometeorol.* **2006**, *7*, 937–952. [CrossRef]
36. Swenson, S.C.; Lawrence, D.M.; Lee, H. Improved simulation of the terrestrial hydrological cycle in permafrost regions by the Community Land Model. *J. Adv. Model. Earth Syst.* **2012**, *4*, M08002. [CrossRef]
37. Swenson, S.C.; Lawrence, D.M. A new fractional snow-covered area parameterization for the Community Land Model and its effect on the surface energy balance. *J. Geophys. Res. Atmos.* **2012**, *117*, D21107. [CrossRef]
38. Wang, C.; Yang, K. A new scheme for considering soil water-heat transport coupling based on Community Land Model: Model description and preliminary validation. *J. Adv. Model. Earth Syst.* **2018**, *10*, 927–950. [CrossRef]
39. Li, S.; Yang, K.; Wang, C. Bias characteristics of land surface model (CLM4. 5) over the Tibetan Plateau during soil freezing-thawing period and its causes. *J. Glaciol. Geocryol.* **2018**, *40*, 322–334.
40. Gardelle, J.; Berthier, E.; Arnaud, Y. Slight mass gain of Karakoram glaciers in the early twenty-first century. *Nat. Geosci.* **2012**, *5*, 322–325. [CrossRef]
41. Gao, J.; Xie, Z.; Wang, A.; Liu, S.; Zeng, Y.; Liu, B.; Li, R.; Jia, B.; Qin, P.; Xie, J. A new frozen soil parameterization including frost and thaw fronts in the Community Land Model. *J. Adv. Model. Earth Syst.* **2019**, *11*, 659–679. [CrossRef]
42. Gao, Y.; Xu, J.; Chen, D. Evaluation of WRF mesoscale climate simulations over the Tibetan Plateau during 1979–2011. *J. Clim.* **2015**, *28*, 2823–2841. [CrossRef]
43. Liu, X.; Cheng, Z.; Yan, L.; Yin, Z.Y. Elevation dependency of recent and future minimum surface air temperature trends in the Tibetan Plateau and its surroundings. *Glob. Planet. Change* **2009**, *68*, 164. [CrossRef]
44. Guo, D.; Sun, J.; Yang, K.; Pepin, N.; Xu, Y. Revisiting recent elevation-dependent warming on the Tibetan Plateau using satellite-based data sets. *J. Geophys. Res. Atmos.* **2019**, *124*, 8511–8521. [CrossRef]
45. IPCC; Masson-Delmotte, V.; Zhai, P.; Pirani, A.; Connors, S.L.; Péan, C.; Berger, S.; Caud, N.; Chen, Y.; Goldfarb, L.; et al. Climate change 2021: The physical science basis. In *Contribution of Working Group I to the Sixth Assessment Report of the Intergovernmental Panel on Climate Change*; Cambridge University Press: Cambridge, UK, 2021.
46. Zhou, Y.; Gao, X.; Zhang, K.; Li, Y.; Yang, L. Spatiotemporal variations in 3.2 m soil temperature in China during 1980–2017. *Clim. Dyn.* **2020**, *54*, 1233–1244. [CrossRef]
47. Wang, T.; Yang, D.; Fang, B.; Yang, W.; Qin, Y.; Wang, Y. Data-driven mapping of the spatial distribution and potential changes of frozen ground over the Tibetan Plateau. *Sci. Total Environ.* **2019**, *649*, 515–525. [CrossRef] [PubMed]
48. Gao, Y.; Chen, F.; Lettenmaier, D.P.; Xu, J.; Xiao, L.; Li, X. Does elevation-dependent warming hold true above 5000 m elevation? Lessons from the Tibetan Plateau. *Npj Clim. Atmos. Sci.* **2018**, *1*, 19. [CrossRef]
49. Qiu, G.; Liu, J.; Liu, H. *Geocryological glossary*; Gansu Science and Technology Press China: Lanzhou, China, 1994.
50. Guo, D.; Yang, M.; Wang, H. Characteristics of land surface heat and water exchange under different soil freeze/thaw conditions over the central Tibetan Plateau. *Hydrol. Process.* **2011**, *25*, 2531–2541. [CrossRef]
51. Yang, K.; Wang, C. Water storage effect of soil freeze-thaw process and its impacts on soil hydro-thermal regime variations. *Agric. For. Meteorol.* **2019**, *265*, 280–294. [CrossRef]
52. Wang, C.; Yang, K.; Li, Y.; Wu, D.; Bo, Y. Impacts of spatiotemporal anomalies of Tibetan Plateau snow cover on summer precipitation in eastern China. *J. Clim.* **2017**, *30*, 885–903. [CrossRef]

remote sensing

MDPI

Article

Evaluation of the Spatial and Temporal Variations of Condensation and Desublimation over the Qinghai–Tibet Plateau Based on Penman Model Using Hourly ERA5-Land and ERA5 Reanalysis Datasets

Hongyuan Li [1,2], Rensheng Chen [1,3,*], Chuntan Han [1,2] and Yong Yang [1]

1 Qilian Alpine Ecology and Hydrology Research Station, Northwest Institute of Eco-Environment and Resources, Chinese Academy of Sciences, Lanzhou 730000, China
2 University of Chinese Academy of Sciences, Beijing 100049, China
3 College of Urban and Environmental Sciences, Northwest University, Xi'an 710000, China
* Correspondence: crs2008@lzb.ac.cn

Abstract: Condensation and desublimation are important processes of nocturnal land–atmosphere interactions, energy transfer, and the water cycle, and have important ecological and hydrological roles in mitigating physiological water deficits caused by low temperatures and reducing the risk of frost damage to plants, animals, and microorganisms near the surface in the Alpine Region. The aim of the present study is to evaluate the spatial and temporal variations of condensation and desublimation from 1950 to 2020 based on Penman model using hourly ERA5-Land and ERA5 reanalysis datasets on the Qinghai–Tibet Plateau (QTP), where condensation and desublimation occur frequently but lack quantitative evaluation. The results showed that: (1) Condensation showed a decreasing trend from southeast to northwest, with annual mean condensation ranging from 0 mm to 72.8 mm, while desublimation showed regional enrichment rather than zonal variation, with the annual mean desublimation ranging from 0 mm to 23.6 mm; (2) At 95% confidence level, condensation showed a significant increasing trend in the central and western QTP, while desublimation showed a significant decreasing trend in most regions of the QTP, and the decreasing trend of desublimation was more obvious than the increasing trend of condensation; (3) Both condensation and desublimation showed significant seasonal characteristics; the maximum monthly condensation was 2.37 mm and the monthly mean condensation was 0.70 mm, while the maximum monthly desublimation was 1.45 mm and the monthly mean desublimation was 0.95 mm; (4) The annual mean condensation was 8.45 mm, with an increasing trend of 0.24 mm/10a, the annual mean desublimation was 11.45 mm, with a decreasing trend of −0.26 mm/10a, and the total annual mean condensation and desublimation was 19.89 mm, with a weak decreasing trend on the QTP; (5) The increase in condensation is most associated with the increase in precipitation, while the decrease in desublimation is most associated with the increase in air temperature on the QTP.

Keywords: condensation; desublimation; land–atmosphere interactions; latent heat flux; ERA5-Land; Qinghai–Tibet Plateau

Citation: Li, H.; Chen, R.; Han, C.; Yang, Y. Evaluation of the Spatial and Temporal Variations of Condensation and Desublimation over the Qinghai–Tibet Plateau Based on Penman Model Using Hourly ERA5-Land and ERA5 Reanalysis Datasets. *Remote Sens.* **2022**, *14*, 5815. https://doi.org/10.3390/rs14225815

Academic Editor: Alexander Kokhanovsky

Received: 30 August 2022
Accepted: 14 November 2022
Published: 17 November 2022

1. Introduction

Condensation and desublimation are both phase transition processes of gaseous water after radiation cooling, where condensation is the process of supersaturation of gaseous water to liquid water [1,2], while desublimation is the process of supersaturation of gaseous water to solid water directly [3]. The main difference between condensation and desublimation is whether the dew point temperature is above or below 0 °C when the gaseous water is supersaturated. That is, when the dew point temperature is equal to or greater than 0 °C, the supersaturation of gaseous water leads to condensation, while when the dew point temperature is below 0 °C, the supersaturation of gaseous water leads to desublimation [4]. Both condensation and desublimation are processes of water transfer from the air to the surface (downward water transfer), as opposed to evaporation and sublimation, which are processes of water transfer from the surface to the air (upward water transfer) [5]. Evaporation and sublimation occur mainly during the day and result in water loss near the surface, while condensation and desublimation occur mainly at night and result in water input near the surface [6,7]. Due to the large difference in the intensity of daytime and nighttime land–atmosphere interactions, results in condensation and desublimation are of a much smaller order of magnitude than evaporation and sublimation [5]. Therefore, condensation and desublimation are often neglected in the study of land–atmosphere interactions.

Although much smaller in order of magnitude than evaporation and sublimation, condensation and desublimation have specific ecological and hydrological roles. Condensation and desublimation are often specifically referred to as dew and hoarfrost, respectively, and are often considered to be important components of non-rainfall water inputs (NRWI) [8–11]. The ecological and hydrological roles of condensation and desublimation can usually be summarized in two aspects: on the one hand, condensation (dew) is an effective source of water for plants, animals, and microorganisms in arid and semi-arid regions and in other regions during the dry season, and is ecologically important for alleviating water stress and maintaining ecosystem functioning in these regions [4,11]. On the other hand, the release of large amounts of latent heat when condensation (dew) and desublimation (hoarfrost) occur can alleviate the temperature stress caused by low temperatures, and plays an ecological role in protecting plants, animals, and microorganisms from frost damage [4,12]. Therefore, the study of condensation and desublimation not only provides a clearer understanding of land–atmosphere interactions, energy transfer, and water cycles near the surface, but also contributes to the understanding of ecosystem functioning mechanisms.

The main limitations to the study of condensation and desublimation are not only their small magnitude, but also their difficulty of measurement [7,13,14]. For the measurement of condensation and desublimation, direct weighing of condensation and desublimation amounts is the more commonly used method [15–17], but this method is not suitable for long periods of time and large spatial scales. Due to condensation and desublimation both being phase transitions of gaseous water, which are accompanied by the release of latent heat, the observation or calculation of latent heat flux to estimate the condensation and desublimation iares the most effective indirect methods at present [4,13,18,19]. The common methods for estimating condensation and desublimation based on the observation or calculation of latent heat flux include the Eddy Covariance method [20,21], the Bowen Ratio Energy Balance method [7,18], the Aerodynamic method [22] and the Penman model [12,23,24]. Among these methods, the Penman model was chosen for the present study to calculate the latent heat flux because of its applicability and easy access to parameters. When the latent heat flux is negative, it indicates the occurrence of condensation or desublimation, while when the latent heat flux is positive, it indicates the occurrence of evaporation or sublimation.

Located in South-Central Asia, the QTP is the highest plateau in the world with its average altitude. The high altitude leads to a unique type of plateau climate on the QTP, characterized by strong radiation, low temperatures and greater diurnal temperature difference, which result in condensation and desublimation occurring more frequently [25]. However, the harsh environment and complex topography have led to the sparse meteorological stations, limiting the study of land–atmosphere interactions, such as condensation and desublimation on the QTP [26–28]. With the development of remote sensing and the data assimilation principle, the produced reanalysis datasets compensate for the scarcity of meteorological stations on the QTP, providing sufficient data support for large-scale, high-precision studies of land–atmosphere interactions, energy transfer and water cycle on the QTP [29]. In view of this, this study uses the ERA5-Land and ERA5 reanalysis datasets with high spatial and temporal resolution to conduct condensation and desublimation evaluation, which is useful for improving the understanding of land–atmosphere interactions, energy transfer, and water cycles on the QTP. Climate change and global warming are already indisputable facts nowadays, with the changes in temperature and precipitation having attracted widespread attention [30–32]. Climate change and global warming have caused changes in land surface processes such as evaporation and sublimation, however, the effects of climate change and global warming on condensation and desublimation processes are not yet known. The QTP is a sensitive area for climate change and global warming, and the effects of climate change and global warming on condensation and desublimation may be more significant, so it is necessary to quantitatively evaluate the variations of condensation and desublimation on the QTP under climate change.

In the present study, the spatial and temporal variations of condensation and desublimation on the QTP during 1950~2020 were evaluated based on the Penman model using the hourly meteorological variables from ERA5-Land and ERA5 reanalysis datasets. The main objectives of this study are (1) to evaluate the magnitude of condensation and desublimation, (2) to analyze the spatial variations of condensation and desublimation, (3) to quantify the temporal variations of condensation and desublimation, and (4) to discuss the impact of condensation and desublimation on alpine ecosystem on the QTP under climate change.

2. Materials and Methods

2.1. Study Area

Located in South-Central Asia, the QTP is the highest plateau in the world by average altitude, and has long been known as the Roof of the World. With an average altitude of over 4000 m, the QTP has developed an alpine climate characterized by strong radiation, low temperatures, and large temperature differences between day and night. The extremely cold climate has led to the widespread glaciers, snow cover, and permafrost on the QTP, which is the source of many rivers in Asia, so it is thus also known as the Water Tower of Asia [33].

The annual mean temperature of the QTP is approximately −2.5 °C and the annual mean precipitation is approximately 380 mm. Both temperature and precipitation show significant seasonal differences, with high temperature and high precipitation in summer and low temperature and low precipitation in winter. In addition, both temperature and precipitation have a decreasing trend from southeast to northwest, and the temperature has a clear tendency to decrease with increasing altitude, while precipitation has the characteristic of increasing with increasing altitude [34]. Due to the spatial pattern of temperature and precipitation, the vegetation cover also has a decreasing trend from southeast to northwest and decreasing with increasing altitude on the QTP [35].

Due to the complex topography and harsh climatic conditions, the meteorological stations are sparsely distributed within the QTP; therefore, the reanalysis datasets produced by the data assimilation principle based on models and observations, effectively complements the lack of data for land surface process studies, such as condensation, desublimation, evaporation, and sublimation.

The unique climatic characteristics lead to frequent water and heat exchange, especially influenced by climate warming in recent decades, which has accelerated the energy transfer and water cycle within the QTP, while the frequency and rate of condensation and desublimation, as well as evaporation and sublimation, are increasing. Therefore, it is important to evaluate the spatial and temporal variations of condensation and desublimation to improve the understanding of the variations in the land–atmosphere interactions, energy transfer, and water cycles on the QTP.

2.2. Datasets

2.2.1. ERA5-Land and ERA5 Reanalysis Datasets

Two reanalysis datasets, ERA5-Land and ERA5, were used as data input for this study. Among these, ERA5 is the fifth generation of the European Centre for Medium-range Weather Forecasts (ECMWF) reanalysis for the global climate and weather for the past 4 to 7 decades, which combines model data with observations from across the world into a globally complete and consistent dataset using data assimilation principle [36,37]. Compared to ERA5, ERA5-Land provides a consistent view of the evolution of land variables over several decades at an enhanced resolution [38–40]. ERA5 provides hourly estimates for a large number of atmospheric, ocean-wave, and land-surface quantities, while ERA5-Land has been produced by replaying the land component of the ERA5 climate reanalysis. Although no additional data assimilation was performed in ERA5-Land compared to ERA5, ERA5-Land produced at a higher resolution and forceed by ERA5 atmospheric parameters with lapse rate correction. Therefore, ERA5-Land has better applicability than ERA5 in the analysis of land surface processes [28]. According to the method of Penman model parameter calculation, eight variables from ERA5-Land and one variable from ERA5 were selected as meteorological variables needed for model parameter calculation in this study, and the details of all selected meteorological variables are listed in Table 1. In order to keep the spatial resolution uniform for all meteorological variables, the friction velocity from ERA5 was resampled to a spatial resolution of 0.1°. In addition, the total precipitation from ERA5-Land was used in order to assess the variations of precipitation on the QTP.

Table 1. Meteorological variables required from ERA5-Land and ERA5 reanalysis datasets.

Meteorological Variables	Symbols	Units	Spatial Resolution	Temporal Resolution	Datasets
2 m temperature	T_a	K	0.1° × 0.1°	Hourly	ERA5-Land
2 m dewpoint temperature	T_d	K	0.1° × 0.1°	Hourly	ERA5-Land
10 m u-component of wind	u	m s^{-1}	0.1° × 0.1°	Hourly	ERA5-Land
10 m v-component of wind	v	m s^{-1}	0.1° × 0.1°	Hourly	ERA5-Land
Surface pressure	P_a	Pa	0.1° × 0.1°	Hourly	ERA5-Land
Surface net solar radiation	R_s	J m^{-2}	0.1° × 0.1°	Hourly	ERA5-Land
Surface net thermal radiation	R_t	J m^{-2}	0.1° × 0.1°	Hourly	ERA5-Land
Skin temperature	T_s	K	0.1° × 0.1°	Hourly	ERA5-Land
Friction velocity	u_*	m s^{-1}	0.25° × 0.25°	Hourly	ERA5
Total precipitation	*pre*	m	0.1° × 0.1°	Hourly	ERA5-Land

2.2.2. Observed Meteorological Variables

To verify the accuracy of the Penman model for estimating condensation and desublimation using nine meteorological variables as inputs from the ERA5-Land and ERA5 reanalysis datasets, the monthly and annual condensation and desublimation estimated by the Penman model were compared with those observed by the Eddy Covariance at nine flux stations on the QTP, respectively. The locations of the nine flux stations on the QTP are shown in Figure 1, and the basic information of each station is shown in Table 2. The meteorological variables observed by the nine flux stations used include temperature, relative humidity, and latent heat flux, which obtained from the datasets of a long-term dataset of integrated land-atmosphere interaction observations on the Tibetan Plateau (2005–2016) [41,42] and the datasets of Heihe integrated observatory network [43–45], at the National Tibetan Plateau Data Center.

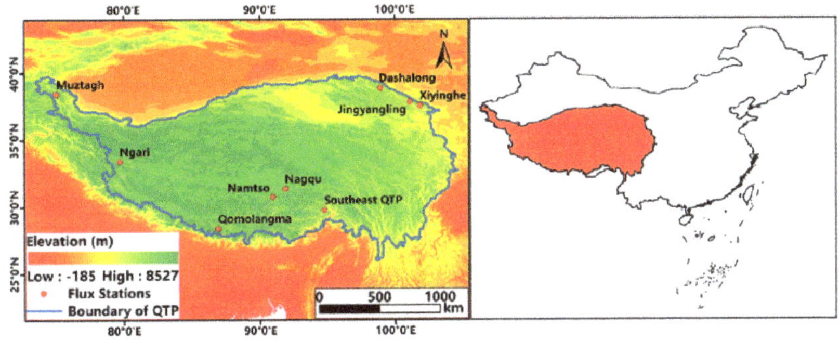

Figure 1. Overview of the Qinghai–Tibet Plateau and the location of nine flux stations.

For comparison, the latent heat flux observed by Eddy Covariance collected from the nine flux stations on the QTP was converted into condensation and desublimation as measured condensation and desublimation, based on the relationship between latent heat flux and moisture transport (Equation (5)). The reason why we choose the observed condensation and desublimation by Eddy Covariance as the measured condensation and desublimation is that the Eddy Covariance technique is currently recognized as a flux observation technique with a high degree of confidence [46]. The dew point temperature was used to distinguish between measured condensation and measured desublimation. When the dew point temperature is below 0 °C, the downward water transfer is condensation, while when the dew point temperature is greater than or equal to 0 °C, the downward water transfer is desublimation. The dew point temperature was calculated according to the method recommended by the FAO [47], which has the general form of:

$$T_d = \frac{116.91 + 237.3 \cdot \ln(e_a)}{16.78 - \ln(e_a)} \tag{1}$$

where e_a is the actual water vapor pressure (kPa), which is calculated as follows:

$$e_s = 0.6108 \cdot \exp\left(\frac{17.27 \cdot T_a}{237.3 + T_a}\right) \tag{2}$$

$$e_a = \frac{e_s \cdot RH}{100} \tag{3}$$

where e_s is the saturation water vapor pressure (kPa), T_a is the air temperature (°C), and RH is the relative humidity (100%).

Table 2. Information of the nine flux stations and the required meteorological variables on the QTP.

Station Name	Longitude	Latitude	Elevation	Meteorological Variables	Temporal Resolution	Period
	°	°	m	°C, %, W m^{-2}		
Nagqu	91.90	31.37	4509	T_a, RH, λE	Hourly	2005~2016
Qomolangma	86.95	28.36	4298	T_a, RH, λE	Hourly	2005~2016
Southeast QTP	94.74	29.77	3327	T_a, RH, λE	Hourly	2005~2016
Ngari	79.70	33.39	4270	T_a, RH, λE	Hourly	2005~2016
Muztagh	75.03	38.42	3668	T_a, RH, λE	Hourly	2005~2016
Namtso	90.96	30.77	4730	T_a, RH, λE	Hourly	2005~2016
Xiyinghe	101.86	37.56	3616	T_a, RH, λE	Half-hour	2016~2020
Jingyangling	101.12	37.84	3750	T_a, RH, λE	Half-hour	2016~2020
Dashalong	98.94	38.84	3739	T_a, RH, λE	Half-hour	2016~2020

2.3. Methods

In the present study, the above-mentioned eight meteorological variables from the ERA5-Land reanalysis dataset and one meteorological variable from the ERA5 reanalysis dataset (Table 1) were used as inputs to calculate the latent heat flux based on the Penman model. According to the rules of the Penman model, when the latent heat flux is negative, it indicates the transport of moisture from the air to the surface, which represents the occurrence of condensation or desublimation.

The premise of calculating latent heat flux based on the Penman model is that the surface of ground or feature is in a sufficiently wet state; otherwise, both negative and positive latent heat flux represent potential latent heat flux, such as potential evaporation [48]. When the surface temperature is equal to or lower than the dew point temperature, the near-surface air reaches saturation and supersaturation, at which point the surface of the ground or feature can be considered to be in a wet state, satisfying the conditions for the applicability of the Penman model [4]. Therefore, the negative latent heat flux calculated based on the Penman model can represent the actual condensation and desublimation only when the surface temperature is equal to or lower than the dew point temperature. Where the negative latent heat flux represents the occurrence of condensation when the dew point temperature is greater than or equal to 0 °C, and the negative latent heat flux represents the occurrence of desublimation when the dew point temperature is lower than 0 °C.

Based on the results of condensation and desublimation calculated by the Penman model, the MK trend test and Sen's slope analysis, as well as other common numerical statistical methods, were used to quantify the condensation and desublimation at different time scales, thus providing a more comprehensive understanding of the spatial and temporal variations of condensation and desublimation on the QTP during 1950~2020. The specific flow of this study is given by Figure 2.

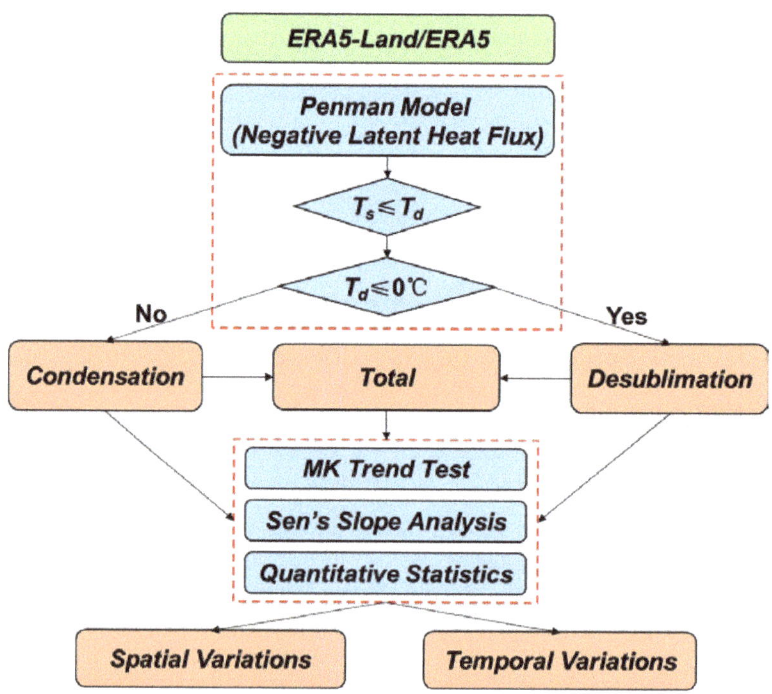

Figure 2. Flow chart of this study.

2.3.1. Penman Model

The Penman model has a good physical basis with two parts, the thermal term on the left and the dynamic term on the right [48,49], with the form:

$$\lambda E = \frac{\Delta(R_n - G_0)}{\Delta + \gamma} + \frac{\rho_a c_p (e_s - e_a)/r_a}{\Delta + \gamma} \tag{4}$$

where λE (W m^{-2}) is the latent heat flux, Δ (kPa °C^{-1}) is the slope of the saturation vapor pressure curve, R_n (W m^{-2}) is the net radiation flux, G_0 (W m^{-2}) is the surface soil heat flux, ρ_a (kg m^{-3}) is the air density, c_p (J kg^{-1} °C^{-1}) is the specific heat of air at constant pressure, here the value of 1013 J kg^{-1} °C^{-1} is used. e_s (kPa) is the saturated vapor pressure, e_a (kPa) is the actual vapor pressure, r_a (m s^{-1}) is the aerodynamic resistance of vapor transport, and γ (kPa °C^{-1}) is the psychrometric constant. All the model parameters are on the hourly timescale. The negative latent heat flux is converted to condensation or desublimation by:

$$E = \frac{A}{\rho}\frac{\lambda E}{\lambda} \tag{5}$$

where E (mm) is the hourly water equivalent of condensation or desublimation, λ (MJ kg^{-1}) is the latent heat of vaporization or sublimation, A (s) is the time interval of hourly meteorological variables, and ρ (kg m^{-3}) is the water density. By accumulating the hourly condensation and desublimation, the monthly and annual condensation and desublimation used for quantitative analysis can be obtained. The parameters for the Penman model and their calculation methods are listed in detail by Table 3, and the required meteorological variables are listed in detail by Table 1.

Table 3. Parameters for the Penman model and their calculation method.

Model Parameters	Symbols	Units	Calculation Methods	References
Slope of saturation vapor pressure curve	Δ	kPa °C^{-1}	$\Delta = \frac{4098 \cdot e_s}{(237.3+T_a)^2}$	[50,51]
Net radiation	R_n	W m^{-2}	$R_n = R_s + R_t$	[47,52]
Surface soil heat flux	G_0	W m^{-2}	$G_0 = 0.5 \cdot R_n \ (R_n \leq 0)$	[47,52]
Air density	ρ_a	kg m^{-3}	$\rho_a = 1.293 \frac{P_a}{P_{atm}} \frac{273.15}{273.15+T_a}$	[53,54]
Specific heat of air at constant pressure	c_p	J kg^{-1} °C^{-1}	1013	[47]
Saturated vapor pressure	e_s	kPa	$e_s = 0.6108 \cdot \exp(\frac{17.27 \cdot T_a}{237.3+T_a})$	[50,55]
Actual vapor pressure	e_a	kPa	$e_a = 0.6108 \cdot \exp(\frac{17.27 \cdot T_d}{237.3+T_d})$	[50,55]
Aerodynamic resistance of vapor transport	r_a	m s^{-1}	$r_a = \frac{\sqrt{u^2+v^2}}{u_*^2}\left(\frac{4.87}{\ln(67.8z-5.42)}\right) \ (R_n \leq 0)$	[47,52,56]
Height of wind component	z	m	10	[38–40]
Psychrometric constant	γ	kPa °C^{-1}	$\gamma = \begin{cases} 0.665 \times 10^{-3} P_a \ T_d > 0 \\ 0.588 \times 10^{-3} P_a \ T_d \leq 0 \end{cases}$	[47,52]
Latent Heat of Vaporization/Sublimation	λ	MJ kg^{-1}	$\lambda = \begin{cases} 2.501 - (2.361 \times 10^{-3})T_a \ T_d > 0 \\ 2.835 - (2.361 \times 10^{-3})T_a \ T_d \leq 0 \end{cases}$	[47,52]
Time interval	A	s	3600	Constant
Water density	ρ	kg m^{-3}	1000	Constant

2.3.2. MK Trend Test

To analyze the trend of condensation and desublimation, the Mann–Kendall (MK) trend test [57–60] was used to quantify the trend and its significance of condensation and desublimation on the QTP during 1950~2020. The MK trend test is performed as follows:

$$S = \sum_{i=1}^{n-1} \sum_{j=i+1}^{n} \text{sgn}(x_j - x_i) \tag{6}$$

$$\text{sgn}(x_j - x_i) = \begin{cases} 1 & (x_j - x_i) > 0 \\ 0 & (x_j - x_i) = 0 \\ -1 & (x_j - x_i) < 0 \end{cases} \tag{7}$$

$$\text{var}(S) = \frac{n(n-1)(2n+5) - \sum_{k=1}^{m} t_k(t_k - 1)(2t_k + 5)}{18} \tag{8}$$

$$Z_c = \begin{cases} (S-1)/\sqrt{\text{var}(S)} & S > 0 \\ 0 & S = 0 \\ (S+1)/\sqrt{\text{var}(S)} & S < 0 \end{cases} \tag{9}$$

where S is the statistic of the dataset, n is the length of the dataset, x_i and x_j are the sequential data values in time series i and j, m is the number of tied groups, and t_k denotes the number of ties of extent k and a tied group is a set of sample data having the same value, Z_c is the standardized statistics of the dataset, and the positive Z_c indicates an increasing trend of the dataset, while the negative values Z_c indicate a decreasing trend of the dataset. If $|Z_c| > Z_{1-\alpha/2}$, the trend is statistically significant, otherwise, the trend is not statistically significant. Trend test of condensation and desublimation was done at the significance level of $\alpha = 0.05$ ($|Z_{1-\alpha/2}| = 1.96$, 95% confidence level), i.e., when the Z_c of condensation or coagulation was greater than 1.96, it indicates that condensation or desublimation shows a significant increasing trend, while when the Z_c of condensation or desublimation was lower than −1.96, it indicates that condensation or desublimation shows a significant decreasing trend; otherwise, there was no significant trend of condensation or desublimation.

2.3.3. Sen's Slope Analysis

The magnitude of the trend (i.e., slope in variation per unit time) of condensation and desublimation was determined using a non-parametric method known as Sen's slope analysis [59–61], and the slope is expressed as:

$$\beta = \text{Median}\left(\frac{x_j - x_i}{j - i}\right) 1 < i < j < n \tag{10}$$

where β is the slope of the dataset, a positive β denotes an increasing trend of condensation or desublimation, while a negative β means a decreasing trend of condensation or desublimation. A larger value of $|\beta|$ means a greater increase or decrease in condensation and desublimation.

3. Results

3.1. Accuracy of Estimated Condensation and Desublimation

The comparison of the monthly condensation and desublimation measured by the Eddy Covariance with those estimated by the Penman model at nine flux stations on the QTP is shown in Figures 3 and 4, respectively. As can be seen from Figure 3, the coefficient of determination (R^2) between estimated and measured monthly condensation is greater than or equal to 0.55 for all stations, indicating that the Penman model with nine meteorological variables from ERA5-Land and ERA5 reanalysis datasets as inputs has good applicability in estimating monthly condensation on the QTP. Among the nine flux stations on the QTP, the largest R^2 between estimated and measured monthly condensation is 0.94 for Xiyinghe station, the smallest R^2 between estimated and measured monthly condensation is 0.55 for Ngari station, and the average R^2 between estimated and measured monthly condensation is 0.82 for all stations.

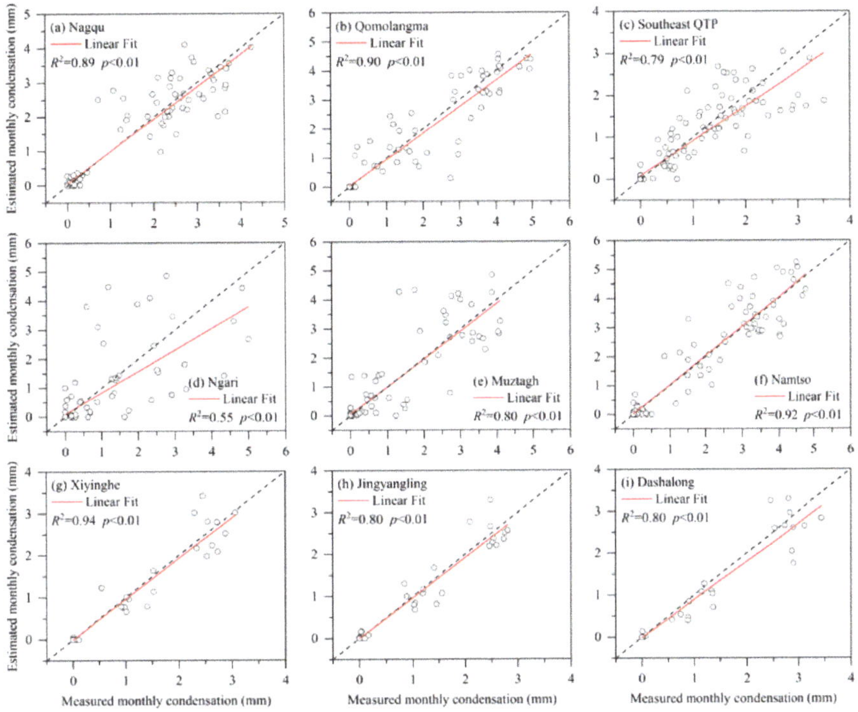

Figure 3. Comparison of estimated and measured monthly condensation at the nine flux stations on the QTP.

Compared with the R^2 between estimated and measured monthly condensation, the R^2 between estimated and measured monthly desublimation is much smaller at the nine flux stations on the QTP (Figure 4). As can be seen from Figure 4, the R^2 between estimated and measured monthly desublimation is less than or equal to 0.71 for all stations, indicating that the Penman model with nine meteorological variables from ERA5-Land and ERA5 reanalysis datasets as inputs is less applicable in estimating monthly desublimation than in estimating monthly condensation on the QTP. Among the nine flux stations on the QTP, the largest R^2 between estimated and measured monthly desublimation is 0.71 for Qomolangma station, the smallest R^2 between estimated and measured monthly desublimation is 0.46 for Xiyinghe station, and the average R^2 between estimated and measured monthly desublimation is 0.59 for all stations.

Although the R^2 between estimated and measured monthly condensation is higher than the R^2 between estimated and measured monthly desublimation, the R^2 between estimated and measured annual condensation is much smaller than the R^2 between estimated and measured annual desublimation, as shown in Figure 5. The average R^2 between estimated and measured annual condensation for all stations is 0.56, while the average R^2 between estimated and measured annual desublimation is 0.91.

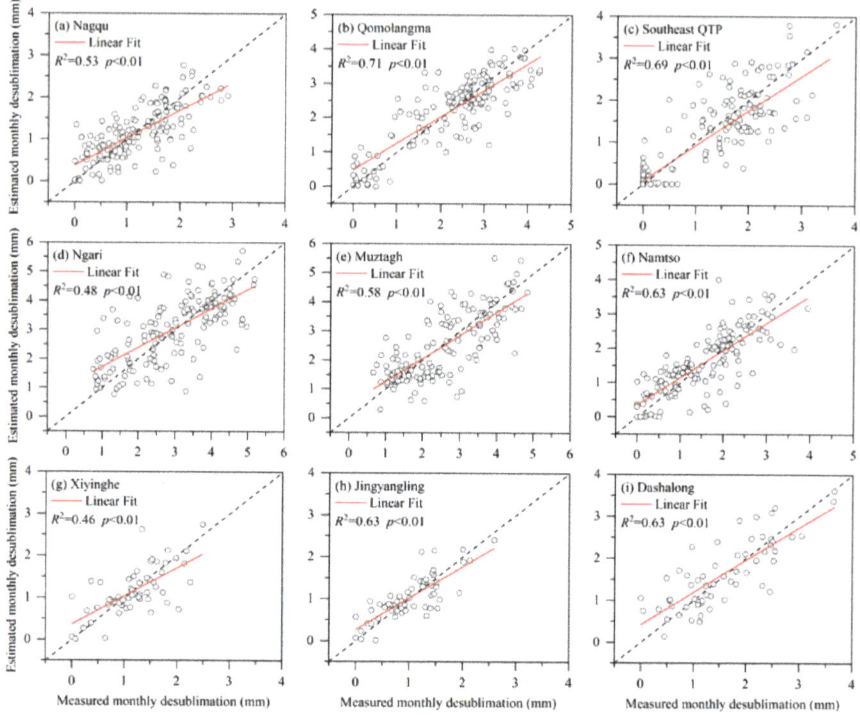

Figure 4. Comparison of estimated and measured monthly desublimation at the nine flux stations on the QTP.

Clearly, the Penman model with nine meteorological variables from ERA5-Land and ERA5 as inputs is more accurate for estimating monthly condensation than for estimating annual condensation, and less accurate for estimating monthly desublimation than for estimating annual desublimation. Overall, the accuracy of the estimated monthly and annual condensation and desublimation is adequate for the study.

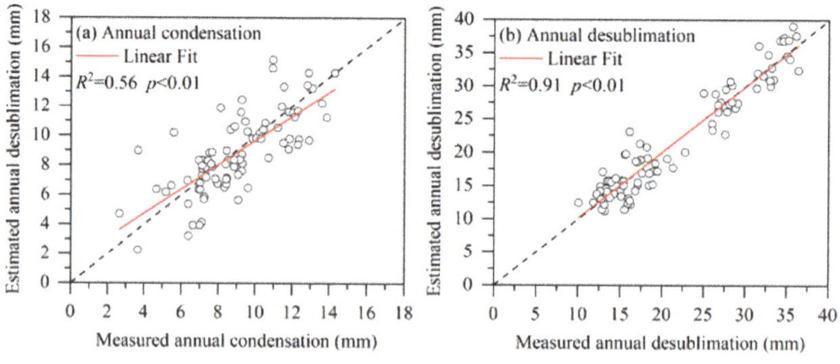

Figure 5. Comparison of estimated and measured annual condensation, annual desublimation at all flux stations on the QTP.

3.2. Spatial Distribution of Condensation and Desublimation

The spatial distribution characteristics of annual mean condensation, annual mean desublimation, and annual mean total condensation and desublimation on the QTP from 1950 to 2020 and their standard deviations are shown in Figure 6. As can be seen in

Figure 6a, the annual mean condensation on the QTP showed a decreasing trend from southeast to northwest, with a large difference between the annual condensation in the southeast and northwest, and the annual mean condensation ranges from 0 mm to 72.8 mm. However, areas with condensation between 36 mm and 72.8 mm account for less than 5% of the total area of the QTP. The areas with the highest annual condensation were in the southeastern river valleys, and the areas with the lowest annual condensation were in the Northwest QTP and the Tsaidam Basin. The spatial distribution of the standard deviation of annual condensation was basically consistent with that of annual mean condensation, and the greater the annual mean condensation, the greater the standard deviation of annual condensation, and vice versa (Figure 6d).

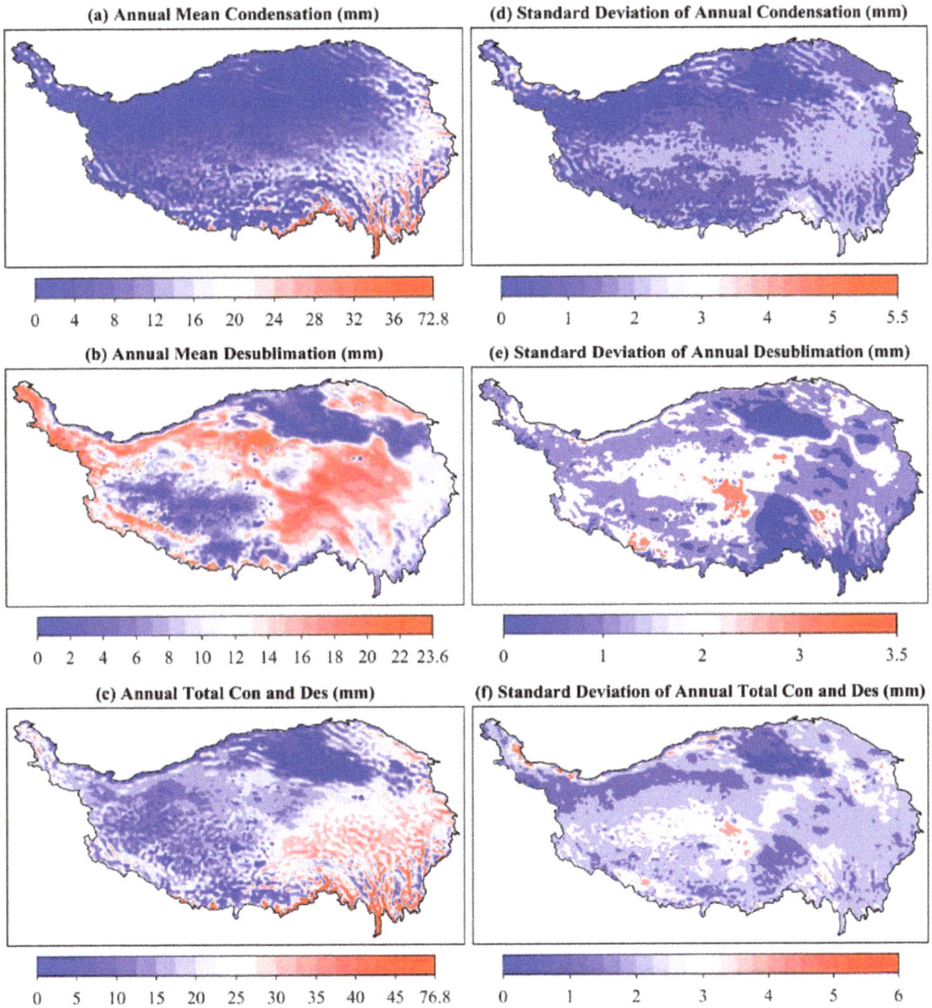

Figure 6. Spatial distribution characteristics of annual mean condensation, annual mean desublimation and annual total mean condensation and desublimation from 1950 to 2020 and their standard deviations. (**a**) Annual mean condensation; (**b**) Annual mean desublimation; (**c**) Annual mean total condensation (Con) and desublimation (Des); (**d**) the Standard divisions of annual condensation; (**e**) the Standard divisions of annual desublimation; (**f**) the Standard divisions of annual total condensation and desublimation.

Unlike the spatial distribution characteristics of annual mean condensation, the spatial distribution of annual mean desublimation did not have zonal characteristics seen in Figure 6b. The annual mean desublimation in the Qiangtang Plateau, Tsaidam Basin, and the eastern edge of the QTP were significantly lower than those in other regions. The annual mean desublimation on the QTP as a whole ranges from 0 mm to 23.6 mm, and the difference in annual mean desublimation between regions is much smaller than that of annual mean condensation. The spatial distribution of the standard deviation of the annual desublimation was not consistent with the spatial distribution of the annual mean desublimation, and the standard deviations were small overall (Figure 6e).

Influenced by the spatial distribution of annual mean condensation and annual mean desublimation, the spatial distribution of total annual mean condensation and desublimation showed obvious regional differences, i.e., the southeastern and northeastern parts of the QTP were the regions with the largest total condensation and desublimation, followed by the Qiangtang Plateau and the lowest in the Tsaidam Basin (Figure 6c). The range of the total annual mean condensation and desublimation for the whole QTP was from 0 mm to 76.8 mm, but only less than 5% of the areas have the total annual mean condensation and desublimation exceeding 45 mm. In addition, the spatial distribution characteristics of the standard deviation of the total annual condensation and desublimation were consistent with the spatial distribution of the annual condensation, with the Qiangtang Plateau being the region with the larger standard deviation of the total annual condensation and desublimation, and the Tsaidam Basin being the region with the smaller standard deviation of the total annual condensation and desublimation (Figure 6f).

3.3. *Spatial Trends of Condensation and Desublimation*

In order to quantify the spatial trends of condensation, desublimation, and total condensation and desublimation on the QTP, a significance analysis based on MK trend test and magnitude of the trends based on Sen's slope analysis were conducted for annual condensation, annual desublimation, and total annual condensation and desublimation as shown in Figure 7. As can be seen from Figure 7a, the annual condensation showed a clear increasing trend in the central part of the QTP, and only a small part of the northeastern QTP showed a clear decreasing trend, while the annual condensation in other regions did not show a clear trend. In addition, it can also be seen from Figure 7d that the annual condensation showed an increasing trend in most regions of the QTP and a decreasing trend in the Tsaidam Basin and the eastern and southern margins of the QTP, with a maximum increase rate of 1.70 mm/10a and a maximum decrease rate of −1.16 mm/10a.

Figure 7b shows that the annual desublimation showed a significant decreasing trend in most regions of the QTP, and only in a very small part of the eastern QTP showed a significant increasing trend. The magnitude of the annual desublimation trend in Figure 7e also indicates that desublimation showed a decreasing trend in the majority of the QTP, with a maximum decrease rate of −0.9 mm/10a and a maximum increase rate of 0.43 mm/10a.

Since the decreasing trend of desublimation on the QTP was more significant than the increasing trend of condensation, the decreasing trend of total condensation and desublimation was more obvious than the increasing trend shown in Figure 7c. The total condensation and desublimation showed a regionally significant decreasing trend in the northeastern QTP, and southeastern QTP and the Qiangtang Plateau, while the increasing trend of total condensation and desublimation was not regional, and the maximum decreasing rate of annual condensation and desublimation was −1.49 mm/10a and the maximum increasing rate was 1.62 mm/10a.

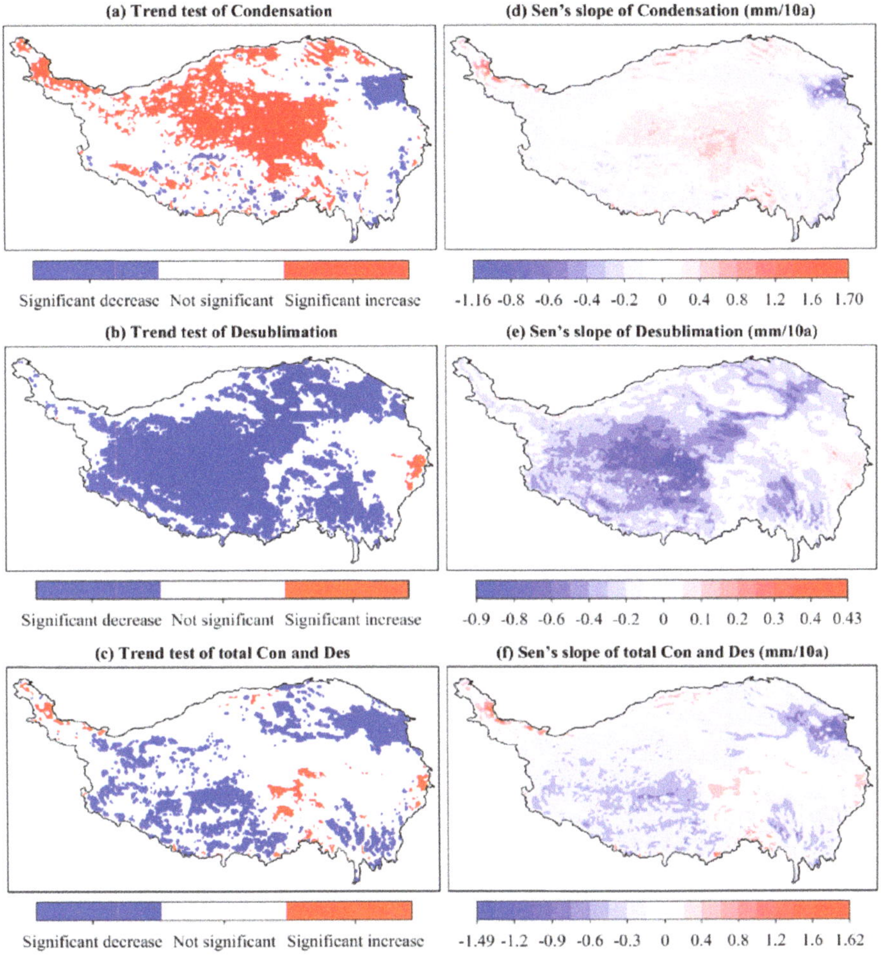

Figure 7. Significance and magnitude of the trend of annual condensation, annual desublimation, and annual total condensation and desublimation from 1950 to 2020 based on MK trend test (95% confidence level) and Sen's slope analysis. (**a**) Trend test of condensation; (**b**) Trend test of desublimation; (**c**) Trend test of total condensation (Con) and desublimation (Des); (**d**) Sen's slope of condensation; (**e**) Sen's slope of desublimation; (**f**) Sen's slope of total condensation and desublimation.

3.4. Spatial Variations in Condensation and Desublimation

To better visualize the spatial variation characteristics of condensation and desublimation on the QTP, the percentage change of condensation and desublimation in the last two decades compared to the first two decades of the study period from 1950 to 2020 was quantified, as shown in Figure 8. From Figure 8a,b, it is clear that the increasing proportion of annual condensation on the QTP has an increasing trend from southeast to northwest, while the decreasing proportion of annual desublimation and the decreasing proportion of total annual condensation and desublimation showed a regional rather than a zonal pattern. In general, in the last two decades compared to the first two decades of the QTP from 1950 to 2020, the mean percentage increase of annual condensation was 19.6%, the mean percentage decrease of annual desublimation was 11.5%, and the mean percentage decrease of annual total condensation and desublimation was 4.6%.

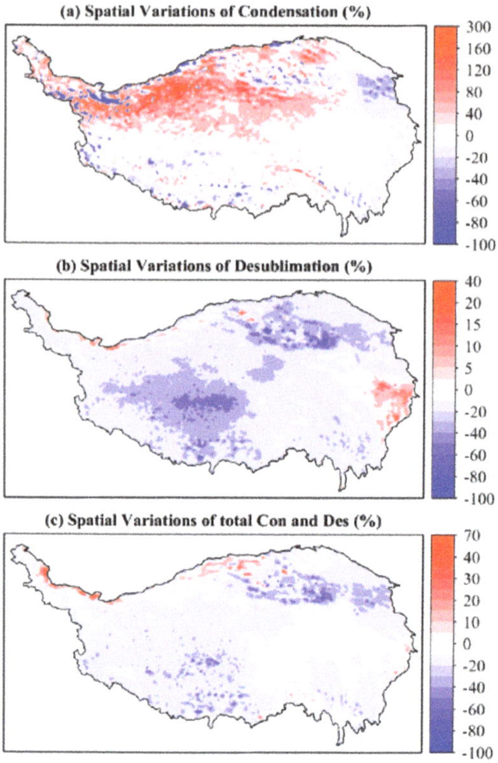

Figure 8. Proportional variations in annual condensation, annual desublimation and annual total condensation and desublimation on the QTP in the last two decades compared to the first two decades, from 1950 to 2020.

3.5. Monthly Variations in Condensation and Desublimation

Figure 9 shows the monthly variations of condensation, desublimation, and the total condensation and desublimation averaged over 1950~2020. From Figure 9a, it can be seen that condensation showed obvious seasonal differences, i.e., the summer was the season with the highest amount of condensation, while the winter was the season with the lowest amount of condensation. The August was the month with the highest amount of condensation of 2.37 ± 0.44 mm, and January was the month with the lowest amount of condensation of 0.02 mm, which was almost negligible. As can be seen from Figure 9b, the desublimation also showed a clear seasonality, but the seasonality of desublimation was different from the seasonality of condensation, i.e., there were two seasons in the year when desublimation was enriched, namely, spring and autumn, in which the month with the highest amount of desublimation was October of 1.45 ± 0.2 mm, and the month with the lowest amount of desublimation was July of 0.5 ± 0.15 mm. The monthly variations of the total condensation and desublimation was influenced by the monthly variations of condensation and desublimation, with July to September being the period of maximum total condensation and desublimation, while December to February of the following year was the period of minimum total condensation and desublimation (Figure 9c). The maximum amount of the total condensation and desublimation was in August, with an average value of 2.97 ± 0.37 mm, and the minimum amount of the total condensation and desublimation was in February, with an average value of 0.77 ± 0.10 mm.

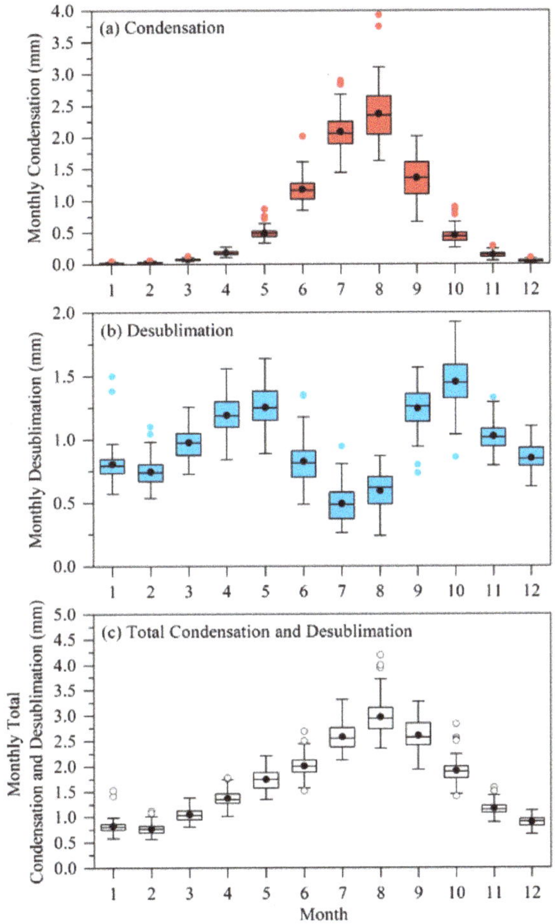

Figure 9. Variations of monthly condensation, monthly desublimation, and monthly total condensation and desublimation. (**a**) Monthly condensation; (**b**) Monthly desublimation; (**c**) Monthly total.

Although the monthly maximum amount of condensation was much larger than the monthly maximum amount of desublimation, the monthly mean amount of desublimation was larger than the monthly mean amount of condensation. The monthly mean amount of condensation was 0.70 mm, the monthly mean amount of desublimation was 0.95 mm, and the monthly mean total amount of condensation and desublimation was 1.66 mm.

3.6. Annual Variations in Condensation and Desublimation

The variations of annual condensation, annual desublimation, and their sums for the entire region of the QTP are shown in Figure 10. From Figure 10a, it is clear that the annual desublimation showed a continuous decreasing trend, while the annual condensation fluctuated more in the initial years and then showed a continuous increasing trend. The rate of decrease of desublimation was −0.26/10a and the rate of increase of condensation in the later years was 0.24/10a. Since the absolute value of the decrease rate of annual desublimation was greater than the increase rate of annual condensation, the total annual condensation and desublimation also showed a weak decreasing trend, as shown in Figure 10b. Furthermore, it can be found from Figure 10a that the increase in annual condensation will exceed the decrease in annual desublimation in recent years, and the increase in condensation will dominate in the future, and the annual total condensation and desublimation will change from a decreasing trend to an increasing trend. It can also be found in Figure 10a that the annual mean desublimation was greater than the annual mean condensation, with annual mean desublimation of 11.45 ± 0.85 mm and annual mean condensation of 8.45 ± 0.94 mm, for annual mean total condensation and desublimation of 19.89 ± 1.2 mm. The higher annual mean desublimation than the annual mean condensation reflects the cold climate characteristics of the QTP, while the increasing trend of condensation and decreasing trend of desublimation also reflect the warming of the regional climate.

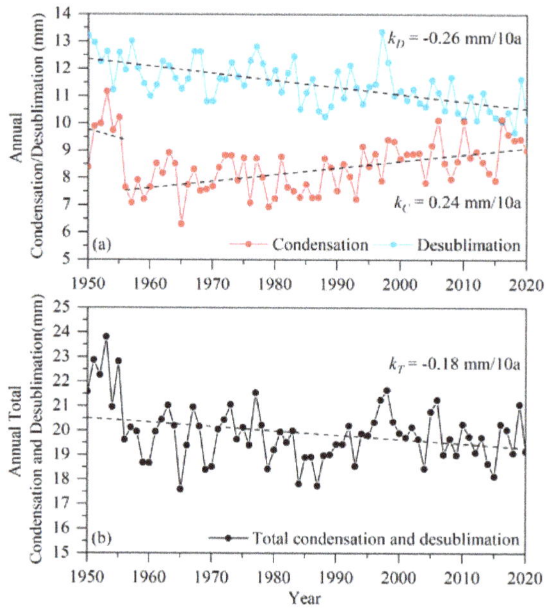

Figure 10. Variations of annual condensation, annual desublimation, and annual total condensation and desublimation. (**a**) Annual variations of condensation and desublimation; (**b**) Annual variations of the total condensation and desublimation. k_C represents the rate of variation in annual condensation, k_D represents the rate of variation in annual condensation, and k_T represents the rate of variation in annual total condensation and condensation.

3.7. Influencing Factors of Condensation and Desublimation Variations

The occurrence of condensation and condensation is the result of a combination of meteorological factors, among which temperature and relative humidity are the most important meteorological variables affecting the occurrence of condensation and desublimation. Since the relative humidity is closely related to variations in air temperature and precipitation, the current study focuses on analyzing the variations in air temperature, precipitation, and the relationship between variations in condensation and desublimation and variations in air temperature and precipitation on the QTP.

As shown in Figure 11a,d, the annual mean temperature and annual precipitation showed an overall increasing trend, but both trends were phased, i.e., both the annual mean temperature and annual precipitation showed a decreasing trend followed by an increasing trend on the QTP from 1950 to 2020. Since the 1970s, the average annual temperature has been increasing at a rate of 0.25 °C/10a, and the annual precipitation has been increasing at a rate of 13.02 mm/10a. From 1950 to 2020, the mean annual temperature was −3.95 °C and the mean annual precipitation was 388 mm of the QTP.

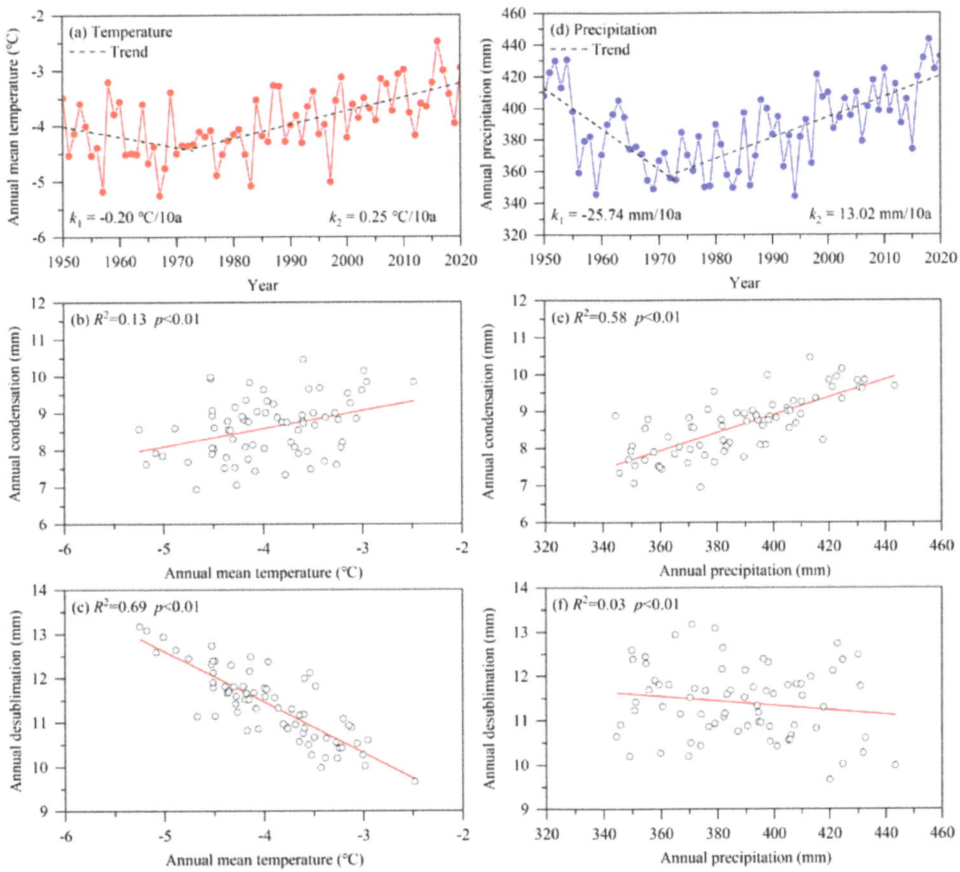

Figure 11. Variations in annual mean temperature and annual precipitation, and the correlation between annual condensation, annual desublimation, and annual mean temperature, annual precipitation on the QTP from 1950 to 2020. (**a**) Annual mean temperature; (**b**) The correlation between annual condensation and annual mean temperature; (**c**) The correlation between annual desublimation and annual mean temperature; (**d**) Annual precipitation; (**e**) The correlation between annual condensation and annual precipitation; (**f**) The correlation between annual desublimation and annual precipitation.

From Figures 10a and 11a,b, it can be seen that the variations in annual mean temperature and annual precipitation are more consistent with the variations in annual condensation. In addition, from Figure 11b,c,e,f, it can be seen that there is a positive correlation between annual mean temperature, annual precipitation, and annual condensation, and a negative correlation with annual desublimation, indicating that the increase in annual mean temperature and annual precipitation is favorable to the occurrence of condensation and unfavorable to the occurrence of desublimation on the QTP.

The above results show that the increase in annual condensation is most correlated with the increase in annual precipitation, while the decrease in desublimation is most correlated with the increase in temperature, and the increase in precipitation is the dominant meteorological variable in the increasing trend of annual condensation, while the increase in temperature is the dominant meteorological variable in the decreasing trend of annual desublimation.

4. Discussion

4.1. Uncertainty in the Evaluation of Condensation and Desublimation

In the present study, nine meteorological variables from the ERA5-Land and ERA5 reanalysis datasets were used as inputs to generalize the occurrence of condensation and desublimation based on the Penman model, which inevitably resulted in systematic errors in the evaluation of condensation and desublimation. Therefore, the main uncertainties in this study arise from the two following aspects: the applicability of ERA5-Land and ERA5 reanalysis datasets for land surface process studies on the one hand, and the applicability of the Penman model in evaluating condensation and desublimation occurrence studies on the other hand.

ERA5-Land and ERA5 are currently more recognized reanalysis datasets with high spatial and temporal resolution, and they have been widely used in studies of land surface processes including land–atmosphere interactions, meteorology, and hydrology [28]. The applicability evaluation of ERA5-Land on the northeastern QTP shows that the correlation between the assimilated temperature in ERA5-Land and the observed temperature at meteorological stations was very high, and ERA5-Land reproduced the spatial distribution of temperature more accurately; however, ERA5-Land underestimated the temperature of different degrees, and this phenomenon was also found in the applicability evaluation of ERA5-Land on a national scale [28,62]. In addition, the evaluation of different variables in ERA5-Land in different regions also shows that ERA5-Land has better performance in reproducing corresponding meteorological elements [63–65]. Compared to ERA5-Land, ERA5 is more widely applicable due to its earlier development. Most studies based on ERA5 reanalysis datasets have shown that ERA5 variables have good spatial and temporal consistency with the observed meteorological elements, despite under- and over-estimation of some meteorological variables [66]. Compared with other types of reanalysis data, ERA5-Land and ERA5 have higher spatial and temporal resolution, making them the most widely used reanalysis datasets at present. In general, the ERA5-Land and ERA5 reanalysis datasets have good applicability in land surface process studies, especially for large scale land–atmosphere interactions, and meteorological and hydrological process studies.

The widely accepted method for estimating the latent heat flux is the Eddy Covariance method [67], but the Eddy Covariance equipment is expensive and only applicable to in situ observations, which cannot be used for large scale and regional latent heat flux estimation. Compared with the Eddy Covariance method, the Penman model also has a better physical basis and can be used for large scale and regional latent heat flux estimation [68]. The estimation of condensation (dew) and desublimation (hoarfrost) based on the Penman model shows very good agreement between the total amount of dew and hoarfrost estimated by the Penman model and the total amount of dew and hoarfrost measured by microlysimeters; and the Penman model slightly underestimates the total amount of dew and hoarfrost [4,24].

From Figures 3–5, the average R^2 between the estimated and measured monthly condensation and monthly desublimation are 0.82 and 0.59, respectively, and the average R^2 between the estimated and measured annual condensation and annual desublimation are 0.56 and 0.91, respectively, for the nine flux stations on the QTP, indicating that the accuracy of the monthly and annual condensation and desublimation, estimated by the Penman model with nine meteorological variables from ERA5-Land and ERA5 as inputs, is adequate for the present study. Overall, although there are uncertainties in evaluating condensation and desublimation based on the Penman model using the ERA5-Land and ERA5 reanalysis datasets, the method can effectively reproduce the spatial and temporal variations of condensation and desublimation, which is useful for improving the understanding of condensation and desublimation over the QTP.

4.2. Impact of Condensation and Desublimation on Alpine Ecosystem

Previous studies of condensation and desublimation mainly focused on arid and semi-arid regions [8,18,69–71], and there are less studies of condensation and desublimation in alpine regions [4,12]. However, the magnitude of condensation and desublimation in alpine regions is much larger than that in arid and semi-arid regions [72,73], and the ecological and hydrological roles of condensation and desublimation in alpine regions and arid and semi-arid regions is obviously different [25]. The alpine regions are the generally referred to areas with high altitude, low temperature, and large diurnal temperature differences, as well as areas with short growing periods and low effective accumulation temperatures due to the high altitude [74]. The near-surface atmospheric hydrothermal conditions in alpine regions are highly variable, with frequent evaporation, sublimation, condensation, and desublimation interaction processes, which play an important role in the ecological and water vapor internal circulation processes in the region [75–77].

Observation experiments on condensation and desublimation carried out in recent years in alpine regions have shown that condensation and desublimation have unique ecological and hydrological roles in alpine regions, and that condensation and desublimation are one of the water sources in alpine regions that are no less important than in arid and semi-arid regions [11,78,79]. Observations in the coarse gravel accumulation area of Siberian slopes show that condensation recharge of water vapor exists in the pore space of mountain gravel and gravel layers, the amount of condensation can reach 80 mm in summer [80], and condensation recharge accounts for 15–20% of the total recharge. Kuhle found in the alpine mountains of the northern slopes of the Himalayas that condensation causes alpine rocky slopes to be completely covered by alpine meadows [81]; and in the alpine belt, where 80% of the soil is coarse-boned, the potential soil moisture can last only 13 days under continuous sunny weather but 22 days with surface condensation occurring [82]. The present study shows that the annual mean total condensation and desublimation is 19.89 mm, the annual mean precipitation is 388 mm, and the ratio of total annual condensation and condensation to annual precipitation is 5.13% on the QTP. In fact, the direct contribution of condensation and desublimation to surface water input in alpine regions is low, and condensation and desublimation mainly moderates the periodic physiological water deficit of vegetation caused by low temperatures [25]. In alpine regions with greater diurnal temperature differences and strong radiative cooling at night, the greater the land–atmosphere temperature difference, the higher the risk of frost damage to vegetation; the formation process of condensation and desublimation releases part of the latent heat, which replenishes the energy deficit at the surface, alleviates the low temperature stress on vegetation, and reduces the risk of frost damage to vegetation [12].

In summary, the ecological and hydrological roles of condensation and condensation in alpine regions like the QTP are reflected in the two following aspects: on the one hand, condensation and desublimation are part of surface water input, which alleviate the physiological water deficit of surface plants, animals, and microorganisms due to low temperatures; on the other hand, the occurrence of condensation and desublimation releases latent heat, which replenishes the energy deficit near the surface, alleviates the difference in ground temperature, and serves to reduce the risk of frost damage to surface plants and animals and microorganisms. Therefore, it can be considered that condensation and desublimation play an important role in maintaining the development of alpine ecosystems, with areas of increased condensation and desublimation indicating a trend towards better ecosystem conditions, while areas of decreased condensation and desublimation imply a trend towards worse ecosystem conditions.

5. Conclusions

The present study focuses on evaluating the spatial and temporal variations of condensation and desublimation on the QTP from 1950 to 2020 based on the Penman model using meteorological variables from ERA5-Land and ERA5 reanalysis datasets. Condensation and desublimation were often neglected, but condensation and desublimation were major processes of land–atmosphere interactions, energy transfer, and water cycle at night and have important ecological and hydrological roles in alpine regions to alleviate the physiological water deficit of surface plants, animals, and microorganisms due to low temperature and reduce the risk of frost damage to surface plants, animals, and microorganisms. Therefore, evaluating the spatial and temporal variations of condensation and desublimation not only improves the understanding of land–atmosphere interactions, energy transfer, and water cycles, but also contributes to the understanding of the operational mechanisms of alpine ecosystems on the QTP.

The results showed that there were large regional differences in condensation and desublimation on the QTP, with annual mean condensation ranging from 0 mm to 72.8 mm, annual mean desublimation ranging from 0 mm to 23.6 mm, and annual mean total condensation and desublimation ranging from 0 mm to 76.8 mm. Condensation showed a significant increasing trend in the central and northwestern parts of the QTP, desublimation showed a significant decreasing trend in most parts of the QTP, and the decreasing trend of desublimation was more significant than the increasing trend of condensation.

Both condensation and desublimation have significant seasonal characteristics, with condensation being most abundant in the summer, and desublimation in the spring and autumn. The maximum monthly condensation was 2.37 mm, the monthly mean condensation was 0.70 mm, the maximum monthly desublimation was 1.45 mm, the monthly mean condensation was 0.95 mm, and the total monthly mean condensation and desublimation was 1.66 mm. The annual mean condensation was 8.45 mm, the annual mean desublimation was 11.45 mm, and the total condensation and desublimation was 19.89 mm. The annual condensation hadan increasing trend of 0.24 mm/10a, the annual desublimation had a decreasing trend of −0.26 mm/10a, and the total annual mean condensation and desublimation had a weak decreasing trend. The increase in condensation is most associated with the increase in precipitation, while the decrease in desublimation is most associated with the increase in air temperature on the QTP.

The ratio of total annual condensation and desublimation to precipitation is 5.13%. Although the annual mean desublimation was higher than the annual mean condensation, the increase in condensation will exceed the decrease in desublimation in the future. As condensation exceeds desublimation, the total amount of condensation and desublimation tends to increase, indicating that the ecosystem shows a trend toward improvement on the QTP.

Author Contributions: Conceptualization, H.L. and C.H.; methodology, H.L. and Y.Y.; software, H.L.; investigation, H.L. and C.H.; data curation, C.H. and R.C.; writing—original draft preparation, H.L.; writing—review and editing, C.H., Y.Y. and R.C.; funding acquisition, R.C., C.H. and Y.Y. All authors have read and agreed to the published version of the manuscript.

Funding: This Research was funded by the Joint Research Project of Three-River Headwaters National Park, Chinese Academy of Sciences and the People's Government of Qinghai Province (LHZX-2020-11), the National Natural Sciences Foundation of China (41971041), the Sciences and Technology Plan Project of Gansu Province (21JR7RA056), and the Open Research Fund of the National Cryosphere Desert Data Center (2021kf09).

Data Availability Statement: The ERA5-Land reanalysis dataset from the ECMWF used in this study can be accessed online (https://cds.climate.copernicus.eu/cdsapp#!/dataset/reanalysis-era5-land?tab=overview/ (accessed on 1 March 2022)). The ERA5 reanalysis dataset from the ECMWF used in this study can be accessed online (https://cds.climate.copernicus.eu/cdsapp#!/dataset/reanalysis-era5-single-levels?tab=overview/ (accessed on 1 March 2022)). The observed meteorological variables from the National Tibetan Plateau Data Center used in this study can be accessed online (https://data.tpdc.ac.cn/en/data/b9ab35b2-81fb-4330-925f-4d9860ac47c3/ (accessed on 1 March 2022); https://data.tpdc.ac.cn/en/special/b2f26432-5d99-4126-ab94-883613cba133/?introduce=true/ (accessed on 1 March 2022)).

Acknowledgments: The authors would like to thank European Centre for Medium-range Weather Forecasts (ECMWF) (https://cds.climate.copernicus.eu/, (accessed on 1 March 2022)) for providing the ERA5-Land and ERA5 reanalysis datasets and National Tibetan Plateau Data Center (https://data.tpdc.ac.cn/en/, (accessed on 1 March 2022)) for providing the observed meteorological variables, and the authors also would like to thank all team members in Qilian Alpine Ecology and Hydrology Research Station, Northwest Institute of Eco-Environment and Resources, Chinese Academy of Sciences.

Conflicts of Interest: The authors declare no conflict of interest.

References

1. Agam, N.; Berliner, P. Dew formation and water vapor adsorption in semi-arid environments—A review. *J. Arid. Environ.* **2006**, *65*, 572–590. [CrossRef]
2. Beysens, D. The formation of dew. *Atmos. Res.* **1995**, *39*, 215–237. [CrossRef]
3. Lee, J.; Lee, K.S. The behavior of frost layer growth under conditions favorable for desublimation. *Int. J. Heat Mass Transf.* **2018**, *120*, 259–266. [CrossRef]
4. Li, H.; Han, C.; Yang, Y.; Chen, R. Formation and variations of dew and hoarfrost in the Hulu Catchment on Northeast Qinghai-Tibet Plateau, China. *J. Hydrol. Reg. Stud.* **2022**, *42*, 101179. [CrossRef]
5. Guo, S.; Chen, R.; Li, H. Surface Sublimation/Evaporation and Condensation/Deposition and Their Links to Westerlies During 2020 on the August-One Glacier, the Semi-Arid Qilian Mountains of Northeast Tibetan Plateau. *J. Geophys. Res. Atmos.* **2022**, *127*, e2022JD036494. [CrossRef]
6. De Jong, C. The contribution of condensation to the water cycle under high-mountain conditions. *Hydrol. Process.* **2005**, *19*, 2419–2435. [CrossRef]
7. Malek, E.; McCurdy, G.; Giles, B. Dew contribution to the annual water balances in semi-arid desert valleys. *J. Arid. Environ.* **1999**, *42*, 71–80. [CrossRef]
8. Gao, Z.; Shi, W.; Wang, X.; Wang, Y. Non-rainfall water contributions to dryland jujube plantation evapotranspiration in the Hilly Loess Region of China. *J. Hydrol.* **2020**, *583*, 124604. [CrossRef]
9. Evans, S.; Todd-Brown, K.E.O.; Jacobson, K.; Jacobson, P. Non-rainfall Moisture: A Key Driver of Microbial Respiration from Standing Litter in Arid, Semiarid, and Mesic Grasslands. *Ecosystems* **2019**, *23*, 1154–1169. [CrossRef]
10. Zheng, J.; Peng, C.; Li, H.; Li, S.; Huang, S.; Hu, Y.; Zhang, J.; Li, D. The role of non-rainfall water on physiological activation in desert biological soil crusts. *J. Hydrol.* **2018**, *556*, 790–799. [CrossRef]
11. Wang, L.; Kaseke, K.F.; Seely, M.K. Effects of non-rainfall water inputs on ecosystem functions. *Wiley Interdiscip. Rev. Water* **2017**, *4*, e1179. [CrossRef]
12. Groh, J.; Slawitsch, V.; Herndl, M.; Graf, A.; Vereecken, H.; Pütz, T. Determining dew and hoar frost formation for a low mountain range and alpine grassland site by weighable lysimeter. *J. Hydrol.* **2018**, *563*, 372–381. [CrossRef]
13. Florentin, A.; Agam, N. Estimating non-rainfall-water-inputs-derived latent heat flux with turbulence-based methods. *Agric. For. Meteorol.* **2017**, *247*, 533–540. [CrossRef]
14. Ninari, N.; Berliner, P.R. The role of dew in the water and heat balance of bare loess soil in the Negev Desert: Quantifying the actual dew deposition on the soil surface. *Atmos. Res.* **2002**, *64*, 323–334. [CrossRef]
15. Kidron, G.J.; Kronenfeld, R. Microlysimeters overestimate the amount of non-rainfall water—An experimental approach. *Catena* **2020**, *194*, 104691. [CrossRef]
16. Zhang, Q.; Wang, S.; Yue, P.; Wang, R. A measurement, quantitative identification and estimation method(QINRW) of non-rainfall water component by lysimeter. *MethodsX* **2019**, *6*, 2873–2881. [CrossRef]
17. Hanisch, S.; Lohrey, C.; Buerkert, A. Dewfall and its ecological significance in semi-arid coastal south-western Madagascar. *J. Arid. Environ.* **2015**, *121*, 24–31. [CrossRef]
18. Zhuang, Y.; Zhao, W. Dew formation and its variation in Haloxylon ammodendron plantations at the edge of a desert oasis, northwestern China. *Agric. For. Meteorol.* **2017**, *247*, 541–550. [CrossRef]
19. Madeira, A.C.; Kim, K.S.; Taylor, S.; Gleason, M. A simple cloud-based energy balance model to estimate dew. *Agric. For. Meteorol.* **2002**, *111*, 55–63. [CrossRef]
20. Wang, Z.; Zhang, Z.; Su, Y. A new water balance equation introducing dew amount in arid area. *J. Hydraul. Eng.* **2019**, *50*, 710–720.
21. Hao, X.; Li, C.; Guo, B.; Ma, J.; Ayup, M.; Chen, Z. Dew formation and its long-term trend in a desert riparian forest ecosystem on the eastern edge of the Taklimakan Desert in China. *J. Hydrol.* **2012**, *472*, 90–98. [CrossRef]
22. Jacobs, A.F.G.; Heusinkveld, B.G.; Berkowicz, S.M. Passive dew collection in a grassland area, The Netherlands. *Atmos. Res.* **2008**, *87*, 377–385. [CrossRef]
23. Uclés, O.; Villagarcía, L.; Moro, M.J.; Canton, Y.; Domingo, F. Role of dewfall in the water balance of a semiarid coastal steppe ecosystem. *Hydrol. Processes* **2014**, *28*, 2271–2280. [CrossRef]
24. Jacobs, A.F.G.; Heusinkveld, B.G.; Wichink Kruit, R.J.; Berkowicz, S.M. Contribution of dew to the water budget of a grassland area in the Netherlands. *Water Resour. Res.* **2006**, *42*, W03415. [CrossRef]

25. Chuntan, H.; Hongyuan, L.; Rensheng, C.; Zhangwen, L.; Junfeng, L.; Yong, Y.; Xiqiang, W. Advances in the study on NRWIs in Alpine Mountains. *Plateau Meteorol.* 2022, *in press*.
26. Zhang, X. Dynamics Changes of Typical Inland Lakes on Tibetan Plateau Using Multi-Sensor Remote Sensing Data. Ph.D. Thesis, Northwest A&F University, Xi'an, China, 2015.
27. Zou, H.; Zhu, J. Applicability of reanalysis data in the study of surface temperature in Qinghai-Tibet Plateau. In Proceedings of the 28th Annual Meeting of China Meteorological Society, Xiamen, China, 3–4 November 2011.
28. Li, H.; Liu, G.; Han, C.; Yang, Y.; Chen, R. Quantifying the Trends and Variations in the Frost-Free Period and the Number of Frost Days across China under Climate Change Using ERA5-Land Reanalysis Dataset. *Remote Sens.* **2022**, *14*, 2400. [CrossRef]
29. Yang, J.; Huang, M.; Zhai, P. Performance of the CRA-40/Land, CMFD, and ERA-Interim Datasets in Reflecting Changes in Surface Air Temperature over the Tibetan Plateau. *J. Meteorol. Res.* **2021**, *35*, 663–672. [CrossRef]
30. Ullah, I.; Ma, X.; Ren, G.; Yin, J.; Iyakaremye, V.; Syed, S.; Lu, K.; Xing, Y.; Singh, V.P. Recent Changes in Drought Events over South Asia and Their Possible Linkages with Climatic and Dynamic Factors. *Remote Sens.* **2022**, *14*, 3219. [CrossRef]
31. Ullah, I.; Ma, X.; Asfaw, T.G.; Yin, J.; Iyakaremye, V.; Saleem, F.; Xing, Y.; Azam, K.; Syed, S. Projected Changes in Increased Drought Risks Over South Asia Under a Warmer Climate. *Earth's Future* **2022**, *10*, e2022EF002830. [CrossRef]
32. Ullah, I.; Saleem, F.; Iyakaremye, V.; Yin, J.; Ma, X.; Syed, S.; Hina, S.; Asfaw, T.G.; Omer, A. Projected Changes in Socioeconomic Exposure to Heatwaves in South Asia Under Changing Climate. *Earth's Future* **2022**, *10*, e2021EF002240. [CrossRef]
33. Yao, T.; Bolch, T.; Chen, D.; Gao, J.; Immerzeel, W.W.; Piao, S.; Su, F.; Thompson, L.; Wada, Y.; Wang, L.; et al. The imbalance of the Asian water tower. *Nat. Rev. Earth Environ.* **2022**, *3*, 618–632. [CrossRef]
34. Xianru, L. Characteristics of Temperature and Precipitation Change on the Tibet Plateau under the Background of Global Warming. *Adv. Geosci.* **2019**, *9*, 1042–1049.
35. Ga, Z.; Sirong, C.; Bing, Z. Spatio-temporal variation of vegetation coverage over the Tibetan Plateau and its responses to climatic factors. *Acta Ecol. Sin.* **2018**, *38*, 3208–3218.
36. Hersbach, H.; Bell, B.; Berrisford, P.; Biavati, G.; Horányi, A.; Muñoz Sabater, J.; Nicolas, J.; Peubey, C.; Radu, R.; Rozum, I.; et al. ERA5 Hourly Data on Single Levels from 1959 to Present—Copernicus Climate Change Service (C3S) Climate Data Store (CDS). 2018. Available online: https://cds.climate.copernicus.eu/cdsapp#!/dataset/reanalysis-era5-single-levels-monthly-means?tab=overview (accessed on 10 March 2022).
37. Bell, B.; Hersbach, H.; Berrisford, P.; Dahlgren, P.; Horányi, A.; Muñoz Sabater, J.; Nicolas, J.; Radu, R.; Schepers, D.; Simmons, A.; et al. ERA5 Hourly Data on Single Levels from 1950 to 1978 (Preliminary Version)—Copernicus Climate Change Service (C3S) Climate Data Store (CDS). 2020. Available online: https://cds.climate.copernicus.eu/cdsapp#!/dataset/reanalysis-era5-pressure-levels-preliminary-back-extension?tab=overview (accessed on 10 March 2022).
38. Muñoz-Sabater, J. ERA5-Land Hourly Data from 1950 to 1980—Copernicus Climate Change Service (C3S) Climate Data Store (CDS). 2021. Available online: https://cds.climate.copernicus.eu/cdsapp#!/dataset/reanalysis-era5-land?tab=overview/ (accessed on 10 March 2022).
39. Muñoz-Sabater, J. ERA5-Land Hourly Data from 1981 to Present—Copernicus Climate Change Service (C3S) Climate Data Store (CDS). 2019. Available online: https://cds.climate.copernicus.eu/cdsapp#!/dataset/reanalysis-era5-land?tab=overview/ (accessed on 10 March 2022).
40. Muñoz-Sabater, J.; Dutra, E.; Agustí-Panareda, A.; Albergel, C.; Arduini, G.; Balsamo, G.; Boussetta, S.; Choulga, M.; Harrigan, S.; Hersbach, H.; et al. ERA5-Land: A state-of-the-art global reanalysis dataset for land applications. *Earth Syst. Sci. Data* **2021**, *13*, 4349–4383. [CrossRef]
41. Ma, Y.; Hu, Z.; Xie, Z.; Ma, W.; Wang, B.; Chen, X.; Li, J.; Zhong, L.; Sun, F.; Gu, L.; et al. A long-term (2005–2016) dataset of hourly integrated land-atmosphere interaction observations on the Tibetan Plateau. *Earth Syst. Sci. Data* **2020**, *12*, 2937–2957. [CrossRef]
42. Ma, Y. Long-Term Dataset of Integrated Land-Atmosphere Interaction Observations on the Tibetan Plateau (2005–2016)—National Tibetan Plateau Data Center. 2020. Available online: https://doi.org/10.11888/Meteoro.tpdc.270910 (accessed on 1 March 2022).
43. Liu, S.; Xu, Z.; Che, T.; Xiao, Q.; Ma, M.; Liu, Q.; Jin, R.; Guo, J.; Wang, L.; Wang, W.; et al. The Heihe Integrated Observatory Network: A Basin-Scale Land Surface Processes Observatory in China. *Vadose Zone J.* **2018**, *17*, 180072. [CrossRef]
44. Liu, S.; Xu, Z.; Wang, W.; Jia, Z.; Zhu, M.; Bai, J.; Wang, J. A comparison of eddy-covariance and large aperture scintillometer measurements with respect to the energy balance problem. *Hydrol. Earth Syst. Sci.* **2011**, *15*, 1291–1306. [CrossRef]
45. Liu, S.; Che, T.; Xu, Z.; Ren, Z.; Tan, J.; Zhang, Y. Qilian Mountains Integrated Observatory Network: Dataset of Heihe Integrated Observatory Network—National Tibetan Plateau Data Center. 2021. Available online: https://doi.org/10.11888/Geogra.tpdc.271440 (accessed on 1 March 2022).
46. Shuhai, G. Observation and Simulation of Sublimation/Evaporation at the Top of Midlatitude August-One Glacier, Qilian Mountains. Ph.D. Thesis, University of Chinese Academy of Sciences, Beijing, China, 2019.
47. Allen, R.G.; Pereira, L.S.; Raes, D.; Smith, M. *Crop Evapotranspiration-Guidelines for Computing Crop Water Requirements*; FAO: Rome, Italy, 1998.
48. Huang, H. *Principle and Simulation of Soil-Vegetation-Atmosphere Interaction*; China Meteorological Press: Beijing, China, 1997; pp. 66–72.
49. Penman, H.L. Natural Evapoation from Open Water, Bare Soil and Grass. *Proc. R. Soc. Lond. Ser. A Math. Phys.* **1948**, *193*, 120–145.
50. Murray, F.W. On the Computation of Saturation Vapor Pressure. *J. Appl. Meteorol. Climatol.* **1967**, *6*, 203–204. [CrossRef]

51. Allen, R. Penman-Monteith equation. In *Encyclopedia of Soils in the Environment*; Elsevier: Oxford, UK, 2005; pp. 180–188.
52. Brunt, D. *Physical and Dynamical Meteorology, 2nd ed*; Cambridge University Press: Cambridge, UK, 1952; 428p.
53. Xiaoqing, L. Design and improvement of the experiment for determining air density. *Exp. Teach. Appar.* **2021**, *38*, 25–26.
54. Jiumin, Y.; Lili, W.; Haibin, K.; Yuhua, Z.; Jinzhong, L.; Lili, Z. Improvement of air density and gas universal constant measurement experiment. *Phys. Exp.* **2011**, *31*, 24–26.
55. Tetens, O. Uber einige meteorologische Begriffe. *Z. Geophys.* **1930**, *6*, 297–309.
56. Brown, K.W.; Rosenberg, N.J. A Resistance Model to Predict Evapotranspiration and Its Application to a Sugar Beet Field. *Agron. J.* **1973**, *65*, 341–347. [CrossRef]
57. Mann, H. Non-Parametric Test Against Trend. *Econometrica* **1945**, *13*, 245–259. [CrossRef]
58. Kendall, M.G. *Rank Correlation Methods*, 4th ed.; Charles Griffin: London, UK, 1975.
59. Agarwal, S.; Suchithra, A.S.; Singh, S.P. Analysis and Interpretation of Rainfall Trend using Mann- Kendall's and Sen's Slope Method. *Indian J. Ecol.* **2021**, *48*, 453–457.
60. Jagadeesh, P.; Agrawal, S. Investigation of trends and its magnitude by non-parameteric Mann-Kendall and Sen's slope methods. *Int. J. Hydrol. Sci. Technol.* **2015**, *5*, 83–94. [CrossRef]
61. Sen, P.K. Estimates of the Regression Coefficient Based on Kendall's Tau. *J. Am. Stat. Assoc.* **1968**, *63*, 1379–1389. [CrossRef]
62. Li, Y.; Qin, X.; Liu, Y.; Zizhen, J.; Liu, J.; Wang, L.; Jizu, C. Evaluation of Long-Term and High-Resolution Gridded Precipitation and Temperature Products in the Qilian Mountains, Qinghai–Tibet Plateau. *Front. Environ. Sci.* **2022**, *10*, 906821. [CrossRef]
63. Bonshoms, M.; Ubeda, J.; Liguori, G.; Krner, P.; Navarro, L.; Cruz, R. Validation of ERA5-Land temperature and relative humidity on four Peruvian glaciers using on-glacier observations. *J. Mt. Sci.* **2022**, *19*, 1849–1873. [CrossRef]
64. Khadka, A.; Wagnon, P.; Brun, F.; Shrestha, D.; Lejeune, Y.; Arnaud, Y. Evaluation of ERA5-Land and HARv2 reanalysis data at high elevation in the upper Dudh Koshi basin (Everest region, Nepal). *J. Appl. Meteorol. Climatol.* **2022**, *61*, 931–954. [CrossRef]
65. Shuping, Y.; Yechao, Y.; Shuwen, Z.; Jiuchun, Y.; Wenjuan, W. Spatiotemporal variations of soil freeze-thaw state in Northeast China based on the ERA5-LAND dataset. *Acta Geogr. Sin.* **2021**, *76*, 2765–2779.
66. Qing, S.; Sun, X.; Li, Y. Evaluation of ERA5 reanalysis soil moisture over inner mongolia. *Sci. Technol. Eng.* **2020**, *20*, 2161–2168.
67. Kai, Z.; Qiong, G.; Zhicai, Z.; Chan, Z. A Calibration problem in applying the Eddy Covariance technique to measure Bowen ratio. *J. Beijing Norm. Univ. (Nat. Sci.)* **2008**, *44*, 207–211.
68. Yang, Y.; Chen, R.; Song, Y.; Han, C.; Liu, J.; Liu, Z. Sensitivity of potential evapotranspiration to meteorological factors and their elevational gradients in the Qilian Mountains, northwestern China. *J. Hydrol.* **2019**, *568*, 147–159. [CrossRef]
69. Zhuang, Y.; Zhao, W.; Luo, L.; Wang, L. Dew formation characteristics in the gravel desert ecosystem and its ecological roles on Reaumuria soongorica. *J. Hydrol.* **2021**, *603*, 126932. [CrossRef]
70. Zhang, Q.; Wang, S.; Yue, P.; Wang, S. Variation characteristics of non-rainfall water and its contribution to crop water requirements in China's summer monsoon transition zone. *J. Hydrol.* **2019**, *578*, 124039. [CrossRef]
71. Uclés, O.; Villagarcía, L.; Cantón, Y.; Lázaro, R.; Domingo, F. Non-rainfall water inputs are controlled by aspect in a semiarid ecosystem. *J. Arid. Environ.* **2015**, *113*, 43–50. [CrossRef]
72. Genxu, W.; Guangsheng, L.; Chunjie, L. Effects of changes in alpine grassland vegetation cover on hillslope hydrological processes in a permafrost watershed. *J. Hydrol.* **2012**, *444*, 22–33. [CrossRef]
73. Genxu, W.; Shengnan, L.; Hongchang, H.; Yuanshou, L. Water regime shifts in the active soil layer of the Qinghai–Tibet Plateau permafrost region, under different levels of vegetation. *Geoderma* **2009**, *149*, 280–289. [CrossRef]
74. Dawen, Y.; Yuanrun, Z.; Bing, G.; Hongyi, L.; Pengtao, Y. *Ecohydrological Process and Coupling Simulation in Alpine Mountains*; Science Press: Beijing, China, 2020; pp. 1–10.
75. Yongjian, D.; Shiqiang, Z. Study on water internal recycle process and mechanism in typical mountain areas of inland basins, northwest China: Progress and challenge. *Adv. Earth Sci.* **2018**, *33*, 719–727.
76. Mcguire, A.D. Environmental Variation, Vegetation Distribution, and Carbon Dynamics in High Latitudes. *AGU Fall Meet. Abstr.* **2001**, *2001*, B12F-01.
77. Jorgenson, M.T.; Racine, C.; Walters, J.; Osterkamp, T. Permafrost Degradation and Ecological Changes Associated with a Warming Climate in Central Alaska. *Clim. Chang.* **2001**, *48*, 551–579. [CrossRef]
78. Wehren, B.; Weingartner, R.; Schädler, B.; Viviroli, D. *General Characteristics of Alpine Waters*; Springer: Berlin/Heidelberg, Germany, 2010; pp. 17–58.
79. Uclés, O.; Villagarcía, L.; Cantón, Y.; Domingo, F. Microlysimeter station for long term non-rainfall water input and evaporation studies. *Agric. For. Meteorol.* **2013**, *182*, 13–20. [CrossRef]
80. Dongxin, G.; Tieliang, L.; Weixin, Z. *General Geocryology*; Science Press: Beijing, China, 1988; pp. 60–68.
81. Kuhle, M. The cold deserts of high Asia (Tibet and contiguous mountains). *GeoJournal* **1990**, *20*, 319–323. [CrossRef]
82. Körner, C. *Alpine Plant Life: Functional Plant Ecology of High Mountain Ecosystems*; Springer: Berlin/Heidelberg, Germany, 2003; pp. 103–125.

remote sensing

MDPI

Article

The Formation of Yardangs Surrounding the Suoyang City Ruins in the Hexi Corridor of Northwestern China and Its Climatic–Environmental Significance

Qingbin Fan [1,2], Jie Liao [1,*], Yan Li [3], Wei Ye [4], Tao Wang [1], Lupeng Yu [5], Xiao Feng [1,2] and Weixiao Han [6]

1 Key Laboratory of Desert and Desertification, Northwest Institute of Eco-Environment and Resources, Chinese Academy of Sciences, Lanzhou 730000, China
2 University of Chinese Academy of Sciences, Beijing 100049, China
3 College of Forestry, Nanjing Forestry University, 159 Longpan Road, Nanjing 210037, China
4 College of Geography and Environmental Sciences, Zhejiang Normal University, Jinhua 321004, China
5 School of Resource and Environmental Sciences, Linyi University, Linyi 276000, China
6 Key Laboratory of Remote Sensing of Gansu Province, Heihe Remote Sensing Experimental Research Station, Northwest Institute of Eco-Environment and Resources, Chinese Academy of Sciences, Lanzhou 730000, China
* Correspondence: liaojie@lzb.ac.cn

Abstract: The yardangs surrounding the Suoyang City ruins are proven to be wind-eroded landforms developed in an oasis which was used for agriculture in history. According to OSL and ^{14}C dating, as well as historical records of local human activities, we suggest that the formation of yardangs in the Suoyang City oasis probably started in the mid-Yuan Dynasty of China (AD 1291). After being abandoned, the Suoyang City oasis quickly evolved into desert land with yardangs and nebkhas under the background of desertification enlargement in a cold, dry climate in the Hexi Corridor. Although human factors are considered to have played an important role in the process of desertification, the effect imposed by climatic changes should not be ignored. Desertification constitutes a serious threat to human survival and development, we should reasonably develop and utilize water and land resources, effectively prevent and control desertification, and promote the harmonious development between man and nature in arid areas.

Keywords: yardangs; wind erosion; climatic–environmental significance

Citation: Fan, Q.; Liao, J.; Li, Y.; Ye, W.; Wang, T.; Yu, L.; Feng, X.; Han, W. The Formation of Yardangs Surrounding the Suoyang City Ruins in the Hexi Corridor of Northwestern China and Its Climatic–Environmental Significance. *Remote Sens.* **2022**, *14*, 5628. https://doi.org/10.3390/rs14215628

Academic Editors: Massimo Menenti, Yaoming Ma, Li Jia and Lei Zhong

Received: 13 September 2022
Accepted: 1 November 2022
Published: 7 November 2022

1. Introduction

The word yardangs are used in geomorphology to describe elongated, streamlined landforms that are produced by wind erosion [1–4]. Yardangs are present in many of the world's major deserts, including the Sahara Desert in Africa, the Namib Desert in Namibia, the Lut Desert in Iran, and the Taklimakan Desert in China, as well as on planets such as Mars and Venus [3,4]. Early explorers and geographers were attracted by the unique morphology and mysterious origin of these landforms and referred to them as Dragon City or Demon Castle. For example, yardangs in the Lop Nur region of China were described as "White Dragon Dunes" in the *Book of the Han Dynasty* 2000 years ago [3,5]. Li Daoyuan, a Chinese geographer during the Wei Dynasty 1500 years ago, proposed that the yardangs in the Lop Nur region were formed by water erosion and subsequently sculpted by wind [5]. The Swedish explorer Sven Hedin proposed the name "yardangs" (a Turkmen word that means "steep hills") in the Lop Nur region of Northwest China in 1903 [1]. This word subsequently became the standard terminology that is widely accepted and used by scholars in geomorphology.

Previous studies have mostly focused on yardangs formed in Gobi desert [3–8] rather than in agricultural areas. Suoyang City and its surrounding oasis are typical representatives of the ancient cities and oases that suffered from severe desertification in the Hexi

Corridor. The occurrence of yardangs indicates an important climatic and environmental change in this region. In this study, we investigated the yardangs around the Suoyang City ruins in the Hexi Corridor, which was a famous agricultural region in history. Our aim was to explore the formation process of the yardangs and their climatic–environmental significance, as well as provide a reference for the development of agriculture, the prevention and control of desertification, and the harmonious development between man and nature in arid areas.

2. Study Area

The yardangs around the Suoyang City ruins (Figure 1) are located in Guazhou basin of the Hexi Corridor. The Guazhou basin (39°52′N–41°53′N and 94°45′E–97°00′E) is located in Northwestern China. Since the Cenozoic, the collision and compression between the Indian and Eurasian plates have resulted in rapid uplift of the Tibetan Plateau, forming a wide range of basin mountain tectonic patterns in both the interior and the edge of the plateau. The Guazhou basin (Figure 1) is located in the basin mountain tectonic pattern on the northeast edge of the Tibetan Plateau, with the Qilian Mountains to the south and the Beishan Mountains to the north. Therefore, the terrain has a high elevation in the south and north, is low in the middle, and gradually inclines toward the center of the basin. The Guazhou basin is located in the hinterland of the Eurasian continent, with an arid climate, scarce precipitation, and low vegetation coverage. The main river in this region is the Shule River. The annual average temperature is 6.79 °C (1998–2018) (Figure 2), with a warm temperate continental arid climate. The annual average precipitation is 64.96 mm (1998–2018), of which 60% occurs during the summer [9]. The zonal soil is gray-brown desert soil.

Guazhou was located at a bottleneck of the old Silk Road and was a hub for economic and cultural exchange between China and the West in history. The Suoyang City ruins (Figure 3a) were listed as a world cultural heritage site in June 2014.

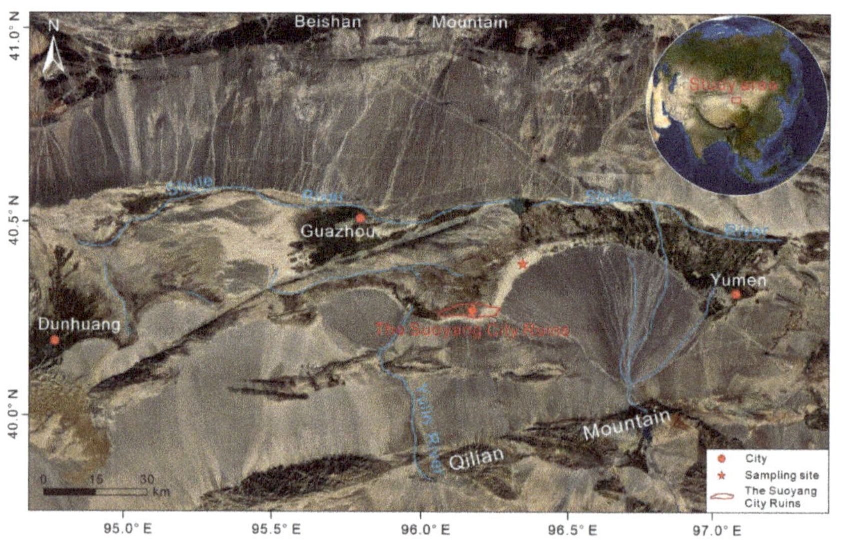

Figure 1. Locations of the Suoyang City ruins in Guazhou basin and the sample site.

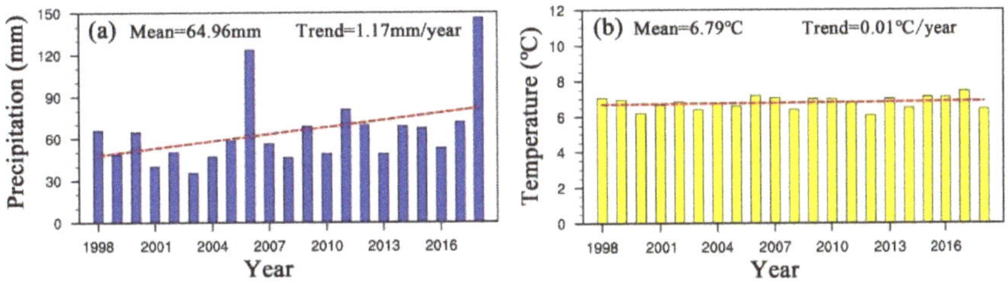

Figure 2. Average annual precipitation (**a**) and temperature (**b**) of the study area in 1998–2018.

In the past, the oasis reclamation area around the Suoyang City ruins (Figure 3a) was a prosperous agricultural region. Even today, the remains of ancient irrigation channels and field ridges are still visible. A dam of about 100 m wide was found 8 km to the southeast of the Suoyang City ruins, and there is an irrigation channel connected to the dam and the upstream mountain pass of the Shule River. In the downstream, the channel is divided into several branches leading to the ancient oasis reclamation area around the Suoyang City ruins. On the basis of remote sensing images of field ridges and irrigation channels, we calculated that the ancient reclamation area around the Suoyang City ruins covered roughly 3.42×10^2 km^2. Although the Suoyang City oasis was a prosperous region in history, the geomorphic landscape of this area is currently dominated by vast yardangs (Figures 3b and 4a), nebkhas (Figure 4b), and saline–alkaline land; in some areas, the nebkhas are as high as the ramparts.

Figure 3. The Ta er Temple (**a**) in Suoyang City ruins, and the surrounding yardangs (**b**).

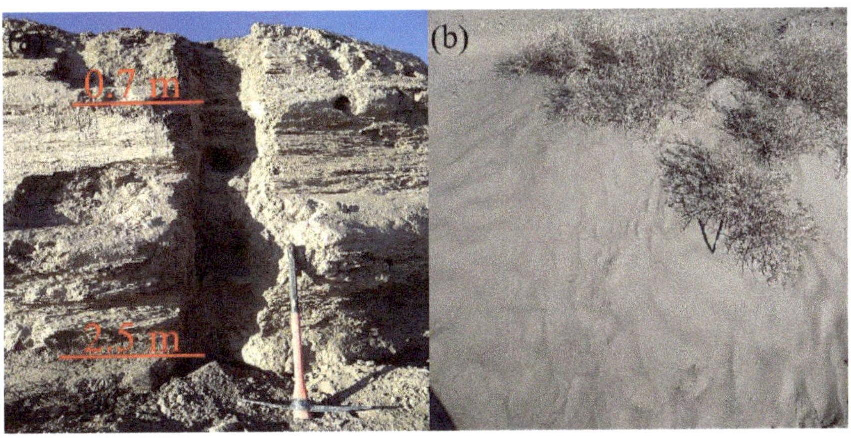

Figure 4. The dating sample profile (**a**) and nebkhas (**b**)surrounding the Suoyang City ruins.

3. Materials and Methods

3.1. Temperature and Precipitation

The annual mean precipitation and temperature used to describe climatic conditions in the study area were derived from the TRMM_3B43 and CRU TS4.05 climate datasets, respectively. As shown in Figure 2, the mean annual precipitation from 1998 to 2018 was 64.96 mm, much less than 200 mm (the thresholds for defining arid land), indicating an extreme arid climate in the study area. The average annual temperature in the last 20 years (1998–2018) was low and very stable. Additionally, although the average annual temperature remains low (6.79 °C) and shows little long-term trend, climate change also imposes a great influence on this region for a significant increasing trend ($p < 0.05$) of precipitation, as detected by the Mann–Kendall (MK) trend test, which may have resulted from the impact of global warming [10–13].

3.2. The Age of Yardangs Surrounding the Suoyang City Ruins

Yardangs are wind-eroded landforms that are different from wind accumulation landforms. The strata age is not the same as the age of the yardangs, the age of the yardangs is the time when wind erosion first began to cut into the uppermost strata of the land surface. Therefore, it is very difficult to determine the age of the yardangs. However, the occurrence of yardangs indicates significant environmental changes, especially for the development of yardangs in oasis agricultural areas. Therefore, it is very important to determine the age of the yardangs surrounding the Suoyang City oasis.

In this paper, we attempted to date the existing top stratum of the yardangs using the optically stimulated luminescence (OSL) dating method to obtain an approximate age of the yardangs. Samples were dated using the OSL dating method in Luminescence Research Laboratory, Shandong Provincial Key Laboratory of Water and Soil Conservation and Environmental Protection, School of Resource and Environmental Sciences, Linyi University. We collected a 2.8 m deep yardang profile (Figure 4a) and dated the age at the depths of 0.7 m and 2.5 m; the 0.7 m section showed an age of 0.87 ± 0.04 ka (Table 1), and the 2.5 m section showed an age of 1.62 ± 0.08 ka (Table 1). As the Suoyang City oasis is located at the extreme end of the alluvial fan, the deposition rate is relatively stable, and the sedimentary facies (Figure 4a) also showed that the yardang sedimentary stratum is very uniform; thus, we assume that this section was relatively uniformly deposited, 0.7–2.5 m apart by 1.8 m. It took 750 years to deposit this section; thus, we can calculate that the deposition rate of this section is 0.24 cm/year. According to this deposition rate, we can calculate that the age of the top stratum of the yardang profile seen at present is about 0.578 ka; hence, we can calculate that the top stratum of the yardang formed at about AD 1444. Radiocarbon ^{14}C dating of branches in the top stratum of the yardang in previous studies showed that the yardangs surrounding the Suoyang City oasis started to form at around AD 1410–AD 1460 [14]. Both methods assume that the top strata of the yardangs are the original final sedimentary layers and that the yardangs began to form immediately after the original final sedimentary layer deposit. The two methods also assume that the top stratum of the yardang seen at present has not been eroded. However, since the formation of yardangs, it is impossible for the top strata of the yardangs to have not suffered from erosion over time. Therefore, the stratigraphic ages determined using the two methods are slightly later than the real formation time of the yardangs.

Table 1. K-feldspar pIRIR dating results for samples.

Samples	Depth (m)	Over-Dispersion (%)	U ppm	Th ppm	K (%)	Moisture (%)	Dose Rate (Gy/ka)	CAM De (Gy)	CAM Age (ka)
SYC-02	0.7	8	2.32	9.20	1.59	5 ± 2	3.70 ± 0.10	3.23 ± 0.1	0.87 ± 0.04
SYC-01	2.5	10	2.34	9.00	1.55	5 ± 2	3.59 ± 0.11	5.8 ± 0.2	1.62 ± 0.08

4. Characteristics and Controlling Factors of the Yardangs

4.1. Characteristics of the Yardangs

The yardangs around the Suoyang City ruins are located on the western edge of the alluvial fan of the Shule River in Guazhou basin. Unlike the yardangs reported in previous studies [3,6–8], this region was an oasis reclamation area in the past. Through investigation, we found that the yardangs were mainly yellow in appearance, elongated and streamlined in morphology, and developed on fluvial and lacustrine sediments [11]. It is generally believed that the length of large yardangs is >1000 m, of medium yardangs is 10–1000 m, and of small yardangs is 1–10 m [14]. In this study, 100 yardangs were measured in the study area, and we found that their length was 1.6–25.8 m, width was 0.7–9.2 m, and height was 0.8–3.2 m; therefore, the yardangs around the Suoyang City ruins are small–medium in size. The most important characteristic is that the yardangs surrounding the Suoyang City ruins show obviously directional, parallel, and streamlined characteristics (Figures 3b and 5). The majority of the yardangs extend in the east–west direction (Figures 3b and 5), consistent with the local prevailing wind direction (Figures 5 and 6), but inconsistent with the direction of gullies formed by flooding (north–south) in this region [9]. In Guazhou basin, the North Mountain lies in the north, and Qilian Mountain lies in the south; thus, the terrain is high on both sides and low in the middle. Affected by the terrain, the gullies scoured by local rivers and floods formed by precipitation extend in the north–south direction. A meteorological station was installed in the center of the yardang distribution area surrounding the Suoyang City ruins; the meteorological data (2016–2018) show that the annual sand-driving wind direction in the study area is mainly easterly (Figure 6), followed by westerly, whereas other wind directions are rare [9,15]. Therefore, it is easy to understand that the trend of the yardangs was consistent with the local prevailing wind direction (Figures 5 and 6), but inconsistent with the direction of gullies formed by rivers and flooding (north–south) in this region. The strong local prevailing wind transports a large amount of dust and sand, which grinds and erodes the land surface into long parallel ridges and grooves, forming large areas of yardangs.

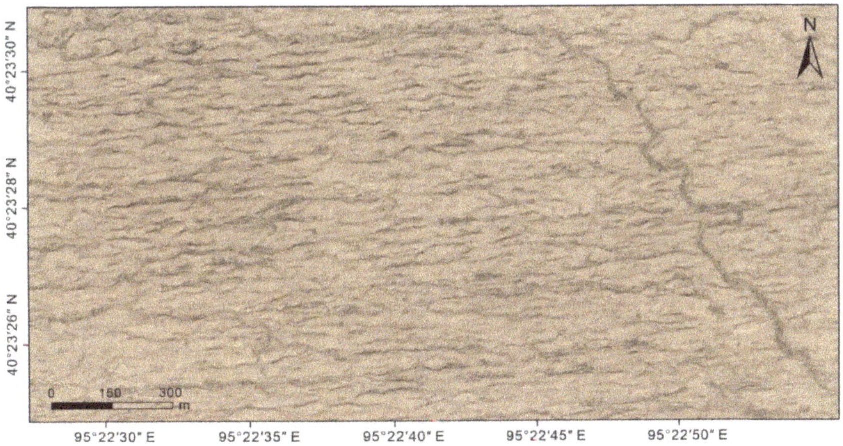

Figure 5. Remote sensing image of the yardangs surrounding the Suoyang City ruins.

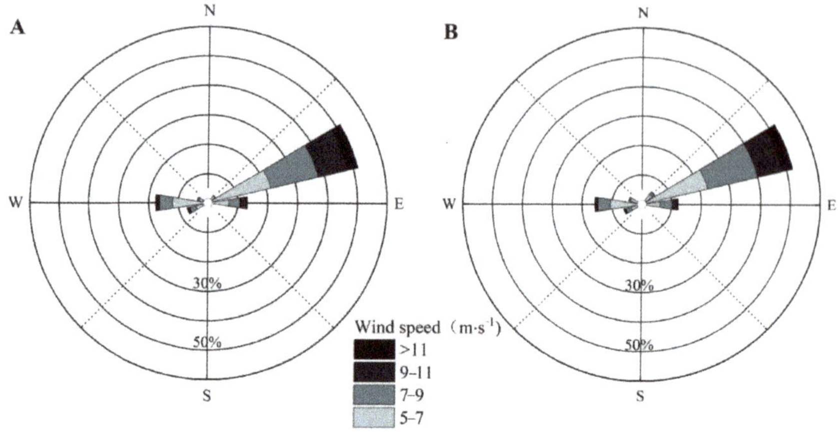

Figure 6. Annual sand-driving wind rose chart: (**A**) 2016–2017; (**B**) 2017–2018.

4.2. Controlling Factors of the Yardangs

4.2.1. Sediment Characteristics

The collision and compression between the Indian and Eurasian plates since the Cenozoic caused rapid uplift of the Tibetan Plateau, forming a range of basin and mountain tectonic patterns. The Guazhou basin is just located in a basin mountain tectonic pattern on the northeastern edge of the Tibetan Plateau, with the Qilian Mountains to the south and the Beishan Mountains to the north. The Shule River, one of the big three inland rivers in the Hexi Corridor, runs through the basin. The Shule River sediments and the proluvial deposits brought from the Beishan and Qilian mountains formed a huge piedmont alluvial fan. These thick Quaternary fluvial and lacustrine sediments provide material conditions for the development of agriculture and the formation of yardangs.

4.2.2. Fluvial Erosion

Yardangs are considered to be among the most typical wind-eroded landforms, but the role of water flow should not be ignored. Despite an extreme arid climate, the intermittent water flow formed by precipitation plays an important role in the development of yardangs. Although the average annual precipitation in Guazhou is only 64.96 mm (Figure 2), the precipitation during rainstorms is sufficient to form surface runoff and erode the surface, forming dense gully systems, which provides conditions that are conducive to enhance wind erosion.

4.2.3. Wind Erosion

Wind erosion is the most significant agent responsible for the formation of yardangs in the Suoyang City oasis reclamation area. As Guazhou basin is located in the hinterland of the Eurasian continent and is far from the sea, the Kunlun Mountains, Qilian Mountains, Himalayas, and Qinling Mountains block the flow of humid air from the Indian and Pacific Oceans, resulting in an extremely dry climate, highlighting wind as the most important external force in this region. Guazhou basin is located in the Shule River valley between the Beishan Mountains and the Qilian Mountains. When airflow enters the valley, the air duct narrows, and the wind speed increases; thus, Guazhou is known as the "wind reservoir of the world" [15]. Accordingly, wind erosion in Guazhou basin is very strong. Previous studies have shown that the threshold velocity for sand movement in Guazhou basin is mainly composed of two groups of wind with opposite directions. The main wind direction is northeast, accounting for 68.86% of the annual threshold velocity for sand movement, followed by southwest wind, accounting for 27.67% of the annual threshold velocity for sand movement (Figure 6) [9]. Through field investigations, we found that the long axis of yardangs in the Suoyang City oasis reclamation area mostly extends in the east–west direction, consistent with the local prevailing wind direction but inconsistent with the north–south direction of ravines scoured by rainstorms and floods in this area [9]. Wind erosion is obviously the main external force for the formation of yardangs surrounding the Suoyang City ruins.

4.2.4. Weathering and Collapse

Physical weathering and collapse are also important factors in the development of yardangs. Physical weathering caused by annual and diurnal temperature variations greatly influences the fluvial and lacustrine sediments, thereby impacting yardangs. Guazhou is deeply inland, with big annual and diurnal temperature variations, meaning that physical weathering is significant to the yardangs in the Suoyang City oasis reclamation area. The temperature changes create mechanical stress that generates horizontal and vertical fractures, potentially leading to the collapse of sediments. Wind and water erosion of the underlying soft and loose sediments can also lead to the collapse of overlying strata.

5. Results and Discussion

5.1. Desertification Process of the Oasis in the Hexi Corridor

The Hexi Corridor is located in Northwest China, with an extreme arid climate, low precipitation, and sparse vegetation. The agricultural production is mainly concentrated in oasis areas in the lower reaches of rivers. In the past, desertification of oasis in the Hexi Corridor occurred frequently, causing the loss of agricultural land resources, a sharp reduction in biomass, and even the destruction of the whole oasis ecosystem. Therefore, the Hexi Corridor is one of the most severe desertification areas in Northwest China [16–18].

There are ten large desertification areas that evolved from the oasis in the Hexi Corridor. These desert areas are all located in the lower reaches of rivers, with poor water resources and potential instability in ecosystems, meaning that they are prone to desertification. For thousands of years, strong winds swept across the ground like a comb, eroding the farmland into yardangs and nebkhas; some areas were almost completely swallowed by quicksand, and people's homes became places where wind and sand raged. The Hexi Corridor in Gansu Province, through which the famous old Silk Road used to pass, is one of the areas with the most ancient city ruins. These ruins are considered to be the best historical specimens of China's ancient civilization and Silk Road culture, as well as the most direct historical evidence of environmental changes in the ancient oasis.

Over the last 2000 years, the agricultural production and social development in the Hexi Corridor made brilliant achievements, especially in the Han, Sui, and Tang Dynasties (202 BC–AD 907). However, there have been stages of environmental changes with serious desertification in the Hexi Corridor. The desertification process is mainly concentrated in the Wei, Jin, and North–South Dynasties (AD 220–AD 589), the late Tang and Five Dynasties (AD 907–AD 960), and the Ming and Qing Dynasties (AD 1644–AD 1912). The desertification area was 1070 km^2 in the Wei, Jin, and North–South Dynasties, 1765 km^2 in the late Tang and Five Dynasties, and 6884 km^2 in the Ming and Qing Dynasties [18]. Moreover, over the past 2000 years, among the 38 ancient cities abandoned through desertification in the Hexi Corridor, 21.05% were abandoned during the Wei, Jin, and North–South Dynasties, 21.05% were abandoned during the end of the Tang and Five Dynasties, and 57.9% were abandoned during the Ming and Qing dynasties [16]. It has been reported that there is a good relationship between climate changes and desertification in history of the Hexi Corridor [19]. The general characteristics of climate changes over the last 2000 years in China show that there have been four warm periods since the Qin Dynasty, namely, the western and eastern Han Dynasties (200 BC–AD 180), the Sui and Tang dynasties (AD 541–AD 810), the Song and Yuan dynasties (AD 931–AD 1320), and the 20th century (AD 1921–AD 2000). Moreover, there were three cold phases including the Wei, Jin, and North–South Dynasties (AD 181–AD 540), the late Tang and Five Dynasties (AD 811–AD 930), and the Ming and Qing dynasties (AD 1321–AD 1920) [19].

Throughout China's history, the warm period often brought a warm, wet climate, social stability, and prosperity, as well as a good ecological environment, while the cold period was often accompanied by a cold, dry climate. In addition, climate changes could bring about serious social and ecological problems, such as war and desertification. The relationship between climate and wars in the past 2000 years of China show that most of the wars occurred in low-temperature periods. There were seven relatively cold periods (AD 180–AD 360, AD 420–AD 540, AD 840–AD 960, AD 1110–AD 1200, AD 1290–AD 1500, AD 1560–AD 1680, and AD 1830–AD 1890) in Chinese history. Except for the period of AD 420–AD 540, the other six period all correspond to the high-frequency period of wars [19]. The Wei, Jin, and North–South Dynasties and the Ming and Qing Dynasties were the two coldest periods in Chinese history, as well as the periods with the highest incidence of wars in Chinese history. In the former period, 595 wars occurred, while, in the Ming and Qing Dynasties, 810 wars occurred. Moreover, in cold periods, the ethnic minorities in border areas often moved inward, leading to wars breaking out [19]. During the years of wars, people either died in the war or fled and moved to other places, while the nomadic lifestyle of the ethnic minorities usually replaced the agricultural lifestyle of the Han people. Therefore, large areas of farmlands in the oasis were abandoned and exposed to the surface, without the protection of vegetation; the ancient oasis then became desert land under strong wind erosion in cold periods, and the surrounding cities usually declined or were abandoned.

5.2. Desertification Process of the Suoyang City Oasis

The Suoyang City ruins are among the most representative desertified ancient city ruins in the Hexi Corridor. Suoyang City was built in the Western Jin Dynasty (AD 295) and was the capital of Guazhou county from the Tang Dynasty to Yuan Dynasty (AD 618–AD 1291). The Suoyang City oasis was a very prosperous area in history, especially flourishing during the Tang Dynasty [17]. On 1 August 2022, People's Daily Online of China reported that the latest archaeological evidence confirmed that the Ta er Temple in the Suoyang City ruins was a high-level temple on the old Silk Road, which was built during the Sui and Tang Dynasties (AD 581–AD 907) and flourished in the Western Xia Dynasty (AD 1038–AD 1227), indicating that the Suoyang City oasis should be a relatively prosperous area without desertification in this period, whereas Guazhou experienced some changes during the Yuan Dynasty. In the early Yuan Dynasty, Guazhou still existed. However, soon after, the residents of Guazhou were ordered to move out. In AD 1288, the government ordered the residents of Guazhou to move to Gan State, and, in AD 1291, the Guazhou residents were ordered to move to Su State. After these large migrations, the residents of Guazhou were very few, and historical document reported that Guazhou only existed in name [17]. From then on, Suoyang City became empty; the surrounding farmland was abandoned and suffered from strong wind erosion. Therefore, we can infer that the desertification process of the Suoyang City oasis probably started in the mid-Yuan Dynasty (AD 1291).

In the early Ming Dynasty (AD 1368–AD 1644), the decline of Suoyang City and its surrounding oasis was very serious; the government reopened Suoyang City and repaired it twice during 1435–1494 [17]. In 1472 AD, the Hami Wei (a military organization) moved to Suoyang City. In the third year of Jiajing (AD 1524), the government could not resist the attack of border ethnic minorities; hence, it officially abandoned the large areas west of Jiayuguan State (including Dunhuang and Guazhou) and moved all residents inland [13]. In the following 200 years, Dunhuang and Guazhou were repeatedly occupied by nomadic tribes from Turpan, Hami, and Mongolia, and the oasis surrounding Suoyang City no longer operated [17]. In 1738 of the early Qing Dynasty, it was recorded that there was little arable land around Suoyang City, and the former irrigation channels were all dry and covered with sand [15], indicating that the Suoyang City oasis had evolved into desert land.

In the Qing Dynasty (AD 1644–AD 1912), the government gradually recovered vast areas west of Jiayuguan State and resumed the management of Guazhou, whereas the development of the drainage area of the Shule River focused on the east and north of the alluvial fan, and the waste land in the east and north of the alluvial fan was extensively reclaimed [17]. It was recorded in the 58th year of Kangxi (AD 1719) that the government built Yumen City, blocked the Shule River estuary, and drove the water to the southeast to irrigate the newly reclaimed waste land. From then on, the river channel that originally flowed to Suoyang City in the west of the alluvial fan was cut off and dried up [17]. The limited discharge of the Shule River was used to irrigate the newly reclaimed waste land to the east. Therefore, the Suoyang City oasis was completely dried up, and it quickly evolved into desert land with yardangs and nebkhas under strong wind erosion.

In conclusion, Suoyang City was first built in the Western Jin Dynasty (AD 295) and was the capital of Guazhou county from the Tang Dynasty to the Yuan Dynasty (AD 618–AD 1291). In 1291, the residents of Guazhou moved to Su State, after which Guazhou only existed in name. Hence, we suggest that the desertification of the Suoyang City oasis started in the mid-Yuan Dynasty (AD 1291), accelerated following the abandonment of Guazhou in the mid-Ming Dynasty (AD 1524), and completely evolved into desert land after the diversion of the Shule River in the early Qing Dynasty (1719 AD). Coupled with the OSL dating results, we suggest with some confidence that the occurrence of yardangs in the Suoyang City oasis started in the mid-Yuan Dynasty (AD 1291).

5.3. Environmental Significance of Desertification and the Occurrence of Yardangs Surrounding the Suoyang City Ruins

The Hexi Corridor is an important agricultural production base and a densely populated area in Northwest China. Therefore, desertification of the oasis and the ancient cities in the Hexi Corridor has important environmental significance. Over the last 2000 years, desertification of the Hexi corridor mainly occurred in the Wei, Jin, and North–South Dynasties, the late Tang and Five Dynasties, and the Ming and Qing Dynasties [16]. Moreover, the abandonment of all 38 ancient cities through desertification in the Hexi Corridor in history also occurred in these cold periods [16]. The reasons can be divided into natural factors and human factors. In terms of natural factors, climate change, especially a cold, dry climate, can easily cause desertification. With regard to human factors, war, river diversion, the increasing intensified human activities such as over-deforestation, overgrazing, reclaiming waste land, and building reservoirs for water storage are responsible for the desertification in the Hexi Corridor [16]. In addition, population is also an important factor, especially for an arid area located in NW China with a weak environmental carrying capacity. In history, the population of the Hexi Corridor rarely exceeded 4×10^5 from the Han to Tang dynasties (202 BC–AD 907); even in the mid-Ming Dynasty, the total population was about 3.5×10^5, and water resources were mainly used to meet the needs of irrigation and everyday life. Thus, water and land resource exploitation did not play a leading role in environment changes of the Hexi Corridor [16]. However, by the early Qing Dynasty, a great deal of immigrants came to the Hexi Corridor. It was reported that, during the Year of Jiaqing (AD 1796–AD 1820) in Qing Dynasty, the total population of the Hexi corridor was about 1.274×10^6 [16]. For the first time, the population density in the Hexi region rose to 8.8 per square kilometer and broke through the critical index of population pressure in an arid region (seven persons per square kilometer set by the United Nations in 1977) [16]. Human activities gradually replaced natural factors and became the primary factor on environmental changes in the Hexi corridor. Subsequently with the development of science and technology, human activities become the principal cause during the process of desertification over the last 300 years [16,19].

In history, Guazhou was rich in water resources, such as the Han Dynasty (202 BC–AD 220) and the Tang Dynasty (AD 618–AD 907) [20]. It was reported that, during the Han and Tang Dynasty, a large tributary of the Shule River (named Ming River) flowed to the Suoyang City oasis area, forming a big swamp (namely a lake) 130 km long and 30 km wide in the downstream [17]. The large tributary and the big swamp provided sufficient irrigation water for the Suoyang City oasis. By the mid-Yuan Dynasty, the ecological environment of Guazhou was much worse than the Tang Dynasty, with infinite dunes [21]. Moreover, when the Swedish Explorer Sven Hedin passed through Guazhou in 1933, local people reported that the discharge of the Shule River was much larger 100 years ago; Sven Hedin believed that this change was not completely caused by the increase in irrigation [21]. In history, the Shule River transported a large amount of water and would terminate in Lop Nur in Xinjiang. The Shule River became a seasonal river below Guazhou in the first half of the 19th century at the latest [21], because Qi Yunshi and Lin Zexu recorded that the Shule River was dry near Guazhou when they were demoted to Ili of Xinjiang in 1810 and 1842, respectively, indicating a reduction in river volume and a more arid climate [21].

Both natural factors and human factors are responsible for the desertification of the Suoyang City oasis and the occurrence of yardangs. In terms of natural factors, in the last 1000 years, the climate in China gradually turned cold and dry, and the discharge of rivers and lakes decreased as well [16,19], especially during the Ming and Qing dynasties, which occurred in the Little Ice Age. Historical records also show an obvious increase in sand and wind disasters all over the world in this period [14]. On the other hand, the increasing intensified human activities also played an important role in the process of desertification [16–18]. For example, after the construction of Shuangta reservoir in 1960, the river volume in the lower reaches of the Shule River was greatly reduced. Furthermore, once the oasis turned into a desert, people often left the oasis and moved to a new oasis [22]. Over time, the area of desertification became larger and larger, as land resources in arid areas are very limited; desertification not only leads to a reduction in people's living space but also seriously threatens human survival and sustainable development of the society. For example, the reduction in river discharge in the lower reaches of the Shule River caused the retreat of downstream lakes, the shrinkage of wetlands, a reduction in biodiversity, an increase in soil salinization, and other serious ecological and environmental problems [23]. Therefore, we should reasonably develop and use land and water resources, effectively prevent and control desertification, and promote harmonious development between man and nature in arid areas.

6. Conclusions

The yardangs surrounding the Suoyang City ruins were proven to be wind-eroded landforms developed in an oasis which was used for agriculture in history. According to OSL and ^{14}C dating, as well as historical records of local human activities, we suggest that the formation of yardangs in the Suoyang City oasis probably started in the mid-Yuan Dynasty of China (AD 1291). After being abandoned, the Suoyang City oasis quickly evolved into desert land with yardangs and nebkhas under the background of desertification enlargement in the cold, dry climate in the Hexi Corridor. Although human factors are considered to have played an important role in the process of desertification, the effect imposed by climatic changes should not be ignored. Desertification constitutes a serious threat to human survival and development; we should reasonably develop and utilize water and land resources, effectively prevent and control desertification, and promote the harmonious development between man and nature in arid areas.

Author Contributions: Conceptualization, J.L. and Q.F.; methodology, Q.F., L.Y. and Y.L.; investigation, Q.F., W.H. and W.Y.; data curation, Q.F. and X.F.; writing—original draft preparation, Q.F.; writing—review and editing, Q.F. and J.L.; visualization, Q.F. and T.W.; supervision, W.Y.; project administration, T.W.; funding acquisition, W.Y. and T.W. All authors have read and agreed to the published version of the manuscript.

Funding: This research was funded by the Second Tibetan Plateau Scientific Expedition and Research Program, grant number 2019QZKK0305.

Acknowledgments: The authors thank the editor and three anonymous reviewers for their laborious work and the valuable comments on the manuscript.

Conflicts of Interest: The authors declare no conflict of interest.

References

1. Hedin, S. *Lop Nor: Scientific Results of a Journey in Central Asia (1889–1902)*; Lithographic Institute of the General Staff of the Swedish Army: Stockholm, Sweden, 1905; Volume 2.
2. Bristow, C.S.; Drake, N.; Armitage, S. Deflation in the dustiest place on Earth: The Bodélé Depression, Chad. *Geomorphology* **2009**, *105*, 50–58. [CrossRef]
3. Dong, Z.; Lv, P.; Lu, J.; Qian, G.; Zhang, Z.; Luo, W. Geomorphology and origin of yardangs in the Kumtagh Desert, Northwest China. *Geomorphology* **2012**, *139*, 145–154. [CrossRef]
4. Liang, X.; Niu, Q.; Qu, J.; Liu, B.; Liu, B.; Zhai, X.; Niu, B. Applying end-member modeling to extricate the sedimentary environment of yardang strata in the Dunhuang Yardang National Geopark, northwestern China. *Catena* **2019**, *180*, 238–251. [CrossRef]
5. Xia, X. Formation of the yardangs in Lop Nor region. In *Scientific Expedition in Lop Nor*; Xia, X., Ed.; Science Press: Beijing, China, 1987; pp. 52–59.
6. Breed, C.S.; Grolier, M.J.; McCauley, J.F. Eolian features in the western desert of Egypt and some applications to Mars. *Geophys. Res.* **1979**, *84*, 8205–8221.
7. Brookes, I.A. Aeolian erosional lineations in the Libyan Desert, Dakhla Region, Egypt. *Geomorphology* **2001**, *39*, 189–209. [CrossRef]
8. Al-Dousari, A.M.; Al-Elaj, M.; Al-Enezi, E.; Al-Shareeda, A. Origin and characteristics of yardangs in the Um Al-Rimam depressions (N Kuwait). *Geomorphology* **2009**, *104*, 93–104. [CrossRef]
9. Xiaolei, L.; Qinghe, N.; Zhishan, A.; Jianjun, Q.; Yaping, S.; Liang, W. Sand-driving wind regime and sand drift potential in the yardang landform areas of southern Suoyang Town, Guazhou, Gansu, China. *J. Desert Res.* **2019**, *39*, 48–55.
10. Wang, Q.; Zhai, P.M.; Qin, D.H. New perspectives on 'warming-wetting' trend in Xinjiang, China. *Adv. Clim. Chang. Res.* **2020**, *11*, 252–260. [CrossRef]
11. Li, C.; Lai, C.; Peng, F.; Xue, X.; You, Q.; Liu, F.; Guo, P.; Liao, J.; Wang, T. Dominant Plant Functional Group Determine the Response of the Temporal Stability of Plant Community Biomass to 9-Year Warming on the Qinghai-Tibetan Plateau. *Front. Plant Sci.* **2021**, *12*, 704138. [CrossRef] [PubMed]
12. Li, C.; Peng, F.; Xue, X.; You, Q.; Lai, C.; Zhang, W.; Cheng, Y. Productivity and quality of alpine grassland vary with soil water availability under experimental warming. *Front. Plant Sci.* **2018**, *9*, 1790. [CrossRef] [PubMed]
13. Ren, Y.; Liu, J.; Liu, S.; Wang, Z.; Liu, T.; Shalamzari, M. Effects of Climate Change on Vegetation Growth in the Yellow River Basin from 2000 to 2019. *Remote Sens.* **2022**, *14*, 687. [CrossRef]
14. Ma, L. A Study on the Historical Environment of the Suo Yangcheng Area during Ming and Qing Dynasty. Master's Thesis, Lanzhou University, Lanzhou, China, 2018.
15. Huang, K.; Tang, L. The Yardangs in Anxi County. *J. Northwest Norm. Univ. Nat. Sci. Ed.* **1986**, *22*, 43–47.
16. Wang, N.; Zhang, C.; Li, G.; Cheng, H. Historical desertification process in Hexi Corridor, China. *Chin. Geogr. Sci.* **2005**, *15*, 245–253. [CrossRef]
17. Li, B. Investigation and Study on the Relics of Suoyang City and Process of Desertification in Its Surrounding Reclaimed Land. *J. Desert Res.* **1991**, *11*, 211–215.
18. Tang, X.; Li, S. An Analysis on the oasis evolution of Hexi Corridor in historical period. *J. Arid. Land Resour. Environ.* **2021**, *35*, 48–55.
19. Ge, Q.; Zheng, J.; Hao, Z.; Liu, H. General characteristics of climate changes during the past 2000 years in China. *Sci. China Earth Sci.* **2013**, *56*, 321–329. [CrossRef]
20. Li, B. Notes on Mingshui (Jiduanshui) River and Mingze lake of the Han and Tang Dynasties. *Duanghuang Res.* **2001**, *2*, 60–67.
21. Zhang, J. Research on the Shule River's Water System Evolution in Historical Peoriods and Related Issues. *J. Chin. Hist. Geogr.* **2010**, *25*, 15–30.
22. Li, B. Several theortical problems on historical geography of desert. *Sci. Geogr. Sin.* **1999**, *19*, 211–215.
23. Yang, G.S.; Qu, Y.G.; Dong, G.R.; Chen, G.T.; Li, D.L.; Zhang, J.G.; Wu, G.H.; Hu, S.X. Study on ecology protection in lower reaches area of the Shulehe River. *J. Desert Res.* **2005**, *25*, 472–482.

remote sensing

MDPI

Article

A Strict Validation of MODIS Lake Surface Water Temperature on the Tibetan Plateau

Lazhu [1,2], Kun Yang [2,3,*], Jun Qin [4], Juzhi Hou [2], Yanbin Lei [2], Junbo Wang [2], Anning Huang [5], Yingying Chen [2], Baohong Ding [2] and Xin Li [2]

1 Research Center for Ecology, College of Science, Tibet University, Lhasa 850011, China
2 Institute of Tibetan Plateau Research, Chinese Academy of Sciences, Beijing 100101, China
3 Ministry of Education Key Laboratory for Earth System Modeling, Department of Earth System Science, Tsinghua University, Beijing 100084, China
4 State Key Laboratory of Resources and Environmental Information System, Institute of Geographic Sciences and Natural Resources Research, Chinese Academy of Sciences, Beijing 100101, China
5 CMA-NJU Joint Laboratory for Climate Prediction Studies, School of Atmospheric Sciences, Nanjing University, Nanjing 210023, China
* Correspondence: yangk@tsinghua.edu.cn

Abstract: Lake surface water temperature (LSWT) is a key parameter in understanding the variability of lake thermal conditions and evaporation. The MODIS-derived LSWT is widely used as a reference for lake model validations and process studies in data-scarce regions. In this study, the accuracy of the MODIS LSWT was examined on the Tibetan Plateau (TP). In-situ subsurface temperatures were collected at five large lakes. Although the observation period covers from summer to winter, only the observations during the lake turnover period (from October to freeze-up), when the lakes are well mixed, can be used as ground truth. The MODIS LSWT agrees well with the selected in-situ data for the five large lakes, with root mean square error (RMSE) < 1 °C at nighttime and <2 °C in the daytime, indicating a high accuracy of the MODIS LSWT data. Before the turnover period, the water is thermally stratified and the surface water is warmer than the subsurface water, and thus the in-situ subsurface water temperature data and the MODIS LSWT have different representativeness. In this case, if the observations are used as a validation basis, the MODIS errors could be much magnified. This in turn indicates the importance of period selection for the validation.

Keywords: Tibetan Plateau; MODIS; lake surface water temperature; lake turnover

Citation: L.; Yang, K.; Qin, J.; Hou, J.; Lei, Y.; Wang, J.; Huang, A.; Chen, Y.; Ding, B.; Li, X. A Strict Validation of MODIS Lake Surface Water Temperature on the Tibetan Plateau. *Remote Sens.* **2022**, *14*, 5454. https://doi.org/10.3390/rs14215454

Academic Editor: Catherine Ottlé

Received: 30 August 2022
Accepted: 25 October 2022
Published: 30 October 2022

1. Introduction

The contrasting properties between the lake and land surfaces (e.g., their surface albedo, heat capacity and roughness length) can cause local air circulations that affect the local climate [1–5]. In particular, lake surface evaporation after the onset of the dry and cold season can increase greatly, causing heavy rainfall or snow downwind of large lakes at high latitudes and high elevations [2,6,7]. Therefore, lake–atmosphere interactions are included in the land surface schemes used in high-resolution weather/climate modeling [1,8–13]. The lake surface water temperature (LSWT) is a crucial parameter in calculating evaporation and sensible heat flux at the lake–air interface [14–18]. In addition, LSWT is widely used in lake biology and climate change studies [19,20]. However, in-situ LSWT data are often not available in remote regions such as the Tibetan Plateau.

Instead of in-situ LSWT data, satellite remote sensing data, if proven reliable, can be used as an alternative [10,11,21–23]. The Moderate Resolution Imaging Spectroradiometer (MODIS) LSWT product has been broadly validated worldwide, showing acceptable errors [24–28]. For the two largest Swedish lakes, the MODIS retrieval showed good agreement with in-situ measurements at a depth of 0.5 m, with a root mean square error (RMSE) of less than 0.50 °C [29]. At the Great Salt Lake, Utah, the MODIS-derived LSWT showed a cold bias (−1.5 °C) and an RMSE of 1.6 °C compared to the in-situ measurements at a depth of 0.5 m [30]. A cold bias was also found at Lake Taihu, China, and the RMSE was less than 1.8 °C when compared with field measurements at depths of 0.5 to 1 m [31]. These validation studies do not pay attention to the representativeness of in-situ measurements, which could be different for different lake thermal phases.

The Tibetan Plateau (TP) has more than 1400 inland lakes with areas greater than 1 km^2, and the total area of TP lakes represents more than 50% of the inland water area in China [32]. Most of the lakes have expanded since the mid-1990s; the expansion is continuing [33,34] and represents one of the most outstanding environmental change events [35]. To better understand the response of the lake water and energy budgets to climate change on the data-scarce TP, the MODIS LSWT has been used to investigate the lake warming trends [20,36–38] and to validate the simulated bulk temperature of the mixed layer in lake modeling [22,39–42].

However, caution must be exercised regarding the accuracy of the MODIS LSWT when it is used for lake process studies on the TP. Xiao et al. [43] showed that the daytime MODIS LSWT had an RMSE of 1.5 °C at Lake Qinghai compared to the in-situ measurements at a depth of 0.5 m, but the comparison was based only on a short period of lake thermal stratification (July of 2010 and 2011). Zhang et al. [36] found that the nighttime MODIS LSWT had an unexpected high RMSE (4.5 °C) in another large lake (Lake Nam Co) using radiometric skin temperature collected near the lake shoreline, but the comparison period contained the lake ice phase during which the difference between the two data sets was large. These studies did not consider the impacts of lake stratification and attributed the difference between the MODIS LSWT and in-situ data to retrieving errors. Recently, Wan et al. [44] released a 15-year (2001–2015) time series of the daytime and nighttime MODIS LSWT for TP lakes, but the accuracy of the MODIS LSWT was not evaluated by the in-situ observations. Therefore, validation studies of the MODIS LSWT in the region are limited and even questionable; specifically, they ignored the conditions under which in-situ subsurface temperature data can be used to validate the satellite-retrieved surface temperature.

A comprehensive evaluation of the MODIS LSWT data should take the TP's unique climate and environment into account. The solar radiation over the TP is very strong [45,46] and the solar heat on the water surface may cause obvious thermal stratification. This phenomenon should be accounted for carefully, otherwise it will lead to artificially magnified errors in the MODIS LSWT. The goal of this study is to consider this phenomenon in the MODIS LSWT evaluation process and to present criteria for validations and applications of the MODIS LSWT over the TP.

In this study, buoy data at five lakes during the open-water period of a full year were used to investigate the accuracy of the MODIS LSWT data during two periods (lake stratification and turnover). The lakes and data sets (buoy data and the MODIS LSWT) are introduced in Section 2, and the comparison method is described in Section 3. The discrepancies between the MODIS LSWT and in-situ data are presented in Section 4, and the reason for the substantial underestimates in the nighttime MODIS LSWT in summer is discussed in Section 5. Finally, the results of this study are summarized in Section 6.

2. Study Area and Data Sets

The climate in the TP is dominated by the Asian monsoon during summer and early autumn, and by westerlies during other seasons. The influence of the monsoon is most prominent during July and August, and the monsoon season in this study refers to the period during July and August. The climate in northwest TP is mainly influenced by the westerlies throughout all seasons. Most lakes in the TP are located in the central and western TP and are influenced by both the monsoon and the westerlies.

2.1. Lake Locations

Subsurface water temperature data were collected for a full annual cycle at five large lakes which are located in different climate regions of the TP (Figure 1 and Table 1). The Lake Bangong Co (33.68°N, 79.22°E, 671 km^2) on the arid western TP, is an endorheic and dimictic lake. According to the meteorological data observed in 2013 at Ngari Station for Desert Environment Observation and Research, CAS (about 10 km away from Bangong Co), the mean annual air temperature is 1.2 °C with the lowest value of −26.1 °C in January, the mean annual relative humidity is 33% and the annual precipitation is 125 mm. Lake Zhari Namco (30.93°N, 85.61°E, 997 km^2) is a salt semiarid lake on the southern TP, around which the mean annual air temperature is 1.2 °C and the annual precipitation is 384 mm in 2017 [47]. Lake Dagze Co (31.89°N, 87.52°E, 295 km^2) is a brackish lake on the central TP and its salinity is 18 g/L [48]. Based on observation from Xianza Meteorological station 150 km away from the lake, the mean annual air temperature is 0.6 °C and annual precipitation is 316 mm in 2012 [48]. Lake Nam Co (30.74°N, 90.61°E, 2021 km^2) is a dimictic, deep lake, and it is the third largest lake in the TP. The weather station on the northeastern shore of Lake Nam Co (distance of about 1.5 km) shows that the mean annual air temperature and relative humidity are −1.2 °C and 55%, and annual precipitation is 488 mm in 2013, respectively [40]. Located in the central Himalayas, Lake Peiku Co (28.89°N, 85.59°E, 280 km^2) is a deep alpine lake with a mean annual air temperature of 4.4 °C in 2015/2016 and an annual precipitation varying between 150 and 200 mm [49].

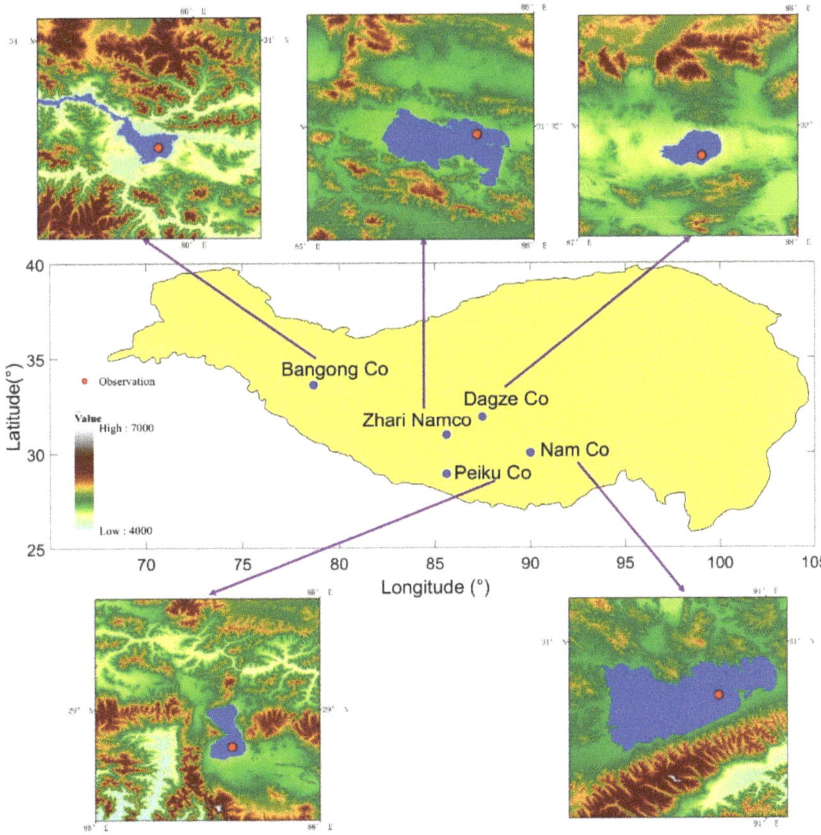

Figure 1. Location of five Tibetan Plateau lakes (Bangong Co, Zhari Namco, Dagze Co, Paiku Co, and Nam Co) and their surrounding topography (the red dots indicate water temperature monitoring stations).

Table 1. Geographical information and measurement information on the buoy stations for the five lakes. The lake depths are at the buoy stations.

Lake	Co-Ordinates	Area (km^2)	Elevation (m a.s.l)	Lake Depth (m)	Measurement Depth (m)	Motoring Sensor	Acquisition Period
Bangong Co	33.2–34.0°N 78.4–80.0°E	671	4241	36.6	5	HOBO U22-001	1 Aug 2012–31 Jul 2013
Zhari Namco	30.75–31.1°N 85.3–85.9°E	1000	4613	59.3	0.7	HOBO U22-001	1 Aug 2016–31 Jul 2017
Dagze Co	31.8–32.0°N 87.4–87.7°E	311	4450	37.3	4	HOBO U22-001	1 Sep 2012–31 Aug 2013
Nam Co	30.5–30.95°N 90.2–91.05°E	2024	4710	92.0	3	Hydrolab DS5	1 Aug 2012–31 Jul 2013
Peiku Co	28.75–29.05°N 85.45–85.7°E	280	4590	42	0.5	HOBO U22-001	1 Jun 2016–31 May 2017

2.2. In-Situ Measurements of Lake Temperature

Table 1 provides the water temperature measurement information on the buoy stations. Buoy data at Lake Nam Co were measured using a Hydrolab DS5 [50]. This probe has an accuracy of ±0.1 °C between −5 °C and 50 °C, with a resolution of 0.01 °C. Water temperatures in the other four lakes were recorded via a HOBO water temperature probe V2/U22-001 (Onset Corp., Cape Cod, USA) [48], which has a temperature range of −10 °C to 50 °C with an accuracy of ±0.2 °C and a resolution of 0.02 °C.

The measuring depths and intervals were set according to the following considerations in addition to specific reasons for the individual lakes. In previous studies [29,30], the subsurface temperatures were measured at depths shallower than 1 m. We measured temperatures at depths deeper than 1 m in some lakes to avoid destruction by ice in winter due to the lack of lake ice thickness information at the buoy sites. Specifically, subsurface temperatures were measured at depths of 5 m, 0.7 m, 4 m, 3 m and 0.5 m at these five lakes in the order shown in Table 1. The measured temperatures are able to represent the bulk temperature of the mixed layer of the lakes [48,50]. In addition, there are considerable diurnal variations in the lake surface temperature due to solar heating in the lakes' stratified period [20,26,51]. Minnett [52] recommended that validation should be conducted within ±2 h of the satellite overpass, as followed in previous studies [31,53,54]. Therefore, the buoy data in this study were collected at intervals of 1 or 2 h, and then the average within ±1 h (2 h) of the satellite overpass was sub-sampled to enable comparison between the two data sets.

According to the measurements, four of the five large lakes were completely frozen between approximately December–early January and late April–May [55,56], while the remaining one (Peiku Co) [49] was ice-free in the studied year. After the ice break-up, the lake water temperature started to increase, and the lakes became thermally stratified. The maximum water temperature occurred around late August–early September, then the temperature started to decrease, but the lakes were still stratified. From October, the upper mixed layer drastically deepened, and the lake turnover appeared before the ice cover. In this study, the lake stratified period is referred to as the period from May to September, while the lake turnover period is defined as the period from October to the freeze-up date.

2.3. MODIS Lake Surface Temperature

The MODIS level 3, 1 km nominal resolution at nadir, daily land surface temperature products (MO/YD11A1) were obtained through the NASA Level 1 and Atmosphere Archive and Distribution System (https://ladsweb.nascom.nasa.gov/data/ (accessed on 20 August 2020)). The inland lake surface temperature contained in these products was derived using thermal infrared (TIR) bands 31 and 32, and a generalized split-window algorithm. LSWT was retrieved from both platforms (Terra and Aqua) at different times on the same day. In both platforms, two instantaneous observations were collected every day (Terra: approximately 10:30 and 22:30 local time, Aqua: approximately 13:30 and 01:30 local time). The MODIS LSWT data are pre-processed to account for atmospheric and surface emissivity effects. The cloud mask (MOD35) used for inland water provides a surface temperature measurement when there is a 66% or greater confidence of clear-sky conditions [57], otherwise no temperature measurement is produced. More details on this product were provided by Savtchenko et al. [58] and Wan et al. [59].

Satellite-derived LSWT measurements represent the instantaneous water temperature of the uppermost ~10–20 μm deep molecular layer; this is known as the skin temperature [60,61]. However, thermometers are installed well below the lake surface and thus the measured ones are usually the bulk temperature of the mixed layer. Differences between the skin and bulk temperature are called skin effects [62–64]. As the lake thermal structure shown in Wilson et al. [64] demonstrates, the skin temperature of the upper-most layer whose thickness is less than 1 mm is controlled by heat and the momentum flux at the lake–air interface; the bulk temperature, at just a few centimeters of depth, is warmer than the skin temperature by several tenths of 1 °C due to solar heating [64,65]; for the depth of no less than 1 m, the difference between bulk temperature and skin temperature can be several degrees [27,53]. The skin effect can be impacted by solar radiation which varies with latitude and altitude, and lake thermal phases for different thermal regimes in the lake's stratified and turnover periods [27,31,66]. On the Tibetan Plateau, the stronger solar radiation in the daytime and intense longwave cooling at nighttime due to the thin air might enlarge the skin effect.

3. Methods

The MODIS level 3 daily LSWTs derived from both Terra and Aqua platforms were compared with the in-situ measurements. The MODIS LSWT at a specific lake was obtained using the MODIS Reprojection Tool (MRT, https://lpdaac.usgs.gov/tools/modis_reprojection_tool (accessed on 20 July 2019)), which involved three steps as follows.

The first step was to identify the lake water body using the MODIS land mask product at 250 m resolution (http://www.landcover.org/data/watermask/(accessed on 15 July 2019)). The land mask image was then resampled to 0.01° spatial resolution to match the resolution of the MODIS LSWT product, which was resampled to 0.01° using MRT.

The second step was data quality assurance to select LSWT data free of cloud contamination and uses the quality flag stored in quality control (QC) scientific data sets. Only pixels with quality flags designated as good quality and the average LSWT errors less than or equal to 2 K were retained.

The third step was to obtain the spatially representative MODIS LSWT data. For each lake water temperature station, a 3 × 3 pixel array centered on the monitoring station was extracted from the MODIS LSWT data, and its average LSWT was calculated for subsequent comparison with the in-situ data. Such spatial averaging is widely used for removing spatial divergence [25,29,65–67].

The selected daily MODIS LSWT data sets from Aqua and Terra were compared with the in-situ measurements at the five lakes. Given that the agreement between the observation and the MODIS data may depend on lake thermal conditions, we evaluated the daytime and nighttime MODIS LSWT in the lake stratified period and the lake turnover period, respectively. The performance of the MODIS LSWT was assessed via mean difference (MD) and root mean square difference (RMSD) between the MODIS retrievals and in-situ measurements. Finally, we investigated the impact of in-situ data from different periods and times for the evaluation of the MODIS LSWT.

4. Results

The MODIS LSWT is the lake skin temperature, whereas the in-situ data represent the bulk temperature of the mixed layer of the lakes. The MODIS LSWT was evaluated for the open-water period of a full year. When lakes were frozen (usually from January to April in large Tibetan lakes), the MODIS retrieved the ice surface temperature while the in-situ data were obtained from the subsurface water layer, so they are not comparable. Four lakes (Bangong Co, Zhari Namco, Dagze Co and Nam Co) had ice cover during the study year, and the other one (Peiku Co) remained ice-free. The ice-covered period in the four lakes can be identified with the MODIS LSWT consequent of most of the satellite data being missing when the lake was ice-covered [68].

Figure 2 shows the comparisons between the four MODIS products (Terra and Aqua, in daytime and nighttime) and the buoy data in the five large lakes, and Figure 3 shows the error metrics of the evaluation. The open-water period consists of the lake stratified period (from May to September) and the lake turnover period (from October to the freeze-up date). As shown in Figures 2 and 3, the differences between the MODIS LSWT and in-situ data at the five large lakes are quite variable during the two periods. They were analyzed in detail in the following sub-sections.

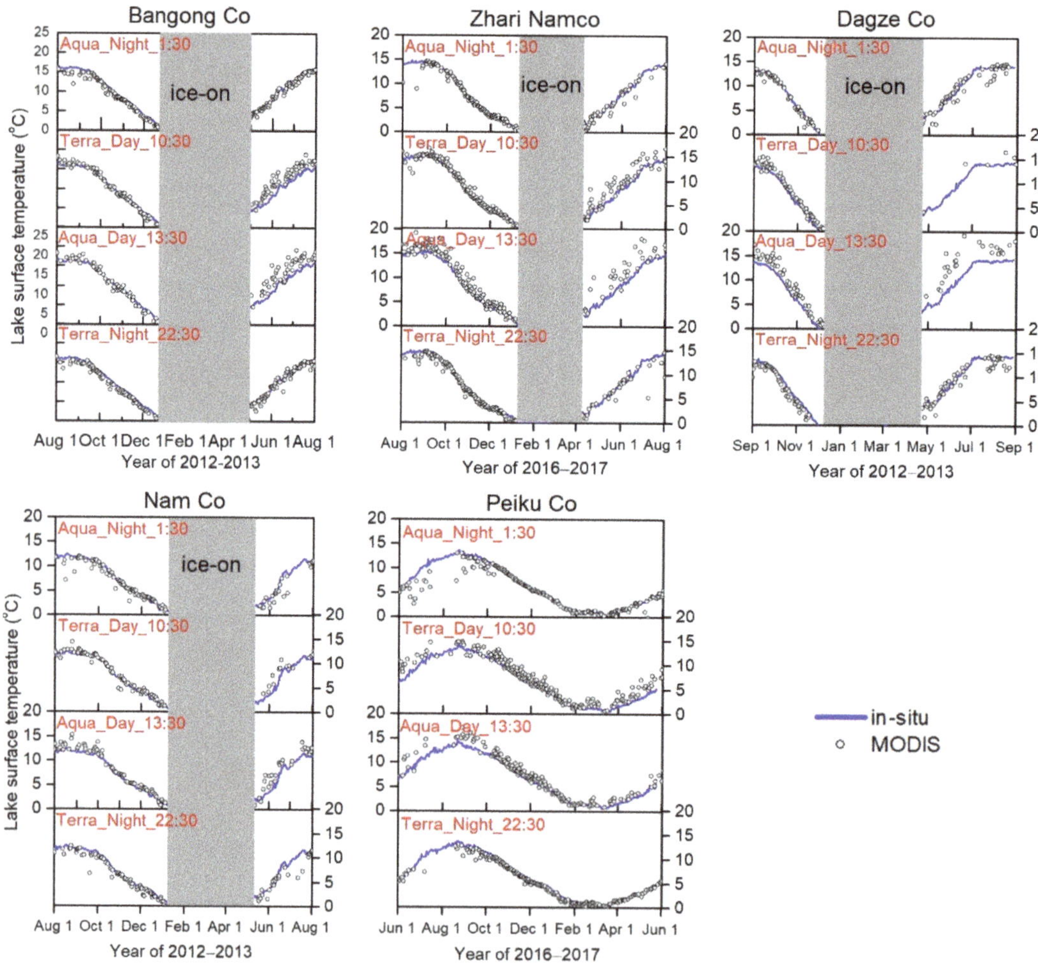

Figure 2. Comparison of daily MODIS-retrieved lake surface temperature and in-situ observed bulk temperature in the five large lakes. The four data layers for each lake represent daytime and nighttime for Terra and Aqua.

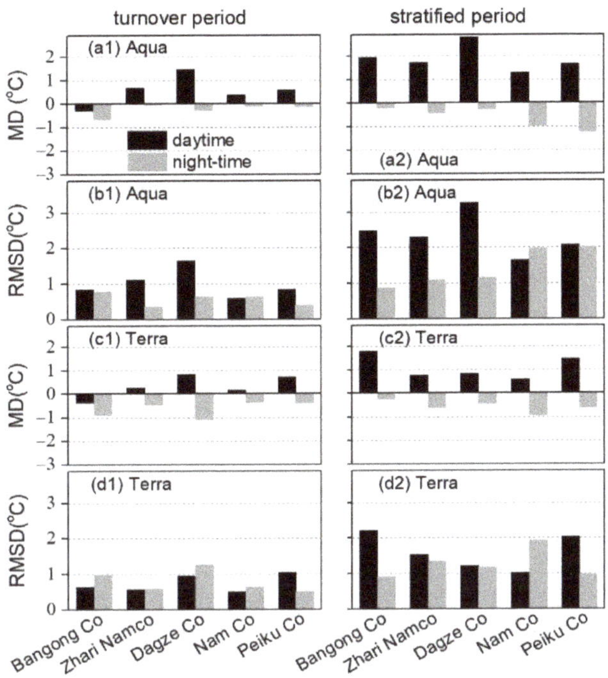

Figure 3. Mean difference (MD (**a1,a2,c1,c2**)) and root mean square difference (RMSD (**b1,b2, d1,d2**)) between MODIS LSWT and in-situ data in the daytime and at night, (1) during the lake turnover period, (2) during the lake stratified period, for Aqua (**a1,a2,b1,b2**) and Terra (**c1,c2,d1,d2**), in the five large lakes on the TP.

4.1. Results during the Lake Turnover Period

During the lake turnover period, the increased air–lake temperature gradient on the TP combined with strong wind [69,70] results in a considerable energy release from the lake to the overlying air, in terms of large latent and sensible heat fluxes as well as longwave radiation [40,71]. The lake heat loss makes the lake surface water temperature rapidly decrease, causing a sharp increase in the water density of the upper water layer, which induces vertical water convection to a large extent and eventually develops into lake turnover. As an example, Figure 4 shows that the temperature is uniform in this mixed layer. The turnover greatly reduces the difference between the surface skin temperature and the bulk water temperature, and, therefore, the subsurface temperature data can be used to validate the MODIS LSWT.

As shown in Figures 2 and 3, the MODIS LSWT is consistent with the in-situ data during this period for the five lakes, in terms of both magnitude and temporal variability. The RMSD is less than <1 °C at night and <2 °C in the daytime, indicating a high accuracy of the MODIS LSWT. The difference between the MODIS LSWT and in-situ data in Terra is relatively larger than that in Aqua, in terms of both MD and RMSD. Among the four MODIS LSWT data sets, retrieval at night from the Aqua sensor performed best in terms of consistency with the bulk temperature.

While the turnover makes the vertical water mix, the difference between the skin and bulk temperature remains. In general, the daytime MODIS LSWT is greater than the in-situ values, while the nighttime MODIS LSWT shows negligible differences from the in-situ data. The larger difference between the MODIS LSWT and the in-situ data may be due to solar heating and should not be attributed to the errors in the MODIS LSWT, as solar radiation on the TP is stronger than that along the same latitude [46,72]. To understand this, we analyzed the difference in lake temperature between the daytime and nighttime.

During this turnover period, there is strong longwave emission from the lake surface while strong solar radiation is received by the lake surface. This may cause a cooling of the surface layer at night but a warming in the daytime. This diurnal change in skin temperature can be detected by the MODIS via the Aqua platform, as shown in Figure 5a. However, the subsurface temperature of these deep lakes has a negligible response to the diurnal thermal conditions (Figure 5b) because of their huge heat capacity and high level of vertical mixing. The diurnal variation in subsurface temperature is less than that in skin temperature. Therefore, the larger MD and RMSD of the MODIS LSWT in the daytime is due to its greater representativeness in vertical differing than that of the in-situ data.

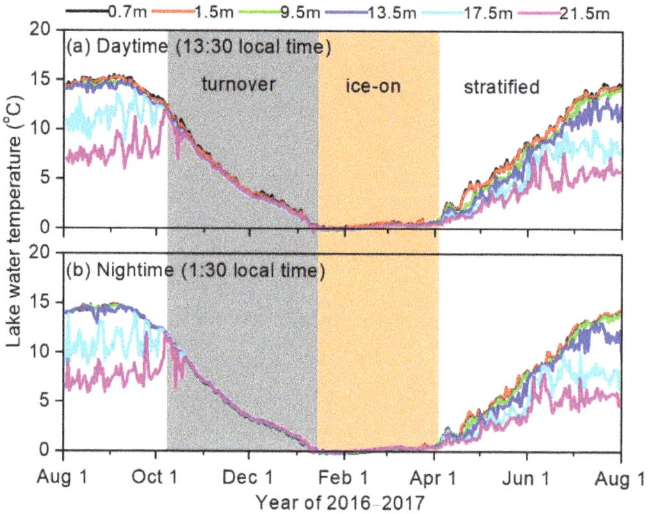

Figure 4. Observed annual cycle of thermal structure of Lake Zhari Namco in the year of 2016–2017. Water temperature was measured at five depths.

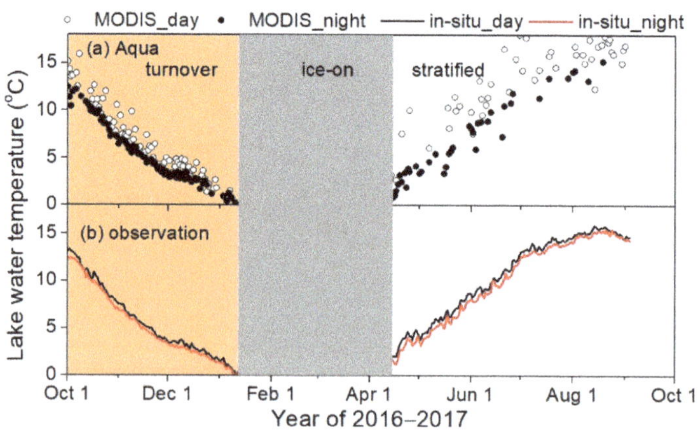

Figure 5. Time series of (**a**) MODIS retrievals and (**b**) in-situ data in the daytime and nighttime for Lake Zhari Namco in the year of 2016–2017.

Overall, the MODIS retrievals perform well during the lake turnover period for these large lakes, and using the nighttime MODIS LSWT as the ground truth for the evaluation is the most reliable.

4.2. Results during the Lake Stratified Period

At night in the lake stratified period, the MODIS LSWT from both Aqua/Terra platforms is slightly lower than the in-situ data due to the skin cooling effect of the lake, with a negative MD of less than 1 °C (Figures 2 and 3). This phenomenon is seen in both the turnover period and the stratified period. However, the RMSD in the lake stratified period is clearly larger than that in the lake turnover period (Figure 3). According to Figures 2 and 3, the MODIS Aqua has an RMSD that ranges between 0.9 °C and 2.0 °C and the MODIS Terra has an RSMD range between 0.9 °C and 1.9 °C. This is associated with the different water layer representativeness between the MODIS retrieval and the in-situ data during the summer, as discussed below.

In the daytime of the lake stratified period, the RMSD may exceed 3 °C and is larger than that at night, so our focus is the discrepancy between the MODIS retrievals and in-situ measurement in the daytime. As shown in Figure 3a2,c2, the daytime MODIS LSWT from both Aqua/Terra platforms is clearly higher than the in-situ data, as indicated by the positive MD. There are two causes for the larger errors in the daytime.

The first is the strong solar radiation in the TP. The solar heating at midday reaches its diurnal peak with a magnitude greater than 1000 Wm^{-2} [45,73]. According to the observations [74,75], the water clarity of lakes on the TP, as defined by the Secchi depth, is high: the mean is about 5 m. Therefore, most of these lakes are relatively transparent, allowing solar radiation to penetrate into deep layers. Solar radiation exponentially attenuates with water depth, which makes a large proportion of solar energy, particularly in the near-infrared band, be absorbed by the upper-most thin layer during the summer. However, as shown in Figure 4, there is a strong thermal stratification. This causes a much higher daytime MODIS LSWT than the subsurface temperature at the Aqua daytime overpass time (13:30 local time). As shown in Figure 3a2,c2, the daytime MD in Aqua is larger than 1.5 °C while in Terra it ranges between 0.6 °C and 1.8 °C.

The second is the depth of the in-situ measurements. According to the water temperature profile proposed by Wilson et al. [64], the water temperature in the subsurface layer varies rapidly with depth when the lake is thermally stratified. The depths of in-situ measurements in some lakes are considerably larger than that in other regions [25,27,31,53], therefore the measured subsurface temperature may be considerably cooler than the skin temperature due to the strong thermal stratification in the daytime.

Therefore, the representativeness in the vertical profile is different between the MODIS and the subsurface data. During the lake turnover period, the water is well mixed and the difference in representativeness is small and the evaluation result during this period is credible. However, during the stratification period, the water temperature varies with depth and the difference in representativeness is considerable; therefore, the evaluation during this period is not credible, and the difference between the in-situ data and the MODIS LSWT cannot be harshly attributed to the errors in the MODIS data.

Based on the above evaluation and reasoning, the MODIS LSWT has good accuracy for the TP lakes, with a root mean square error (RMSE) of less than 1.6 °C.

5. Discussions

Although the MODIS LSWT has good accuracy, there are still some distinct errors. Figure 2 shows some substantial underestimates in the nighttime MODIS LSWT when compared to the in-situ data during the monsoon season (summer), as also reported by Wan et al. [44]. Theoretically, the skin cooling effect at night causes the skin temperature to be slightly lower than the subsurface temperature, but it cannot explain such large cold biases at night during the monsoon season. We noticed that the large cold biases strongly depend on climate zones. For the lakes in the monsoon-controlled region (e.g., Lake Peiku Co and Lake Nam Co), some MODIS LSWTs during the monsoon season were severely underestimated. On the contrary, this situation was not so apparent for the lakes in the westerlies-dominated region (e.g., Lake Bangong Co).

During the monsoon season (from June to August), water vapor content in the monsoon-controlled region is considerably higher than in the westerly-dominated region. We speculate that the large cold biases may be related to the water vapor and cloud processes. As an example, we analyzed the relative humidity data from the observations of Lake Peiku Co, a typical lake in the monsoon region. The relative humidity data were measured at the north shoreline [76]. Figure 6a shows the temporal variations of the MODIS LSWT and in-situ measurements, Figure 6b shows the diurnal variations in relative humidity in two days with large cold biases in the MODIS LSWT and Figure 6c shows the same in the adjacent two days with small biases. It can be seen that the nighttimes with large cold biases are more humid than the ones with small biases. So, the large cold biases may be associated with the local water vapor circulation and clouds. On the TP, due to the thin air, the surface energy loss through upwelling longwave radiation is strong, causing rapid cooling of the land surface but weaker cooling of the lake surface. The resulting land–lake thermal contrast can drive cold airflow from the land to the lake surface at night. Meanwhile, evaporation from the warmer lake surface at night can replenish the water vapor in the cold air. The air convergence over the lake surface due to land wind causes upward flow. If the relative humidity is high, the upward flow may result in the condensation of water vapor. The large lakes are still at the warming stage during monsoon season (e.g., Lake Zhari Namco shown in Figure 4) and the land–lake thermal contrast is not so strong at night. Thus, upward motion over the TP's lakes is not strong enough to cause deep convections. Instead, it may cause the condensation of water vapor to form low-level fog or shallow clouds, which are usually undetectable by the MODIS sensor.

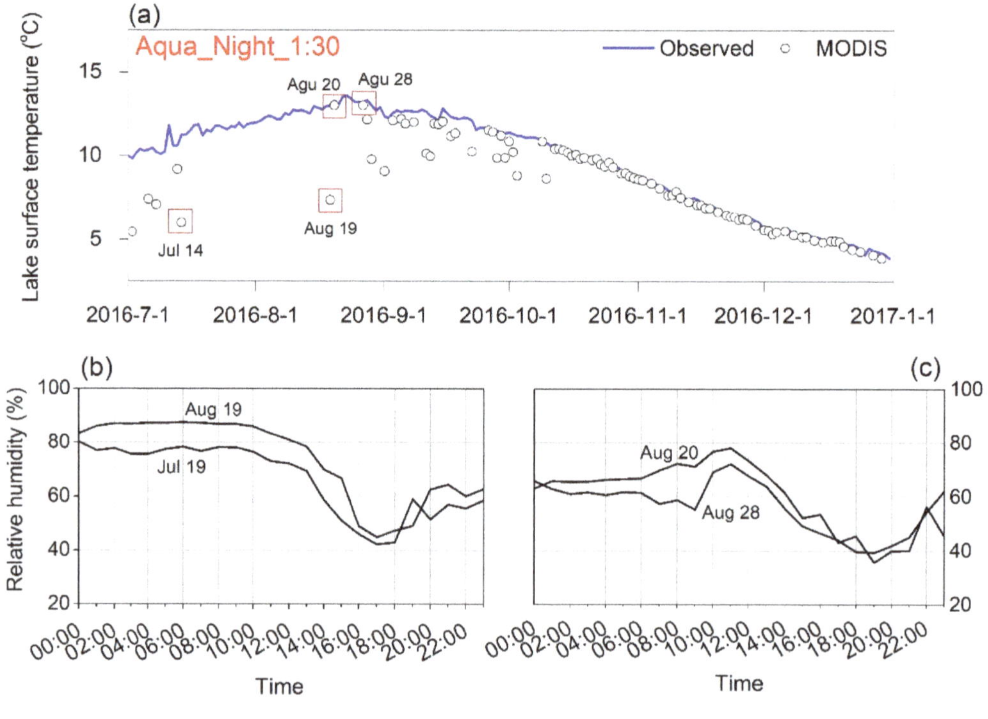

Figure 6. Comparison of MODIS Aqua and in-situ measurements of nighttime LSWT on Lake Peiku Co, (located in Himalaya region) (**a**), the diurnal variation in relative humidity on 14 July 2016 and 29 August 2016 (**b**) and 20 August 2016 and 28 August 2016 (**c**). On these two days in panel (**b**), MODIS LSWTs showed large negative deviations from the in-situ measurements, and the opposite in panel (**c**), indicated by the four red rectangles in part (**a**).

At the same time, the cloud mask criterion for the MO/YD11A1 inland water surface temperature retrieval is relatively relaxed compared to that for retrieving the land surface temperature, i.e., a water surface temperature is retrieved when the confidence of a clear-sky condition is 66% or greater, but the confidence must be greater than 99% for the land temperature retrieval [57]. In other words, the MODIS algorithm's detection of near-surface shallow clouds or fog has lower confidence. Therefore, shallow clouds and fog missing from the cloud mask might result in a considerable underestimate of nighttime LSWT during the monsoon season. Due to the lack of observational data, this speculation is subject to further confirmation in the future.

Given the frequent occurrence of large cold biases, the use of nighttime MODIS LSWT data with large cold biases during the monsoon season should be avoided in the monsoon-controlled region of the TP.

6. Conclusions

There are more than 1400 lakes with an area greater than 1 km^2 on the TP. Due to the lack of in-situ measurements, the MODIS LSWT product has been used to assess changes in lake temperature and to calculate evaporation, but the applicability of this product has remained unclear. Early limited evaluations for this region did not consider the representativeness of in-situ subsurface data, and thus may have caused uncertainties in the evaluation. In this study, we deployed buoys to measure subsurface water temperature in five large lakes on the TP and assessed the advantages and limitations of the MODIS LSWT (both Terra and Aqua) based on the buoy data. Particularly, we pay attention to the representativeness of in-situ data that varies with the seasons.

The agreement between the MODIS-derived LSWT and the in-situ observed subsurface temperature highly depends on the lake's thermal structure. During the turnover period (i.e., October to the freeze-up date), the MODIS LSWT generally shows good agreement with the in-situ data, indicating its high accuracy. Due to water mixing, the in-situ data during the nighttime of this period can be used as the ground truth in the validation of the MODIS LSWT, and the result confirms that the MODIS skin temperature can be representative of the subsurface temperature of the mixed layer of the lakes. During the lake-stratified period (from May to September), the in-situ data measured below the surface cannot represent the skin temperature detected by the MODIS LSWT and cannot be used as the ground truth for the validation of the latter.

Despite good performance in large Tibetan lakes, the MODIS lake surface temperatures can be questionable in the monsoon-controlled region. Due to the influence of the Asian monsoon, the nighttime LSWT can be heavily underestimated in the MODIS retrievals on some humid days in summer. This severely restricts the use of the MODIS LSWT on the Tibetan Plateau.

In summary, the MODIS LSWT has a high accuracy (RMSE < 2 °C) for large lakes in the TP. However, if the MODIS LSWT is assumed to represent the subsurface water temperature of a lake, it is preferable to use its nighttime data during the lake turnover period, but not to use them during the lake stratified period (especially the monsoon period).

Author Contributions: Conceptualization, L. and K.Y.; Investigation, J.H., Y.L., J.W. and Y.C.; Methodology, L., K.Y. and J.Q.; Writing—original draft, L. and K.Y.; Writing—review & editing, J.Q., A.H., Y.C., B.D. and X.L. All authors have read and agreed to the published version of the manuscript.

Funding: This work was supported by the Second Tibetan Plateau Scientific Expedition and Research Program (Grant No. 2019QZKK0206), the National Natural Science Foundation of China (Grant No. 41975125) and the Natural Science Foundation of Tibet Autonomous Region (Grant No. XZ202201ZR0046G).

Data Availability Statement: The MODIS lake surface temperature product (MO/YD11A1) was obtained through the NASA Level 1 and Atmosphere Archive and Distribution System (https://ladsweb.nascom.nasa.gov (accessed on 20 August 2020)). The in-situ measurements of lake water temperature for Lake Bangong Co, Lake Dagze Co and Lake Nam Co can be obtained from National Tibetan Plateau Data Center (http://data.tpdc.ac.cn/en/ (accessed on 20 August 2020)).

Acknowledgments: We are grateful to everyone who participated in the field campaign establishing lake water temperature stations. We are also grateful to Ngari Station for Desert Environment Observation and Research for providing the station data and the reviewers for their invaluable comments on the manuscript.

Conflicts of Interest: The authors declare that they have no conflict of interest.

References

1. Dutra, E.; Stepanenko, V.M.; Balsamo, G.; Viterbo, P.; Miranda, P.; Mironov, D.; Schär, C. An offline study of the impact of lakes on the performance of the ECMWF surface scheme. *Boreal Environ. Res.* **2010**, *15*, 100–112.
2. Lofgren, B.M. Simulated Effects of Idealized Laurentian Great Lakes on Regional and Large-Scale Climate. *J. Clim.* **1997**, *10*, 2847–2858. [CrossRef]
3. Long, Z.; Perrie, W.; Gyakum, J.; Caya, D.; Laprise, R. Northern lake impacts on local seasonal climate. *J. Hydrometeorol.* **2007**, *8*, 881–896. [CrossRef]
4. Thiery, W.; Davin, E.L.; Panitz, H.J.; Demuzere, M.; Lhermitte, S.; Van Lipzig, N. The impact of the African Great Lakes on the regional climate. *J. Clim.* **2015**, *28*, 4061–4085. [CrossRef]
5. Yao, X.; Yang, K.; Zhou, X.; Wang, Y.; Lazhu; Chen, Y.; Lu, H. Surface friction contrast between water body and land enhances precipitation downwind of a large lake in Tibet. *Clim. Dyn.* **2021**, *56*, 2113–2126. [CrossRef]
6. Li, M.; Ma, Y.; Hu, Z.; Ishikawa, H.; Oku, Y. Snow distribution over the Namco lake area of the Tibetan Plateau. *Hydrol. Earth Syst. Sci.* **2009**, *13*, 2023–2030. [CrossRef]
7. Zhao, Z.; Huang, A.; Ma, W.; Wu, Y.; Wen, L.; La, Z.; Gu, C. Effects of Lake Nam Co and surrounding terrain on extreme precipitation over Nam Co basin, Tibetan Plateau: A case study. *J. Geophys. Res. Atmos.* **2022**, *127*, e2021JD036190. [CrossRef]
8. Balsamo, G.; Salgado, R.; Dutra, E.; Boussetta, S.; Stockdale, T.; Potes, M. On the contribution of lakes in predicting near-surface temperature in a global weather forecasting model. *Tellus A Dyn. Meteorol. Oceanogr.* **2012**, *64*, 15829. [CrossRef]
9. Dai, Y.; Wei, N.; Huang, A.; Zhu, S.; Shangguan, W.; Yuan, H.; Zhang, S.; Liu, S. The lake scheme of the Common Land Model and its performance evaluation. *Chin. Sci. Bull.* **2018**, *63*, 3002–3021. [CrossRef]
10. Gu, H.; Jin, J.; Wu, Y.; Ek, M.B.; Subin, Z.M. Calibration and validation of lake surface temperature simulations with the coupled WRF-lake model. *Clim. Chang.* **2013**, *129*, 471–483. [CrossRef]
11. Stepanenko, V.; Jöhnk, K.D.; Machulskaya, E.; Perroud, M.; Subin, Z.; Nordbo, A.; Mammarella, I.; Mironov, D. Simulation of surface energy fluxes and stratification of a small boreal lake by a set of one-dimensional models. *Tellus A Dyn. Meteorol. Oceanogr.* **2014**, *66*, 21389. [CrossRef]
12. Subin, Z.M.; Riley, W.J.; Mironov, D. An improved lake model for climate simulations: Model structure, evaluation, and sensitivity analyses in CESM1. *J. Adv. Model Earth Syst.* **2012**, *4*, M02001. [CrossRef]
13. Wen, L.; Lv, S.; Li, Z.; Zhao, L.; Nagabhatla, N. Impacts of the two biggest lakes on local temperature and precipitation in the Yellow River source region of the Tibetan Plateau. *Adv. Meteorol.* **2015**, *2015*, 248031. [CrossRef]
14. Dutton, J.A.; Bryson, R.A. Heat Flux in Lake Mendota 1. *Limnol. Oceanogr.* **1962**, *7*, 80–97. [CrossRef]
15. Lewis, W.M., Jr. Temperature, heat, and mixing in Lake Valencia, Venezuela 1. *Limnol. Oceanogr.* **1983**, *28*, 273–286. [CrossRef]
16. Magee, M.R.; Wu, C.H. Response of water temperatures and stratification to changing climate in three lakes with different morphometry. *Hydrol. Earth Syst. Sci.* **2017**, *21*, 6253–6274. [CrossRef]
17. Pour, H.K.; Rontu, L.; Duguay, C.; Eerola, K.; Kourzeneva, E. Impact of satellite-based lake surface observations on the initial state of HIRLAM. Part II: Analysis of lake surface temperature and ice cover. *Tellus A Dyn. Meteorol. Oceanogr.* **2014**, *66*, 21395. [CrossRef]
18. Verburg, P.; Antenucci, J.P. Persistent unstable atmospheric boundary layer enhances sensible and latent heat loss in a tropical great lake: Lake Tanganyika. *J. Geophys. Res.* **2010**, *115*, D11109. [CrossRef]
19. Woolway, R.I.; Kraemer, B.; Zscheischler, J.; Albergel, C. Compound hot temperature and high chlorophyll extreme events in global lakes. *Environ. Res. Lett.* **2021**, *16*, 094013. [CrossRef]
20. Ke, L.; Song, C. Remotely sensed surface temperature variation of an inland saline lake over the central Qinghai–Tibet Plateau. *ISPRS J. Photogramm. Remote Sens.* **2014**, *98*, 157–167. [CrossRef]
21. Fang, X.; Stefan, H.G. Long-term lake water temperature and ice cover simulations/measurements. *Cold Reg. Sci. Technol.* **1996**, *24*, 289–304. [CrossRef]
22. Huang, A.; Lazhu; Wang, J.; Dai, Y.; Yang, K.; Wei, N.; Wen, J.; Wu, Y.; Zhu, X.; Zhang, X.; et al. Evaluating and improving the performance of three 1-D Lake models in a large deep Lake of the central Tibetan Plateau. *J. Geophys. Res. Atmos.* **2019**, *124*, 3143–3167. [CrossRef] [PubMed]
23. Pour, H.K.; Duguay, C.R.; Martynov, A.; Brown, L.C. Simulation of surface temperature and ice cover of large northern lakes with 1-D models: A comparison with MODIS satellite data and in situ measurements. *Tellus A Dyn. Meteorol. Oceanogr.* **2012**, *64*, 17614. [CrossRef]
24. Chavula, G.; Brezonik, P.; Thenkabail, P.; Johnson, T.; Bauer, M. Estimating the surface temperature of Lake Malawi using AVHRR and MODIS satellite imagery. *Phys. Chem. Earth* **2009**, *34*, 749–754. [CrossRef]

25. Hook, S.J.; Vaughan, R.G.; Tonooka, H.; Schladow, S.G. Absolute radiometric in-flight validation of mid infrared and thermal infrared data from ASTER and MODIS on the Terra spacecraft using the Lake Tahoe, CA/NV, USA, automated validation site. *IEEE Trans. Geosci. Remote Sens.* **2007**, *45*, 1798–1807. [CrossRef]
26. Pareeth, S.; Salmaso, N.; Adrian, R.; Neteler, M. Homogenised daily lake surface water temperature data generated from multiple satellite sensors: A long-term case study of a large sub-Alpine lake. *Sci. Rep.* **2016**, *6*, 31251. [CrossRef]
27. Pour, H.K.; Duguay, C.R.; Solberg, R.; Rudjord, Ø. Impact of satellite-based lake surface observations on the initial state of HIRLAM. Part I: Evaluation of remotely-sensed lake surface water temperature observations. *Tellus A Dyn. Meteorol. Oceanogr.* **2014**, *66*, 21534. [CrossRef]
28. Tavares, M.H.; Cunha, A.H.F.; Motta-Marques, D.; Ruhoff, A.L.; Cavalcanti, J.R.; Fragoso, C.R.; Bravo, J.M.; Munar, A.M.; Fan, F.M.; Rodrigues, L.H.R. Comparison of Methods to Estimate Lake-Surface-Water Temperature Using Landsat 7 ETM+ and MODIS Imagery: Case Study of a Large Shallow Subtropical Lake in Southern Brazil. *Water* **2019**, *11*, 168. [CrossRef]
29. Reinart, A.; Reinhold, M. Mapping surface temperature in large lakes with MODIS data. *Remote Sens. Environ.* **2008**, *112*, 603–611. [CrossRef]
30. Crosman, E.T.; Horel, J.D. MODIS-derived surface temperature of the Great Salt Lake. *Remote Sens. Environ.* **2009**, *113*, 73–81. [CrossRef]
31. Liu, G.; Ou, W.; Zhang, Y.; Wu, T.; Zhu, G.; Shi, K.; Qin, B. Validating and mapping surface water temperatures in Lake Taihu: Results from MODIS land surface temperature products. *IEEE J. Sel. Top. Appl. Earth Obs. Remote Sens.* **2015**, *8*, 1230–1244. [CrossRef]
32. Zhang, G.; Luo, W.; Chen, W.; Zheng, G. A robust but variable lake expansion on the Tibetan Plateau. *Sci. Bull.* **2019**, *64*, 1306–1309. [CrossRef]
33. Zhu, L.; Xie, M.; Wu, Y. Quantitative analysis of lake area variations and the influence factors from 1971 to 2004 in the Nam Co basin of the Tibetan Plateau. *Chin. Sci. Bull.* **2010**, *55*, 1294–1303. [CrossRef]
34. Lei, Y.; Yao, T.; Bird, B.W.; Yang, K.; Zhai, J.; Sheng, Y. Coherent lake growth on the central Tibetan Plateau since the 1970s: Characterization and attribution. *J. Hydrol.* **2013**, *483*, 61–67. [CrossRef]
35. Yang, K.; Lu, H.; Yue, S.; Zhang, G.; Lei, Y.; La, Z.; Wang, W. Quantifying recent precipitation change and predicting lake expansion in the Inner Tibetan Plateau. *Clim. Chang.* **2018**, *147*, 149–163. [CrossRef]
36. Zhang, G.; Yao, T.; Xie, H.; Qin, J.; Ye, Q.; Dai, Y.; Guo, R. Estimating surface temperature changes of lakes in the Tibetan Plateau using MODIS LST data. *J. Geophys. Res. Atmos.* **2014**, *119*, 8552–8567. [CrossRef]
37. Song, K.; Wang, M.; Du, J.; Yuan, Y.; Ma, J.; Wang, M.; Mu, G. Spatiotemporal variations of lake surface temperature across the Tibetan Plateau using MODIS LST product. *Remote Sens.* **2016**, *8*, 854. [CrossRef]
38. Wan, W.; Zhao, L.; Xie, H.; Liu, B.; Li, H.; Cui, Y.; Ma, Y.; Hong, Y. Lake Surface Water Temperature Change Over the Tibetan Plateau From 2001 to 2015: A Sensitive Indicator of the Warming Climate. *Geophys. Res. Lett.* **2018**, *45*, 11–177. [CrossRef]
39. Haginoya, S.; Fujii, H.; Kuwagata, T.; Xu, J.; Ishigooka, Y.; Kang, S.; Zhang, Y. Air-lake interaction features found in heat and water exchanges over Nam Co on the Tibetan Plateau. *Sci. Online Lett. Atmos.* **2009**, *5*, 172–175. [CrossRef]
40. Lazhu; Yang, K.; Wang, J.; Lei, Y.; Chen, Y.; Zhu, L.; Ding, B.; Qin, J. Quantifying evaporation and its decadal change for Lake Nam Co, central Tibetan Plateau. *J. Geophys. Res. Atmos.* **2016**, *121*, 7578–7591. [CrossRef]
41. Zhang, Q.; Jin, J.; Zhu, L.; Lu, S. Modelling of Water Surface Temperature of Three Lakes on the Tibetan Plateau using a Physically Based Lake Model. *Atmos. Ocean* **2018**, *56*, 289–295. [CrossRef]
42. Zhou, J.; Wang, L.; Zhang, Y.; Guo, Y.; Li, X.; Liu, W. Exploring the water storage changes in the largest lake (Selin Co) over the Tibetan Plateau during 2003–2012 from a basin-wide hydrological modeling. *Water Resour. Res.* **2015**, *51*, 8060–8086. [CrossRef]
43. Xiao, F.; Ling, F.; Du, Y.; Feng, Q.; Yan, Y.; Chen, H. Evaluation of spatial-temporal dynamics in surface water temperature of Qinghai Lake from 2001 to 2010 by using MODIS data. *J. Arid Land* **2013**, *5*, 452–464. [CrossRef]
44. Wan, W.; Li, H.; Xie, H.; Hong, Y.; Long, D.; Zhao, L.; Han, Z.; Cui, Y.; Liu, B.; Wang, C.; et al. A comprehensive data set of lake surface water temperature over the Tibetan Plateau derived from MODIS LST products 2001–2015. *Sci. Data* **2017**, *4*, 170095. [CrossRef]
45. Yang, K.; He, J.; Tang, W.; Qin, J.; Cheng, C.C. On downward shortwave and longwave radiations over high altitude regions: Observation and modeling in the Tibetan Plateau. *Agric. For. Meteorol.* **2010**, *150*, 38–46. [CrossRef]
46. Tang, W.; Yang, K.; Qin, J.; Min, M. Development of a 50-year daily surface solar radiation dataset over China. *Sci. China-Earth Sci.* **2013**, *56*, 1555–1565. [CrossRef]
47. Pubu, C.; Baima, Y.; Laba, Z.; Luosang, Q.; Zheng, W. Lake Area Variation of Tharinamtso during 2002 to 2017. *Plateau Mt. Meteorol. Res.* **2018**, *4*, 67–70. (In Chinese)
48. Wang, M.; Hou, J.; Lei, Y. Classification of Tibetan lakes based on variations in seasonal lake water temperature. *Chin. Sci. Bull.* **2014**, *59*, 4847–4855. [CrossRef]
49. Lei, Y.; Yao, T.; Yang, K.; Bird, B.W.; Tian, L.; Zhang, X.; Wang, W.; Xiang, Y.; Dai, Y.; Lazhu; et al. An integrated investigation of lake storage and water level changes in the Paiku Co basin, central Himalayas. *J. Hydrol.* **2018**, *562*, 599–608. [CrossRef]
50. Wang, J.; Huang, L.; Ju, J.; Daut, G.; Wang, Y.; Ma, Q.; Zhu, L.; Haberzettl, T.; Baade, J.; Mäusbacher, R. Spatial and temporal variations in water temperature in a high-altitude deep dimictic mountain lake (Nam Co), central Tibetan Plateau. *J. Great Lakes Res.* **2019**, *45*, 212–223. [CrossRef]
51. Sharma, S.; Gray, D.K.; Read, J.S.; O'Reilly, C.M.; Schneider, P.; Qudrat, A.; Gries, C.; Stefanoff, S.; Hampton, S.E.; Hook, S.; et al. A global database of lake surface temperatures collected by in situ and satellite methods from 1985–2009. *Sci. Data* **2015**, *2*, 150008. [CrossRef] [PubMed]

52. Minnett, P.J. Consequences of sea surface temperature variability on the validation and applications of satellite measurements. *J. Geophys. Res. Oceans* **1991**, *96*, 18475–18489. [CrossRef]
53. Oesch, D.C.; Jaquet, J.M.; Hauser, A.; Wunderle, S. Lake surface water temperature retrieval using advanced very high resolution radiometer and Moderate Resolution Imaging Spectroradiometer data: Validation and feasibility study. *J. Geophys. Res. Oceans* **2005**, *110*, C12014. [CrossRef]
54. Oesch, D.; Jaquet, J.M.; Klaus, R.; Schenker, P. Multi-scale thermal pattern monitoring of a large lake (Lake Geneva) using a multi-sensor approach. *Int. J. Remote Sens.* **2008**, *29*, 5785–5808. [CrossRef]
55. Cai, Y.; Ke, C.; Li, X.; Zhang, G.; Duan, Z.; Lee, H. Variations of lake ice phenology on the Tibetan Plateau from 2001 to 2017 based on MODIS data. *J. Geophys. Res. Atmos.* **2019**, *124*, 825–843. [CrossRef]
56. Kropácek, J.; Maussion, F.; Chen, F.; Hoerz, S.; Hochschild, V. Analysis of ice phenology of lakes on the Tibetan Plateau from MODIS data. *Cryosphere* **2013**, *7*, 287. [CrossRef]
57. Wan, Z. *Collection-6 MODIS Land Surface Temperature Products Users' Guide*; ICESS, University of California: Santa Barbara, CA, USA, 2013.
58. Savtchenko, A.; Ouzounov, D.; Ahmad, S.; Acker, J.; Leptoukh, G.; Koziana, J.; Nickless, D. Terra and Aqua MODIS products available from NASA GES DAAC. *Adv. Space Res.* **2004**, *34*, 710–714. [CrossRef]
59. Wan, Z.; Zhang, Y.; Zhang, Q.; Li, Z. Quality assessment and validation of the MODIS global land surface temperature. *Int. J. Remote Sens.* **2004**, *25*, 261–274. [CrossRef]
60. Robinson, I.S.; Wells, N.C.; Charnock, H. The sea surface thermal boundary layer and its relevance to the measurement of sea surface temperature by airborne and spaceborne radiometers. *Int. J. Remote Sens.* **1984**, *5*, 19–45. [CrossRef]
61. Schluessel, P.; Emery, W.J.; Grassl, H.; Mammen, T. On the bulk-skin temperature difference and its impact on satellite remote sensing of sea surface temperature. *J. Geophys. Res. Oceans* **1990**, *95*, 13341–13356. [CrossRef]
62. Donlon, C.J.; Nightingale, T.J.; Sheasby, T.; Turner, J.; Robinson, I.S.; Emergy, W.J. Implications of the oceanic thermal skin temperature deviation at high wind speed. *Geophys. Res. Lett.* **1999**, *26*, 2505–2508. [CrossRef]
63. Donlon, C.J.; Minnett, P.J.; Gentemann, C.; Nightingale, T.J.; Barton, I.J.; Ward, B.; Murray, M.J. Toward improved validation of satellite sea surface skin temperature measurements for climate research. *J. Clim.* **2002**, *15*, 353–369. [CrossRef]
64. Wilson, R.C.; Hook, S.J.; Schneider, P.; Schladow, S.G. Skin and bulk temperature difference at Lake Tahoe: A case study on lake skin effect. *J. Geophys. Res. Atmos.* **2013**, *118*, 10–332. [CrossRef]
65. Hook, S.J.; Prata, F.J.; Alley, R.E.; Abtahi, A.; Richards, R.C.; Schladow, S.G.; Pálmarsson, S. Retrieval of lake bulk and skin temperatures using Along-Track Scanning Radiometer (ATSR-2) data: A case study using Lake Tahoe, California. *J. Atmos. Ocean. Technol.* **2003**, *20*, 534–548. [CrossRef]
66. Schneider, P.; Hook, S.J.; Radocinski, R.G.; Corlett, G.K.; Hulley, G.C.; Schladow, S.G.; Steissberg, T.E. Satellite observations indicate rapid warming trend for lakes in California and Nevada. *Geophys. Res. Lett.* **2009**, *36*, L22402. [CrossRef]
67. Schneider, P.; Hook, S.J. Space observations of inland water bodies show rapid surface warming since 1985. *Geophys. Res. Lett.* **2010**, *37*, L22405. [CrossRef]
68. Lazhu; Yang, K.; Hou, J.; Wang, J.; Lei, Y.; Zhu, L.; Chen, Y.; Wang, M.; He, X. A new finding on the prevalence of rapid water warming during lake ice melting on the Tibetan Plateau. *Sci. Bull.* **2021**, *66*, 2358–2361. [CrossRef]
69. Li, Z.; Lyu, S.; Ao, Y.; Wen, L.; Zhao, L.; Wang, S. Long-term energy flux and radiation balance observations over Lake Ngoring, Tibetan Plateau. *Atmos. Res.* **2015**, *155*, 13–25. [CrossRef]
70. Wang, B.; Ma, Y.; Ma, W.; Su, Z. Physical controls on half-hourly, daily, and monthly turbulent flux and energy budget over a high-altitude small lake on the Tibetan Plateau. *J. Geophys. Res. Atmos.* **2017**, *122*, 2289–2303. [CrossRef]
71. Li, X.; Ma, Y.; Huang, Y.; Hu, X.; Wu, X.; Wang, P.; Li, G.Y.; Zhang, S.Y.; Wu, H.W.; Jiang, Z.Y.; et al. Evaporation and surface energy budget over the largest high-altitude saline lake on the Qinghai-Tibet Plateau. *J. Geophys. Res. Atmos.* **2016**, *121*, 10–470. [CrossRef]
72. Tang, W.; Yang, K.; Qin, J.; Li, X.; Niu, X. A 16-year dataset (2000–2015) of high-resolution (3 h, 10 km) global surface solar radiation. *Earth Syst. Sci. Data* **2019**, *11*, 1905–1915. [CrossRef]
73. Li, Z.; Lyu, S.; Wen, L.; Zhao, L.; Ao, Y.; Wang, S. Effect of a cold, dry air incursion on atmospheric boundary layer processes over a high-altitude lake in the Tibetan Plateau. *Atmos. Res.* **2017**, *185*, 32–43. [CrossRef]
74. Liu, C.; Zhu, L.; Li, J.; Wang, J.; Ju, J.; Qiao, B.; Ma, Q.; Wang, S. The increasing water clarity of Tibetan lakes over last 20 years according to MODIS data. *Remote Sens. Environ.* **2021**, *253*, 112199. [CrossRef]
75. Liu, C.; Zhu, L.; Wang, J.; Ju, J.; Ma, Q.; Qiao, B.; Wang, Y.; Xu, T.; Chen, H.; Kou, Q.Q.; et al. In-situ water quality investigation of the lakes on the Tibetan Plateau. *Sci. Bull.* **2021**, *66*, 1727–1730. [CrossRef]
76. Lei, Y.; Yao, T.; Yang, K.; Lazhu; Ma, Y.; Bird, B.W. Contrasting hydrological and thermal intensities determine seasonal lake-level variations—A case study at Paiku Co on the southern Tibetan Plateau. *Hydrol. Earth Syst. Sci.* **2021**, *25*, 3163–3177. [CrossRef]

remote sensing

MDPI

Article

Applicability Assessment of Coherent Doppler Wind LiDAR for Monitoring during Dusty Weather at the Northern Edge of the Tibetan Plateau

Meiqi Song [1,2,3], Yu Wang [1,2,3], Ali Mamtimin [1,2,3,4,5,*], Jiacheng Gao [1,2,4], Ailiyaer Aihaiti [1,2,3], Chenglong Zhou [1,2,4,5], Fan Yang [1,2,4,5], Wen Huo [1,2,4,5], Cong Wen [1,2,3] and Bo Wang [6]

1 Institute of Desert Meteorology, China Meteorological Administration, Urumqi 830002, China
2 National Observation and Research Station of Desert Meteorology, Taklimakan Desert of Xinjiang, Urumqi 830002, China
3 Taklimakan Desert Meteorology Field Experiment Station of China Meteorological Administration, Urumqi 830002, China
4 Xinjiang Key Laboratory of Desert Meteorology and Sandstorm, Urumqi 830002, China
5 Key Laboratory of Tree-Ring Physical and Chemical Research, China Meteorological Administration, Urumqi 830002, China
6 Xinjiang Meteorological Service Center, Urumqi 830002, China
* Correspondence: ali@idm.cn

Abstract: Wind profile light detection and ranging (LiDAR) is an important tool for observing features within the atmospheric boundary layer. Observations of the wind field and boundary layer height from coherent Doppler wind LiDARs (CDWLs) under sandy and dusty weather conditions were evaluated using observations from two CDWLs and one GTS radio sounding located at the northern edge of the Tibetan plateau from 1 May to 30 August 2021. The results showed that CDWL has good applicability in reproducing wind fields in dust, precipitation, and in clear-sky conditions, and that it is superior to the v wind field for real measurements of the u wind fields. In terms of the planetary boundary layer height (PBLH), the validity of the inversion of PBLH in dusty weather was higher than that under clear-sky conditions. It was found that the PBLH retrieved by the CDWL at 20:00 (BJT) was better than that at 08:00 (BJT). The diurnal variation amplitude of the PBLH before the occurrence of a sandstorm was larger than the diurnal variation amplitude of the PBLH occurring during a sandstorm.

Keywords: coherent doppler wind LiDAR; northern edge of Tibetan plateau; dusty weather; monitoring application assessment

Citation: Song, M.; Wang, Y.; Mamtimin, A.; Gao, J.; Aihaiti, A.; Zhou, C.; Yang, F.; Huo, W.; Wen, C.; Wang, B. Applicability Assessment of Coherent Doppler Wind LiDAR for Monitoring during Dusty Weather at the Northern Edge of the Tibetan Plateau. *Remote Sens.* **2022**, *14*, 5264. https://doi.org/10.3390/rs14205264

Academic Editors: Massimo Menenti, Yaoming Ma, Li Jia and Lei Zhong

Received: 29 August 2022
Accepted: 19 October 2022
Published: 21 October 2022

1. Introduction

Sandy weather has a significant impact on arid regions, with the immediate effect of causing air pollution and mesoscale to large-scale climate change [1–3]. It was discovered that dust storms can affect the heat balance of planetary radiation, which in turn leads to climate change [4].

In recent years, scientists have conducted numerous studies on dust cycles, dust properties, and the environmental effects of dust using ground-based light detection and ranging (LiDAR) information such as the optical intensity, backscatter intensity, depolarization ratio, extinction coefficient, and dual-wavelength signal ratio [5–7]. Laser wind measurement techniques have developed rapidly during the past decades and are mostly used in wind field observation [8], aircraft wake measurements [9], turbulence measurements [10], cloud and atmospheric boundary layer characterization [11], and atmospheric aerosol optical characterization [12]. The fluctuation of boundary layer height with time and the effect of the entrainment layer and vertical wind speed on the boundary layer height were found using micro–pulse LiDAR observations [13]. The boundary layer height retrieved by direct detection LiDAR and coherent Doppler wind LiDAR (CDWL) were correlated with PM2.5 to study a precipitation event [14]. Ground–based and air–based LiDARs combined with ground–based aerosol mass concentrations were used to analyze the optical and physical properties of a dust process [15]. The characteristics of the dust aerosol backscattering coefficient, extinction coefficient, and depolarization ratio have been gradually studied by using polarization LiDAR to monitor pollution and dust storm weather in cities [16–19]. The direct observation of seasonal dust weather can be used to study the frequency and intensity of dust [20] and effectively evaluate the effect of air pollution control [21]. LiDAR was used to analyze the aerosol extinction coefficients for the inversion of dusty weather processes and to obtain correlations between the extinction coefficients and ground-level PM10 concentrations [22]. During the observation of the atmospheric boundary layer of an urban area using 3D scanning coherent Doppler LiDAR, multiple dust-devil-like vortices were detected in the area, and the temporal evolution of the precise 3D structure and vortex intensity was observed [23]. Detection means such as ground-based radar combined with satellite remote-sensing LiDAR can be used to analyze the transport characteristics and optical properties of sand and dust [24]. More novel experiments have been used in the past to quantify changes in aerosol transport and aerosol properties from the Sahara Desert to the Caribbean Sea by means of airborne coherent Doppler wind LiDAR experiments [25]. Observations of dust events in Iceland have confirmed the possibility of using Doppler wind LiDAR to monitor volcanic and sedimentary aerosols [26]. The simultaneous 3D monitoring of wind and pollution is performed using coherent Doppler wind LiDAR, which then generates a high-resolution wind field to track local air pollution sources and their dispersal, as well as to analyze transboundary air pollution events [27].

Previously, there have been no relevant observations and studies on the long duration and continuity of dust storms and floating dust weather at the northern edge of the Tibetan plateau using the coherent Doppler wind LiDAR (CDWL). The vertical evolution pattern of dust aerosol concentration and meteorological elements during the maintenance of dusty weather is not clear. This project proposes the use of CDWL installed at the Minfeng and Yeyik stations on the northern side of the Tibetan plateau region, combined with GTS ratio soundings, to conduct a study on the evolution of the atmospheric boundary layer before and after sand and dust storms and during persistent dusty weather. The objective was to evaluate the applicability of CDWL in the observation of boundary layer elements under sandy and dusty weather to provide new observational support for the development of currently stagnant sand and dust studies. This study was conducted as a basis for the quantitative assessment of the contribution of dusty weather to regional atmospheric dust aerosols and its impact on regional and global changes.

This paper is organized as follows: The site and data resources are described in Section 2. Section 3 provides a comparative analysis of the wind field data observed by CDWL and GTS soundings and evaluates the performance of CDWL wind field observations. Section 4 compares the effect of CDWL on the planetary boundary layer height (PBLH) inversion under different weather conditions. Finally, a discussion and conclusions are provided in Sections 5 and 6. If not specified, Beijing time (BJT) is used in this paper.

2. Materials and Methods

2.1. Data and Information

In this paper, ground-based coherent Doppler wind LiDAR data were acquired from 1 May to 31 August 2021 at Minfeng station, hereafter referred to as MF (82°43′E, 37°04′N, 1410.7 m) and Yeyik station, hereafter referred to as YYK (83°10′E, 36°42′N, 2499.0 m) on the northern edge of the Tibetan plateau (Figure 1). CDWL was obtained using the Wind3D 6000 produced by the Ocean University of China (OUC) and Qingdao Leice Transient Technology Co., Ltd., Qingdao, China. [28].

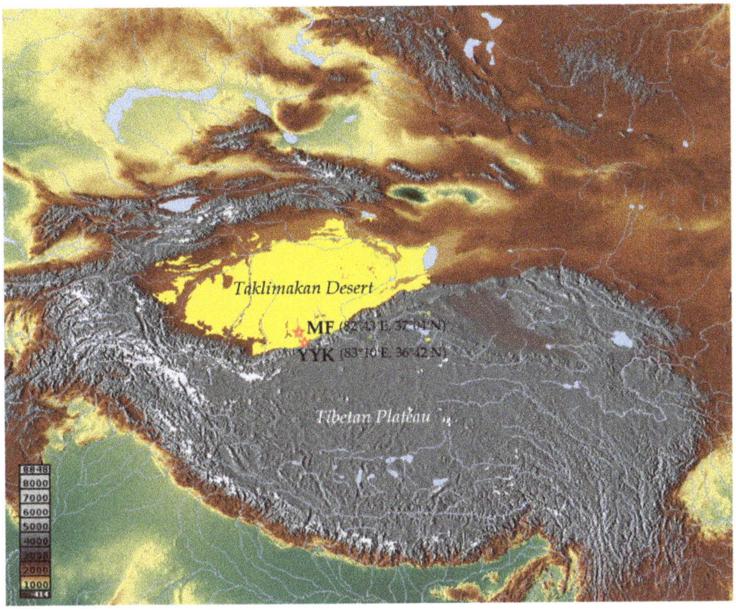

Figure 1. Location map of field experiment station at northern edge of Tibetan plateau.

CDWL performs continuous observations, 24 h a day, with an average of 17 s for one set of observations, and the 10 min average data were used for the analysis in this paper. The maximum detection height of CDWL is around 6 km, the minimum detection height is around 45–100 m, the vertical resolution is 10–30 m, the wind speed measurement accuracy is below 0.1 m/s, and the wind direction measurement accuracy is below 3° [29]. CDWL can detect the horizontal wind direction, wind speed, vertical velocity, atmospheric refractive index structure constant, and atmospheric extinction coefficient, and can obtain more reliable information on the wind field and turbulence intensity in the boundary layer through the electromagnetic wave signals emitted and received back by CDWL. Level 0 is the radial wind speed data, Level 1 is the second level wind profile data, Level 2 is the 10 min average wind profile data, and Level 3 includes the PBLH and cloud height data products. The vertical height of the CDWL observation at the MF station is 51–5017 m with a vertical resolution of 26 m.

The GTS sounding observation data were taken from the Minfeng County Meteorological Bureau (82°43′E, 37°04′N), and the sounding instrument used was a GTS13 digital sounding instrument, which received data through the GFE (L)1 secondary wind radar. In order to facilitate the comparative analysis of the two types of data, the same vertical resolution of 26 m was chosen to match that of the CDWL in MF. The CDWL was selected based on the set of samples with the closest time to the GTS sounding release.

2.2. Research Methods

2.2.1. DBS Wind Field Inversion Method for CDWL

CWDL uses the Doppler effect of light to measure wind. This means that, when the laser propagates in the atmosphere, particles such as aerosols cause the laser to scatter, and the movement of the particles such as aerosols causes the received scattered echo to produce a differential frequency signal, i.e., the Doppler shift. The Doppler shift can be inverted by Formula (1) to show the component of aerosol motion in the direction of the laser beam, i.e., radial velocity or radial wind speed [28].

$$\Delta f_D = \frac{2V_{los}}{\lambda} \tag{1}$$

where Δf_D is the Doppler frequency shift, V_{los} is the radial velocity, and λ is the wavelength of the emitted laser.

The Doppler Beam Swinging (DBS) method is a five-beam method (DBS-5) for the inversion of wind profiles, which requires radial data in five directions—east, west, south, north, and vertical—which can be used to invert horizontal and vertical wind fields [29].

$$\begin{cases} u = \frac{V_N - V_S}{2\cos\theta} \\ v = \frac{V_E - V_W}{2\cos\theta} \\ V_h = \sqrt{u^2 + v^2} \\ w = V_z \end{cases} \tag{2}$$

The wind direction is $\alpha = \left| tan^{-1}\frac{u}{v} \right|$

2.2.2. PBLH Inversion Method for CDWL and GTS Soundings

The inversion of PBLH for CDWL uses a wavelet covariance transform based on the Haar function, which is as follows [14,30]:

$$h\left(\frac{z-b}{a}\right) = \begin{cases} -1,\ b - \frac{a}{2} \le z < b \\ 1,\ b \le z < b + \frac{a}{2} \\ 0,\ \text{other} \end{cases} \tag{3}$$

where z is the height, b is the location where the function is located, and a is the spatial extent or calculation step, which can be formulated according to local characteristics. The covariance function W_f of the Haar function is also defined and denoted as:

$$W_f(a,b) = \frac{1}{a}\int_{z_b}^{z_t} f(z)h\left(\frac{z-b}{a}\right)dz \tag{4}$$

In this paper, the vertical resolution of CDWL Δz is 26 m, and a is 10 times the minimum height difference, which is 260 m. The lowest altitude observed by CDWL z_{min} is 51 m. Therefore, the theoretical CDWL inversion of the PBLH is no less than $z_{min} + a/2$, which is 181 m.

The determination of the PBLH in this study using GTS soundings is a method of determining the height of the boundary layer through thermal differences in the vertical direction. The specific method is the potential temperature gradient method used by Liu and Liang [31].

2.2.3. The k-Means Clustering Analysis

The k-means algorithm is a commonly used clustering method. This study used k-means clustering to classify the PBLH of CDWL and GTS sounding inversions and the difference between them. The aim was to analyze what weather type and other characteristics were present in clusters where the PBLH of the CDWL inversion was close to that of the soundings' inversion. This facilitates an understanding of the conditions under which the CDWL inversion of the PBLH is applicable.

The k-means algorithm works by dividing n sample points into k clusters, with sample points within each cluster having a high degree of similarity and sample points between clusters having a low degree of similarity; moreover, the similarity is calculated based on the average of the sample points in a cluster.

3. Comparison of Wind Field Observations

The samples were analyzed and compared based on the results of u and v wind fields as well as on the wind direction and speed data recorded by CDWL and GTS sounding data at MF from May to August 2021.

3.1. CDWL and GTS Soundings' Observation Wind Field Assessment

Based on the hour-by-hour weather phenomenon records from the MF station, all available samples were classified into three categories: dusty weather (dust storms, blowing sand, and floating dust), precipitation, and a clear sky. All available samples were used for analysis. The CDWL and the horizontal wind field observations at the effective height of the soundings were compared for each of the three weather phenomena. The scatter plot is shown in Figure 2. There was a greater proportion of dusty weather and clear-sky days, with fewer precipitation weather samples. In terms of the linear fit results, the linear coefficients of determination R^2 for wind speed during dusty weather, precipitation, and clear skies were 0.9361, 0.9715, and 0.8747, respectively, and the coefficients of determination R^2 for wind direction were 0.9937, 0.9937, and 0.9592, respectively, all of which passed the significance test ($p < 0.01$, results omitted). The root mean square error (RMSE) of u and v recorded by CDWL and GTS sounding (Table 1) were approximately 2.0 m/s, while the wind speed and direction were less than 1.5 m/s and 40.0°, respectively. The above results are similar to those of previous studies [8,32–34]. The comparison results for the wind direction were also satisfactory and better than previous studies [35].

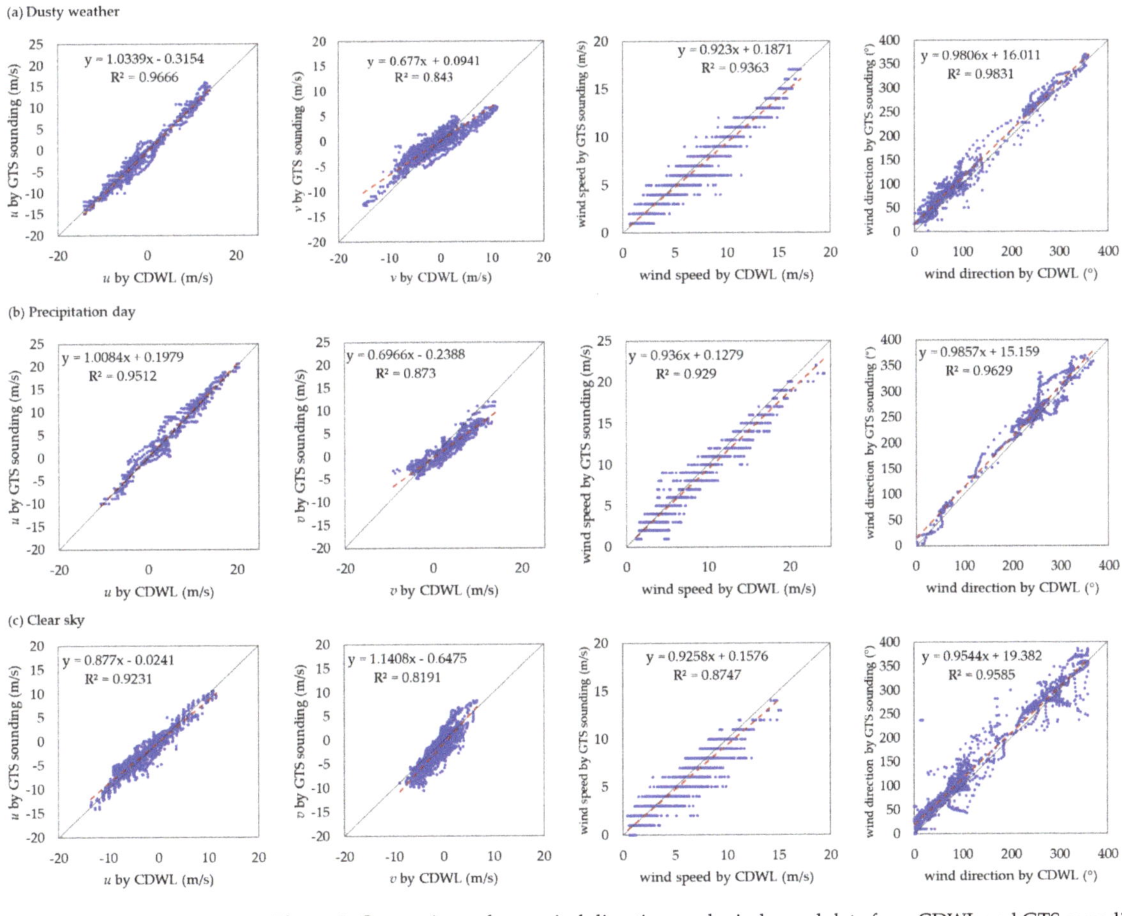

Figure 2. Comparison of u, v, wind direction, and wind speed data from CDWL and GTS sounding samples during (**a**) dusty weather, (**b**) precipitation, and (**c**) a clear sky in MF in August 2021. The regression line (red line) and reference line y = x (black line) are also shown.

Table 1. The root mean square error (RMSE) of each element observed by CDWL and GTS sounding for different weather types.

	u (m/s)	v (m/s)	Wind Speed (m/s)	Wind Direction (°)
Dusty weather	1.18	1.79	1.00	15.04
Precipitation	1.53	1.90	1.05	13.85
Clear sky	1.34	1.97	1.46	21.09

The atmosphere in desert areas is dry and, when there is no precipitation or dusty weather, the aerosol content in the atmosphere is low. When the air is too "clean", the signal returned to the LiDAR is weakened and the signal-to-noise ratio of the LiDAR observation is reduced, which affects the accuracy of the observation to a certain extent. This is also a result of the fact that observations during dusty weather and precipitation are better than when there are clear sky events.

A further comparative analysis of the horizontal wind fields at various heights in the vertical direction at MF was carried out (Figure 3). Below, albeit close to 2300 m, the correlation between the u and v wind fields observed by both instruments passed the significance test ($p < 0.01$). The overall correlation of u was significantly better than the correlation of v. Due to the special nature of the soundings' balloon observations, the balloon will drift with the wind field, and the higher the wind speed and height, the further the balloon will drift. This results in some differences between the balloon observations and the horizontal wind field over CDWL at MF. This also causes a difference in the correlation between u and v.

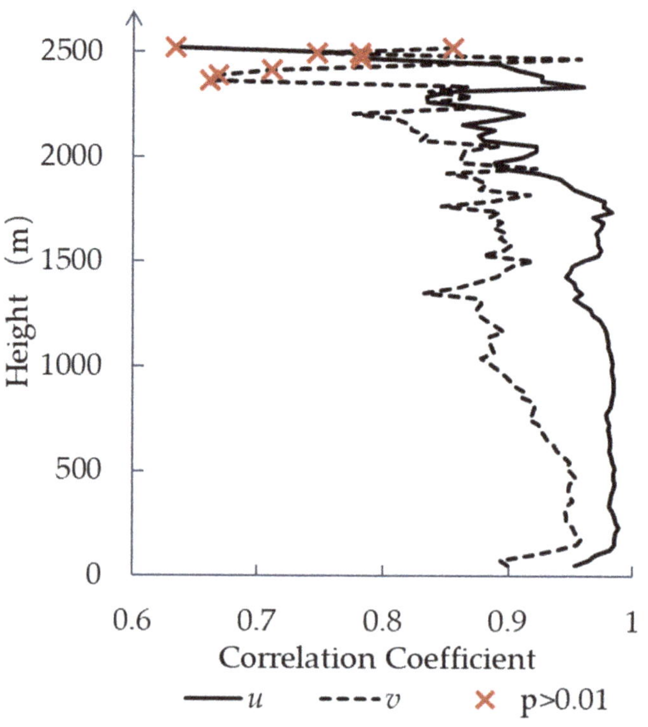

Figure 3. Correlation curves of the CDWL observed u and v wind fields with the GTS sounding observations ($p > 0.01$ does not pass significance test).

3.2. Quantile Distribution and Probability Density Analysis of Horizontal Wind Fields from CDWL and GTS Sounding Observations

The quantile–quantile plot (hereafter referred to as QQ plot) visualizes whether the distribution of the two variables is skewed or not. The probability density distribution plot allows a direct observation of the number of samples distributed in different intervals. Continuing the bias and probability distribution analysis of the horizontal wind field in MF, the QQ plots of the u and v wind fields show (Figure 4, left) that the two instrumental observations are the distribution of u around the 1:1 line and that the probability density distribution is consistent. Moreover, v shows a certain angle to the baseline and the sounding's observation; the sounding's observation was larger than the LiDAR observation when the wind was southerly, and the sounding's observation was smaller than the LiDAR observation when the wind was northerly. The probability density was more concentrated around 0 m/s.

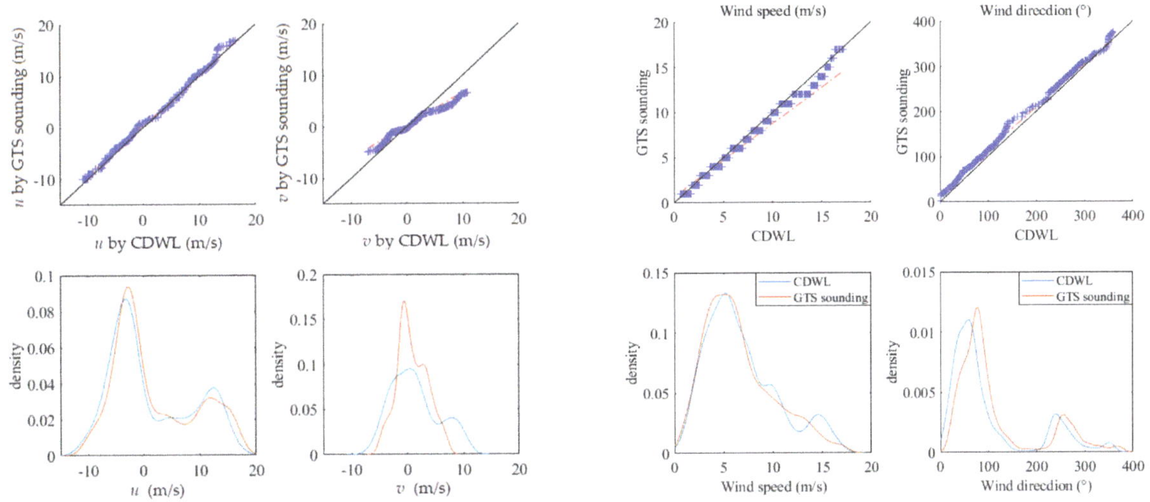

Figure 4. Quantile-quantile plot and probability density distribution of u and v wind fields and wind direction and speed from CDWL and GTS soundings' observations. The regression line (red line) and reference line y = x (black line) are also shown.

The QQ plots of wind direction and wind speed show (Figure 4 right) that the distributions of wind speed and wind direction observed by both instruments coincide more perfectly around the 1:1 line. The CDWL wind direction probability density distribution is, overall, around 10° smaller than the soundings' observations. Therefore, errors were calculated for the wind directions of −9.69°, −9.96°, and −14.44° for CDWL and GTS soundings during dusty weather, precipitation, and clear skies, respectively. This error occurs because we cannot avoid the fact that the horizontal position of the GTS sounding changes with the increase in altitude in accordance with the wind field. This problem has also existed in previous studies [34].

4. Evaluation of PBLH Inversion Results

4.1. Analysis of PBLH Clustering Results

PBLH from soundings data via temperature contour inversions were used as the basis for clustering, based on the PBLH from CDWL and GTS sounding inversions from May to August 2021 and the difference between the two. The samples were classified into three and four types, respectively (Figure 5). The QQ plot of the classification results show that the distribution of the PBLH results from the CDWL inversions of Type 1 and Type 2 is closer to that of the sounding inversions when the first classification case is divided into three types. Type 3 shows that the PBLH of CDWL are generally smaller than GTS sounding inversions. The true atmospheric boundary layer heights of the GTS sounding inversions are in the range of 2000–6000 m, with a concentration of around 4000 m, which exceeds the effective detection height of CDWL.

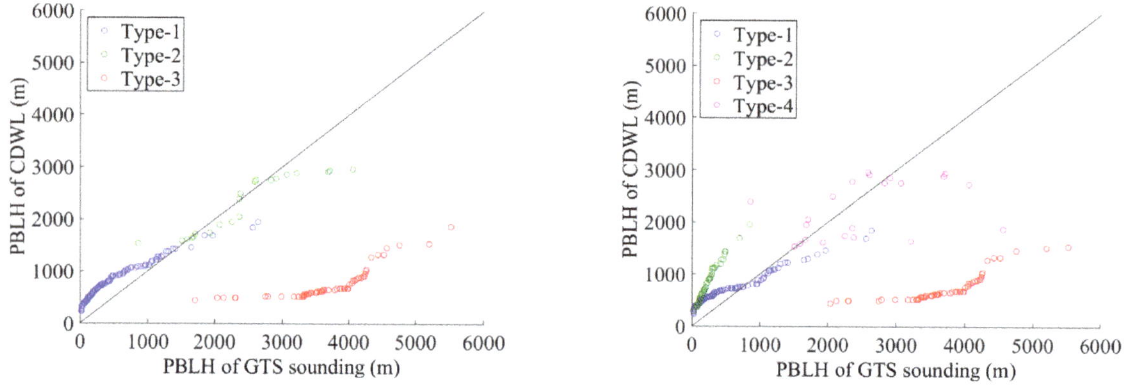

Figure 5. Quantile–quantile plot of the inversion of the PBLH of CDWL and GTS soundings after k-means clustering into three (**left**) and four (**right**) types.

The statistical results of each sample size for the first classification are shown in Table 2. Type 1 and Type 2 had a higher percentage of sample sizes where dusty weather and precipitation occurred, with 54.77% and 47.62% of the categories, respectively. In contrast, Type 3 had the largest proportion of results with clear-sky days, at 70.49%. PBLH in Type 1 was mainly concentrated below 2000 m, and most of the results were recorded at 8:00 am, accounting for 77.27% of the results of Type 1. The overall height of Type 2 was between 2000 and 4000 m, with a higher percentage, 88.89%, at 20:00. Almost all of the results for Type 3 were inversions from 20:00.

Table 2. Weather conditions and percentage of observed moments when PBLH inversion results of CDWL and GTS soundings were divided into three types.

Type	1	2	3
Sample Size	199	21	61
Dust	109 (54.77%)	10 (47.62%)	12 (19.67%)
Precipitation	29 (14.57%)	5 (23.81%)	6 (9.84%)
Clear Sky	61 (30.65%)	6 (28.57%)	43 (70.49%)
Sample Size	154	18	54
8:00	199 (77.27%)	2 (11.11%)	1 (1.85%)
20:00	35 (22.73%)	16 (88.89%)	53 (98.15%)

The PBLH of Type 1 was low overall, with a high proportion of cases occurring at 8:00 and during dusty weather. The PBLH of Type 2 was higher, mainly concentrated in the afternoon during dusty weather and when PBLH development was more vigorous. Type 3 accounted mainly for when PBLH development exceeded the effective height for CDWL detection, and this occurred more often in the late afternoon on clear-sky days.

The results of the second classification case in four types are similar to but different from the case in the three types. The QQ plot of the results of the four classification types shows that Type 1 and Type 4, with a better distribution, are similar to the results of Type 1 and Type 3 when classified into three types. Type 3, with an overall low PBLH from CDWL, is identical to the results of the three types. The difference is that Type 2 of the GTS soundings' inversion of PBLH is also clustered as a whole below 1000 m, and the CDWL inversion is higher than that of the GTS soundings.

Combining the analysis of weather conditions and observation times after classification (Table 3), the characteristics of each type of PBLH were evident in different weather conditions or in the time of the day.

Table 3. PBLH inversion results of CDWL and GTS soundings by weather conditions and time of the day when divided into four types.

Type	1	2	3	4
Sample Size	130	70	59	22
Dust	89 (68.46%)	21 (30.00%)	10 (16.95%)	13 (59.09%)
Precipitation	18 (13.85%)	11 (15.71%)	6 (10.17%)	5 (22.73%)
Clear Sky	23 (17.69%)	38 (54.29%)	43 (72.88%)	4 (18.18%)
Sample Size	101	54	52	19
8:00	70 (69.31%)	49 (90.74%)	0 (0.00%)	3 (15.79%)
20:00	31 (30.69%)	5 (9.26%)	52 (100.00%)	16 (84.21%)

Combining the statistical results of the two classifications, it can be found that dusty weather often occurs when the PBLH inversions of the two are close to each other. When there is a clear-sky day, the surface can receive more solar shortwave radiation and the atmospheric boundary layer develops more vigorously, usually to heights above 4000 m; however, it is often beyond the effective and ineffective detection height of CDWL. Similarly, the PBLH in desert areas is very low before sunrise—below the minimum height for the PBLH inversion by CDWL. This provides the basis for the subsequent analysis of the PBLH inversion results in this study. For example, the analysis can be carried out for different weather types or within the effective height range.

4.2. Evaluation of PBLH Inversion Results

In order to give a comprehensive assessment of the PBLH inversion by CDWL, the samples in the effective height range were divided into three weather types—dust (D), precipitation (P) and clear skies (S)—based on the analysis of the results of the clustering in Section 3.1, and the PBLHs were compared separately (see Section 4.2.1.). The PBLHs were also compared separately for the two times of the day 8:00 and 20:00 (representing day and night, respectively). The inversions are shown in Section 4.2.2.

4.2.1. Evaluation of PBLH Inversion Results for Different Weather Types

In the QQ plot (Figure 6), which divides the samples in the effective height range into three weather types—dust (D), precipitation (P), and clear skies (S)—the PBLH distribution was closer to the 1:1 baseline for dusty weather, followed by the inversion results for precipitation, and again for clear skies. The probability density distribution graph shows that the PBLH is usually below 1000 m. Below 1000 m, the results of the CDWL inversions were higher than those of the GTS soundings for all three weather types, and the difference was more obvious for clear-sky days.

In order to quantify the effect of the PBLH inversions for the different weather types in the effective height range, a linear fit of the CDWL and GTS sounding inversions of the PBLH under different weather conditions was distributed (Figure 7). It is clear that the PBLH inversions for dust and precipitation were better than those for clear weather. The coefficients of determination R^2 were 0.7666 and 0.6906 for dust and precipitation, respectively, and was 0.4926 for clear skies; the PBLH distributions for all three weather conditions passed the significance test ($p < 0.001$). The values of R^2 are consistent in magnitude compared to past studies [30].

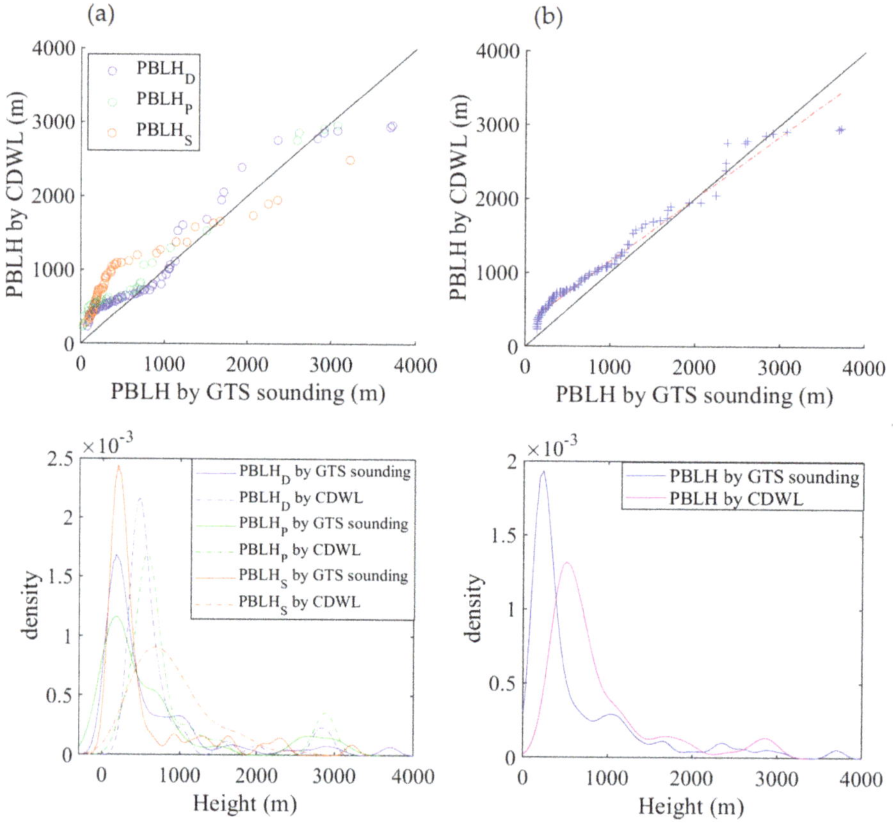

Figure 6. Quantile–quantile plot and probability density distribution of the PBLH inversions of CDWL and GTS soundings by (**a**) weather type and (**b**) overall sample for dust (D), precipitation (P), and clear skies (S).

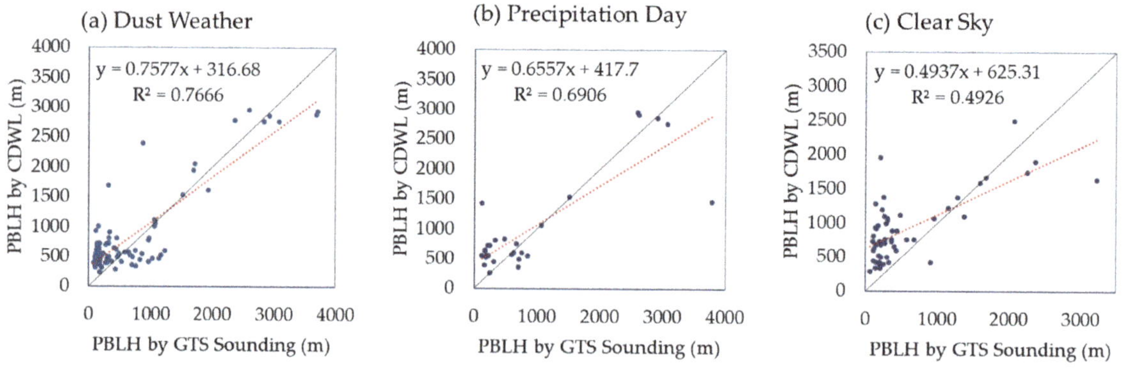

Figure 7. Comparison of CDWL and GTS sounding PBLH inversion results for (**a**) dusty weather, (**b**) precipitation, and (**c**) clear skies in the effective height range. Also shown is the regression line (red line) and reference line y = x (black line).

4.2.2. Evaluation of PBLH Inversion Results for Different Observation Times during Dusty and Precipitation Days

Figure 8 shows the correlation between the CDWL and GTS sounding inversions of the PBLH at 08:00 and 20:00 during dusty and precipitation days. The test samples were screened for the effective height of the CDWL inversion of the PBLH. The time around sunrise is represented by 8:00, when the PBLH is low and the observations within the effective height range are mainly concentrated below 1000 m, with an R^2 of 0.5165, which passes the significance test ($p < 0.01$). The time when the PBLH has developed and accumulated over the day is represented by 20:00. The inversion of the PBLH (Figure 8b) within the effective height range was a better fit with an R^2 of 0.8977, which passes the significance test ($p < 0.001$). The inversion of PBLH by CDWL has been effective in the past [10,30]. The inversion effect of this study on PBLH within the effective inversion height in the desert area is similar to the above study, which also proves the inversion ability of CDWL on the PBLH within the effective height, especially at 20:00.

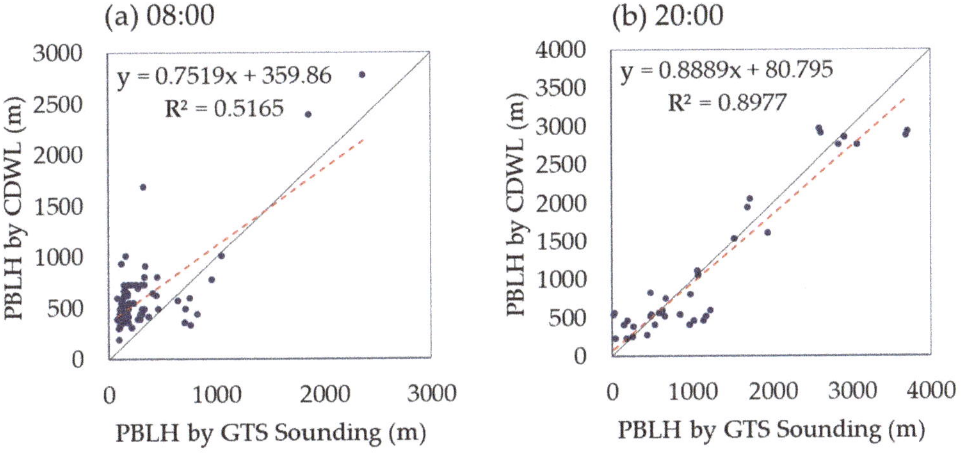

Figure 8. Comparison of CDWL and GTS sounding PBLH inversions at (**a**) 08:00 BJT and (**b**) 20:00 BJT in the effective altitude range. The regression line (red line) and reference line y = x (black line) are also shown.

In combination with the above analysis, we can obtain PBLH results very close to the true value through CDWL when the actual PBLH is within the effective height of the PBLH inversion by CDWL.

5. Discussion

Our comparison and analysis of the accuracy of CDWL observations in this particular region on the northern edge of the Tibetan plateau, where no similar studies have been carried out in the past, show that CDWL is suitable for the accurate observation of wind fields in this region under a variety of weather conditions; moreover, it is also efficient for the acquisition of the PBLH within the effective inversion height.

The detection of wind fields by CDWL is already a relatively mature observation technique. The advantages of this instrument are its large scanning volume, its mobility, and the fact that it provides three-dimensional wind measurements, as well as its relatively high temporal and spatial resolution compared to other measurement devices [36]. In the past, a wind profile radar was used to monitor a variety of weather processes and collect 3D wind field data. Typically, for urban air pollution in Beijing [8] and Binzhou [27], for the monitoring of Typhoon Lekima [32], and for Xiamen on clear and precipitation days [34], it showed good results in terms of wind direction and speed compared with sounding observations. The results of this study, using CDWL for wind field observation in the desert region, are similar to those of the previous studies mentioned above, and both R^2 and RMSE are consistent in magnitude, which confirms that the CDWL used in this study is suitable for wind field detection in this region. The inversion of the PBLH using CDWL data is also a research direction that has been carried out more frequently. The inversion of the PBLH for Hefei city [10,14], Xilingol, Inner Mongolia [30], the seaside city of Xiamen [34], and an ocean island site in China [37] has been effective in the past. The inversion effect of this study on the PBLH within the effective inversion height in the desert area is similar to the above study, which also proves the inversion ability of CDWL on the PBLH within the effective height, especially during dusty weather.

In the meantime, we would like to analyze the inversion of the CDWL on the PBLH by means of an actual example of a dust event. Figure 9 shows a day of four GTS sounding inversions of the PBLH (02:00, 08:00, 14:00, 20:00) and the CDWL continuum inversions of the PBLH, where the full range of samples for the time period is shown, including samples that do not meet the valid inversion height of the CDWL. A correlation analysis for 02:00 and 14:00 was not performed in Section 4.2.2 because the regular GTS soundings only had two moments per day, 08:00 and 20:00, and only in July was an encrypted sounding experiment conducted to add two more moments, 2:00 and 4:00. However, the sample size was still less than 30, which was not enough to complete the correlation analysis. Nevertheless, for this study, 02:00 and 14:00 are also noteworthy moments. The individual case of dusty weather occurred on 17–20 July (Figure 9). The CDWL and GTS sounding inversions of the PBLH for this process showed a continuous elevation of the PBLH during clear-sky days prior to the onset of the dust storm process. It is clear to see that the PBLH inversion results at the time of the dust event are due both before and after the dust event. Moreover, the PBLH inversion results at 02:00 and 14:00 are overall better than 20:00 for PBLH. The heating effect of solar radiation, which is directly absorbed by the surface on clear-sky days when the atmosphere is dry and clean, causes the convective boundary layer to accumulate and thicken. When a dust storm occurs, there is a significant decrease in PBLH during the daytime and an increase in PBLH at night compared to sunny days. The decrease in the difference in the daily variation in the PBLH is due to the fact that the dust aerosols floating in the atmosphere block the solar radiation from reaching the surface and the surface heating atmospheric effect is weakened, which suppresses the elevation of PBLH during the daytime. Furthermore, the relatively stable atmosphere during the maintenance of floating dusty weather allows the residual layer to be maintained at a certain height during the night. When the atmospheric dust aerosols dissipate and the atmosphere returns to a dry and clean state, the daytime PBLH is developed again, and the daily variation in the PBLH varies significantly.

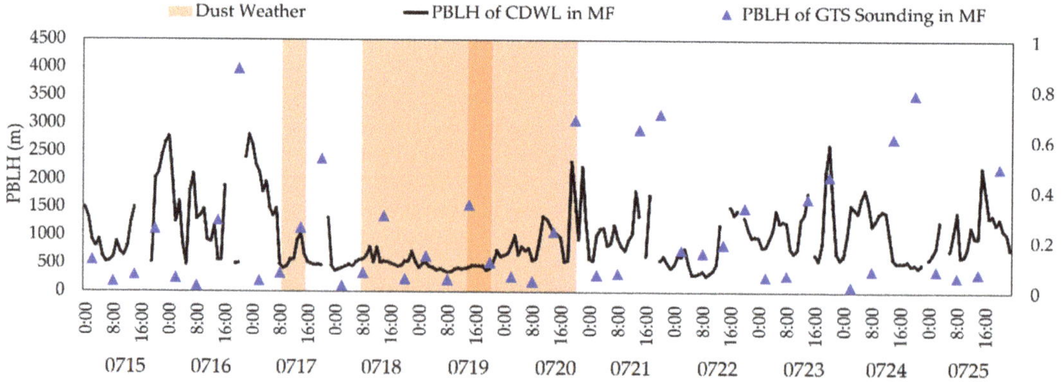

Figure 9. PBLH of dusty weather processes and before and after soundings and the CDWL inversions at the MF station.

The limitation of this study is that, compared with other regions, the effect becomes worse when the effective inversion range of CDWL is exceeded because the PBLH is lower at night and lifts higher during midday hours in desert areas. This is due to the rapid absorption of solar short-wave radiation and the warming of the subsurface in the desert region during clear days; moreover, the atmosphere is subject to surface heating and turbulence, which may make the PBLH develop vigorously in the late afternoon. At night, the PBLH decreases to a lower altitude due to the thin atmospheric aerosol and excessively dry air in the desert area, as well as the rapid cooling of the surface and the large temperature difference formed during the day [38]. This may be the reason why the actual PBLH exceeds the effective inversion height of CDWL. Moreover, the release time of the GTS sounding used for the validation of the PBLH in this study is limited, and, to address this limitation and dilemma, we suggest increasing the number of daily sounding experiments in the next experimental design. In future experiments, to compensate for the shortcomings of the CDWL, the use of microwave wave radar and combined radar technology is proposed, such as, for example, a combination of microwave wave radar and LiDAR. In turn, a more comprehensive and better detection of the area will be carried out.

6. Conclusions

In this paper, ground-based CDWL data from two stations were collected over a 4-month period in 2021. The two stations were the MF station located at Minfeng County and the YYK observation station located at Yeyik Township, Minfeng County, on the northern side of the Tibetan plateau on the southern margin of the Taklimakan Desert. We also cooperated with the GTS sounding observation experiment in the same period, from the Minfeng County National Basic Meteorological Observatory. In order to verify the accuracy of the CDWL observations and analyze the applicability of CDWL monitoring in the Minfeng area, the wind field PBLH observed by CDWL and GTS sounding data were compared and analyzed. The results show that:

(1) CDWL has good applicability for wind field observations in dust storm, floating dust, and clear-sky conditions, and that it is superior to the v wind field for the real measurements of u wind fields;

(2) The results of the cluster analysis for the PBLH inversion show that, when the PBLH inversion results were good, dusty weather accounted for more than the other weather types, at approximately 50–70%, with more samples observed at 08:00. When the PBLH inversion results were not good, clear-sky weather accounted for a higher percentage, with more samples at 20:00. Limited by the effective observation height of CDWL, when the boundary layer developed vigorously during clear-sky days, the PBLH exceeded the highest effective detection height of CDWL. Therefore, this part of the data showed an overall underestimation;

(3) Within the effective inversion height of the PBLH by CDWL, the inversion of the PBLH for dusty weather and precipitation is better than that for sunny days. The inversion of the PBLH at 20:00 is better than that at 08:00, and both passed the significance test.

Author Contributions: Conceptualization, M.S. and A.M.; methodology, M.S. and A.A.; software, M.S. and Y.W.; validation, A.M., A.A. and J.G.; formal analysis, J.G.; investigation, Y.W.; resources, Y.W.; data curation, M.S.; writing—original draft preparation, M.S.; writing—review and editing, M.S.; visualization, M.S.; supervision, A.M. and F.Y.; project administration, C.Z. and W.H.; funding acquisition, C.W. and B.W. All authors have read and agreed to the published version of the manuscript.

Funding: This work was supported by the China Desert Meteorological Research Fund (Sqj2021007); Special funds for basic scientific research business expenses of central-level public welfare scientific research institutes (IDM2021001) and National Natural Science Foundation of China (41875023, 42030612).

Data Availability Statement: The data used in this paper can be provided by A.M. (ail@idm.cn) upon request.

Acknowledgments: The authors would like to acknowledge the GTS sounding data support of Minfeng County Meteorological Bureau, Hotan.

Conflicts of Interest: The authors declare no conflict of interest.

References

1. Qian, W.; Quan, L.; Shi, S. Variations of the Dust Storm in China and its Climatic Control. *J. Clim.* **2002**, *10*, 1216–1229. [CrossRef]
2. Adams, J.W.; Rodriguez, D.; Cox, R.A. The uptake of SO2 on Saharan dust: A flow tube study. *Atmos. Chem. Phys. Discuss.* **2005**, *5*, 2679–2689. [CrossRef]
3. Brunekreef, B.; Forsberg, B. Epidemiological evidence of effects of coarse airborne particles on health. *Eur. Respir. J.* **2005**, *26*, 309–318. [CrossRef]
4. Zakey, A.S.; Solmon, F.; Giorgi, F. Development and testing of a desert dust module in a regional climate model. *Atmos. Chem. Phys.* **2006**, *6*, 4687–4704. [CrossRef]
5. Wang, Y.; Zhuang, G.; Tang, A.; Zhang, W.; Sun, Y.; Wang, Z.; An, Z. The evolution of chemical components of aerosols at five monitoring sites of China during dust storms. *Atmos. Environ.* **2007**, *41*, 1091–1106. [CrossRef]
6. Du, W.; Xin, J.; Wang, M.; Gao, Q.; Wang, Y. Photometric measurements of spring aerosol optical properties in dust and non-dust periods in China. *Atmos. Environ.* **2011**, *42*, 7981–7987. [CrossRef]
7. Zhang, Z.G.; Jiao, M.Y.; Bi, B.G.; Zhao, L.N.; Gao, Q.X.; Guo, J.H.; Ren, Z.H. Analysis of the Heavy Polluting Effects of Sand Dust Weather in Beijing. *Res. Environ. Sci.* **2009**, *22*, 309–314. [CrossRef]
8. Dai, L.; Xin, J.; Zuo, H.; Ma, Y.; Zhang, L.; Wu, X.; Ma, Y.; Jia, D.; Wu, F. Multilevel Validation of Doppler Wind Lidar by the 325 m Meteorological Tower in the Planetary Boundary Layer of Beijing. *Atmosphere* **2020**, *11*, 1051. [CrossRef]
9. Zhai, X.; Wu, S.; Liu, B. Doppler lidar investigation of wind turbine wake characteristics and atmospheric turbulence under different surface roughness. *Opt. Express* **2017**, *25*, A515–A529. [CrossRef] [PubMed]
10. Jiang, P.; Yuan, J.; Wu, K.; Wang, L.; Xia, H. Turbulence Detection in the Atmospheric Boundary Layer Using Coherent Doppler Wind Lidar and Microwave Radiometer. *Remote Sens.* **2022**, *14*, 2951. [CrossRef]
11. Dai, G.; Wu, S.; Song, X.; Liu, L. Optical and Geometrical Properties of Cirrus Clouds over the Tibetan Plateau Measured by LiDAR and Radiosonde Soundings during the Summertime in 2014. *Remote Sens.* **2019**, *11*, 302. [CrossRef]
12. Zhang, Y.; Zheng, Y.; Tan, W.; Guo, P.; Xu, Q.; Chen, S.; Lin, R.; Chen, S.; Chen, H. Two Practical Methods to Retrieve Aerosol Optical Properties from Coherent Doppler Lidar. *Remote Sens.* **2022**, *14*, 2700. [CrossRef]
13. Chen, W.; Kuze, H.; Uchiyama, A.; Suzuki, Y.; Takeuchi, N. One-year observation of urban mixed layer characteristics at Tsukuba, Japan using a micro pulse lidar. *Atmos. Environ.* **2001**, *35*, 4273–4280. [CrossRef]
14. Wang, C.; Jia, M.; Xia, H.; Wu, Y.; Wei, T.; Shang, X.; Yang, C.; Xue, X.; Dou, X. Relationship analysis of PM2.5 and boundary layer height using an aerosol and turbulence detection lidar. *Atmos. Meas. Tech.* **2019**, *12*, 3303–3315. [CrossRef]
15. Kim, S.W.; Yoon, S.C.; Kim, J.; Kang, J.Y.; Sugimoto, N. Asian dust event observed in seoul, korea, during 29-31 may 2008: Analysis of transport and vertical distribution of dust particles from lidar and surface measurements. *Sci. Total Environ.* **2010**, *408*, 1707–1718. [CrossRef] [PubMed]
16. Liu, D.; Qi, F.D.; Jin, C.J.; Yue, G.M.; Zhou, J. Polarization Lidar Observations of Cirrus Clouds and Asian Dust Aerosols over Hefei. *Chin. J. Atmos. Sci.* **2003**, *6*, 1093–1100. [CrossRef]
17. Xie, C.; Nishizawa, T.; Sugimoto, N.; Matsui, I.; Wang, Z. Characteristics of aerosol optical properties in pollution and Asian dust episodes over Beijing, China. *Appl. Opt.* **2008**, *47*, 4945–4951. [CrossRef]
18. Dong, X.H.; Qi, H.; Ren, L.J.; Wang, Y.P.; Di, Y.A.; Sugimoto, N.; Sakmoto, K.; Wang, Q.Y. Application and Data Demonstration of Lidar in Sandstorm Observation. *Res. Environ. Sci.* **2007**, *2*, 106–111. [CrossRef]
19. Guo, B.J.; Liu, L.; Huang, D.P.; Wang, L.L.; Li, X.L. Analysis of Lidar Measurements from a Dust Event. *Meteorol. Mon.* **2008**, *5*, 52–57. [CrossRef]
20. Wang, Y.P.; Chen, Y.; Dong, X.H.; Yin, H.M. Observation of Sand-Dust Weather by Lidar in the Spring of 2005 at Beijing. *Chin. J. Spectrosc. Lab.* **2008**, *25*, 33–35. [CrossRef]
21. Yang, T.; Wang, Z.F.; Zhang, B.; Wang, X.Q.; Wang, W.; Gbaguidi, A.; Gong, Y.B. Evaluation of the effect of air pollution control during the Beijing 2008 Olympic Games using Lidar data. *Chin. Sci. Bull.* **2010**, *55*, 1311–1316. [CrossRef]
22. Huang, T.; Song, Y.; Hu, W.D.; Guo, B.J. Lidar Detection of a Sand-Dust Process in Dalian, Liaoning, China. *J. Desert Res.* **2010**, *30*, 983–988.
23. Fujiwara, C.; Yamashita, K.; Nakanishi, M.; Fujiyoshi, Y. Dust Devil–Like Vortices in an Urban Area Detected by a 3D Scanning Doppler Lidar. *J. Appl. Meteorol. Climatol.* **2011**, *3*, 534–547. [CrossRef]
24. Ma, J.H.; Gu, S.Q.; Chen, M.; Shi, H.; Zhang, G.L. Analysis of a dust case using lidar in Shanghai. *Acta Ecol. Sin.* **2012**, *32*, 1085–1096. [CrossRef]
25. Chouza, F.; Reitebuch, O.; Groß, S.; Rahm, S.; Freudenthaler, V.; Toledano, C.; Weinzierl, B. Retrieval of aerosol backscatter and extinction from airborne coherent Doppler wind lidar measurements. *Atmos. Meas. Tech.* **2015**, *8*, 2909–2926. [CrossRef]
26. Yang, S.; Preißler, J.; Wiegner, M.; von Löwis, S.; Petersen, G.N.; Parks, M.M.; Finger, D.C. Monitoring Dust Events Using Doppler Lidar and Ceilometer in Iceland. *Atmosphere* **2020**, *11*, 1294. [CrossRef]

27. Yuan, J.; Wu, Y.; Shu, Z.; Su, L.; Tang, D.; Yang, Y.; Dong, J.; Yu, S.; Zhang, Z.; Xia, H. Real-Time Synchronous 3-D Detection of Air Pollution and Wind Using a Solo Coherent Doppler Wind Lidar. *Remote Sens.* **2022**, *14*, 2809. [CrossRef]
28. Wang, X.; Dai, G.; Wu, S.; Sun, K.; Song, X.; Chen, W.; Li, R.; Yin, J.; Wang, X. Retrieval and Calculation of Vertical Aerosol Mass Fluxes by a Coherent Doppler Lidar and a Sun Photometer. *Remote Sens.* **2021**, *13*, 3259. [CrossRef]
29. Song, X.; Wang, F.; Yin, J.; Wang, X.; Liu, X.; Ren, H.; Wu, S. Three-Beam Methodology and Observations of Wind Profiling with Doppler Lidar. *Period. Ocean Univ. China* **2020**, *50*, 136–144. [CrossRef]
30. Wang, L.; Wei, Q.; Xia, H.; Wei, T.; Yuan, J.; Jiang, P. Robust Solution for Boundary Layer Height Detections with Coherent Doppler Wind Lidar. *Adv. Atmos. Sci.* **2021**, *38*, 1920–1928. [CrossRef]
31. Liu, S.; Liang, X. Observed diurnal cycle climatology of planetary boundary layer height. *J. Clim.* **2010**, *23*, 5790–5809. [CrossRef]
32. Tang, S.; Guo, Y.; Wang, X.; Tang, J.; Li, T.; Zhao, B.; Zhang, S.; Li, Y. Validation of Doppler Wind Lidar during Super Typhoon Lekima (2019). *Front. Earth Sci.* **2022**, *16*, 75–89. [CrossRef]
33. Lin, X.; Wei, Y.; Chen, H.; Wang, Y. The effect assessment of wind field inversion based on WPR in precipitation. *J. Appl. Meteor. Sci.* **2020**, *31*, 361–372. [CrossRef]
34. Ren, Y.; Chen, S.; Yu, A.; Fan, M. Reliability Analysis of Wind Field and Boundary Layer Height in Xiamen Retrieved by Coherent Doppler Wind Lidar. *J. Arid Meteorol.* **2021**, *39*, 514–523. [CrossRef]
35. Wildmann, N.; Päschke, E.; Roiger, A.; Mallaun, C. Towards improved turbulence estimation with Doppler wind lidar velocity-azimuth display (VAD) scans. *Atmos. Meas. Tech.* **2020**, *13*, 4141–4158. [CrossRef]
36. Liu, Z.; Barlow, J.F.; Chan, P.-W.; Fung, J.C.H.; Li, Y.; Ren, C.; Mak, H.W.L.; Ng, E. A Review of Progress and Applications of Pulsed Doppler Wind LiDARs. *Remote Sens.* **2019**, *11*, 2522. [CrossRef]
37. Chen, Y.; Jin, X.; Weng, N.; Zhu, W.; Liu, Q.; Chen, J. Simultaneous Extraction of Planetary Boundary-Layer Height and Aerosol Optical Properties from Coherent Doppler Wind Lidar. *Sensors* **2022**, *22*, 3412. [CrossRef]
38. Wang, Y.; Sayit, H.; Ali, M.; Zhu, J.; Zhou, C.; Huo, W.; Yang, F.; Yang, X.; Gao, J.; Zhao, X. Evaluation of five planetary boundary layer schemes in WRF over China's largest semi-fixed desert. *Atmos. Res.* **2021**, *256*, 105567. [CrossRef]

remote sensing

Article

Analysis of Drought Characteristics Projections for the Tibetan Plateau Based on the GFDL-ESM2M Climate Model

Yu Liu [1,2], Zhifeng Jia [3,4,*], Xiaoyi Ma [1,2], Yongqiang Wang [1,2], Ronghao Guan [1,2], Zilong Guan [3,5], Yuhui Gu [1,2] and Wei Zhao [6]

1 College of Water Resources and Architectural Engineering, Northwest A&F University, Xianyang 712100, China
2 Key Laboratory of Agricultural Soil and Water Engineering in Arid and Semiarid Areas, Ministry of Education, Northwest A&F University, Xianyang 712100, China
3 School of Water and Environment, Chang'an University, Xi'an 710054, China
4 Key Laboratory of Subsurface Hydrology and Ecological Effects in Arid Region of the Ministry of Education, Chang'an University, Xi'an 710054, China
5 Northwest Engineering Corporation Limited, Power China, Xi'an 710065, China
6 Environmental Engineering Evaluation Center in Tibet Autonomous Region, Lhasa 850000, China
* Correspondence: 409538088@chd.edu.cn; Tel.: +86-15829012186

Citation: Liu, Y.; Jia, Z.; Ma, X.; Wang, Y.; Guan, R.; Guan, Z.; Gu, Y.; Zhao, W. Analysis of Drought Characteristics Projections for the Tibetan Plateau Based on the GFDL-ESM2M Climate Model. *Remote Sens.* **2022**, *14*, 5084. https://doi.org/10.3390/rs14205084

Academic Editors: Massimo Menenti, Yaoming Ma, Li Jia and Lei Zhong

Received: 16 August 2022
Accepted: 8 October 2022
Published: 12 October 2022

Abstract: Under conditions of continuous global warming, research into the future change trends of regional dry-wet climates is key for coping with and adapting to climate change, and is also an important topic in the field of climate change prediction. In this study, daily precipitation and mean temperature datasets under four representative concentrative pathway (RCP) scenarios in the geophysical fluid dynamics laboratory Earth system model with modular ocean model (GFDL-ESM2M) version 4 were used to calculate the standardized precipitation-evapotranspiration index of the Tibetan Plateau (TP) at different time scales. Using a multi-analytical approach including the Mann–Kendall trend test and run theory, the spatiotemporal variation characteristics of drought in the TP from 2016 to 2099 were studied. The results show that the overall future climate of the TP will develop towards warm and humid, and that the monthly-scale wet–dry changes will develop non-uniformly. As the concentration of carbon dioxide emissions increases in the future, the proportion of extremely significant aridification and humidification areas in the TP will significantly increase, and the possibility of extreme disasters will also increase. Moreover, influenced by the increase of annual TP precipitation, the annual scale of future drought in the region will tend to decrease slightly, and the spatial distributions of the frequency and intensity of droughts at all levels will develop uniformly. Under all four RCP scenarios, the drought duration of the TP was mainly less than 3 months, and the drought cycle in the southern region was longer than that in the northern region. The results of this study provide a new basis for the development of adaptive measures for the TP to cope with climate change.

Keywords: GFDL-ESM2M; RCPs; drought characteristics projections; standardized precipitation-evapotranspiration index; Tibetan Plateau

1. Introduction

Drought is one of the most common and widely distributed natural disasters, and is often responsible for serious losses [1]. It often causes problems such as reduced agricultural production and exacerbated ecological deterioration [2,3]. Especially under the influences of global climate change and rapid urbanization, frequent drought disasters seriously restrict the sustainable development of economies and societies [4,5] and lead to ecological and environmental problems, such as water shortages, land degradation, and desertification [6–9]. Over the past 40 years, about 12 million hectares of land have been lost to drought and desertification each year, and this trend will continue to expand in the 21st century, especially in the mid-latitudes [10,11].

According to the Drought Numbers Report, from 1998 to 2017, the global economic loss caused by drought was as high as $124 billion. Since 2000, the number and duration of droughts globally has increased by 29%, affecting about 1.4 billion people. Over the past century, more than 10 million people have died from major drought events, costing the global economy hundreds of billions of dollars, and both numbers are rising [12–14]. Frequent drought disasters have become important in restricting regional sustainable development and ecological protection [15]. In addition, the AR6 Working Group I report released by the United Nations Special Assessment Committee on Climate Change pointed out that in the future, global warming will intensify, and the frequency of extreme high temperature events and marine heat waves will increase accordingly. Additional studies have also shown that future drought duration, frequency, and intensity will increase to varying degrees [16]. Therefore, analyzing the spatiotemporal variation characteristics of future drought is not only beneficial to regional disaster prevention and mitigation, but is also important for rationally managing and distributing regional water resources, as well as for improving regional economic and social development planning.

However, most studies focus on the social, agricultural, and environmental impacts of droughts on a global or regional scale for historical periods [17]. For example, Zhu et al. [18] constructed a dynamic evaluation model and applied it to the distribution and development trend of comprehensive drought disaster risk in Xuzhou, China; Hu et al. [19] used the crop water deficit index for evaluation of agricultural drought, described the spatiotemporal variation of drought in the growth period of winter wheat in the Huang-Huai-Hai plain, and constructed a drought disaster risk index of winter wheat in each growth period; and Orimoloye et al. [20] used the enhanced vegetation index and standardized precipitation index (SPI) to study drought disaster events and their temporal and spatial patterns in Free State Province, South Africa. Overall, there are relatively few studies on the changing characteristics of drought under different climate scenarios in the future, considering the increase of carbon dioxide concentration. In particular, the Tibetan Plateau (TP) region, which is sensitive to climate change and has an active hydrological cycle, presents significant knowledge gaps [21].

In addition, existing studies on meteorological drought mainly focus on arid and semi-arid regions [22]. For example, Annette et al. [23] used the ParFlow-CLM model to study the driving factors of drought-related changes in the southern United States; Wang et al. [24] evaluated the drought monitoring effect of remote sensing precipitation products based on a proposed grid drought index, revealing the drought characteristics of the Yellow River Basin from 1998 to 2016; and Yang et al. [25] constructed a composite drought index to determine the duration, peak, and severity of drought in the Weihe River Basin, to assess multivariate drought risk. However, little is known about the variation characteristics of drought in the TP, which is rich in glacier resources, especially in terms of the variation characteristics of drought at different time scales in the future [26]. TP glaciers are important buffers for regional drought resistance [27]. They can significantly affect the climate model in East Asia and atmospheric circulation in the northern hemisphere through dynamic, thermal, and frictional atmospheric effects; affect the regional and surrounding land-atmosphere interactions and dry–wet changes; and also play a greater role in adapting to climate change pressures [28,29]. In the context of global warming, the temperature of the TP has increased significantly, accelerating the melting of ice and snow and increasing permafrost activities [30,31]. How will the dry and wet conditions of the TP change under global warming in the future? This is a very important scientific question in the field of climate science.

This study is aimed at addressing this question. Considering TP as the study area, according to the climate change prediction results obtained using the geophysical fluid dynamics laboratory Earth system model with modular ocean model (GFDL-ESM2M) version 4, and based on the annual scale standardized precipitation-evapotranspiration index (SPEI), the spatiotemporal variation trends of meteorological drought in the TP under various future emissions scenarios were predicted. In addition, based on the monthly SPEI, the spatiotemporal evolution characteristics of future meteorological droughts in the TP were systematically analyzed with respect to the duration, frequency, intensity, and cycle of drought events, and the characteristics of changes in drought under different climate scenarios were compared and analyzed. The results contribute to a better understanding of the evolution of meteorological drought on the TP and provide a scientific basis for relevant agencies to further improve drought prevention systems and formulate appropriate drought disaster prevention measures and countermeasures.

2. Materials and Methods

2.1. Study Area

The TP, located in southwest China, has a total area of ~2.5 million km^2 (26°00′–39°47′N, 73°19′–104°47′E). It is the largest plateau in China and the highest plateau in the world, sometimes called "the roof of the world". The TP has a mean elevation of more than 4000 m and its 12 major rivers flow to East Asia, Southeast Asia, and South Asia. The major outflow rivers include the Yangtze, Yellow, Lancang, Nujiang, and Yarlung Zangbo rivers, and the lake area of the TP comprises more than 1500 large and small lakes. The more famous lakes are the Nam Co, Qinghai, Qarhan Salt, and Eling lakes (Figure 1a). The TP belongs to a plateau climate zone with distinct dry and wet conditions and frequent nighttime rains. The mean annual precipitation in the southern part of the TP is more than 1500 mm, whereas the mean annual precipitation in the Qaidam Basin (QB) in the northeast is less than 200 mm (Figure 1b). In addition, the TP experiences strong radiation, a high degree of sunshine, low temperatures, and low cumulative temperatures. The temperatures decrease with increasing altitude and latitude. The daily temperature difference is large. The mean annual temperature of the plateau hinterland is below 0 °C, and the mean temperature of the warmest month in large regions is less than 10 °C (Figure 1c).

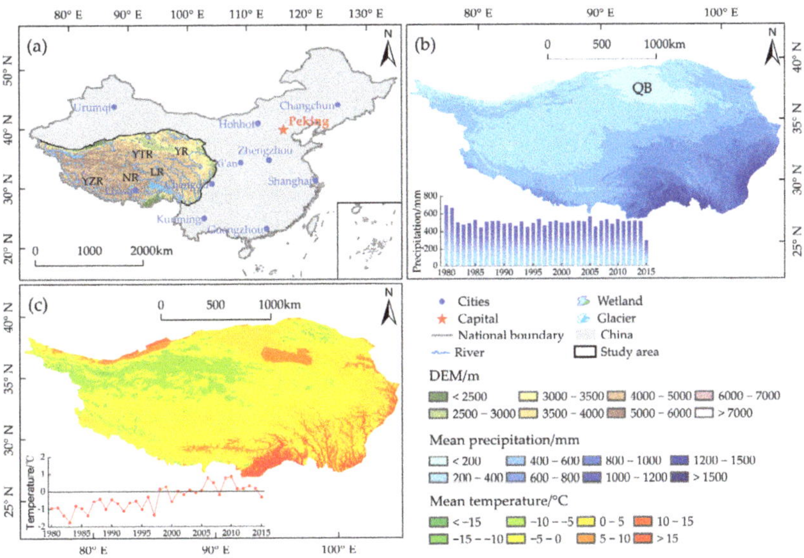

Figure 1. Main environmental characteristics of the TP: (**a**) geographical location, topography and river system of the study area, (**b**) spatiotemporal variation characteristics of precipitation in the study area from 1985 to 2015, (**c**) spatiotemporal variation characteristics of mean temperature in the study area from 1985 to 2015. River name abbreviations: LR-Lancang; NR-Nujiang; YR-Yellow; YTR-Yangtze; YZR-Yarlung Zangbo.

2.2. Data Source and Climate Model

A precipitation and temperature dataset for the TP from 1980 to 2015 was obtained from the Resources and Environment Science and Data Center of the Chinese Academy of Sciences (http://www.resdc.cn/, accessed on 11 November 2021) [32]. The dataset was based on daily observation data from more than 2400 meteorological stations in China and was generated by sorting, calculation, and spatial interpolation processing. In this study, the dataset for precipitation and temperature on the TP from 1980 to 2015 was obtained by mask extraction.

The future climate model selected for this study uses the GFDL-ESM2M. GFDL-ESM2M is a coupled model intercomparison project phase 5 (CMIP5) global coupled carbon-climate model proposed by the US National Oceanic and Atmospheric Administration (NOAA) that is similar to GFDL's previous generation climate model 2.1 (CM2.1) [33]. CM2.1 has exhibited efficient performance in simulating and reproducing global climate interannual variability and climate characteristics [34]. In contrast to the CM2.1 model, the GFDL-ESM2M uses modular ocean model version 4.1, with vertical pressure layers and new biogeochemical algorithms and stoichiometric phytoplankton functional group dynamics. GFDL-ESM2M includes a revised land model to simulate competing vegetation distributions and functions, including carbon cycling among vegetation, soil, and the atmosphere [35]. Therefore, GFDL-ESM2M has a wide range of applications and can be adapted to various climatic conditions around the world. Jia et al. [36] conducted a comprehensive evaluation of GFDL–ESM2M by comparing the performance of 33 CMIP5 general circulation models (GCMs) in a temperature simulation of the TP. Their results showed that GFDL-ESM2M exhibits better temperature simulation performance than other GCMs, which indicates that GFDL-ESM2M is well suited to the simulation of future climate changes in the TP. In addition, China's special program for climate change science and technology development during the 12th five-year plan period has simulated the future water resources situation in China based on the GFDL-ESM2M climate model [37]. Based on that climate model, Ma et al. [38] predicted the future trend of dry–wet area and climate change in China. The study showed that the GFDL-ESM2M climate model has a good application effect in China. Therefore, GFDL-ESM2M was considered in this study to predict future drought changes in the TP.

The GFDL-ESM2M climate model includes the representative concentrative pathways (RCP) 2.6, 4.5, 6.0, and 8.5 climate scenarios. The RCP2.6, RCP4.5, RCP6.0, and RCP8.5 climate scenarios represent very low, low, medium, and high levels of greenhouse gas emissions, respectively, meaning that radiative forcing will stabilize at 2.6, 4.5, 6.0, and 8.5 W/m^2, respectively, by 2100, and the carbon dioxide concentration will reach 490, 650, 850, and 1370 ppm, respectively. In this study, we used daily precipitation and temperature datasets from multiple climate models from the National Tibetan Plateau Data Center (TPDC; http://data.tpdc.ac.cn, accessed on 16 November 2021) [39,40]. In addition, the geographic coordinate system of all datasets was uniformly transformed into GCS_WGS_1984.

2.3. Methods

2.3.1. Standardized Precipitation-Evapotranspiration Index (SPEI)

The meteorological drought index integrates different climatic factors that represent the overall climate and environment of the region [41]. The commonly used meteorological drought indices include the SPI, Palmer drought severity index (PDSI), and SPEI [42]. The SPEI, proposed by Vicente-Serrano et al. [43], is used to characterize meteorological drought. This index inherits the characteristics of the PDSI, considering evapotranspiration to be sensitive to temperature, as well as the advantages of SPI with respect to simplicity of calculation, multiple time scales, and multiple spatial comparisons [44,45]. In the context of global warming, temperature increase has become one of the most important factors affecting regional drought [46–48]. Therefore, SPEI has unique advantages in studying the characteristics of drought on different time scales in the future, which is of great significance to further understand the impact of climate change on drought. Previous studies have shown that the annual-scale SPEI (SPEI-12) is more suitable for long-term drought trend assessment [49], and the monthly-scale SPEI (SPEI-1) is sensitive to short-term dry and wet variation [50]. Therefore, in this study, SPEI was used to analyze the future drought variation characteristics of the TP on both the annual and monthly scales. The calculation method of SPEI refers to Vicente-Serrano et al., but it should be noted that in this study, we used the Thornthwaite approach with a limited range of mean air temperatures to calculate the potential evapotranspiration (PET) for the TP [51,52]. The formula used is as follows:

$$\mathrm{PET} = \begin{cases} 0 & \text{if } T < 0 \\ 16\left(\frac{10T}{I}\right)^{\alpha} & \text{if } 26.5 > T \geq 0 \\ -415.85 + 32.24T - 0.43T^2 & \text{if } T \geq 26.5 \end{cases} \tag{1}$$

where I is the heat index, T is the average air temperature (in °C), and α is estimated using an I-related third-order polynomial:

$$I = \sum_{i=1}^{12} \left(\frac{T_i}{5}\right)^{1.514}, \tag{2}$$

$$\alpha = 0.49239 + 1.792 \times 10^{-2} I - 7.71 \times 10^{-5} I^2 + 6.75 \times 10^{-7} I^3, \tag{3}$$

According to the "Meteorological Drought Grade" classification standard, the SPEI values were divided into nine classes, as shown in Table 1.

Table 1. Categorization according to the SPEI values.

SPEI Value	Category	SPEI Value	Category	SPEI Value	Category
<−2	Extreme drought	−1 to −0.5	Light drought	1 to 1.5	Moderately wet
−2 to −1.5	Severe drought	−0.5 to 0.5	Normal	1.5 to 2	Severely wet
−1.5 to −1	Moderate drought	0.5 to 1	Lightly wet	>2	Extremely wet

2.3.2. Mann–Kendall (M-K) Test

The M-K test is often used to analyze the changing trend and significance of various factors [53,54]. The change trend of SPEI can be expressed as:

$$\mathrm{Slope} = \mathrm{median}\left(\frac{x_j - x_k}{j - k}\right), \tag{4}$$

where Slope is the rate of change, Slope > 0 represents a wetting trend, and Slope < 0 represents a drought trend.

The significance calculation formula is as follows:

$$Z = \begin{cases} \frac{S-1}{\sqrt{n(n-1)(2n+5)/18}} & \text{for } S > 0 \\ 0 & \text{for } S = 0, \\ \frac{S+1}{\sqrt{n(n-1)(2n+5)/18}} & \text{for } S < 0 \end{cases} \tag{5}$$

$$S = \sum_{k=1}^{n-1} \sum_{j=k+1}^{n} \mathrm{sgn}(x_j - x_k), \tag{6}$$

where S is the Kendall sum statistic, sgn is a symbolic function, and x_j and x_k are the parameter values at times j and k, respectively. When $|Z| \geq Z_{1-\alpha/2}$, the null hypothesis that the trend is not significant is not rejected. In this study, α = 0.01, α = 0.05, and α = 0.1 significance levels were considered, and the corresponding value of $Z_{1-\alpha/2}$ were 2.58, 1.96, and 1.64, respectively [55,56]. Based on the SPEI trend, it is further divided into six levels, as detailed in Table 2.

Table 2. Index Classification.

Data Type	Slope	∣Z∣	Description	Class	Slope	∣Z∣	Description	Class
SPEI	>0	[2.58, +∞)	Extremely significant humidification	EW	<0	[2.58, +∞)	Extremely significant aridification	ED
		[1.64, 2.58)	Significant humidification	SW		[1.64, 2.58)	Significant aridification	SD
		[0, 1.64)	Insignificant humidification	IW		[0, 1.64)	Insignificant aridification	ID

2.3.3. Run Theory

The run theory is typically used to identify drought events requiring a cutoff level k (k = drought level corresponding to SPEI) [57,58]. In this study, combined with the drought characteristics of each grade of the TP, the value of k is −1 (i.e., moderate drought and above). When the value of the random variable (SPEI) is greater than −1, it has a positive run; otherwise, it has a negative run. In drought research, the length of a negative run is called the drought duration (*D*), and the drought intensity (*S*) is the area encompassed by the drought duration and the intercept level (Figure 2).

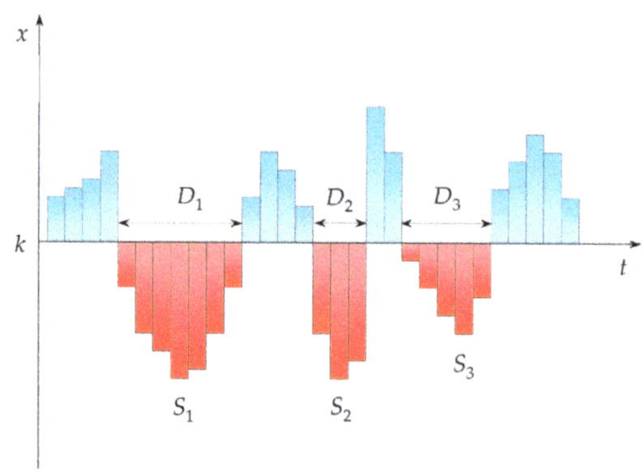

Figure 2. Schematic diagram of drought characteristics recognition process.

According to the run theory, to calculate the return periods of different run lengths [59,60], the calculation steps are as follows:

$$P = \frac{1}{n}\sum_{t=1}^{n} xf(x) \quad x = 1, 2, \cdots, n, \tag{7}$$

where P is the mean probability of occurrence of different drought durations, x is the drought duration, n is the number of months in the series of monthly precipitation for many years, and $f(x)$ is the number of occurrences of drought duration x.

The expected length (E) of the runs of different drought durations can be expressed by the following formula:

$$E = \sum_{t=1}^{n} xf(x) / \sum_{t=1}^{n} f(x), \tag{8}$$

After eliminating the influence of dependent effects, the migration probability (P_z) of different drought durations can be expressed as follows:

$$P_z = n(E-1)/(E(n-1)). \tag{9}$$

Hence, the one-dimensional run probability distribution function ($F(x)$) with parameters including the drought duration and intensity can be expressed as follows:

$$F(x) = \frac{1 + (1 - P_z)(n - x)}{1 + (1 - P_z)(n - 1)} {P_z}^{x-1}. \tag{10}$$

According to the definition of the run-length return period ($T(x)$), $T(x)$ can be expressed by the following formula:

$$T(x) = \frac{x + ({P_z}^x - 1)/(1 - P_z)}{12}. \tag{11}$$

In order to reflect the main research content more clarity, we have drawn a flowchart (Figure 3), which mainly illustrates the input datasets, calculations, methods, and output results of this study.

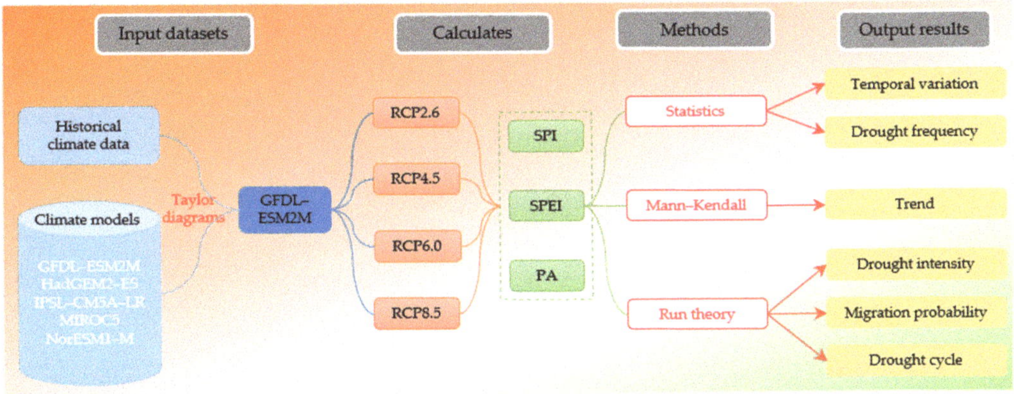

Figure 3. Flowchart including the main steps of input dataset, computation, method, and output result. The climate models are defined as follows: HadGEM2-ES-Hadley Centre global environment model version 2-Earth system configuration, IPSL-CM5A-LR-Institut Pierre-Simon Laplace to study natural climate variability and climate response to natural and anthropogenic forcings as part of the 5th phase of the coupled model intercomparison project, MIROC5-model of interdisciplinary research on climate version 5, and NorESM1-M-Norwegian Earth system model.

3. Results

3.1. Model Simulation Capability Assessment

Because the CMIP5 model shows different simulation performances for different climate elements in different regions, the model with the poorest performance will have a serious impact on the results [61]. Therefore, according to the simulation ability of different models in the TP for precipitation and mean temperature, this study selected the best model, and on this basis, analyzed the future drought variation characteristics of the TP. In order to more comprehensively and intuitively analyze the precipitation and mean temperature simulation capabilities of the five CMIP5 models in the TP, the Taylor diagrams analysis method is introduced here [62]. The Taylor diagrams comprehensively examine the matching degree of model simulation and observation data from three aspects: the spatial correlation coefficient, ratio of standard deviation, and centralization root mean square error (RMSD).

Figure 4 shows the Taylor diagrams of precipitation and mean temperature over the TP for five CMIP5 models under the RCP2.6 climate scenario from 2007 to 2015. The Taylor diagram results show that the precipitation simulation capabilities of different models differ to a certain degree, and the simulation results of the HadGEM2-ES, GFDL-ESM2M, and NorESM1-M climate models are significantly better than the IPSL-CM5A-LR and MIROC5 climate simulations. In addition, the variation amplitudes of the GFDL-ESM2M, NorESM1-M, and IPSL-CM5A-LR climate model simulation results are close to the observed data (Figure 4a). The five CMIP5 models have a similar ability to simulate the mean temperature time series, and the spatial correlation coefficients are all above 0.9. Among them, the GFDL-ESM2M climate model is the closest to the observational data, indicating that this model can better simulate the variation characteristics of mean temperature (Figure 4b). Overall, the GFDL-ESM2M model is better than the other four CMIP5 models in terms of its ability to comprehensively simulate precipitation and mean temperature in the TP.

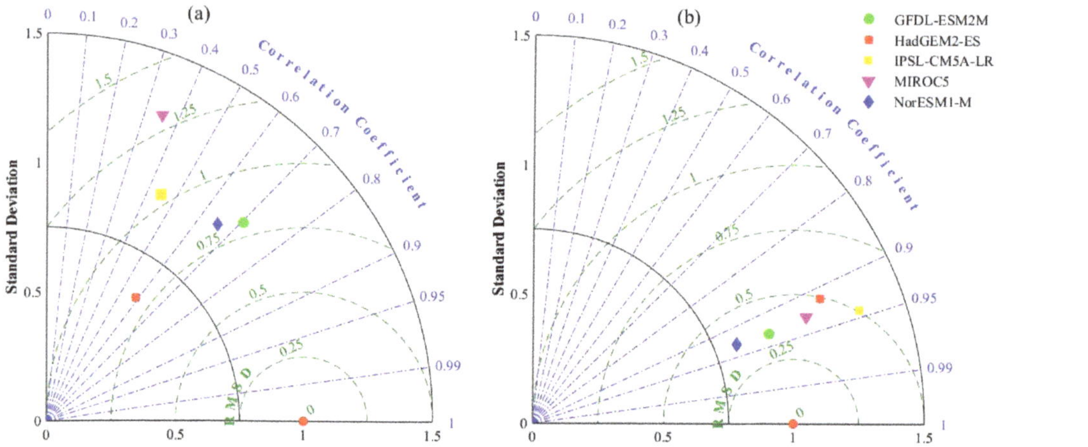

Figure 4. Taylor diagrams of different parameterization schemes under the RCP2.6 climate scenario: (**a**) precipitation and (**b**) mean temperature from 2007 to 2015.

3.2. Spatiotemporal Variation of SPEI

3.2.1. Temporal Variation Characteristics of SPEI

To explore the characteristics of the future drought evolution of the TP, the trends in SPEI-12 from 2016 to 2099 under four climatic scenarios were analyzed (Figure 5). Under the RCP2.6, RCP4.5, and RCP8.5 climate scenarios, the mean values of SPEI-12 for the TP exhibit insignificant increases ($p > 0.1$), i.e., 8.93×10^{-5}, 7.21×10^{-4}, and 1.52×10^{-3}, respectively. Under the RCP6.0 climate scenario, the mean value of SPEI-12 for the TP exhibits an insignificant ($p > 0.1$) decreasing trend of 2.07×10^{-4}. In addition, under RCP2.6, RCP4.5, RCP6.0, and RCP8.5, the annual drought probabilities for the TP from 2016 to 2099 are 13.10%, 9.52%, 10.71%, and 14.29%, respectively, and the forecasted drought in each case is mainly light drought.

In general, the mean value of SPEI-12 for the TP from 2016 to 2099 under the RCP2.6, RCP4.5, and RCP8.5 climatic scenarios mainly exhibit an increasing trend, indicating that the drought degree in the study area will be slowing in the future. This is mainly because of the increasing trends of precipitation and temperature over the TP in the future, with precipitation having a more significant effect on wet–dry changes over the TP.

To further research the intra-year distribution characteristics of the future wet–dry variation on the TP, a statistical map of the temporal variation of the TP SPEI-1 from 2016 to 2099 under the four climatic scenarios was drawn (Figure 6). Under the RCP2.6, RCP4.5, RCP6.0, and RCP8.5 climate scenarios, the future dry months (SPEI-1 < −0.5) on the TP will be concentrated in the summer (June to August) and will account for 96.50%, 94.74%, 95.48%, and 92.86% of the total dry months, respectively. The wet months (SPEI-1 > 0.5) will be concentrated in the spring (March to May) and October and will account for 88.67%, 89.80%, 92.35%, and 90.61% of the total wet months, respectively. In addition, under RCP2.6, RCP4.5, RCP6.0, and RCP8.5 for the period of 2016–2099, regarding the monthly extreme difference and standard deviation (Std) of the SPEI-1 of the TP, it can be expected that the maximum and minimum values will occur in May and December, respectively. As the concentration of carbon dioxide emissions increases, the maximum value of the monthly extreme difference (and the Std) will increase, as evidenced by predicted values of 1.87 (0.40), 1.70 (0.38), 2.14 (0.42), and 2.25 (0.45), respectively.

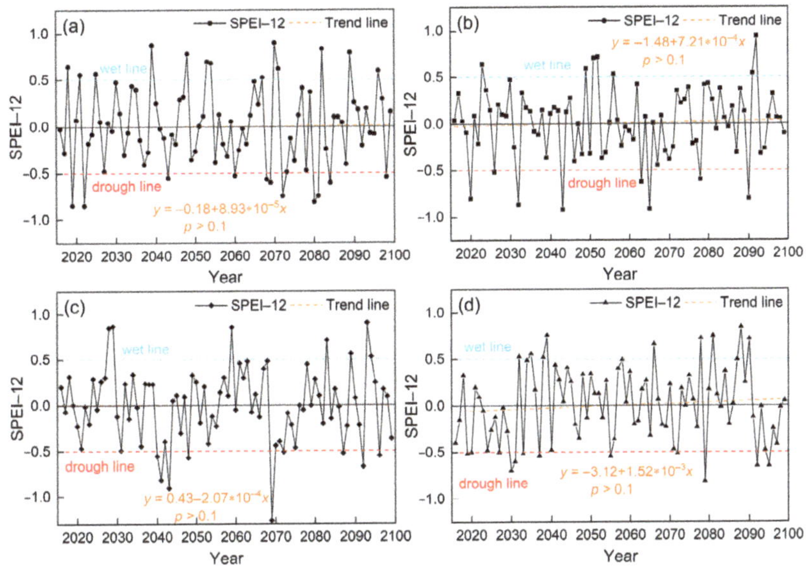

Figure 5. Temporal variation of SPEI-12 in the TP from 2016 to 2099 under the (**a**) RCP2.6, (**b**) RCP4.5, (**c**) RCP6.0, and (**d**) RCP8.5 climate scenarios.

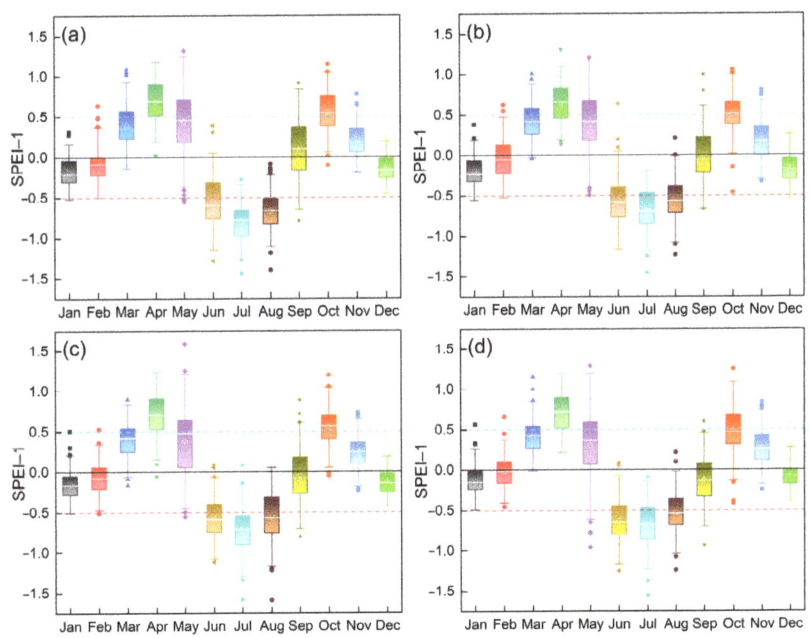

Figure 6. Temporal variation of the SPEI-1 on the TP from 2016 to 2099 under four climate scenarios: (**a**) RCP2.6, (**b**) RCP4.5, (**c**) RCP6.0, and (**d**) RCP8.5.

In general, under the four climate scenarios, the characteristics of monthly wet–dry variation and the annual distribution patterns for the TP are highly similar. However, as the concentration of carbon dioxide emissions increases, the dispersion of intra-year wet–dry variation will gradually increase, that is, towards the direction of non-uniform development. This indicates that as the concentration of carbon dioxide emissions increases, the intensity of climate extremes on the TP may also increase.

3.2.2. Spatial Variation Characteristics of SPEI

In this study, only the spatial variation characteristics of SPEI-12 under four future climate scenarios were analyzed based on Equation (4). Under the RCP2.6, RCP4.5, RCP6.0, and RCP8.5 climate scenarios, there will be a certain extent in the SPEI-12 of the TP from 2016 to 2099 (Figure 7). The SPEI-12/10 years ranges from −0.152 to 0.116, from −0.196 to 0.167, from −0.325 to 0.181, and from −0.374 to 0.269, respectively, for the four climate scenarios, as listed above. These results indicate that as carbon dioxide emissions increase, climate change on the TP will become more extreme in the future, and the trend of increasing extreme drought events will become more obvious. Under RCP2.6, the resulting aridification areas will be mainly concentrated in the Qilian Mountains, the Hengduan Mountains, and Lhasa. Under RCP4.5, RCP6.0, and RCP8.5, the aridification areas will be mainly concentrated in the QB, Kunlun Mountains, and Altun Mountains.

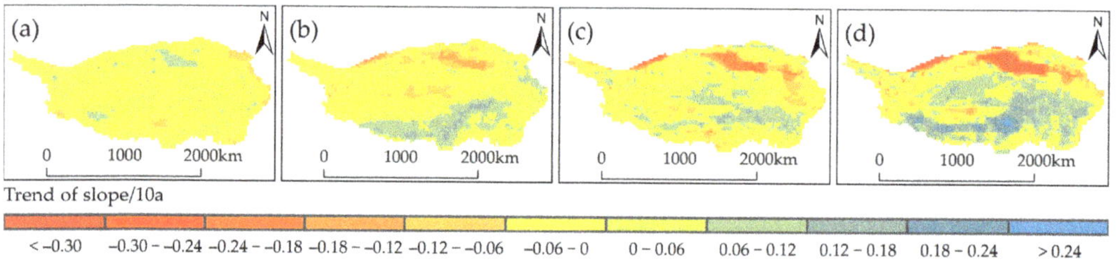

Figure 7. Spatial distribution of SPEI-12 trends on the TP from 2016 to 2099 under four climate scenarios: (**a**) RCP2.6, (**b**) RCP4.5, (**c**) RCP6.0, and (**d**) RCP8.5.

Under the RCP2.6, RCP4.5, RCP6.0, and RCP8.5 climate scenarios, 47.87%, 47.65%, 44.80%, and 35.45%, respectively, of the TP region will exhibit a trend of aridification in the future; that is, the SPEI will exhibit a decreasing trend (Figure 8). Under the four climate scenarios, the aridification of the TP will be dominated by insignificant aridification, and the proportions of insignificant aridification will be 45.60%, 40.58%, 33.39%, and 19.75%, respectively. As the carbon dioxide emissions concentration increases, the proportion of the area of the TP exhibiting extremely significant aridification will also increase in the future, from 0.01% to 2.79% to 6.35% to 10.06% for RCP2.6 to RCP8.5, respectively.

In addition, as the concentration of carbon dioxide emissions increases, the proportion of humidified areas in the TP will increase under the RCP2.6, RCP4.5, RCP6.0, and RCP8.5 climate scenarios, from 52.13% to 52.35% to 55.20% to 65.55%, respectively. Moreover, the proportions of insignificantly humidified area will be 50.05%, 40.11%, 41.74%, and 34.23%, respectively. In the future, the TP will exhibit extremely significant humidified area ratios of 0.05%, 3.39%, 3.04%, and 14.78%, respectively.

In summary, as the concentration of carbon dioxide emissions increases in the future, the proportion of areal aridification on the TP will decrease, but the proportions of extremely significant aridification and humidified areas will increase. This suggests that the TP will be more prone to extreme droughts and floods in the future.

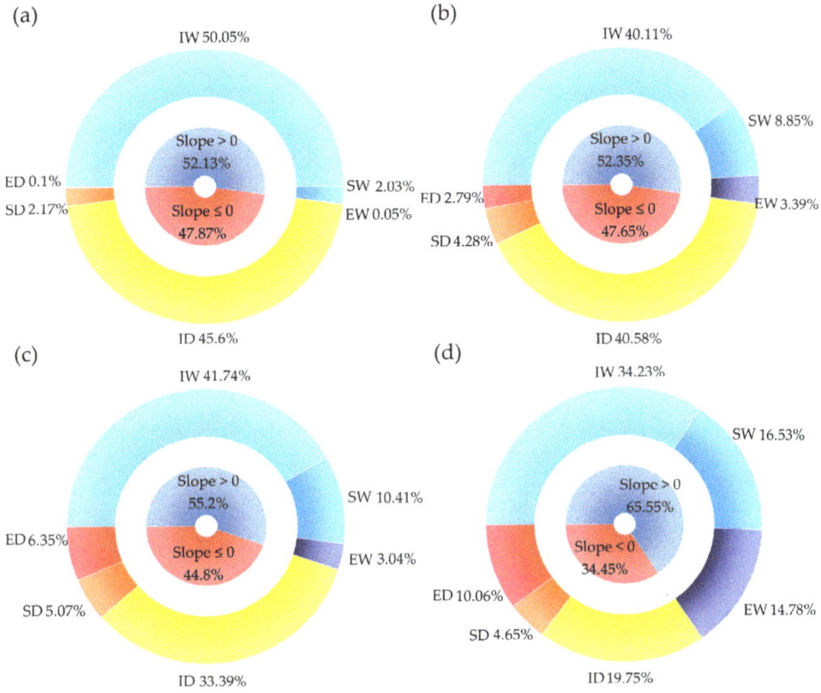

Figure 8. SPEI-12 changes on the TP from 2016 to 2099 under four climate scenarios: (**a**) RCP2.6, (**b**) RCP4.5, (**c**) RCP6.0, and (**d**) RCP8.5. IW-insignificant wetting, SW-significant wetting, EW-extremely significant wetting, ID-insignificant drought, SD-significant drought, ED-extremely significant drought.

3.3. *Analysis of Drought Change Characteristics*

Based on meteorological predictions for the different climatic scenarios, SPEI-1 was used to analyze the characteristics of drought changes for the TP from 2016 to 2099. In this study, we mainly analyzed the changing characteristics of drought frequency, drought intensity, and drought cycles for the TP under the four climate scenarios.

3.3.1. Variation Characteristics of Drought Frequency in Different Grades

According to the drought identification method, combined with the SPEI-1 calculation results for the $0.25° \times 0.25°$ grid of the TP, the droughts in different regions of the TP were distinguished, and the frequencies of drought events predicted from 2016 to 2099 for each grid under the four climate scenarios were summarized and counted (Figure 9). Under the RCP2.6, RCP4.5, RCP6.0, and RCP8.5 climate scenarios, the spatial variation characteristics of the overall drought frequency on the TP in the future will conform to light drought > moderate drought > severe drought > extreme drought ranking. Under the four climatic scenarios, the frequency of light drought will be between 0.11% and 40.84%, mainly in the southern regions of the TP. The frequency of moderate drought will be between 0.08% and 25.50%, mainly in the QB and southern regions of the TP. The frequency of severe drought will be less than 16.94%, mainly in the northwestern regions of the TP. Extreme droughts will occur less frequently, mainly concentrated in the western and central regions of the TP, with a maximum of 6.32%.

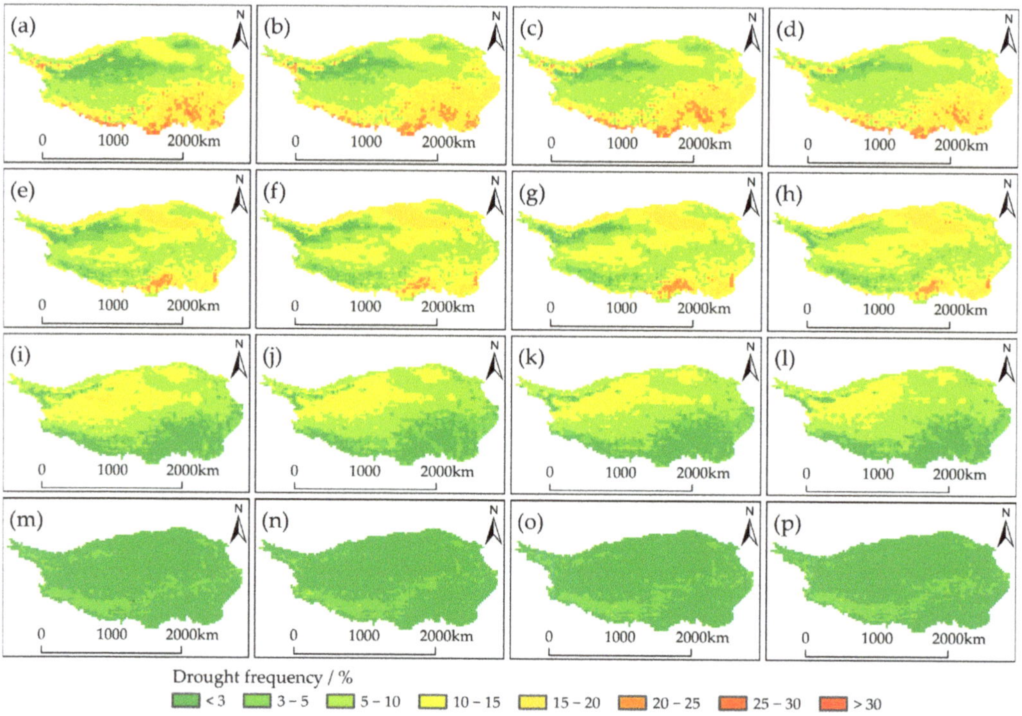

Figure 9. Spatial distributions of the frequency of droughts of various grades on the TP from 2016 to 2099 under four climatic scenarios: Frequency of light drought under the (**a**) RCP2.6, (**b**) RCP4.5, (**c**) RCP6.0, and (**d**) RCP8.5 climatic scenarios; frequency of moderate drought under (**e**) RCP2.6, (**f**) RCP4.5, (**g**) RCP6.0, and (**h**) RCP8.5; frequency of severe drought under (**i**) RCP2.6, (**j**) RCP4.5, (**k**) RCP6.0, and (**l**) RCP8.5; and frequency of extreme drought under (**m**) RCP2.6, (**n**) RCP4.5, (**o**) RCP6.0, and (**p**) RCP8.5.

To examine further the variation characteristics of future drought changes of various categories for the TP, a statistical analysis of the different categories of drought at each grid point was carried out (Figure 10). Under the four climate scenarios, the mean frequencies of monthly droughts on the TP from 2016 to 2099 were relatively similar, but overall, the mean frequency of droughts is predicted to increase at a rate of 2.65% as the concentration of carbon dioxide emissions increase (RCP8.5 (30.27%) > RCP6.0 (30.23%) > RCP4.5 (30.07%) > RCP2.6 (29.49%)). The mean frequencies of light droughts are predicted to be RCP2.6 (10.81%) < RCP8.5 (11.30%) < RCP4.5 (11.42%) < RCP6.0 (11.33%), with the largest rate of increase being 5.72%. The mean frequencies of moderate and extreme droughts are predicted to increase (i.e., RCP2.6 < RCP4.5 < RCP6.0 < RCP8.5), and the rates of increase for moderate and extreme droughts are predicted to be 9.07% and 4.74%, respectively. The mean frequencies of severe droughts are predicted to decrease gradually (i.e., RCP2.6 (7.34%) > RCP4.5 (6.89%) > RCP6.0 (6.77%) > RCP8.5 (6.66%)), with a decreasing rate of 9.25%. In addition, under the four climatic scenarios from RCP2.6 to RCP8.5, the Std of various drought categories in the TP are in the order of: light drought > moderate drought > severe drought > extreme drought, and Cv of various drought categories are in the order of: extreme drought > severe drought > light drought > moderate drought. These results also indicate that, as the concentration of carbon dioxide emissions increases, the Std and Cv of various drought categories on the TP are predicted to exhibit overall decreasing trends.

In general, as the concentration of carbon dioxide emissions increases, the frequency of monthly droughts on the TP from 2016 to 2099 will increase, the increase of the frequency of moderate droughts will become even more pronounced, and all categories of drought will trend toward more uniform spatial distributions. The degree of dispersion of the spatial distribution of droughts at different levels for the four climate scenarios are in the order of: extreme drought > severe drought > light drought > moderate drought.

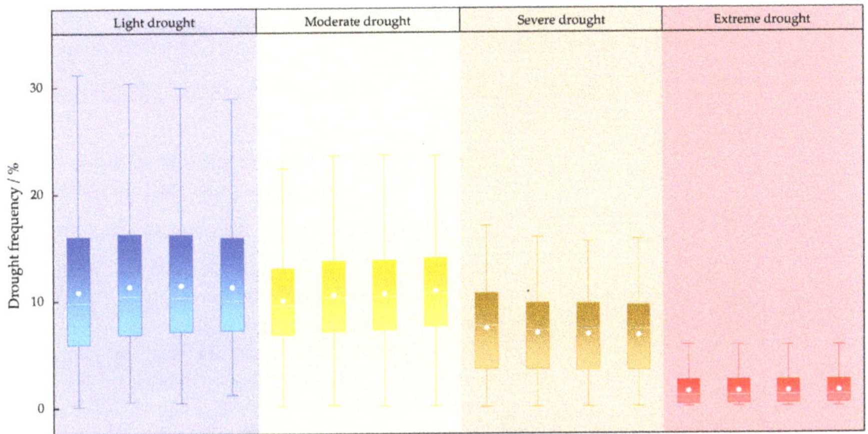

RCP	2.6	4.5	6.0	8.5	2.6	4.5	6.0	8.5	2.6	4.5	6.0	8.5	2.6	4.5	6.0	8.5
Mean	10.81	11.33	11.42	11.30	9.97	10.45	10.60	10.87	7.34	6.89	6.77	6.66	1.38	1.41	1.43	1.44
Median	9.81	10.41	10.27	10.00	9.48	10.18	10.30	10.59	7.61	7.12	7.29	7.24	0.98	1.01	1.06	1.08
Std	5.98	5.52	5.48	5.13	4.39	4.42	4.48	4.28	4.13	3.73	3.71	3.46	1.30	1.29		1.28
Cv	0.55	0.49	0.48	0.45	0.44	0.42		0.39	0.56	0.54	0.55	0.52	0.95	0.92	0.90	0.89

Figure 10. Predicted drought frequencies for various categories at each grid point on the TP from 2016 to 2099 under four climatic scenarios.

3.3.2. Variation Characteristics of Drought Intensity under Different Climatic Scenarios

Based on the calculated sub-mean drought intensity of each grid point, the spatial distribution map of the sub-mean drought intensity for the four climatic scenarios considered for the TP from 2016 to 2099 was drawn (Figure 11). Under the RCP2.6, RCP4.5, RCP6.0, and RCP8.5 climate scenarios, the sub-mean drought intensities on the TP will range from 1.030 to 3.106, 1.042 to 3.044, 1.027 to 2.973, and 1.033 to 2.759, respectively. The region with the strongest drought intensity is located mainly in the northwest and central part of the TP, while the region with the weakest drought intensity is located mainly in the south of the TP. As the concentration of carbon dioxide emissions increases, the spatial means of the sub-mean drought intensity on the TP from 2016 to 2099 are predicted to be 1.549, 1.535, 1.534, and 1.531, respectively; the Std values are predicted to be 0.169, 0.165, 0.168, and 0.163, respectively; and the Cv values are predicted to be 0.109, 0.107, 0.109, and 0.107, respectively.

In general, there are certain differences in the spatial distributions of the sub-mean drought intensities for the four climatic scenarios for the TP from 2016 to 2099, but the spatial distribution of drought intensities is more uniform than that of drought frequency. The spatial mean values of the sub-mean drought intensity are low, not exceeding 1.55, and the Cv values are also low, not exceeding 0.11. As the concentration of carbon dioxide emissions increases, the overall change in the spatial mean of sub-mean drought intensity is predicted to exhibits a decreasing trend, and the spatial distribution of sub-mean drought intensity develops in a more uniform direction.

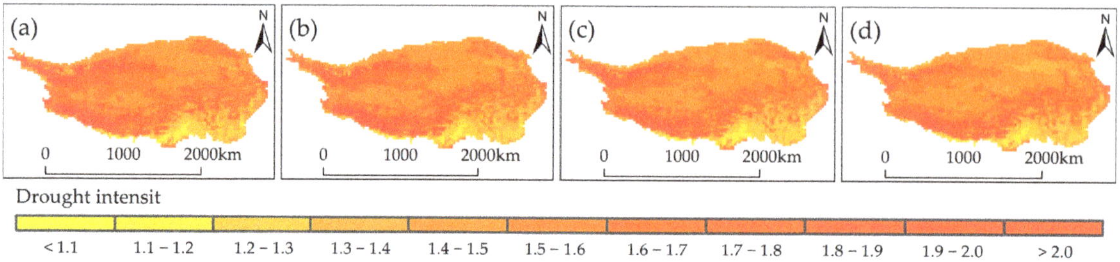

Figure 11. Spatial distribution characteristics of sub-mean drought intensity on the TP from 2016 to 2099 under the (**a**) RCP2.6, (**b**) RCP4.5, (**c**) RCP6.0, and (**d**) RCP8.5 climate scenarios.

3.3.3. Migration Probabilities and Return Periods of Different Drought Durations

In this study, the run theory was used to calculate the migration probabilities and return periods of different drought durations on the TP under the four climatic scenarios. In calculating the migration probabilities of different drought durations, combined with the SPEI drought grade distribution and the impacts of different grades of drought on the TP, k is taken as −1, i.e., this study only examined the migration probabilities of moderate and more severe droughts on the TP (Figure 12). Under the four climatic scenarios, the drought duration of the TP exhibited approximately the same trend as the theoretical probability distribution. As the drought duration increases, the theoretical probability density decreases exponentially, and when the drought duration exceeds 5 months, the theoretical probability density approaches 0. In addition, under the RCP2.6, RCP4.5, RCP6.0, and RCP8.5 climate scenarios, the drought duration on the TP is mainly less than 3 months, accounting for 82.09%, 80.89%, 79.80%, and 78.28% of the total drought duration, respectively. This indicates that as the concentration of carbon dioxide emissions increases, the possibility of long-lasting meteorological drought on the TP gradually increases.

When the TP drought duration exceeds 5 months, the theoretical probability density will approach zero. Therefore, only the spatial distribution map of the TP drought cycle from 1 to 5 months (Figure 13) was drawn in this study. It can be seen from Figure 13 that for the same drought duration conditions, the spatial distribution of the drought cycles on the TP under the four climate scenarios was relatively consistent. However, as the concentration of carbon dioxide emissions increases, the drought cycle of the TP is gradually shortened. In addition, under the four climate scenarios, the drought cycle of drought duration for one month on the TP is less than one year. The distribution of drought cycles of drought durations of 2–5 months shows obvious differences between the north and south, and the drought cycles in the southern region are longer than in the northern region. The drought cycle of 2-month drought duration in the southern region of the TP is between 2 and 10 years, and that in the northern region is between 1 and 2 years. The cycle of 3-month drought duration in the southern region is between 10 and 50 years, and that in the northern region is between 2 and 5 years. The cycle of 4-month drought duration in the southern region is between 50 and 200 years, and that in the northern region is between 2 and 10 years. The cycle of 5-month drought duration in the southern region is between 200 and 500 years, and that in the northern region is between 5 and 20 years.

Based on these results, it can be concluded that the drought duration is different across the different regions of the TP, and the distribution of drought cycles is also different. The greater the drought duration, the longer the drought cycle. From the perspective of the spatial distribution of drought cycles, the southern region of the TP has a longer drought cycle than the northern region, indicating that the northern region is more susceptible to drought, which is consistent with the actual situation. The reason for this is that the southern region of the TP has a subtropical and tropical northern marginal mountain forest climate, with an annual precipitation of more than 1000 mm, while the annual precipitation in the northern region of the TP is less than 600 mm (Figure 1b). In addition, precipitation in the northern region of the TP is more concentrated in summer, therefore, spring and autumn droughts are more likely to occur in the northern region.

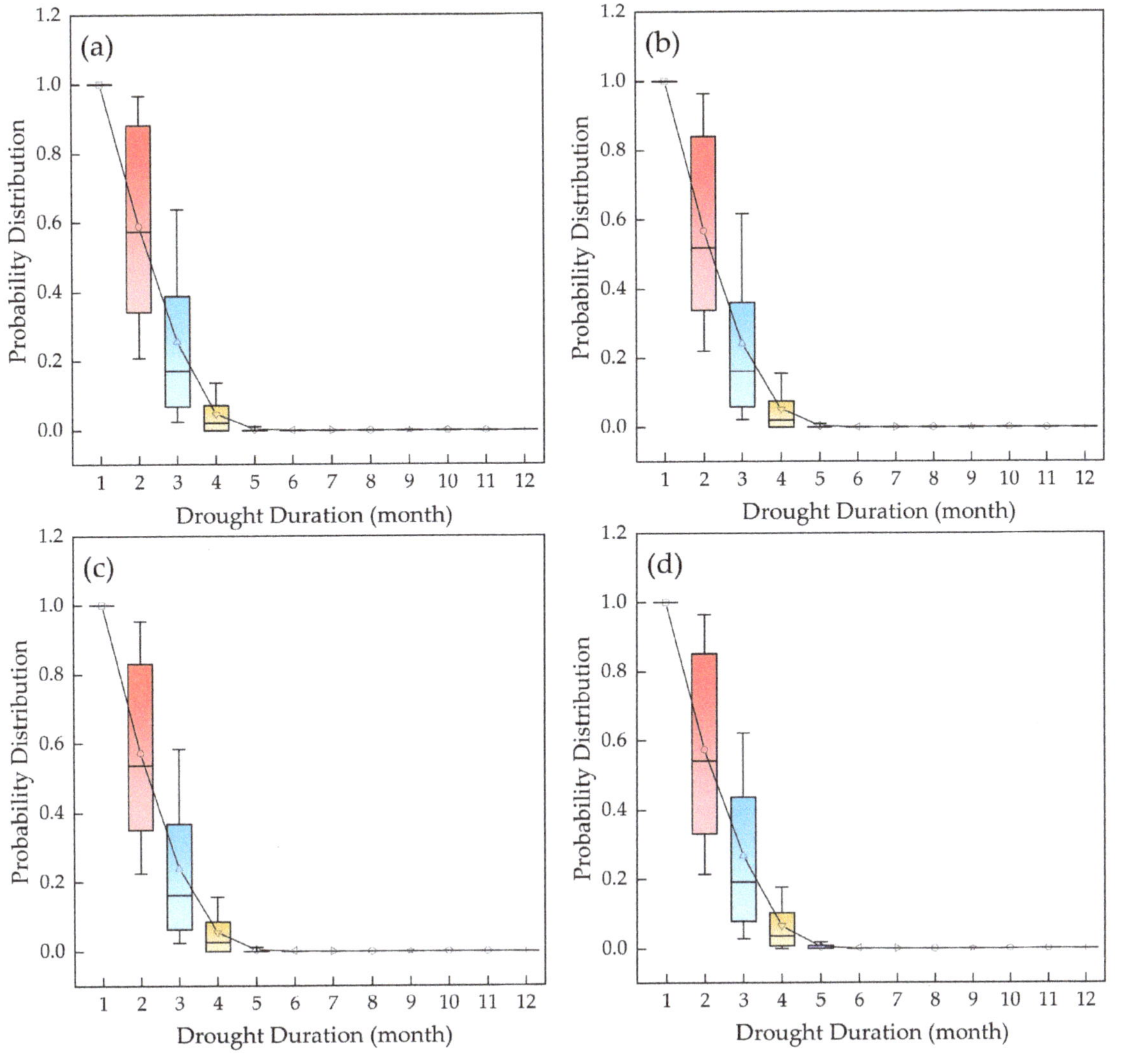

Figure 12. Relationship between drought duration and migration probability distribution on the TP under the (**a**) RCP2.6, (**b**) RCP4.5, (**c**) RCP6.0, and (**d**) RCP8.5 climate scenarios.

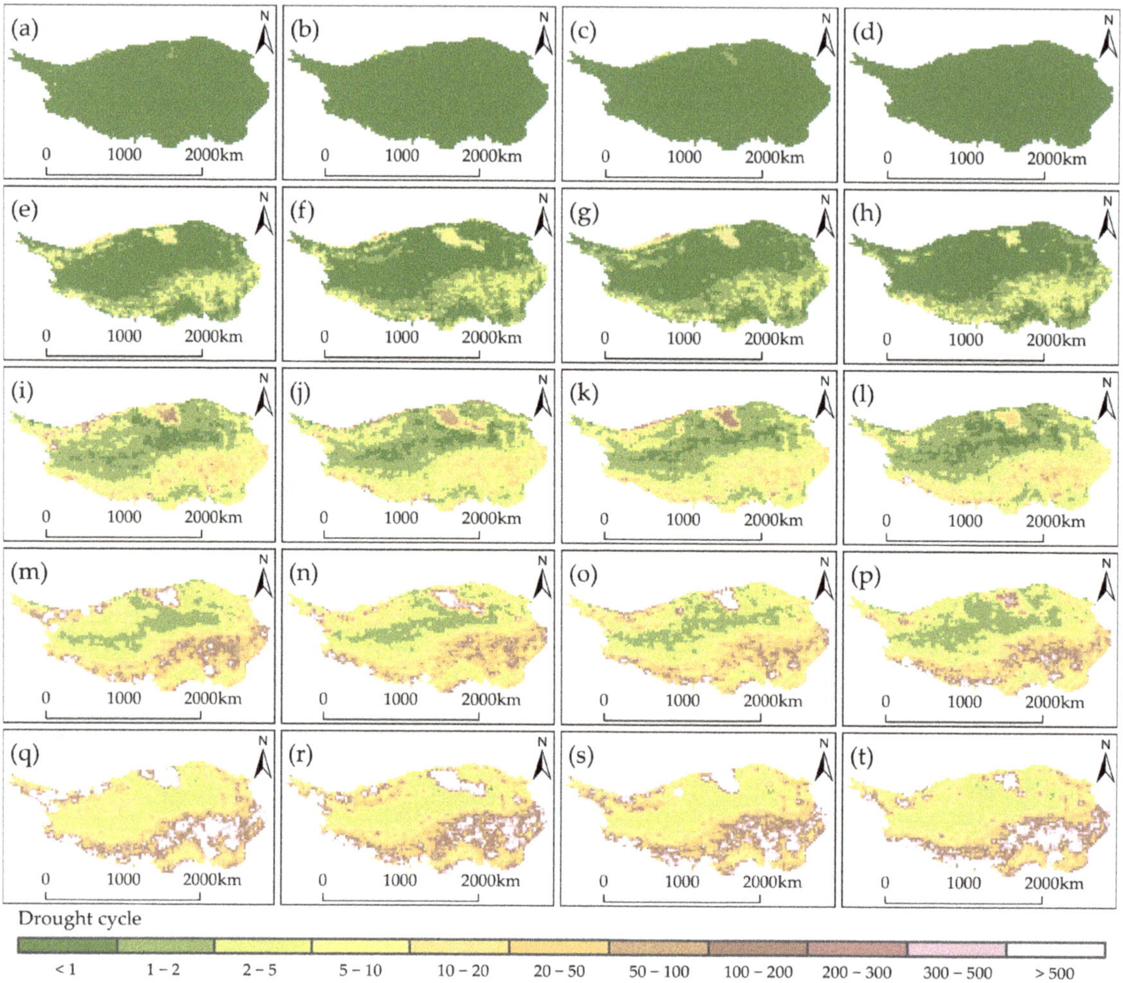

Figure 13. Distribution maps of the drought cycle of the TP for drought durations of 1–5 months under four climatic scenarios: drought duration of 1 month under the (**a**) RCP2.6, (**b**) RCP4.5, (**c**) RCP6.0, and (**d**) RCP8.5 climate scenarios; drought duration of 2 months under (**e**) RCP2.6, (**f**) RCP4.5, (**g**) RCP6.0, and (**h**) RCP8.5; drought duration of 3 months under (**i**) RCP2.6, (**j**) RCP4.5, (**k**) RCP6.0, and (**l**) RCP8.5; drought duration of 4 months under (**m**) RCP2.6, (**n**) RCP4.5, (**o**) RCP6.0, and (**p**) RCP8.5; drought duration of 5 months under (**q**) RCP2.6, (**r**) RCP4.5, (**s**) RCP6.0, and (**t**) RCP8.5.

4. Discussion

4.1. Adaptability Analysis of SPEI Index

To evaluate the ability of the SPEI index to reflect the annual scale of drought on the TP, the SPEI, SPI, and precipitation anomaly in percentage (PA) indices of each grid point under the four climate scenarios from 2016 to 2099 were used to conduct a correlation analysis (Figure 14). The following levels of degree of correlation were defined: greater than 0.60, a strong positive correlation; between 0.40 and 0.60, a significant correlation; between 0 and 0.4, a weak positive correlation; and less than 0, a negative correlation.

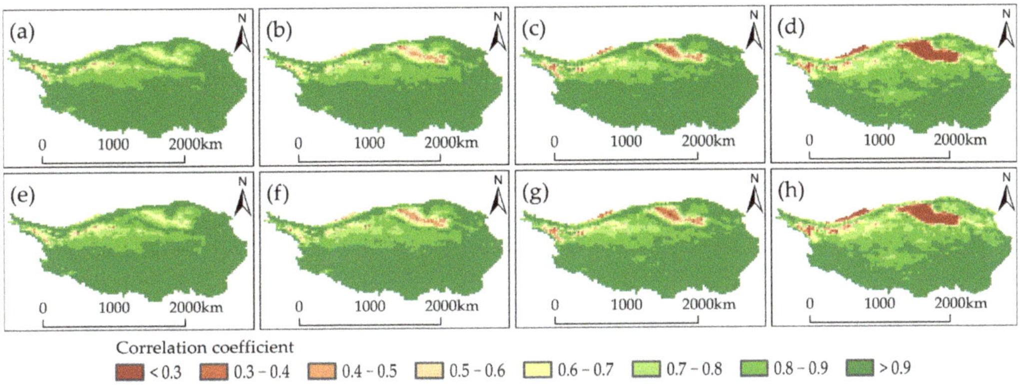

Figure 14. Correlations of three drought indices under different climate scenarios: PA index and SPEI index correlation under the (**a**) RCP2.6, (**b**) RCP4.5, (**c**) RCP6.0, and (**d**) RCP8.5 climate scenarios; SPI index and SPEI index correlation under (**e**) RCP2.6, (**f**) RCP4.5, (**g**) RCP6.0, and (**h**) RCP8.5.

Judging from the spatial distribution results of the correlation coefficients of the three drought indices under the four climate scenarios, the correlations between the SPEI, SPI, and PA indices for the southern part of the TP are stronger than those for the northern part, because the southern region receives more precipitation and is mostly a high-altitude mountainous area, with relatively low temperatures and evaporation [63,64]. The change in precipitation dominates the wet–dry change in the southern part of the TP [65]. Therefore, the SPI and PA indices, which only consider precipitation, can better reflect the wet–dry changes in the southern part of the TP. The SPI, PA, and SPEI indices are all strongly positively correlated. However, the precipitation in the northern part of the TP is relatively low, especially in the QB, where precipitation and evaporation jointly dominate the wet–dry changes in this region [66,67]. Therefore, the correlations between the SPI, PA, and SPEI indices in the northern region are poorer than in the southern region.

In addition, the temperature and evaporation of the TP increase with carbon dioxide emissions, and the correlations between the SPEI index (which considers precipitation and evaporation) and the SPI and PA indices (which only consider precipitation) gradually decrease. This is especially true for the QB under the RCP8.5 climate scenario: the correlation coefficients of the SPI, PA, and SPEI indices were all less than 0.4, indicating weak positive correlations. This indicates that as the temperature increases in the future, the influence of evaporation on climate change on the TP will gradually increase, and the SPEI index (which considers precipitation and evaporation) will better reflect the characteristics of climate change on the TP [68] This is consistent with research results obtained by Xu et al. [69], who found that by using different drought indices to predict the drought characteristics of humid subtropical basins in China in the context of climate warming, the SPI would not reflect the effect of evaporation and would underestimate the frequency of regional droughts.

4.2. *Difference Analysis of Drought Variation at Different Time Scales*

The annual-scale degree of drought in the region will tend to decrease slightly in the future (Figure 5) as a result of the increases in precipitation and temperature in the TP, except under the RCP6.0 climate scenario. Xu et al. [70] analyzed China's past drought trends and found that the TP has exhibited a wetting trend, which is consistent with our research findings on the future drought trends predicted for the TP. In addition, we found that the frequency of monthly-scale droughts in the TP is predicted to increase, which is consistent with the results of Han et al. [57] and Wang et al. [71], who found that southwest China may suffer from more severe drought disasters in the future.

In general, there will be differences in the characteristics of drought change at different time scales on the TP in the future. The main reasons for this phenomenon are that the frequency of severe drought and extreme drought will increase significantly and that extremely humid events will also become more frequent, especially under the RCP8.5 climate scenario (Figure 8). The annual-scale SPEI can reflect the relationship between the long-term trend and interannual variation of drought, but it is less sensitive to extreme events, so it is more suitable for long-term drought trend assessment [49]. The monthly-scale SPEI can reflect the characteristics of regional short-term dry and wet variation and is more sensitive to extreme events, and thus it is more suitable for the analysis of drought characteristics [50]. Therefore, this study analyzed the future drought variation characteristics of the TP on annual and monthly scales, which can better reflect the drought variation characteristics of the study area under different climate scenarios.

4.3. Analysis of Difference in Spatiotemporal Variation of Drought Characteristics

As global carbon dioxide emission concentrations, as well as the extreme difference and Std of the SPEI-1 for the TP, are predicted to increase, this indicates that the dispersion of dry-wet changes in each month will also gradually increase, and that the annual distribution law will gradually develop in a non-uniform direction (Figure 6). The main reasons for this phenomenon are that extreme precipitation in the TP will respond strongly to warming in the future, daily mean precipitation extremes will increase, extreme precipitation events will occur significantly more frequently, and the intra-annual distribution of precipitation will be more uneven than in the past, which is consistent with previous research results [72–74].

Spatially, as the carbon dioxide emission concentration increases, the spatial distributions of drought and sub-mean drought intensity at all levels on the TP will develop in a more uniform direction (Figures 8 and 10). There are two main reasons for this phenomenon. First, in the context of climate warming, the temperature of the TP has increased significantly, which has aggravated the melting of ice and snow and the increase of permafrost active layers [75,76]. These changes have led to soil moisture change and migration (especially in the upper layer), and have affected water and heat exchange between the land and air [77]. Second, the TP has a high altitude and complex terrain, with a large spatial variability of annual and seasonal precipitation [78,79]. As the overall precipitation on the TP increases, the spatial distribution of water differences decreases [80].

In general, the annual precipitation on the TP will increase in the future, and the annual distribution will develop in a more uneven direction, resulting in differences in the spatiotemporal variation of drought on the TP and increasing the frequency of extreme climate events on the TP. Previous studies have shown that ~45% of the world's land is affected by drought disasters, and the annual losses due to drought are as high as 6 to 8 billion dollars [81]. In the future, the global area affected by drought will continue to expand, and there will be a trend of gradual expansion of drought from arid areas to sub-humid and humid areas [11]. By the end of this century, it is expected that dry land will cover half of the Earth's land surface, and the total area of extremely dry land will increase by more than double [82,83]. In addition, studies have shown that future aridification will not only cause huge economic losses but also lead to systematic and abrupt changes in various ecosystem properties [84,85]. Under the scenario of future global warming, the TP, as a sensitive and vulnerable region, will face a greater threat of drought, especially extreme drought events. Therefore, we should continue to pay close attention to climate change in the TP and improve our abilities to predict and provide early warnings of extreme climate disasters [86,87].

5. Conclusions

Annual and monthly scale SPEI drought indices were used with the GFDL-ESM2M climate model to study predicted drought trends for the TP from 2016 to 2099 under four RCP scenarios: RCP2.6, RCP4.5, RCP6.0, and RCP8.5, representing very low, low, medium, and high levels of greenhouse gas emissions, respectively. The duration, frequency, intensity, and cycle of drought events were analyzed based on the variation of SPEI on the TP over different time, and the applicability of different drought indices was examined. The following conclusions are drawn from the results:

(1) The future climate of the TP is predicted to be warmer and more humid than that of the past, and these changes are most obvious under the RCP8.5 climate scenario. As the concentration of carbon dioxide emissions increases, the annual wet–dry variation of the TP will tend to develop in a non-uniform direction, and the proportion of areas of extremely significant aridity and humidification will both increase significantly, which indicates the possibility of increased extreme disasters for the region in the future.

(2) Under all four climate scenarios, the TP will be dominated by light drought in the future. As the carbon dioxide emission concentration increases, the frequency of occurrence of droughts in the TP will gradually increase, yet the spatial average value of sub-mean drought intensity will decrease. However, the spatial distribution of both these factors will tend to develop in a uniform direction.

(3) Under all four climate scenarios, the drought duration of the TP is mainly less than 3 months, and when the drought duration exceeds 5 months, the theoretical probability density will approach 0. As the carbon dioxide emission concentration increases, the drought cycle of the TP will gradually shorten. The southern region of the TP has a longer drought cycle than the northern region, which indicates that the northern region is more susceptible to drought.

Author Contributions: Conceptualization, Y.L., Z.J. and X.M.; methodology, Y.L. and Z.J.; software, Y.L.; validation, Y.L.; formal analysis, Y.L.; investigation, Z.J.; resources, Z.J. and X.M.; data curation, Y.L. and W.Z.; writing—original draft preparation, Y.L., Y.W., R.G. and Z.J.; writing—review and editing, Y.L., Y.W., Z.G., Y.G. and X.M.; visualization, X.M.; supervision, Y.L. and Z.J.; project administration, Z.J. and X.M.; funding acquisition, Z.J. and X.M. All authors have read and agreed to the published version of the manuscript.

Funding: This research was funded by the National Natural Science Foundation of China (52179048, 42001033), the National Key R&D Program of China (2021YFD1900600), and the Natural Science Basic Research Plan in Shaanxi Province of China (2021JQ-237).

Institutional Review Board Statement: Not applicable.

Informed Consent Statement: Not applicable.

Data Availability Statement: Not applicable.

Acknowledgments: The authors gratefully acknowledge the National Tibetan Plateau Data Center and the Institute of Geographic Sciences and Natural Resources Research, CAS for providing support.

Conflicts of Interest: The authors declare no conflict of interest.

References

1. Dai, A. Increasing drought under global warming in observations and models. *Nat. Clim. Chang.* **2013**, *3*, 52–58. [CrossRef]
2. Liu, Z.P.; Wang, Y.Q.; Shao, M.G.; Jia, X.X.; Li, X.L. Spatiotemporal analysis of multiscalar drought characteristics across the Loess Plateau of China. *J. Hydrol.* **2016**, *534*, 281–299. [CrossRef]
3. Duan, W.L.; Zou, S.; Chen, Y.N.; Daniel, N.; Fang, G.H.; Wang, Y. Sustainable water management for cross-border resources: The Balkhash Lake Basin of Central Asia, 1931–2015. *J. Clean. Prod.* **2020**, *263*, 121614. [CrossRef]
4. Lin, Y.; Grimm, N.B. Modelling potential impacts of climate change on water and nitrate export from a mid-sized, semiarid watershed in the US Southwest. *Clim. Chang.* **2013**, *120*, 419–431.
5. Zha, X.B.; Luo, P.P.; Zhu, W.; Wang, S.T.; Lyu, J.Q.; Zhou, M.M.; Huo, A.D.; Wang, Z.H. A bibliometric analysis of the research on Sponge City: Current situation and future development direction. *Ecohydrology* **2021**, *14*, e2328. [CrossRef]
6. Huang, W.J.; Duan, W.L.; Chen, Y.N. Rapidly declining surface and terrestrial water resources in Central Asia driven by socioeconomic and climatic changes. *Sci. Total. Environ.* **2021**, *784*, 147193. [CrossRef]
7. Mu, D.R.; Luo, P.P.; Lyu, J.Q.; Zhou, M.M.; Huo, A.D.; Duan, W.; Nover, D.; He, B.; Zhao, X.L. Impact of temporal rainfall patterns on flash floods in Hue City, Vietnam. *J. Flood Risk Manag.* **2020**, *14*, e12668. [CrossRef]
8. Xie, D.N.; Duan, L.; Si, G.Y.; Liu, W.J.; Zhang, T.; Mulder, J. Long–term 15N balance after single–dose input of 15N–labeled NH4+ and NO3– in a subtropical forest under reducing N deposition. *Glob. Biogeochem. Cycles* **2021**, *35*, e2021GB006959. [CrossRef]

9. Zhang, Y.; Luo, P.P.; Zhao, S.F.; Kang, S.X.; Wang, P.B.; Zhou, M.M.; Lyu, J.Q. Control and remediation methods for eutrophic lakes in the past 30 years. *Water Sci. Technol.* **2020**, *81*, 1099–1113. [CrossRef]
10. Feng, S.; Fu, Q. Expansion of global drylands under a warming climate. *Atmos. Chem. Phys. Discuss.* **2013**, *13*, 14637–14665. [CrossRef]
11. Huang, J.P.; Yu, H.P.; Guan, X.D.; Wang, G.Y.; Guo, R.X. Accelerated dryland expansion under climate change. *Nat. Clim. Chang.* **2016**, *6*, 166–171. [CrossRef]
12. United Nations Convention to Combat Desertification. *The Global Land Outlook*, 1st ed.; UNCCD: Bonn, Germany, 2017.
13. Liu, Y.W.; Liu, Y.B.; Wang, W. Inter-comparison of satellite-retrieved and Global Land Data Assimilation System-simulated so moisture datasets for global drought analysis. *Remote Sens. Environ.* **2019**, *220*, 1–18. [CrossRef]
14. Akhtar-Schuster, M.; Stringer, L.C.; Metternicht, G.; Barger, N.N.; Chotte, J.-L.; Kust, G. Assessing the impact of science in the implementation of the United Nations Convention to Combat Desertification. *Land* **2022**, *11*, 568. [CrossRef]
15. Lkhagvadorj, N.; Zhang, J.H.; Battsetseg, T.; Yang, S.S.; Sonam, R.; Ahmed, P.F.; Pangali, S.T.P. Assessment of drought impact on net primary productivity in the terrestrial ecosystems of Mongolia from 2003 to 2018. *Remote Sens.* **2021**, *13*, 2522.
16. Huang, J.L.; Zhai, J.Q.; Jiang, T.; Wang, Y.J.; Li, X.C.; Wang, R.; Xiong, M.; Su, B.D.; Fischer, T. Analysis of future drought characteristics in China using the regional climate model CCLM. *Clim. Dyn.* **2018**, *50*, 507–525. [CrossRef]
17. Wang, Y.X.; Lv, J.; Jamie, H.; Wang, Y.C.; Sun, H.Q.; Lucy, J.B.; Ma, M.M.; Su, Z.C.; Michael, E. Linking drought indices to impacts to support drought risk assessment in Liaoning province, China. *Nat. Hazards Earth Syst. Sci.* **2020**, *20*, 889–906. [CrossRef]
18. Zhu, Y.N.; Lei, L.; Zhao, Y.; Liang, Z.M.; Li, H.H.; Wang, L.Z.; Wang, Q.M. Regional comprehensive drought disaster risk dynamic evaluation based on projection pursuit clusterin. *Water Policy* **2018**, *20*, 410–428. [CrossRef]
19. Hu, Z.; Wu, Z.; Islam, A.R.M.T.; You, X.Y.; Liu, C.; Li, Q.; Zhang, X.S. Spatiotemporal characteristics and risk assessment of agricultural drought disasters during the winter wheat-growing season on the Huang-Huai-Hai Plain, China. *Theor. Appl Climatol.* **2021**, *143*, 1393–1407. [CrossRef]
20. Orimoloye, I.R.; Ololade, O.O.; Belle, J.A. Satellite-based application in drought disaster assessment using terra MOD13Q1 data across free state province, South Africa. *J. Environ. Manag.* **2021**, *285*, 112112. [CrossRef]
21. Li, L.; Zhang, R. Effect of upper-level air temperature changes over the Tibetan Plateau on the genesis frequency of Tibetan Plateau vortices at interannual timescales. *Clim. Dyn.* **2021**, *57*, 341–352. [CrossRef]
22. Abraham, J.G.; Danielle, C.V.K.; Greg, R.H.; Garry, W. Catchment-scale drought: Capturing the whole drought cycle using multiple indicators. *Hydrol. Earth Syst. Sci.* **2020**, *24*, 1985–2002.
23. Annette, H.; Laura, C.; Reed, M. Evaluating the relative importance of precipitation, temperature and land-cover change in the hydrologic response to extreme meteorological drought conditions over the North American High Plains. *Hydrol. Earth Syst. Sci.* **2019**, *23*, 1931–1950.
24. Wang, F.; Yang, H.B.; Wang, Z.M.; Zhang, Z.Z.; Li, Z.H. Drought evaluation with CMORPH satellite precipitation data in the Yellow River Basin by using gridded standardized precipitation evapotranspiration index. *Remote Sens.* **2019**, *11*, 485. [CrossRef]
25. Yang, J.; Chang, J.X.; Wang, Y.M.; Li, Y.Y.; Hu, H.; Chen, Y.T.; Huang, Q.; Yao, J. Comprehensive drought characteristics analysis based on a nonlinear multivariate drought index. *J. Hydrol.* **2018**, *557*, 651–667. [CrossRef]
26. Immerzeel, W.W.; Beek, L.P.H.v.; Bierkens, M.F.P. Climate change will affect the Asian water towers. *Science* **2010**, *328*, 1382–1385. [CrossRef] [PubMed]
27. Pritchard, H.D. Retraction: Asia's glaciers are a regionally important buffer against drought. *Nature* **2018**, *555*, 274. [CrossRef]
28. Wu, G.X.; Liu, Y.M.; He, B.; Bao, Q.; Duan, A.M.; Jin, F.F. Thermal controls on the Asian summer monsoon. *Sci. Rep.* **2012**, *2*, 404. [CrossRef]
29. Wang, M.R.; Wang, J.; Duan, A.M.; Liu, Y.M.; Zhou, S.W. Coupling of the quasi-biweekly oscillation of the Tibetan Plateau summer monsoon with the arctic oscillation. *Geophys. Res. Lett.* **2018**, *45*, 7756–7764. [CrossRef]
30. Duan, A.M.; Xiao, Z.X. Does the climate warming hiatus exist over the Tibetan Plateau? *Sci. Rep.* **2015**, *5*, 13711. [CrossRef]
31. An, S.; Chen, X.Q.; Zhang, X.Y.; Lang, W.G.; Ren, S.L.; Xu, L. Precipitation and minimum temperature are primary climatic controls of alpine grassland autumn phenology on the Qinghai-Tibet Plateau. *Remote Sens.* **2020**, *12*, 431. [CrossRef]
32. Resources and Environment Data Cloud Platform of the Chinese Academy of Sciences. Available online: http://www.resdc.cn/ (accessed on 11 November 2021).
33. Dunne, J.P.; John, J.G.; Adcroft, A.J.; Griffies, S.M.; Hallberg, R.W.; Shevliakova, E.; Stouffer, R.J.; Cooke, W.; Dunne, K.A.; Harrison, M.J.; et al. GFDL's ESM2 Global coupled climate–carbon Earth system models. Part I. *J. Clim.* **2012**, *25*, 6646–6665. [CrossRef]
34. Ng, B.; Cai, W.J.; Walsh, K. Nonlinear feedbacks associated with the Indian Ocean Dipole and their response to Global warming in the GFDL-ESM2M coupled climate model. *J. Clim.* **2014**, *27*, 3904–3919. [CrossRef]
35. Dunne, J.P.; John, J.G.; Shevliakova, E.; Stouffer, R.J.; Krasting, J.P.; Malyshev, S.L.; Milly, P.C.D.; Sentman, L.T.; Adcroft, A.J.; Cooke, W.; et al. GFDL's ESM2 Global coupled climate–carbon Earth system models. Part II. *J. Clim.* **2013**, *26*, 2247–2267. [CrossRef]
36. Jia, K.; Ruan, Y.F.; Yang, Y.Z.; You, Z. Assessment of CMIP5 GCM simulation performance for temperature projection in the Tibetan Plateau. *Earth Space Sci.* **2019**, *6*, 2362–2378. [CrossRef]
37. Bi, Y.J.; Zhao, J.; Wu, D.; Zhao, Y. Spatiotemporal variations of future potential evapotranspiration in the Beijing-Tianjin-Hebei region under GFDL-ESM2M climate model. *Trans. Chin. Soc. Agric. Eng.* **2020**, *36*, 140–149.

38. Ma, D.Y.; Deng, H.Y.; Yin, Y.H.; Wu, S.H.; Zhang, D. Sensitivity of arid/humid patterns in China to future climate change under a high-emissions scenario. *J. Geogr. Sci.* **2019**, *29*, 29–48. [CrossRef]
39. Pan, X.D.; Zhang, L.; Huang, C.L. Future climate projection in Northwest China with RegCM4.6. *Earth Space Sci.* **2020**, *7*, e2019EA000819. [CrossRef]
40. National Tibetan Plateau Data Center. Future Climate Projection of China Based on Regcm4.6 (2007–2099). Available online: http://data.tpdc.ac.cn (accessed on 16 November 2021).
41. Yao, N.; Li, L.C.; Feng, P.Y.; Feng, H.; Liu, D.L.; Liu, Y.; Jiang, K.T.; Hu, X.T.; Li, Y. Projections of drought characteristics in China based on a standardized precipitation and evapotranspiration index and multiple GCMs. *Sci. Total Environ.* **2020**, *704*, 135245. [CrossRef] [PubMed]
42. Liu, Y.; Tian, J.Y.; Liu, R.H.; Ding, L.Q. Influences of climate change and ruman activities on NDVI changes in China. *Remote Sens.* **2021**, *13*, 4326. [CrossRef]
43. Vicente-Serrano, S.M.; BeguerÃ a, S.; LÃ³pez-Moreno, J.I. A Multiscalar drought index sensitive to Global warming: The standardized precipitation evapotranspiration index. *J. Clim.* **2010**, *23*, 1696–1718. [CrossRef]
44. Ndehedehe, C.E.; Agutu, N.O.; Ferreira, V.G.; Getirana, A. Evolutionary drought patterns over the Sahel and their teleconnections with low frequency climate oscillations. *Atmosp. Res.* **2020**, *233*, 104700. [CrossRef]
45. Zhang, B.Q.; Long, B.; Wu, Z.Y.; Wang, Z.K. An evaluation of the performance and the contribution of different modified water demand estimates in drought modeling over water-stressed regions. *Land Degrad. Develop.* **2017**, *28*, 1134–1151. [CrossRef]
46. Zhang, R.W.; Zhao, C.Y.; Ma, X.F.; Brindha, K.; Han, Q.F.; Li, C.F.; Zhao, X.N. Projected spatiotemporal dynamics of drought under Global warming in Central Asia. *Sustainability* **2019**, *11*, 4421. [CrossRef]
47. Zhang, Y.Z.; Huang, C.C.; Tan, Z.H.; Chen, Y.L.; Qiu, H.J.; Huang, C.; Li, Y.Q.; Zhang, Y.X.; Li, X.G.; Shulmeister, J.; et al. Prehistoric and historic overbank floods in the Luoyang Basin along the Luohe River, middle Yellow River basin, China. *Quat. Int.* **2019**, *521*, 118–128. [CrossRef]
48. Jung, I.W.; Chang, H. Climate change impacts on spatial patterns in drought risk in the Willamette River Basin, Oregon, USA. *Theor. Appl. Climatol.* **2012**, *108*, 355–371. [CrossRef]
49. Chen, S.D.; Zhang, L.P.; Liu, X.; Guo, M.Y.; She, D.X.; Qian, B.D. The use of SPEI and TVDI to assess temporal-spatial variations in drought conditions in the Middle and Lower Reaches of the Yangtze River Basin, China. *Adv. Meteorol.* **2018**, *2018*, 1–11. [CrossRef]
50. Li, X.Z.; Huang, W.R. How long should the pre-existing climatic water balance be considered when capturing short-term wetness and dryness over China by using SPEI? *Sci. Total Environ.* **2021**, *786*, 147575. [CrossRef]
51. Thornthwaite, C.W. An approach toward a rational classification of climate. *Soil Sci.* **1948**, *66*, 77. [CrossRef]
52. Willmott, C.J.; Rowe, C.M.; Mintz, Y. Climatology of the terrestrial seasonal water cycle. *J. Climatol.* **1985**, *5*, 589–606. [CrossRef]
53. Wei, X.D.; Wang, N.; Luo, P.P.; Yang, J.; Zhang, J.; Lin, K.L. Spatiotemporal assessment of land marketization and its driving forces for sustainable urban–rural development in Shaanxi Province in China. *Sustainability* **2021**, *13*, 7755. [CrossRef]
54. Güçlü, Y.S. Improved visualization for trend analysis by comparing with classical Mann-Kendall test and ITA. *J. Hydrol.* **2020**, *584*, 124674. [CrossRef]
55. Hamed, K.H. Exact distribution of the Mann–Kendall trend test statistic for persistent data. *J. Hydrol.* **2008**, *365*, 86–94. [CrossRef]
56. Meng, X.Y.; Gao, X.; Li, S.Y.; Lei, J.Q. Spatial and temporal characteristics of vegetation NDVI changes and the driving forces in Mongolia during 1982–2015. *Remote Sens.* **2020**, *12*, 603. [CrossRef]
57. Han, R.C.; Li, Z.L.; Li, Z.J.; Han, Y.Y. Spatial–temporal assessment of historical and future meteorological droughts in China. *Atmosphere* **2021**, *12*, 787. [CrossRef]
58. Wu, R.N.; Zhang, J.q.; Bao, Y.H.; Guo, E.L. Run theory and copula-based drought risk analysis for songnen grassland in Northeastern China. *Sustainability* **2019**, *11*, 6032. [CrossRef]
59. Moyé, L.A.; Kapadia, A.S. Predictions of drought length extreme order statistics using run theory. *J. Hydrol.* **1995**, *169*, 95–110. [CrossRef]
60. Guo, H.; Bao, A.M.; Liu, T.; Guli·, J.; Felix, N.; Jiang, L.L.; Alishir, K.; Philippe, D.M. Spatial and temporal characteristics of droughts in Central Asia during 1966–2015. *Sci. Total Environ.* **2017**, *624*, 1523–1538. [CrossRef]
61. Gallegati, M. Multiscale evaluation of CMIP5 models using wavelet-based descriptive and diagnostic techniques. *Clim. Chang.* **2022**, *170*, 41. [CrossRef]
62. Giuseppe, M.; Alfonso, S. Evaluation of parametric and statistical approaches for the regionalization of flow duration curves in intermittent regimes. *J. Hydrol.* **2013**, *480*, 19–32.
63. Wang, C.P.; Huang, M.T.; Zhai, P.M. Change in drought conditions and its impacts on vegetation growth over the Tibetan Plateau. *Adv. Clim. Chang. Res.* **2021**, *12*, 333–341. [CrossRef]
64. Li, L.; Zhang, R.H.; Wen, M.; Lv, J.M. Regionally different precipitation trends over the Tibetan Plateau in the warming context: A perspective of the Tibetan Plateau vortices. *Geophys. Res. Lett.* **2021**, *48*, e2020GL091680. [CrossRef]
65. Gao, Y.H.; Li, X.; Leung, L.R.; Chen, D.L.; Xu, J.W. Aridity changes in the Tibetan Plateau in a warming climate. *Environ. Res. Lett.* **2015**, *10*, 034013. [CrossRef]
66. Liu, J.; Zhang, T.J.; Wu, Q.B.; Jiang, G.L. Recent climate changes in the Northwestern Qaidam Basin inferred from geothermal gradients. *Earth Sci. Inform.* **2020**, *13*, 261–270. [CrossRef]

67. Jia, Y.X.; Wu, H.B.; Zhang, W.C.; Li, Q.; Yu, Y.Y.; Zhang, C.X.; Sun, A.Z. Quantitative cenozoic climatic reconstruction and its implications for aridification of the Northeastern Tibetan Plateau. *Palaeogeogr. Palaeoclimatol. Palaeoecol.* **2021**, *567*, 110244. [CrossRef]
68. Zarei, A.R.; Shabani, A.; Moghimi, M.M. Accuracy Assessment of the SPEI, RDI and SPI drought indices in regions of Iran with different climate conditions. *Pure Appl. Geophys.* **2021**, *178*, 1387–1403. [CrossRef]
69. Xu, K.; Wu, C.H.; Zhang, C.; Hu, B.X. Uncertainty assessment of drought characteristics projections in humid subtropical basins in China based on multiple CMIP5 models and different index definitions. *J. Hydrol.* **2021**, *600*, 126502. [CrossRef]
70. Xu, K.; Yang, D.W.; Yang, H.B.; Li, Z.; Qin, Y.; Shen, Y. Spatio-temporal variation of drought in China during 1961–2012: A climatic perspective. *J. Hydrol.* **2015**, *526*, 253–264. [CrossRef]
71. Wang, L.; Chen, W.; Zhou, W. Assessment of future drought in Southwest China based on CMIP5 multimodel projections. *Adv. Atmosp. Sci.* **2014**, *31*, 1035–1050. [CrossRef]
72. Easterling, D.R.; Kunkel, K.E.; Wehner, M.F.; Sun, L. Detection and attribution of climate extremes in the observed record. *Weather. Clim. Extrem.* **2016**, *11*, 17–27. [CrossRef]
73. Donat, M.G.; Lowry, A.L.; Alexander, L.V.; O'Gorman, P.A.; Maher, N. More extreme precipitation in the world's dry and wet regions. *Nat. Clim. Chang.* **2016**, *6*, 508–513. [CrossRef]
74. Allan, R.P.; Soden, B.J. Atmospheric warming and the amplification of precipitation extremes. *Science* **2008**, *321*, 1481–1484. [CrossRef] [PubMed]
75. Wang, C.H.; Wang, Z.L.; Kong, Y.; Zhang, F.M.; Yang, K.; Zhang, T.J. Most of the Northern Hemisphere permafrost remains under climate change. *Sci. Rep.* **2019**, *9*, 3295. [CrossRef]
76. Yang, K.; Wu, H.; Qin, J.; Lin, C.G.; Tang, W.J.; Chen, Y.Y. Recent climate changes over the Tibetan Plateau and their impacts on energy and water cycle: A review. *Global. Planet. Chang.* **2014**, *112*, 79–91. [CrossRef]
77. Kang, S.C.; Xu, Y.W.; You, Q.L.; Flügel, W.A.; Pepin, N.; Yao, T.D. Review of climate and cryospheric change in the Tibetan Plateau. *Environ. Res. Lett.* **2010**, *5*, 015101. [CrossRef]
78. Eyring, V.; Cox, P.M.; Flato, G.M.; Gleckler, P.J.; Abramowitz, G.; Caldwell, P.; Collins, W.D.; Gier, B.K.; Hall, A.D.; Hoffman, F.M.; et al. Taking climate model evaluation to the next level. *Nat. Clim. Chang.* **2019**, *9*, 102–110. [CrossRef]
79. Xu, Y.; Xu, C.H. Preliminary Assessment of simulations of climate changes over China by CMIP5 multi-models. *Atmosp. Ocean. Sci. Lett.* **2012**, *5*, 489–494.
80. Su, F.G.; Duan, X.L.; Chen, D.L.; Hao, Z.C.; Cuo, L. Evaluation of the Global climate models in the CMIP5 over the Tibetan Plateau. *J. Clim.* **2013**, *26*, 3187–3208. [CrossRef]
81. Keyantash, J.; Dracup, J.A. The quantification of drought: An evaluation of drought indices. *Bull. Am. Meteorol. Soc.* **2002**, *83*, 1167–1180. [CrossRef]
82. Cook, B.I.; Mankin, J.S.; Marvel, K.; Williams, A.P.; Smerdon, J.E.; Anchukaitis, K.J. Twenty-first century drought projections in the CMIP6 forcing scenarios. *Earth's Future* **2020**, *8*, e2019EF001461. [CrossRef]
83. Yadu, P.; Farshid, F.; Yusuke, S.; Julien, B.; Peter, B.; Anne, G.; Dieter, G.; Gosling, S.N.; Manolis, G.; Lukas, G.; et al. Global terrestrial water storage and drought severity under climate change. *Nat. Clim. Chang.* **2021**, *11*, 226–233.
84. Miguel, B.; Manuel, D.; Santiago, S.; Rocío, H.; Zhao, Y.C.; Gaitán, J.J.; Nicolas, G.; Hugo, S.; Vincent, M.; Anika, L.; et al. Global ecosystem thresholds driven by aridity. *Science* **2020**, *367*, 787–790.
85. Saleska, S.R.; Wu, J.; Kaiyu, G.; Araujo, A.C.; Huete, A.; Nobre, A.D.; Restrepo-Coupe, N. Dry-season greening of Amazon forests. *Nature* **2016**, *531*, E4–E5. [CrossRef] [PubMed]
86. RonnieJ, A.C.; María, B.; Jerónimo, P. Benchmarking of drought and climate indices for agricultural drought monitoring in Argentina. *Sci. Total Environ.* **2021**, *790*, 148090.
87. Sutanto, S.J.; Lanen, H.A.J.V.; Wetterhall, F.; Llort, X. Potential of Pan-European seasonal hydrometeorological drought forecasts obtained from a multihazard early warning system. *Bull. Am. Meteorol. Soc.* **2020**, *101*, E368–E393. [CrossRef]

remote sensing MDPI

Article

Interannual and Monthly Variability of Typical Inland Lakes on the Tibetan Plateau Located in Three Different Climatic Zones

Weiyao Ma [1,2], Ling Bai [1,2,3], Weiqiang Ma [1,4,5,6,*], Wei Hu [1,2], Zhipeng Xie [1,5], Rongmingzhu Su [1,2], Binbin Wang [1,5] and Yaoming Ma [1,2,4,5,6,7]

1 Land-Atmosphere Interaction and its Climatic Effects Group, State Key Laboratory of Tibetan Plateau Earth System, Resources and Environment (TPESRE), Institute of Tibetan Plateau Research, Chinese Academy of Sciences, Beijing 100101, China
2 College of Earth and Planetary Sciences, University of Chinese Academy of Sciences, Beijing 100049, China
3 Meteorological Bureau of Liangping District in Chongqing, Chongqing 405200, China
4 College of Atmospheric Science, Lanzhou University, Lanzhou 730000, China
5 National Observation and Research Station for Qomolongma Special Atmospheric Processes and Environmental Changes, Tingri 858200, China
6 China-Pakistan Joint Research Center on Earth Sciences, Chinese Academy of Sciences, Islamabad 45320, Pakistan
7 Kathmandu Center of Research and Education, Chinese Academy of Sciences, Beijing 100101, China
* Correspondence: wqma@itpcas.ac.cn; Tel.: +86-10-84097057

Citation: Ma, W.; Bai, L.; Ma, W.; Hu, W.; Xie, Z.; Su, R.; Wang, B.; Ma, Y. Interannual and Monthly Variability of Typical Inland Lakes on the Tibetan Plateau Located in Three Different Climatic Zones. *Remote Sens.* **2022**, *14*, 5015. https://doi.org/10.3390/rs14195015

Academic Editor: Guy J.-P. Schumann

Received: 30 August 2022
Accepted: 30 September 2022
Published: 9 October 2022

Abstract: Changes in lake water volume can reflect variations in regional hydrometeorology and are a sensitive indicator of regional environmental change. The Tibetan Plateau, referred to as the "Asian Water Tower", has a large number of lakes. These lakes are in a natural state and are relatively unaffected by human activities. Understanding the changes to lake water volume is a key issue for the study of lake-atmosphere interactions and the effects of lake expansion and contraction on regional climate. By using multisource remote sensing and water level observations, this study systematically analyzed inter-annual changes from 1970 to 2021 of three typical inland lakes basin (Bamu Co-Peng Co basin, Langa Co-Mapum Yumco basin andLongmu Co-Songmuxi Co basin), which are located in different climatic regions of the Tibetan Plateau and monthly changes from 2019 to 2021 of Bamu Co, Langa Co and Longmu Co in the lake area, water level, and water volume. In addition, the study analyzed the response of lakes in different climate regions to climate change from 1979 to 2018. The main conclusions are as follows. (1) From 1970 to 2021, there were similar trends in lake changes between the primary and twin lakes. (2) The changes to lakes in different climatic regions are different: lakes in the monsoon-dominated region showed a significant trend of expansion from 2000 to 2014, but the trend slowed down and stabilized after 2014; lakes in the westerlies-dominated region showed a small expansion trend; lakes in the region affected by both westerlies and the monsoon showed an overall shrinking trend. (3) The monthly variation of lake water volume showed a trend of first increasing and then decreasing, with the largest relative change of lake water volume in August and September. (4) Precipitation is a dominant factor controlling lake variation during the year. (5) Temperature and precipitation are dominant meteorological elements affecting the decadal variation of the lake, and with the warming of the TP, temperature plays an increasingly important role.

Keywords: Tibetan Plateau; typical inland twin lakes; change of water volume; multisource altimetry data; climate zones

1. Introduction

The Tibetan Plateau (TP) has an average elevation of over 4000 m. This region is known as the "Asian Water Tower" as it contains numerous lakes, covering a total area of 5×10^4 km^2 in 2018 [1], and is the birthplace of many large rivers [2]. Most lakes on the TP are in a natural state and are relatively unaffected by human activities. Under the background of TP warming and wetting in recent decades, the change in lake water volume is one of the most sensitive indicators of regional environmental change [3]. The change in lake water volume is the result of the combined action of variability in water level and lake area. The magnitude of these changes determines the heat absorbed and released by lakes, which affects the energy exchange between the land and the atmosphere. As such, lakes need to be closely monitored and are good sentinels of climate change [4–6].

The rapid development of remote sensing, multispectral mapping, and satellite altimetry has led to great improvements in understanding the variations of TP lakes in recent years [7,8]. Multispectral optical remote sensing images, such as the Landsat MSS/TM/ETM+/OLI, Gaofen series satellites (GF−1 and GF−2), and Sentinel data (Sentinel−2), have enabled lake mapping since the 1970s [7–9]. Many previous studies have analyzed the changes in water volume in a single lake on the TP. For example, Zhu et al., 2010, used remote sensing, meteorological data, and field-measured water depth to analyze the temporal and spatial changes in water volume in Nam Co from 1970 to 2004. This study found that lake water volume increased from 78.32 km^3 to 86.38 km^3 across this period at a rate of 2.37 km^3/a [10]. Zhang et al. used field-measured water depth and the area extracted from remote sensing images of Nam Co to quantify the lake water volume and its variation in 2011. The result shows that the water volume of Nam Co increased by 84.24 km^3 from 1976 to 2009 [11]. Qiao et al. used field bathymetry data, remote sensing images, and satellite altimetry of Chibuzhang Co and Duoersuodong Co in 2019. This study showed that these two lakes increased by 2.4 km^3 and 2 km^3 from 2003 to 2014, accounting for 24.5% and 14.1% of the original lake water, respectively [12].

Additional studies have analyzed changes in the lake water volume across the entire TP. Zhang et al. used SRTM DEM (30 m) data to estimate the water volume change of lakes with an area greater than 1 km^2 from 1976 to 2019 in 2021. The results show that lake water storage on the TP increased by about 170 km^3 mainly in the inflow area (158 km^3) [13]. The increase in water volume of glacial-supplied lakes (about 127 km^3) was much higher than that of non-glacial-supplied lakes (43 km^3). This result was related to the large number and wide area of glacier-supplied lakes. In addition, the increase in the water volume of closed lakes (about 163 km^3) was much higher than that of outflow lakes (about 8 km^3). Luo et al. further estimated the water storage change of 242 lakes on the TP from 2003 to 2019 to be 11.51 ± 2.26 km^3/a by integrating ICESat/ICESat−2, the global surface water dataset, and the HydroLAKES dataset in 2021 [14].

At present, most research on the change of lake water volume on the TP is based on the analysis of long-term interannual variability. There are relatively few studies on the trend of water level changes in lakes and the dynamic characteristics of lake areas during the year in different climate zones. The change in lake water volume can accurately reflect changes to the regional climate and hydrological processes and is a sensitive indicator of environmental change. Lake effects affect climate change mainly by affecting the energy budget. The energy absorbed and released will affect the energy exchange between land and the atmosphere and affect the regional climate [15]. Therefore, records of lake water volume changes provide valuable information for understanding how lakes respond to climate change.

Due to the remote location and a lack of observational data, there is currently limited research about intra-annual water volume changes and spatial differences of lakes on the TP. This paper selects three lakes in the westerlies-dominated, monsoon-dominated, and westerlies–monsoon interaction zones of the TP and analyzes the differences in lake water volume change between these regions (Figure 1). First, the monthly lake water area of the three lakes was extracted using multisource remote sensing satellite images and the Normalized Difference Water Index. Then, the variations in lake water level were calculated from remote sensing altimetry data and in situ observations of water level. Finally, the water volume changes of the three lakes and their spatial differences were analyzed by quantifying the estimated change in the lake area and water level.

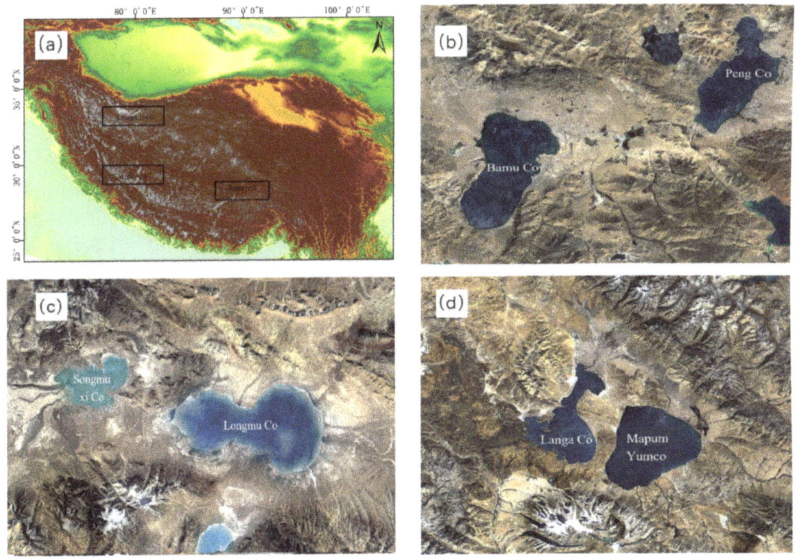

Figure 1. (**a**) Area of the basins studied, (**b**) BMC-PC basin, (**c**) LMC-SMXC basin, (**d**) LAC-MPYC basin.

2. Overview of the Study Area

We analyzed three groups of inland lakes in different climate regions for our study. Firstly, according to the atmospheric circulation field and local dry/wet conditions in the area, we divided the TP into the monsoon region, westerly region, and monsoon–westerly interaction region. Then three groups of lakes with large adjacent lakes and similar topographic conditions in the catchment area were selected from different climate zones, which were Bamu Co (BMC) and Peng Co (PC) in the monsoon zone, the Longmu Co (LMC) and Songmuxi Co (SMXC) in the westerly zone and the Langa Co and Mapum Yumco (MPYC) in the monsoon–westerly interaction zone.

The BMC (31.25°N, 90.58°E)–PC (31.52°N, 90.97°E) basin locate in the east of Bangor Country, in the Tibetan Autonomous Region of China, and it is situated in the north of the Nyenchen Tanglha Mountain. The BMC extends in a north-south direction and is a relatively regular rectangular shape with an average altitude of 4566 m, a lake area of 242.05 km^2 and the PC shows a shape of narrow in the north and wide in the south with an average altitude of 4553 m, the lake area of 175.43 km^2. The basin belongs to the sub-frigid and semi-arid climate region of the TP, which is mainly affected by the monsoon; the mean annual temperature is 0.0~2.0 °C, with the highest average temperature in July and the lowest in January; the annual precipitation is 300~400 mm, and multi-year average precipitation from June to September is 263.30 mm [16]. The area of the catchment is 4839.2 km^2 and 1020.3 km^2 with a supply factor of 25.3 and 7.5, respectively. The two lakes both mainly rely on surface runoff for supply, in which BMC has nine inflow rivers, and among which the Baisangsangqu is mainly a supply river, with a drainage area of 1890.0 km^2, accounting for 38% of the total lake catchment area, and PC has 11 inflow rivers, among which the Dangxiong Co is mainly a supply river [17].

The LAC (30.69°N, 81.22°E)–MPYC (30.67°N, 81.47°E) basin locate in Pulan County in Pulan County in the Ali area of the Tibetan Autonomous Region of China, and there are two high mountains facing each other in the south and north of the basin: The Naimona'nyi Peak on the south side is 7728 m above sea level with developed glaciers; the Kangrinpoche Mountain on the north side is 6656 m above sea level and covered with ice and snow the whole year. There are five tiers of ladder plates in the east and west of the basin, which are 1.5~2.0, 4.0, 8.0~10.0, 13.0~15.0, and 27.0~30.0 m higher than the lake surface, respectively.

The LAC shows a spoon shape, with an average altitude of 4567 m and a lake area of 252.43 km^2, where the northern part is an alluvial fan with a depth of 10 m. The middle part of the lake is narrow and long, with a depth of about 25 m; and the MPYC exhibits a trapezoidal shape that is wide in the north and narrow in the south, with an average altitude of 4572 m, the lake area of 414.23 km^2. The basin locates in a semi-arid region with mountainous shrubs and grasslands, which is a mixed climatic zone that is affected by both the westerly belt and the monsoon, the mean annual temperature is about 2.0 °C, with the highest average temperature in July and the lowest in January, and the lake surface temperature reaches a maximum in August [18]; the annual precipitation is 150~200 mm. The area of the catchment is 2551.5 km^2 and 4148.0 km^2 with a supply factor of 9.5 and 10.0, respectively. The LAC mainly relies on precipitation and the Ganga River and Nagqu River in the north of the lake for the supply, and it also receives water from MPYC in the wet season. The MPYC mainly relies on precipitation and surface runoff supply [17].

The LMC (34.61°N, 80.45°E)–SMXC (34.60°N, 80.25°E) basin locate in the Ritu County of the Tibetan Autonomous Region, and it is situated in the northeast of the lowest part of the Quaternary sedimentary basin on the north side of the Karakoram. The two lakes belonged to the same great lake in the Quaternary Period, and now these two lakes are twin lakes. The LMC shows a gourd shape, with an average altitude of 5012 m and an area of 107.95 km^2; the average altitude of the SMXC is 5015 m, and the area is 32.64 km^2. The climate in the basin is cold and dry and mainly affected by the westerly belt, with an average annual temperature of −8.0 °C and annual precipitation of 75–100 mm. The area of the catchment is 570.0 km^2 and 1605.0 km^2 and the supply factor of 5.9 and 64.2, respectively. The LMC mainly relies on underground runoff for supply, and there are three main rivers in the east of the lake; the SMXC relies on the surface runoff for supply, and there are two main inflow rivers [17].

The main characteristics of the three inland lakes are shown in Table 1.

Table 1. Main characteristics of the lakes in the study region.

	Bamu Co—Peng Co Basin		**Langa Co—Mapum Yumco Basin**		**Longmu Co—Songmuxi Co Basin**	
Climate Region	**Monsoon Region**		**Westerly–Monsoon Interaction Region**		**Westerly Region**	
Mean Annual Temperature (°C)	0.0~2.0		2.0		−8.0	
Annual Precipitation (mm)	300~400		150~200		75~100	
lake	BMC	PC	LAC	MPYC	LMC	SMXC
Latitude	31.25°N	31.52°N	30.69°N	30.67°N	34.61°N	34.60°N
Longitude	90.58°E	90.97°E	81.22°E	81.47°E	80.45°E	80.25°E
Altitude (m)	4566	4553	4567	4572	5012	5015
Lake Area (km^2)	242.05	175.43	252.43	414.23	107.95	36.24
Catchment Area (km^2)	4839.2	1220.3	2551.5	4148.0	570.0	1605.0
Supply Factor	25.3	7.5	9.5	10.0	5.9	64.2
Supply Type	Surface runoff		Surface runoff and Precipitation		Underground runoff	Surface runoff

The climate of the TP is mainly influenced by monsoons (including the Indian monsoon and East Asian monsoon) and westerlies [19,20]. The monsoon brings abundant rainfall to the affected areas during the summer, while the westerlies bring cold and dry weather in winter [21], which will cause different impacts on lakes in different climate zones, resulting in different expansion or contraction trends. The interaction of atmospheric circulation patterns in the TP leads to an uneven distribution of precipitation: The annual precipitation of the BMC-PC basin controlled by the monsoon zone is 300~400 mm; that of the LMC-SMXC basin controlled by the westerly zone is 75–100 mm, and that of LAC-MPYC basin controlled by the westerly–monsoon interaction zone is 150–200 mm, which is between the monsoon zone and the westerly zone.

3. Data and Methods

3.1. Data Source

3.1.1. Satellite Imagery

Multisource remote sensing images were used to extract the annual lake area changes of the six lakes from 1970 to 2021 and monthly lake area changes of the three main lakes from 2019 to 2021 in order to analyze the temporal and spatial variations across the region. Landsat remote sensing images were mainly used to quantify the annual lake area from 1970 to 2021. Sentinel−2 remote sensing images were mainly used to extract monthly lake area data from 2019 to 2021. The Landsat project has completed seven missions since 1972 and provided a long time-series of observations for monitoring water resources on the Earth's surface [9,22–25]. We selected Landsat images with little or no cloud cover and the highest satellite repetition period. Sentinel-1 remote sensing images are supplemented into the dataset. Due to the lack of complete data from 1970 to 2013, we calculated the mean lake area from 1970 to 1990 as the lake area of the 1970s. From 1990 to 2010, we calculated the lake area every five years.

The Landsat images used in this study were collected using the Multispectral Scanner (MSS) of Landsat1-3, Thematic Mapper (TM) of Landsat 5, Enhanced Thematic Mapper Plus (ETM+) of Landsat7 and Operational Land Imager (OLI) of Landsat8. Sentinel−2 is a multispectral imaging satellite developed by the European Space Agency. These data not only have high resolution and a short revisit period but are also freely available to the public [26]. Sentinel−2A and Sentinel−2B satellites were launched and put into use on 23 June 2015 and 7 March 2017, respectively. The two satellites are polar orbit satellites, with an orbital altitude of 786 km and an inclination of 98.62°E. The swath width is 290 km, and the orbital period is 100 min [27].

Under good meteorological conditions, dual-satellite synchronization can shorten the satellite revisit period to 5 days and achieve full coverage in 2–3 days in mid-latitude regions [9]. Sentinel−2A/B satellite is equipped with a multispectral imager MIS (Multi-spectral Instrument). The sensor is arranged with 12 detection elements along the scanning direction. The ground reflection spectrum of the scanning zone is obtained by push-broom scanning. The imager can obtain 13 detection elements. The green light band (B3) and the near-infrared band (B8) are mainly used to calculate the water body index in this study.

Sentinel−1 contains four imaging modes: SM (Strip map), IW (Interferometric Wide swath), EW (Extra-Wide swath), and WV (Wave). The SM, IW, and EM modes include single polarization (HH/VV) and dual polarization ((HH + HV)/(VV + HV)) imaging. The WV mode contains single polarization (HH/VV) imaging. The acronyms VV, HH, HV, and VH refer to different transmit and receive angles, which are vertical transmit/vertical receive, horizontal transmit/horizontal receive, horizontal transmit/vertical receive, and vertical transmit/horizontal receive, respectively, with a single co-polarization. VV + VV and VV + HV refer to dual-frequency cross-polarization with different transmit and receive angles [28]. End users have access to a variety of applications, including surface water monitoring [29]. To ensure the accuracy of the data, the lake area of this study was also compared with a long-term dataset of lake areas on the Tibetan Plateau (1970–2013) [30], and the changing trend was similar.

3.1.2. Satellite Altimetry

Satellite altimetry technology refers to using altimeters carried by satellites to measure the height of the Earth's surface. There are two major categories of satellite altimeters: laser and radar. Laser altimeters are mainly used in the Ice, Cloud, and land Elevation Satellite (ICESat). Laser altimeters have smaller footprints and higher accuracy than radar altimeters [31,32] but a lower time resolution. In this study, we use multisource altimetry data (CryoSat−2, Jason−2, Sentinel−3B, and ICESat−2). These data were combined if available. A brief description of the data is shown in Table 2 below.

Table 2. Summary of the multisource altimetry data used in this study.

Mission	Sensor	Type	Duration	Repetition Period (day)	Footprint Interval (m)	Footprint Diameter (km)	Data Source
CryoSat−2	SIRAL	radar	2010-	369	280	1.65	ESA
Jason−2	Poseidon-3	radar	2008-	10	300	2–4	CNES Aviso+
Sentinel−3B	SRAL	radar	2018-	27	291–306	>2	ESA
ICESat−2	ATLAS	laser	2018-	91	90	0.0175	NASA

3.1.3. Water Level Observational Data

The water level in the lakes was measured using an automatic water level gauge (HOBO−U20). The pressure water level gauge can observe small changes (1 mm) in the lake water level. Two HOBO−U20 loggers were placed on the long axis of the BMC, LAC, and LMC. The field environment restricted measurements of all six lakes. The water level gauges observe hourly water levels and obtain average lake water level data every half hour/day. Due to the climate and geographical environment of the TP, the shallow layers of the lakes are frozen in winter, and the movement of the ice will damage the instruments. Therefore, water level data are mainly concentrated in summer (May, June, and July) and autumn (August, September, and October). Due to the impact of COVID-19 prevention policies, it was not possible to replace the batteries in the water level gauge, resulting in partial data loss. The missing data were filled using satellite altimetry data.

3.1.4. Meteorological Data

This study mainly used the dataset of lake-catchment characteristics for the Tibetan Plateau (LCC−TP v1.0) [33]. The data are the first dataset of lake-catchment characteristics on the TP; it can provide fundamental data for the study of lakes in the TP. The climate variables mainly used the grid-based CMFD dataset to calculate the catchment-level climate characteristics; the CMFD was constructed through the fusion of in situ observation, remote sensing data, and reanalysis datasets, which improved the data quality in the TP. By reprocessing the CMFD dataset, the LCC−TP v1.0 dataset obtained the meteorological data from 1979 to 2018 of lakes over 0.2 km^2 with a temporal resolution of 1 day. We extracted basic meteorological data (temperature, precipitation, wind), which are related to the change of lake for catchments of the BMC, PC, LMC, SMXC, LAC, and MPYC. Moreover, the monthly precipitation from 2019 to 2021 is collected by the automatic weather station and rain gauge set around the BMC, LMC, and LAC.

3.2. *Calculation Method*

3.2.1. Extraction of Water Area

The extraction of water bodies from remote sensing images mainly uses the different spectral characteristics of different ground objects on remote sensing images. There are many satellites remote sensing data available for lake mapping on the TP. The method for extracting the lake area is generally referred to as the Normalized Difference Water Index (*NDWI*), which is a reliable method for the extraction of lake water information on the TP [9]. The normalized difference water body index method was proposed by Mcfeeters [34]. The method highlights the water body and weakens the background environment. The most basic formula (1) is:

$$NDWI = \frac{Green - NIR}{Green + NIR} \tag{1}$$

where *Green* represents the green wave band, and *NIR* represents the near-infrared band.

In this study, the long time-series lake area is mainly extracted by the data of Landsat and Sentinel−1 SAR data, and the monthly lake area from 2019 to 2021 is mainly extracted by the data of Landsat−8 and Sentinel−2 remote sensing images. The data are processed by GEE (Google Earth Engine) platform; the GEE is a comprehensive platform for geographic information data processing and visualization launched by Google. It provides remote sensing data, including Sentinel and Landsat and topographic data, and it can process data online. The Sentinel−1 data provided by the GEE have been preprocessed for thermal noise removal, radiometric calibration, terrain correction, and fringe processing. We adopt the SVM (Support Vector Machine) method to extract the lake water body. The SVM is a kind of generalized linear classifier that classifies data bivariate according to supervised learning. The decision boundary is the maximum-margin hyperplane solved for the learning sample. It can be classified nonlinearly by kernel method, which is one of the common kernel learning methods and is widely used in pattern recognition. In this study, we use the hard margin linear SVM to classify the water body using NDWI; the ultimate goal of this method is to calculate an optimal separating hyperplane based on the calculated NDWI as the identification boundary between land and water bodies (2).

$$m = \frac{2}{\|w\|} \tag{2}$$

where m is the margin of the optimal separating hyperplane, and $\|w\|$ means the two norms of each element. The calculation result of NDWI, which are larger than the optimal separating hyperplane, can be recognized in the water body.

The first step is the processing of remote sensing images, which is carried out on the GEE platform. This preprocessing involves cloud removal and atmospheric correction on the remote sensing images. We then use the NDWI to identify lakes and a machine learning method to classify lake water bodies. This study adopts the SVM method, which enables accurate identification and classification of water bodies. After the water body is determined, the vectorized processing is performed on ArcGIS. The monthly water body area of the lake is finally obtained through manual sight translation.

The second step is the processing of remote sensing images, including a topographic correction and speckle noise removal. The fluctuation of terrain causes significant geometric distortion of remote sensing images and can lead to perspective shrinkage, overlapping, shadows, and other phenomena. In addition, when the echoes of continuous radar pulses are processed over a rough surface, the superposition of reflected electromagnetic waves and the different distances between each scatterer and the sensor means that the echoes are incoherent in phase, resulting in echoes. Pixel-by-pixel variation in intensity, which appears grainy in pattern, resulting in randomly distributed black-and-white specks in the image. In order to address these issues, the images are topographically corrected, and speckle noise is removed. The water body index is then calculated as shown in Equation (3):

$$NDWI = \ln(10 \times VV \times VH) \tag{3}$$

After determining the water body index, we export the image and carry out a vectorization processing in ArcGIS to produce a water area of the lake every month through visual interpretation (VI).

3.2.2. Extraction and Calculation of the Lake Water Level

The first step is the processing of the observational data of the lake water level. The HOBO water level gauge is used to observe the lake water level, producing half-hourly lake water level data. As the gauge use pressure changes to record changes in water level, air pressure data obtained from meteorological observations are subtracted from the water level to remove the influence of atmospheric pressure changes. The monthly average water level is calculated from the data. We mainly focus on the process of lake surface fluctuations and do not consider the absolute elevation of each lake surface. As such, we only use the relative change of lake water level of different months in the summer (May, June, July) and autumn (August, September, October) of the same year, compared to the lake level in May.

The second step involves processing the satellite altimetry data. Using the acquisition of ICESat−2 altimetry data as an example, we use basic geospatial data such as water bodies and lake boundaries to mask the longitude and latitude information of the spot center in the ATL13 product. Second, we extract the sub-satellite track of the ICESat−2 satellite in the lake area and perform photon signal delay and geophysical corrections. Third, we correct the spot heights corresponding to the six pulse signals (gravity level anomaly, level height, and projection deformation) and calculate the corrected instantaneous water level of the lake. Finally, a simple Normalized Median Absolute Deviation method (NMAD) is used to remove outliers and obtain the average lake level [35]. After obtaining the lake water level data from different satellite altimeters for the six lakes, we compare the mean lake water level for the same period to eliminate systematic errors between different satellite altimeters. We also use observational data as a reference for the altimetry-derived water levels to improve the accuracy of the data.

3.2.3. Estimation of Changes in Water Volume

Because the lake area is irregular, the lake water volume can be approximately simplified as an irregular platform, as shown in Equation (4):

$$\mathrm{V} = \frac{1}{3}H \times \left(S' + S + \sqrt{S' \times S}\right) \tag{4}$$

where S' and S are the lake surface area and the lake bottom area, respectively. H is the lake height. The lake water volume change is calculated from the difference between the upper and lower bottom areas of the two lakes, as shown in Equation (5):

$$\Delta V = \frac{1}{3}(H_2 - H_1) \times \left(S_1 + S_2 + \sqrt{S_1 \times S_2}\right) \tag{5}$$

where ΔV is the change of lake water volume from lake surface water level H_1 and area S_1 relative to water level S_2 and area S_2.

A flowchart showing the method for calculating the relative change in lake water volume is shown in Figure 2.

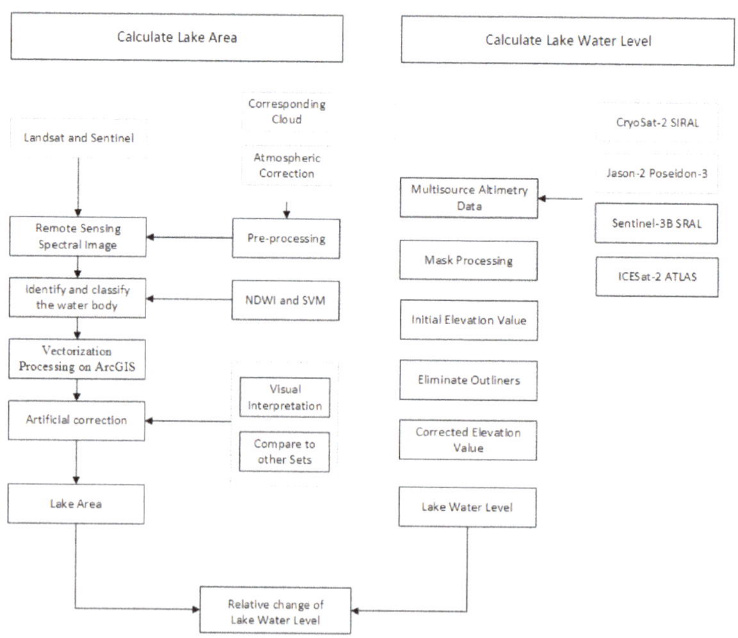

Figure 2. A flowchart illustrating the method used to calculate lake water volume change.

3.2.4. Calculating Lakes Changes Responses to Climate Change

In the study, we mainly used Pearson Correlation to analyze the response of lake changes to climate change. The Pearson correlation coefficient is widely used to measure the degree of correlation between two variables in natural sciences research. Among them, the correlation coefficient of the sample is denoted by r, and p is a probability value. If $p < 0.01$, it means that the null hypothesis is rejected at the 99% confidence level; $p < 0.05$ or 0.1 is the same, and the p-value can be used as an evaluation index of contribution. The formula is shown in formula (6).

$$r = \frac{\Sigma(X-\overline{X})(Y-\overline{Y})}{\sqrt{\Sigma(X-\overline{X})^2\Sigma(Y-\overline{Y})^2}} \quad (6)$$

where the X and Y means lake water area and meteorological elements in the same year.

4. Results and Analyses

4.1. Analysis of Lake Change

Analysis of Area Change

The lake area changes differently in different climate zones from the 1970s to 2021 (Figure 3). Lakes in the monsoon region (BMC and PC) and the westerly region (LMC and SMXC) show an overall expansion trend. The BMC and PC showed a trend of rapid expansion from 1995 to 2005, with growth rates of 4.746 km^2/a and 31.8098 km^2/a, respectively, before the trend stabilizes. The LMC and SMXC showed a steady growth trend from the 1970s to 2021, with growth rates of 1.092 km^2/a and 0.747 km^2/a, respectively. Lakes in the westerly–monsoon interaction zone (LAC and MPYC) showed an overall shrinking trend. Lake areas in this region show a rapid shrinking trend from the 1970s to 2005, with a reduced rate of 1.744 km^2/a and 1.061 km^2/a, after which the reduction slows or stabilizes. The three groups of primary and twin lakes have similar trends, suggesting they are connected by groundwater.

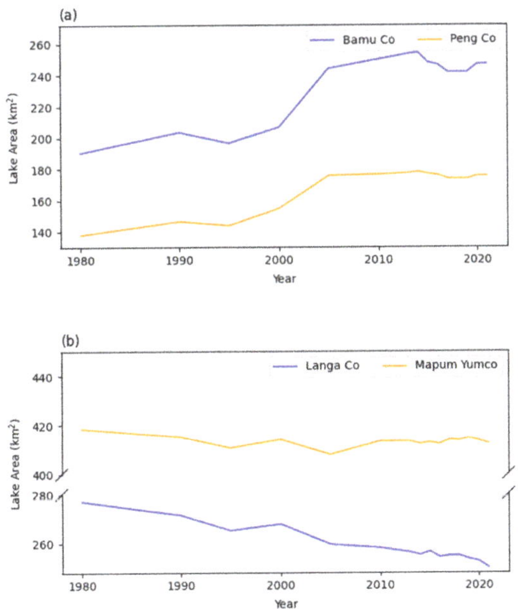

Figure 3. *Cont.*

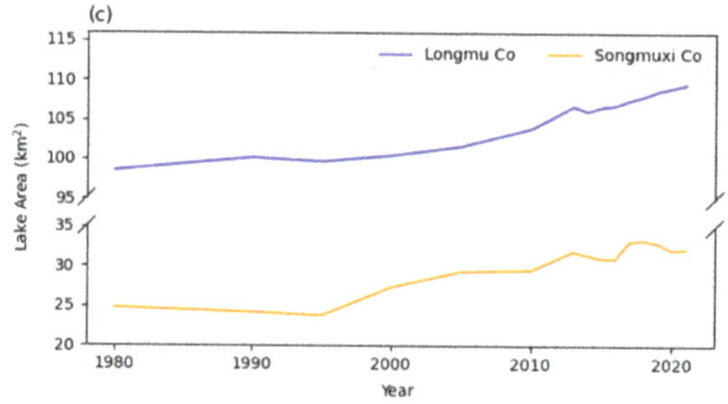

Figure 3. Lake area from 1970 to 2021, (**a**) lakes in the monsoon region, (**b**) lakes in the westerly–monsoon interaction region, (**c**) lakes in the westerly region.

The changes in the lake area of inland lakes on the TP are primarily affected by precipitation, snowfall, and melting of glaciers and permafrost. There are obvious seasonal differences in the supply of these water resources, meaning that changes in lakes within a year cannot be ignored. In order to make the investigation more accurate and make full use of the observed meteorological and lake water level data, the analysis of annual variance in lake water level focuses on 2019–2021. The lakes in the monsoon region and westerly–monsoon interaction region have similar trends during the year. Lakes in these regions both began to expand in May, reached the maximum in August then began to shrink. The lake area expanded slightly in winter after the lake surface froze. The timing of the maximum lake area in the westerly–monsoon interaction region was delayed by one month compared to the monsoon region.

It can also be seen in Figure 4 that the seasonal changes in the lake area from 2019 to 2021 are different. For example, at BMC in 2019, the lake area showed a decreasing trend from January to April. The lake ice started to melt in May, and the lake area expanded and reached a maximum in August before gradually retreating. In 2020, the lake area showed a decreasing trend from January to May. The lake ice began to melt in May, and the lake area began to expand. The lake area reached a maximum in August and then gradually retreated. In 2021, the lake area showed a decreasing trend from January to March, and the lake ice began to melt in May. The lake area began to expand in May and reached a maximum in September before gradually retreating.

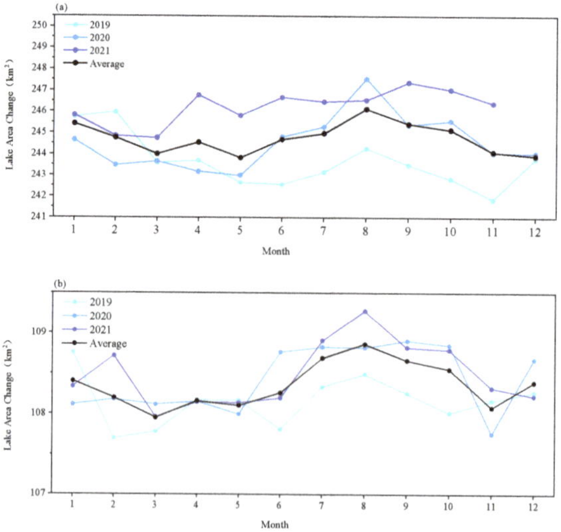

Figure 4. *Cont.*

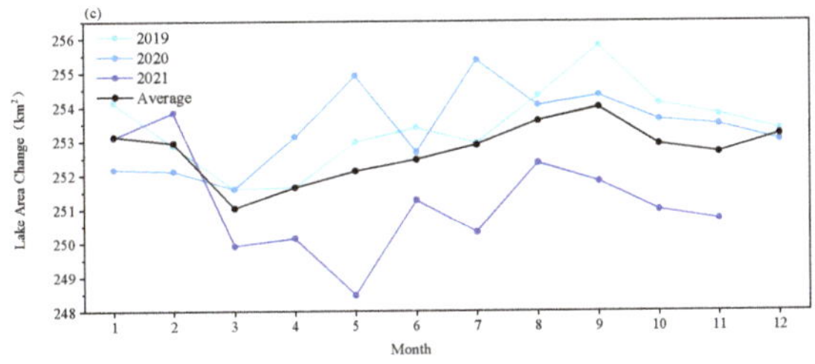

Figure 4. The annual lake area changes from 2019 to 2021, (**a**) BMC, (**b**) LMC, (**c**) LAC.

4.2. Analysis of Lake Water Level Change

The parameter of lake level change is very important for estimating the change of lake water volume and can represent climate change to a certain extent. Lakes located in different climate zones have different trends (Figure 5). The twin lakes have similar interannual variations in lake water level as their respective primary lakes.

In this study, lake water level change is mainly analyzed for seasonal characteristics. The change in lake water level in May is set to zero every year, and then the monthly average value of water level change during the non-freezing period (May to October) relative to that in May is calculated. In order to verify the accuracy of the ICESat−2 ATL13 remote sensing altimetry product, this paper selects 26 repeated ATL13 orbital data passing through the three lakes from May to October every year from 2019 to 2021. We then calculate the change of lake water level obtained by the ATL13 remote sensing altimetry products as the change of lake water level in the month relative to that in May. The changes in the repeater orbit data of ATL13 are not considered due to the relatively small amount of data obtained from BMC (Table 3). Correlation analysis between the relative change of lake water level from ATL13 and the observational data (Figure 6) showed R^2 values of 0.85. Therefore, the ICESat−2/ATL13 laser altimetry product can accurately reflect the relative change in lake water level.

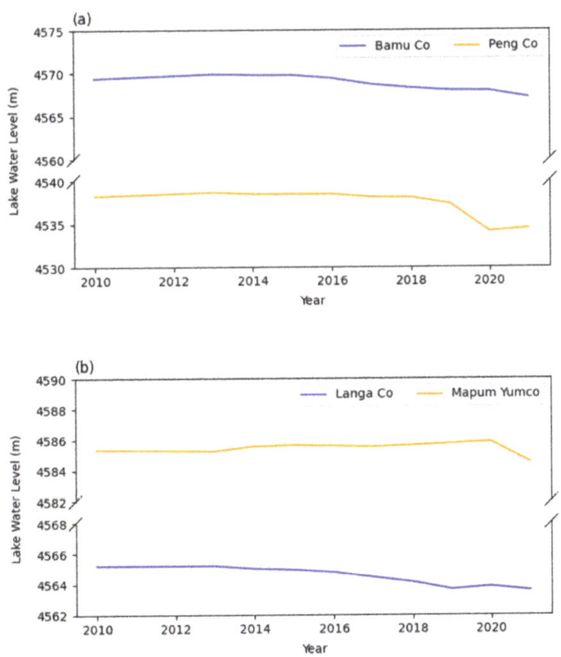

Figure 5. *Cont.*

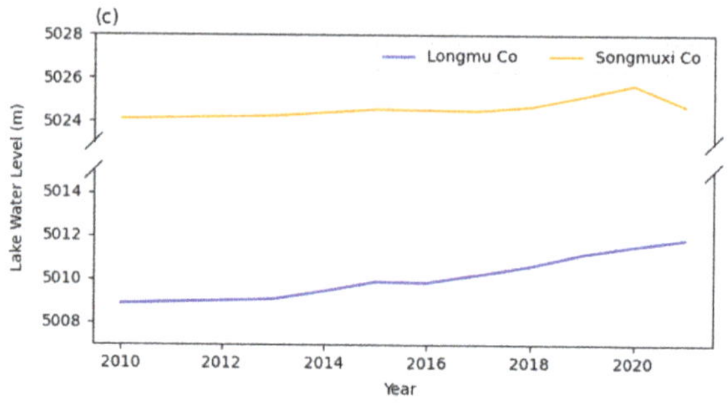

Figure 5. Lake water level from 1970s to 2021, (**a**) lakes in monsoon region, (**b**) lakes in westerly–monsoon interaction region, (**c**) lakes in westerly region.

Table 3. Repeat orbit data from ATL13 of Bamu Co, Langa Co and Longmu Co (m).

Data	Lake	May	June	July	August	September	October
2019	Bamu Co	4566.018	×××	×××	×××	×××	×××
	Langa Co	4567.530	4567.905	×××	4568.220	4568.026	×××
	Longmu Co	5012.203	5012.101	5012.166	5012.238	×××	×××
2020	Bamu Co	×××	×××	×××	×××	4566.523	4566.438
	Langa Co	×××	4567.656	×××	4567.706	4567.698	×××
	Longmu Co	5012.485	×××	5012.507	×××	5012.460	×××
2021	Bamu Co	4569.784	×××	×××	×××	×××	×××
	Langa Co	4567.060	4567.039	×××	×××	4567.147	×××
	Longmu Co	5012.610	5012.689	5012.750	×××	5012.736	5012.688

××× means lack of data.

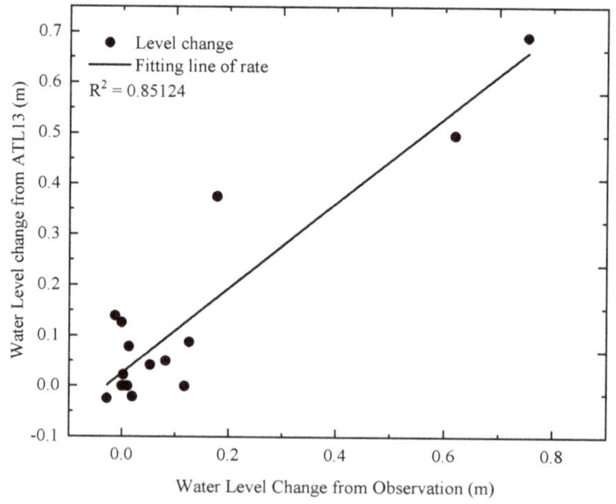

Figure 6. Water level change rate from ICESat−2 and observation.

As is shown in the Figure 7, the relative lake water level change of BMC reached maximum values in September 2019 (0.32 m), August 2020 (0.23 m), and September 2021 (0.42 m). LMC reached maximum values in August 2019 (0.17 m), August 2020 (0.02 m), and September 2021 (0.01 m). LAC reached maximum values in September 2019 (0.75 m), August 2020 (0.07 m), and August 2021 (0.01 m).

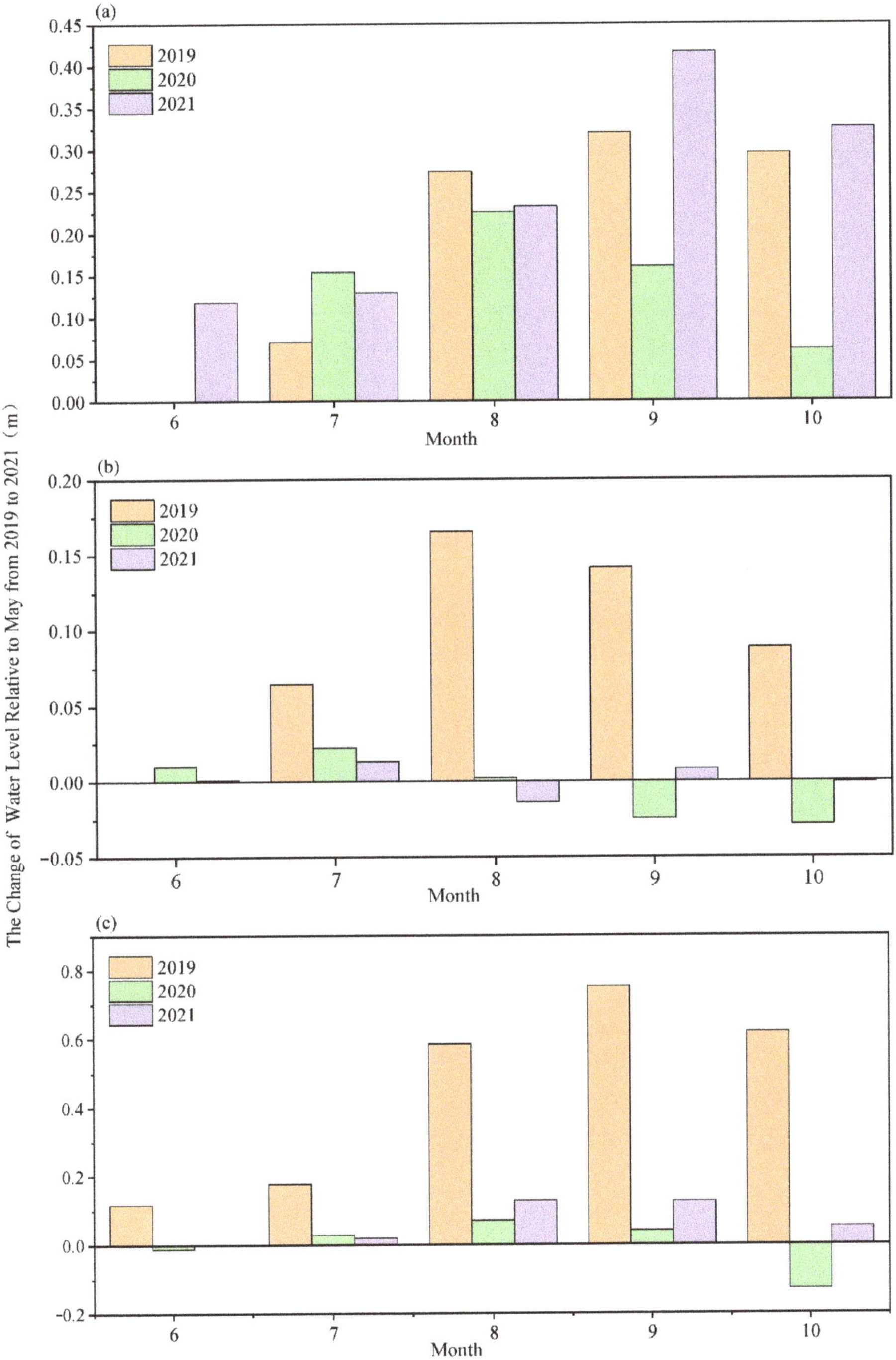

Figure 7. The relative change of water level in (**a**) BMC, (**b**) LMC, (**c**) LAC.

4.3. *Analysis of Relative Change in Lake Water Volume*

The change in lake water volume is a key factor for evaluating the degree of lake impacts on climate. Changes in lake volume are the result of the combined action of water level change and lake area change. By using the obtained lake water level change and lake area change, the lake water volume change can be estimated following Equations (3) and (4) [36,37].

The inter-annual relative change of lake water level from 2013–2021 is calculated relative to 2011 (Figure 8). The inter-annual relative change of lake water level showed a significant difference in the different climatic regions. At the BMC and PC, in the monsoon region, the water volume of the lake was relatively stable before around 2017 but then decreased significantly. LAC in the westerly–monsoon interaction region showed a significant decreasing trend, while MPYC was stable. LMC and SMXC in the westerly region showed a significant increasing trend. However, the growth rate of SMXC is an order of magnitude lower than LMC.

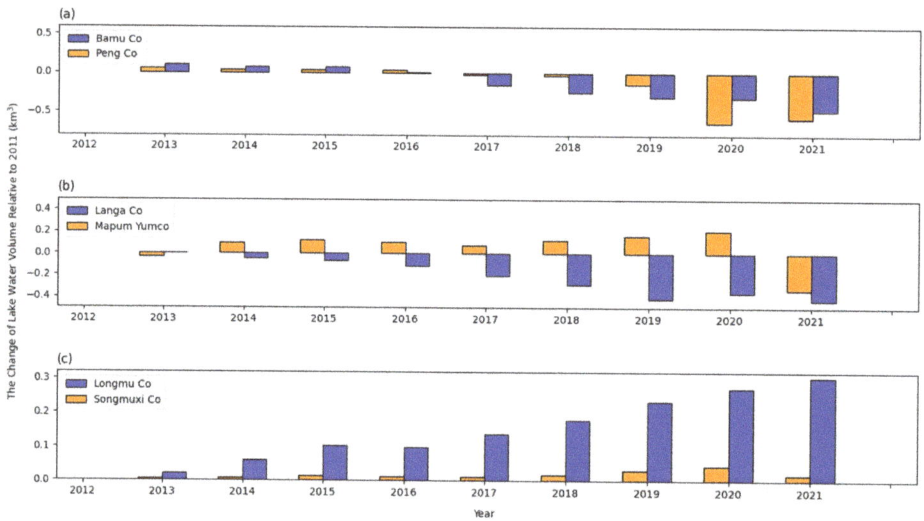

Figure 8. Relative change of lake water volume from 2013 to 2021, (**a**) lakes in the monsoon region, (**b**) lakes in the westerly–monsoon interaction region, (**c**) lakes in the westerly region.

Differences in lake water volume change are significantly related to temporal and spatial variations in precipitation. From 1980–2015, the monsoon region showed a warm-wet tendency ($\Delta T = 0.54$ °C/10a, $\Delta p = -12.44$ mm/10a), the Westerly–monsoon interaction region showed a cold-wet tendency ($\Delta T = -0.71$ °C/10a, $\Delta p = 37.66$ mm/10a) and the westerly region showed a warm–dry tendency ($\Delta T = -0.53$ °C/10a, $\Delta p = -12.44$ mm/a). The combined action of precipitation and melting of glacial snow cover causes lake expansion [38]. In the TP, which is relatively unaffected by human factors, variations in precipitation are the main drivers of inland lake change [39,40].

The annual relative change in 2019–2021 is calculated relative to the first month of the year for which data are available each year (Figure 9). The relative changes in lake water volume in typical inland lakes in different climatic regions are different in different seasons.

The lake water volume of BMC, located in the monsoon region, is relatively stable, and the overall lake water volume shows an increasing trend from 2019 to 2021. Among the annual changes in the lake water volume, the relative change of lake water volume shows a trend of first rising and then falling. The relative change of lake water volume reached maximum values in August 2019 (77.76×10^3 km^{-3}), August 2020 (55.61×10^3 km^3), and September 2021 (102.91×10^3 km^3). The lake water volume in LMC, located in the westerlies area, showed a significant change trend in 2019 but was stable in 2020 and 2021. The annual changes in the lake water volume at LMC reached maximum values in August 2019 (42.16×10^3 km^3), August 2020 (5.66×10^3 km^3), and June 2021 (8.67×10^3 km^3). The lake water volume of LAC, located in the westerly–monsoon interaction area, shows an overall decreasing trend from 2019 to 2021. The relative lake water volume change of LAC reached maximum values in September 2019 (81.44×10^3 km^3), August 2020 (7.66×10^3 km^3), and August 2021 (32.43×10^3 km^3).

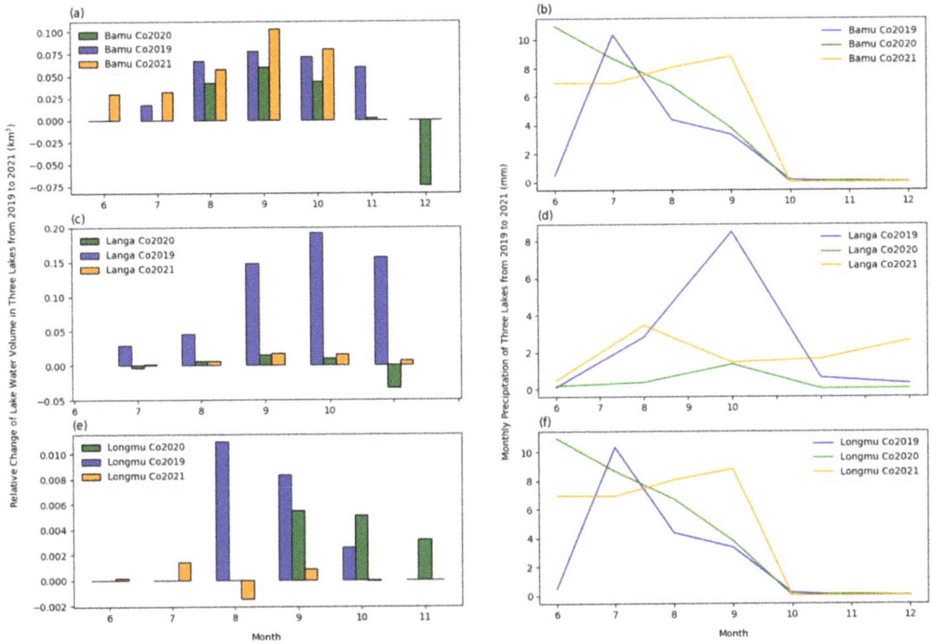

Figure 9. Monthly relative change of lake water volume and precipitation from 2019–2021, (**a**,**b**) lake water volume change and precipitation of BMC, (**c**,**d**) lake water volume change and precipitation of LAC, (**e**,**f**) lake water volume change and precipitation of LMC.

Abnormally high rainfall in the westerly region (LAC and LMC) in 2019 meant that the precipitation was significantly higher compared with other years. This high precipitation shows a good correlation with the obvious increase in lake water volume in 2019. In addition, there is also a good correlation between the peak of precipitation and the increase in lake water volume, indicating that precipitation is an important factor affecting lake change.

4.4. Response of Lake Change to Climate Change

4.4.1. Interdecadal Variation of Meteorological Elements

From 1979 to 2018, the temperature showed an overall increasing trend with different rates: BMC-PC basin in the monsoon region is 0.68 °C/10a; LMC-SMXC basin in the westerly region is 0.24 °C/10a; LAC-MPYC basin in the westerly–monsoon interaction region is 0.064 °C/10a (Figure 10), the annual mean temperature of the three basins are −1.96, −4.26 and −8.48 °C, respectively. Among them, we can see that the BMC-PC basin showed a rapid warming trend, while the LAC-MPYC basin was relatively stable, and it is worth noting that the LMC-SMXC basin was stable before 2000, but it sharply decreased in 2000 then increased significantly.

The annual cumulative precipitation in different climate regions showed an overall increasing trend with a ratio of 36.31, 12.40, and 119.12 mm/10a, respectively, from 1979 to 2018. Among them, the cumulative precipitation of the BMC-PC basin was relatively stable before 2000 but fluctuated significantly after 2000, that of the LAC-MPYC basin increased significantly after 2000, and that of the LMC-SMXC basin was stable before 1998 and then increased rapidly. The increasing trend of the cumulative precipitation rate was mostly bounded by the year 2000, and the cumulative precipitation increased significantly after 2000, which corresponded to the obvious increase in the lake area after 2000 (Figure 11).

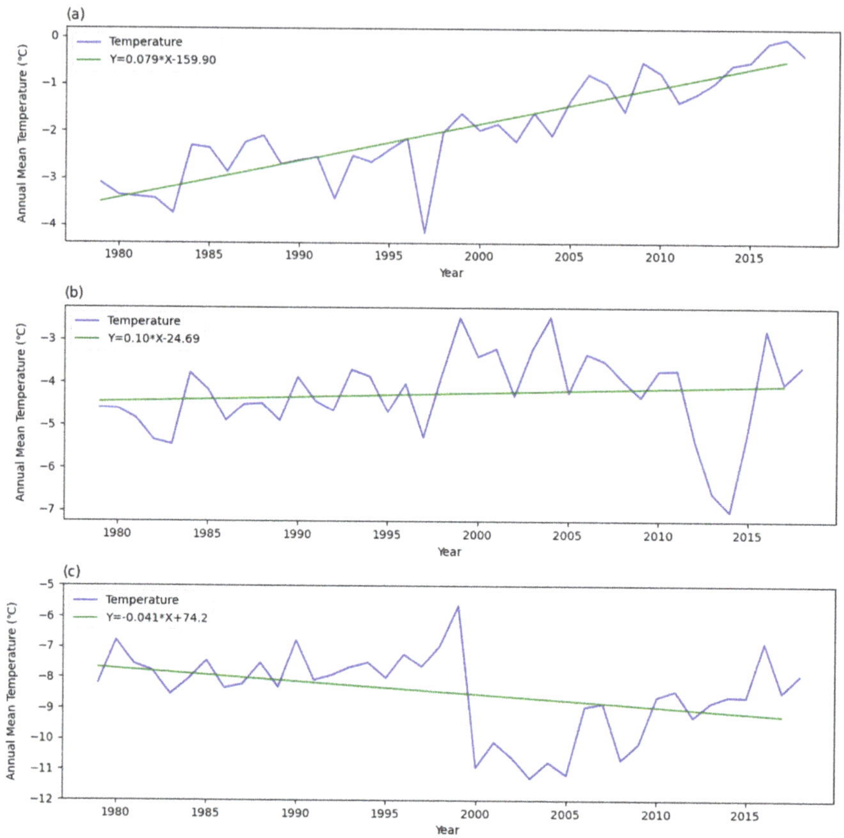

Figure 10. Annual mean temperature of three basins in different climate region, the green lines mean one linear fitting equation, (**a**) the BMC-PC basin, (**b**) the LAC-MPYC basin, (**c**) the LMC-SMXC basin.

The annual mean wind speed from 1979 to 2018 showed different trends in different climate regions: the wind speed of the BMC-PC basin decreased slightly, with the rate of -0.16 m/(s·10a), and the wind speed fluctuated significantly from 1990 to 2000, then stabilized; the wind speed of LAC-MPYC basin decreased with the rate of -0.20 m/(s·10a), the wind speed increased in 2000 then decreased rapidly; the wind speed of LMC-SMXC basin decreased slightly before 1997 then increased rapidly with the rate of 0.39 m/(s·10a) (Figure 12).

The annual average specific humidity is generally stable from 1979 to 2018, but there are abnormally large or small around 2000: the specific humidity of the BMC-PC basin in the monsoon region and the LMC-SMXC basin increased significantly from 1995 to 2000 then decreased gradually while that of the LAC-MPYC basin showed a contrary tendency (Figure 13).

The variation trends of these meteorological elements changed in 2000, and the variation trends of the lake also changed in the same period. Therefore, the response mechanism of lakes to climate change will be analyzed in different periods, and the responses will be discussed over the entire time, before and after 2000.

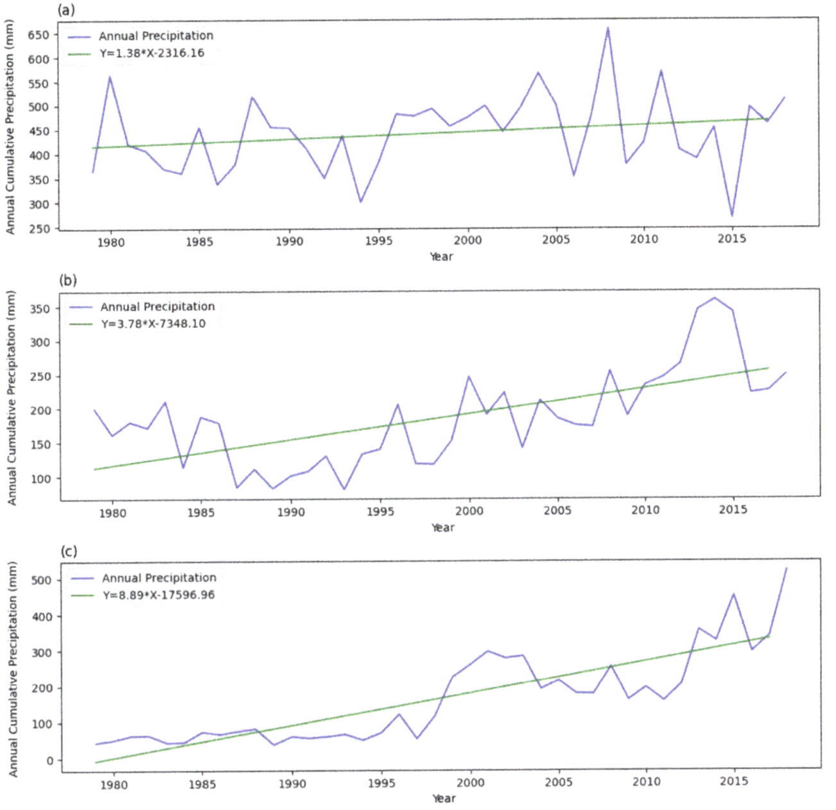

Figure 11. Annual cumulative precipitation of three basins in different climate region, the green lines mean one linear fitting equation, (**a**) the BMC-PC basin, (**b**) the LAC-MPYC basin, (**c**) the LMC-SMXC basin.

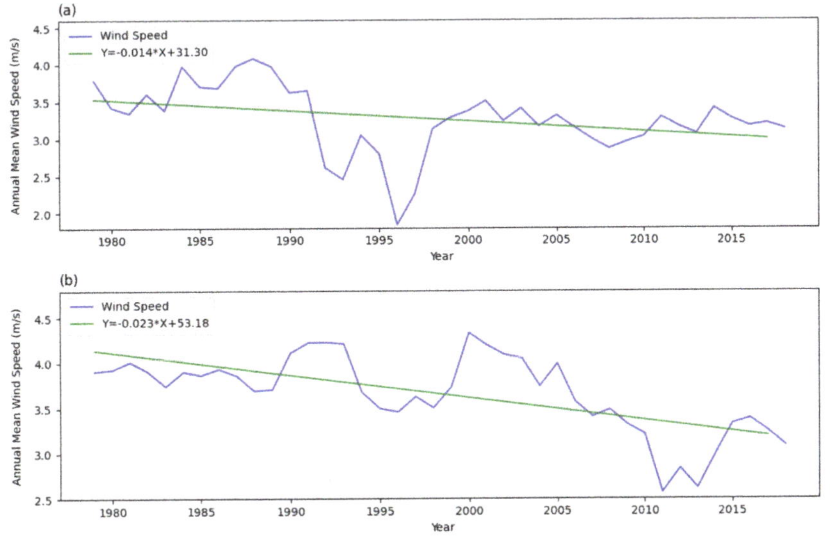

Figure 12. *Cont.*

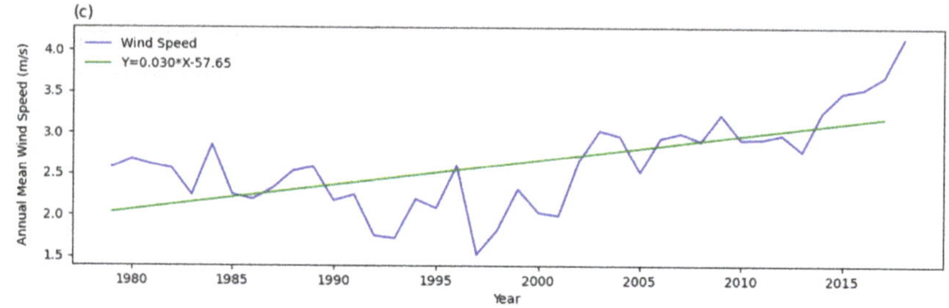

Figure 12. Annual mean wind speed of three basins in different climate region, the green lines mean one linear fitting equation, (**a**) the BMC-PC basin, (**b**) the LAC-MPYC basin, (**c**) the LMC-SMXC basin.

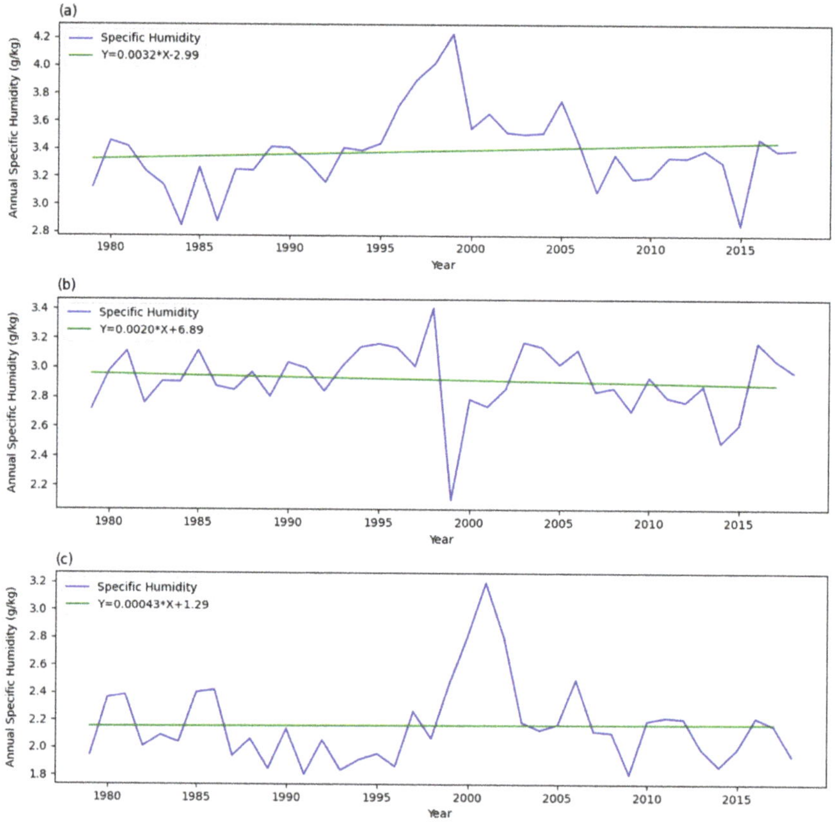

Figure 13. Annual Specific Humidity of three basins in different climate region, the green lines mean one linear fitting equation, (**a**) the BMC-PC basin, (**b**) the LAC-MPYC basin, (**c**) the LMC-SMXC basin.

4.4.2. Response of Lake Change to Climate

In this study, we analyzed the correlation between the lake area and the annual mean temperature, wind speed, specific humidity, and annual cumulative precipitation. Analysis results from 1979 to 2918 show that the main factors affecting the lake change are different from the change in climate zone. For the BMC and PC in the monsoon region, the correlation between temperature and lake area was the highest, with a correlation coefficient of 0.90 ($p < 0.01$), followed by specific humidity, with a correlation coefficient of −0.37 ($p = 0.23$), indicating a strong correlation; for the LMC and SMXC, the correlation between cumulative precipitation and lake area was the highest, with a correlation coefficient of 0.87 ($p < 0.01$) and 0.88 ($p < 0.01$), followed by specific humidity, with a correlation coefficient of −0.36 ($p = 0.24$), indicating a strong correlation; for the LAC and MPYC, the correlation between wind speed, cumulative precipitation, and lake area were both high, with a correlation coefficient of 0.78 ($p < 0.01$) and −0.71 ($p < 0.01$), respectively. In addition, the dominant meteorological factors also changed in different periods. Before 2000, the dominant factors in the monsoon zone, the westerly zone, and the monsoon–westerly interaction zone were temperature, precipitation, and wind speed, while after 2000, which is the period of acceleration of warming on the TP, the dominant factors changed to temperature, precipitation, temperature and wind speed, specific humidity, respectively (Table 4). The wind speed and specific humidity (which can represent evaporation) are the dominant meteorological elements in the LAC-MPYC basin and show a negative correlation; this explains the contraction trend of lake area changes in the LAC and MPYC.

Table 4. Correlation analysis between lake area and meteorological factors.

Year	Climate Region	Dominant Factors	Correlation Coefficient	p-Value
1979–2018	Monsoon Region	Temperature, Specific Humidity	0.90 −0.37	0.001 0.23
	Westerly–monsoon Interaction Region	Temperature Specific Humidity	0.78 −0.71	0.002 0.009
	Westerly Region	Precipitation Specific Humidity	0.88 −0.37	0.002 0.23
1979–2000	Monsoon Region	Temperature	0.69	0.30
	Westerly–monsoon Interaction Region	Wind Speed	0.60	0.28
	Westerly Region	Precipitation	0.62	0.38
2000–2018	Monsoon Region	Temperature	0.49	0.22
	Westerly–monsoon Interaction Region	Wind Speed Specific Humidity	−0.50 −0.42	0.24 0.34
	Westerly Region	Precipitation Temperature	0.80 0.71	0.02 0.05

In conclusion, although the dominant meteorological factors change with different climate regions, temperature and precipitation are always the main factors affecting the change of the lake. With the rapid warming of the TP, the influence of temperature on the expansion of the lake area is becoming more significant. The warming of the lake catchment area leads to the melting of glaciers and snow, which increases runoff and leads to the expansion of lakes. Moreover, with the warming and wetting of the TP, the precipitation gradually increases, which also promotes the expansion of lakes.

5. Discussion

In recent years, the TP has become warmer and wetter. This trend involves a decreasing surface sensible heat and increasing latent heat from the northwest to the southeast, resulting in a significant increase in precipitation in the southeast and a decrease in the northwest [41,42]. In addition to being affected by precipitation, lakes in the TP are also closely related to glacial meltwater and permafrost degradation in the basin. Lake changes in the TP are significantly influenced by water and heat exchange between the lakes and the atmosphere, which in turn affects the regional water cycle. However, quantifying the dominant factors affecting lakes in different regions and how these factors will change under climate change conditions is key to accurately understanding the mechanistic role that lakes play in the water cycle of "Asian Water Tower".

Therefore, in each typical area of the TP, we should strengthen research on lake water balance and its response to climate change. This will involve collecting observations of the spatial and seasonal distribution of precipitation, runoff (including precipitation, meltwater, and underground runoff), and evaporation, and using methods such as a total water balance and isotope segmentation to study the response of lake water balance to changes in different supply sources. Such research would provide valuable information about the processes and mechanisms of how climate change will impact lakes in the future.

The changes in the area, water level, and water volume, as well as the monthly water volume variations of three pairwise lakes from different climate zones, were analyzed using remote sensing data and in situ observations. For the response of lake change to meteorological factors, we mainly focused on qualitative research but lacked quantitative research on the contribution of each meteorological element. In addition, there is still a part of the research that has not been carried out, such as the reason for lake water decline after 2020 due to lack of meteorological data. Therefore, quantifying the contribution of meteorological elements to lake change and using high-frequency meteorological data and eddy covariance data to analyze the reduction of lake level after 2020 will be the focus of the next stage of research. This will be helpful in accurately understanding the mechanism of climate change affecting lakes. Moreover, with the development of tourism in Tibet, the TP has been affected by more and more human activities; consideration of direct human impacts on the TP water supply remains poorly articulated but potentially important to the lake change research [43,44].

6. Conclusions

There is a lack of research on the annual water volume changes and spatial differences of typical lakes in the TP due to the remote location and the lack of observational data. In the study, we have found there is obvious spatial heterogeneity in the seasonal changes of lake water volume in different climate regions. By using Sentine−2 remote sensing images, multisource altimetry, and observational water level data, the following conclusions are drawn.

(1) Inter-annual variations of lakes in different climatic zones are markedly different. From the 1970s to 2021, lakes in the monsoon (BMC and PC) and westerly (LMC and SMXC) regions show an overall expansion trend, while lakes in the westerly–monsoon interaction region (LAC and MPYC) showed an overall shrinking trend [45]. In the westerly–monsoon interaction region, the lake area shows a rapidly shrinking trend from the 1970s to 2005, after which the reduction slows or stabilizes [46]. The three groups of lakes have similar trends.

(2) Monthly variations of the lakes during the year in different climatic zones generally show similar trends. The changes are highly correlated with increases and decreases in monthly rainfall. This correlation is especially strong in 2019, which was a year of abnormal fluctuations in the westerly belt, with increased precipitation and significantly increased monthly changes in lake water volume. In addition, there is a good correlation between the peak of precipitation and lake water volume increase. These findings indicate that precipitation is a dominant factor affecting lake changes in the TP.

(3) The paper focuses on the effects of climate change on lakes from 1979 to 2018. The meteorological factors that dominate lake variation are temperature, precipitation, specific humidity, and wind speed (where specific humidity and wind speed can represent evaporation). Increases in temperature (which promotes melting of glaciers and snow) and precipitation promote the lake expansion, while increases in evaporation cause the lake shrinkage. For lakes in different climate regions, the main impact of meteorological elements is different, but with the accelerated warming on the TP, temperature plays an increasingly important role in accelerating lake expansion, while in the LAC-MPYC basin, evaporation is the leading factor that has caused the lake to shrink over the past decade.

Author Contributions: Conceptualization, W.M. (Weiyao Ma), L.B. and W.M. (Weiqiang Ma); methodology, W.M. (Weiyao Ma) and L.B.; validation, L.B., Z.X., W.H., R.S. and B.W.; formal analysis, W.M. (Weiyao Ma), W.H. and Z.X.; investigation, L.B.; resources, W.M. (Weiyao Ma), L.B., W.H. and R.S.; data curation, W.M. (Weiyao Ma) and L.B.; writing—original draft preparation, W.M. (Weiyao Ma) and L.B.; writing—review and editing, W.M. (Weiqiang Ma), Z.X., W.H., R.S., B.W., and Y.M.; visualization, W.M. (Weiyao Ma); supervision, W.M. (Weiqiang Ma); All authors have read and agreed to the published version of the manuscript.

Funding: This research has been funded by the National Natural Science Foundation of China (Grant No. 41830650); the Second Tibetan Plateau Scientific Expedition and Research Program (STEP) (Grant No. 2019QZKK0103) and the Strategic Priority Research Program of Chinese Academy of Sciences (Grant No. XDA20060101).

Data Availability Statement: The ICESat−2 data can be availability in the National Snow and ICE Data Center (https://nsidc.org/data/atl13/versions/5); The CryoSat−2 data can be availability in European Space Agency Earth Online (https://earth.esa.int/eogateway/missions/cryosat); The Sentinel 3B data can be availability in the Copernicus Open Access Hub (https://scihub.copernicus.eu/); The Jason−2 data can be availability in NOAA NCEI (https://www.ncei.noaa.gov/products/jason-satellite-products).

Acknowledgments: The authors would like to acknowledge the National Aeronautics and Space Administration (NASA) and European Space Agency (ESA) for providing remote-sensing and multisource altimetry data, the Institute of Tibetan Plateau Research, Chinese Academy of Science for providing the lake water level and precipitation data from the Bamu Co, Langa Co and Longmu Co, and the students who went to Tibet to collect data.

Conflicts of Interest: The authors declare no conflict of interest.

References

1. Zhang, G.; Luo, W.; Chen, W.; Zheng, G. A robust but variable lake expansion on the Tibetan Plateau. *Sci. Bull.* **2019**, *64*, 1306–1309. [CrossRef]
2. Immerzeel, W.W.; van Beek, L.P.H.; Bierkens, M.F.P. Climate change will affect the Asian Water Towers. *Science* **2010**, *328*, 1382–1385. [CrossRef]
3. Zhang, G.; Yao, T.; Xie, H.; Yang, K.; Zhu, L.; Shum, C.; Bolch, T.; Yi, S.; Allen, S.; Jiang, L.; et al. Response of Tibetan Plateau lakes to climate change: Trends, patterns, and mechanisms. *Earth Sci. Rev.* **2020**, *208*, 103269. [CrossRef]
4. Carpenter, S.R.; Benson, B.J.; Biggs, R.; Chipman, J.W.; Foley, J.A.; Golding, S.A.; Hammer, R.B.; Hanson, P.C.; Johnson, P.T.J.; Kamarainen, A.M.; et al. Understanding regional change: A comparison of two lake districts. *Bioscience* **2007**, *57*, 323–335. [CrossRef]
5. Pham, S.V.; Leavitt, P.R.; McGowan, S.; Peres-Neto, P. Spatial variability of climate and land-use effects on lakes of the northern Great Plains. *Limnol. Oceanogr.* **2008**, *53*, 728–742. [CrossRef]
6. Williamson, C.E.; Dodds, W.; Kratz, T.K.; Palmer, M.A. Lakes and streams as sentinels of environmental change in terrestrial and atmospheric processes. *Front. Ecol. Environ.* **2008**, *6*, 247–254. [CrossRef]
7. Wang, W.Z.; Huang, D.; Liu, J.F.; Liu, H.W.; Wang, H. Patterns and causes of changes in water level and volume in Tangra Yumco from 1988 to 2018 based on Landsat images and Sentinel-3A synthetic aperture radar. *J. Lake Sci.* **2020**, *32*, 1552–1563.
8. Zhang, G.Q.; Wang, M.M.; Zhou, T.; Chen, W.F. Progress in remote sensing monitoring of lake area, water level, and volume changes on the Tibetan Plateau. *Natl. Remote Sens. Bull.* **2022**, *26*, 115–125.
9. Zhang, G.Q.; Yao, T.D.; Shum, C.K.; Yi, S.; Yang, K.; Xie, H.J.; Feng, W.; Bolch, T.; Wang, L.; Behrangi, A.; et al. Lake volume and groundwater storage variations in Tibetan Plateau's endorheic basin. *Geophys. Res. Lett.* **2017**, *44*, 5550–5560. [CrossRef]
10. Zhu, L.P.; Xie, M.P.; Wu, Y.H. Quantitative analysis of lake area variations and the influence factors from 1971 to 2004 in the Nam Co basin of the Tibetan Plateau. *Chin. Sci. Bull.* **2010**, *55*, 1294–1303. [CrossRef]
11. Zhang, B.; Wu, Y.; Zhu, L.; Wang, J.; Li, J.; Chen, D. Estimation and trend detection of water storage at Nam Co Lake, central Tibetan Plateau. *J. Hydrol.* **2011**, *405*, 161–170. [CrossRef]
12. Qiao, B.J.; Zhu, L.P.; Yang, R.M. Temporal-spatial differences in lake water storage changes and their links to climate change throughout the Tibetan Plateau. *Remote Sens. Environ.* **2019**, *222*, 232–243. [CrossRef]
13. Zhang, G.; Bolch, T.; Chen, W.; Crétaux, J. Comprehensive estimation of lake volume changes on the Tibetan Plateau during 1976–2019 and basin-wide glacier contribution. *Sci. Total Environ.* **2021**, *772*, 145463. [CrossRef] [PubMed]
14. Luo, S.; Song, C.; Zhan, P.; Liu, K.; Chen, T.; Li, W.; Ke, L. Refined estimation of lake water level and storage changes on the Tibetan Plateau from ICESat/ICESat-2. *Catena* **2021**, *200*, 105177. [CrossRef]
15. Changnon, J.; Jones, D. Review of the influences of the Great Lakes on weather. *Water Resour. Res.* **1972**, *8*, 360–371. [CrossRef]
16. Bhasang, T.; Liu, J.S.; Niu, J.F.; Dhawa, T.; Ci, B. Area Variation and Its Causes of Bamu Co Lake in the Central Tibet. *J. Nat. Resour.* **2012**, *27*, 9.

17. Wang, S.M.; Dou, H.S. *The Volume of Chinese Lake*; Science Press: Beijing, China, 1998.
18. Su, R.M.Z.; Ma, W.Q.; Ma, Y.M.; Xie, Z.P.; Wang, B.B.; Hu, W.; Liu, J.S. Investigation of thermal stratification and mixed layer depth in La'ang Co in the Tibetan Plateau. *J. Lake Sci.* **2021**, *33*, 550–560.
19. Krause, P.; Biskop, S.; Helmschrot, J.; Flügel, W.-A.; Kang, S.; Gao, T. Hydrological system analysis and modelling of the Nam Co basin in Tibet. *Adv. Geosci.* **2010**, *27*, 29–36. [CrossRef]
20. Fan, Z.-X.; Thomas, A. Spatiotemporal variability of reference evapotranspiration and its contributing climatic factors in Yunnan province, SW China, 1961–2004. *Clim. Chang.* **2013**, *116*, 309–325. [CrossRef]
21. Yao, T.D.; Thompson, L.; Yang, W.; Yu, W.S.; Gao, Y.; Guo, X.J.; Yang, X.X.; Duan, K.Q.; Zhao, H.B.; Xu, B.Q.; et al. Different glacier status with atmospheric circulations in Tibetan Plateau and surroundings. *Nat. Clim. Chang.* **2012**, *2*, 663–667. [CrossRef]
22. Alsdorf, D.E.; Rodríguez, E.; Lettenmaier, D.P. Measuring surface water from space. *Rev. Geophys.* **2007**, *45*, 24. [CrossRef]
23. Pekel, J.F.; Cottam, A.; Gorelick, N.; Belward, A.S. High-resolution mapping of global surface water and its long-term changes. *Nature* **2016**, *540*, 418–422. [CrossRef]
24. Crétaux, J.-F.; Calmant, S.; Romanovski, V.; Shabunin, A.; Lyard, F.; Bergé-Nguyen, M.; Cazenave, A.; Hernandez, F.; Perosanz, F. An absolute calibration site for radar altimeters in the continental domain: Lake Issykkul in Central Asia. *J. Geod.* **2009**, *83*, 723–735. [CrossRef]
25. Verpoorter, C.; Kutser, T.; Seekell, D.A.; Tranvik, L.J. A global inventory of lakes based on high-resolution satellite imagery. *Geophys. Res. Lett.* **2014**, *41*, 6396–6402. [CrossRef]
26. Li, D.; Wu, B.S.; Chen, B.W.; Xue, Y.; Zhang, Y. Review of water body information extraction based on satellite remote sensing. *J. Tsinghua Univ. Sci. Technol.* **2020**, *60*, 15.
27. Gong, R. Sentinel-2A optical imaging satellite launched. *Space Int.* **2015**, *8*, 36–40.
28. Attema, E.; Cafforio, C.; Gottwald, M.; Pietro, G.; Guarnieri, A.M.; Rocca, F.; Snoeij, P. Flexible dynamic block adaptive quantization for Sentinel-1 SAR missions. *IEEE Geosci. Remote Sens. Lett.* **2010**, *7*, 766–770. [CrossRef]
29. Binh, P.D.; Catherine, P.; Filipe, A. Surface Water Monitoring within Cambodia and the Vietnamese Mekong Delta over a Year, with Sentinel-1 SAR Observations. *Water* **2017**, *9*, 366.
30. Zhang, G.Q.; Chen, W.F.; Xie, H.J. Tibetan Plateau's Lake Level and Volume Changes From NASA's ICESat/ICESat-2 and Landsat Missions. *Geophys. Res. Lett.* **2019**, *46*, 13107–13118. [CrossRef]
31. Neckel, N.; Kropáček, J.; Bolch, T.; Hochschild, V. Glacier mass changes on the Tibetan Plateau 2003–2009 derived from ICESat laser altimetry measurements. *Environ. Res. Lett.* **2014**, *9*, 014009. [CrossRef]
32. Yuan, C.; Gong, P.; Bai, Y. Performance assessment of ICESat-2 laser altimeter data for water-level measurement over lakes and reservoirs in China. *Remote Sens.* **2020**, *12*, 770. [CrossRef]
33. Liu, J.; Fang, P.; Que, Y.; Zhu, L.-J.; Duan, Z.; Tang, G.; Liu, P.; Ji, M.; Liu, Y. A dataset of lake-catchment characteristics for the Tibetan Plateau. *Earth Syst. Sci. Data Discuss.* **2022**, *14*, 1–21. [CrossRef]
34. Mcfeeters, S.K. The use of the Normalized Difference Water Index (NDWI) in the delineation of open water features. *Int. J. Remote Sens.* **1996**, *17*, 1425–1432. [CrossRef]
35. Höhle, J.M. Accuracy assessment of digital elevation models by means of robust statistical methods. *ISPRS J. Photogramm. Remote Sens.* **2009**, *64*, 398–406. [CrossRef]
36. Song, C.; Huang, B.; Ke, L. Modeling and analysis of lake water storage changes on the Tibetan Plateau using multi-mission satellite data. *Remote Sens. Environ.* **2013**, *135*, 25–35. [CrossRef]
37. Zhang, G.; Yao, T.; Xie, H.; Zhang, K.; Zhu, F. Lakes' state and abundance across the Tibetan Plateau. *Chin. Sci. Bull.* **2014**, *59*, 3010–3021. [CrossRef]
38. Lv, L.; Zhang, T.B.; Yi, G.H.; Miao, J.Q.; Li, J.J.; Bie, X.J.; Huang, X.L. Changes of lake areas and its response to the climatic factors in Tibetan Plateau since 2000. *J. Lake Sci.* **2019**, *31*, 573–589.
39. Lei, Y.; Yang, K.; Wang, B.; Sheng, Y.; Bird, B.W.; Zhang, G.; Tian, L. Response of inland lake dynamics over the Tibetan Plateau to climate change. *Clim. Chang.* **2014**, *125*, 281–290. [CrossRef]
40. Li, B.; Zhang, J.; Yu, Z.; Liang, Z.; Chen, L.; Acharya, K. Climate change driven water budget dynamics of a Tibetan inland lake. *Glob. Planet. Chang.* **2017**, *150*, 70–80. [CrossRef]
41. Yang, K.; Wu, H.; Qin, J.; Lin, C.; Tang, W.; Chen, Y. Recent climate changes over the Tibetan Plateau and their impacts on energy and water cycle: A review. *Glob. Planet. Chang.* **2014**, *112*, 79–91. [CrossRef]
42. Han, Y.Z.; Ma, W.Q.; Wang, B.Y.; Ma, Y.M.; Tian, R.X. Climatic Characteristics of Rainfall Change over the Qinghai-Tibetan Plateau from 1980 to 2013. *Plateau Meteorol.* **2017**, *36*, 1477–1486.
43. Cooley, S.W.; Ryan, J.C.; Smith, L.C. Human alteration of global surface water storage variability. *Nature* **2021**, *591*, 78–81. [CrossRef] [PubMed]
44. Vörösmarty, C.J.; Green, P.; Salisbury, J.; Lammers, R.B. Global water resources: Vulnerability from climate change and population growth. *Science* **2000**, *289*, 284–288. [CrossRef] [PubMed]
45. La, B.; Bian, D.; Ci, Z.; Laba, Z.; Chen, T. Study on the Change of Lake Area and Its Causes in the Mapangyong Co Basin in Tibet. *Arid. Zone Res.* **2012**, *29*, 992–996.
46. Yang, R.; Zhu, L.; Wang, J.; Ju, J.; Ma, Q.; Turner, F.; Guo, Y. Spatiotemporal variations in volume of closed lakes on the Tibetan Plateau and their climatic responses from 1976 to 2013. *Clim. Chang.* **2017**, *140*, 621–633. [CrossRef]

remote sensing

MDPI

Article

The Seasonal and Diurnal Variation Characteristics of Soil Moisture at Different Depths from Observational Sites over the Tibetan Plateau

Hongyi Li [1], Ziniu Xiao [2], Junhong Wei [3,4,5] and Ge Wang [6,*]

[1] China Meteorological Administration Training Centre, Beijing 100081, China
[2] State Key Laboratory of Numerical Modeling for Atmospheric Sciences and Geophysical Fluid Dynamics, Institute of Atmospheric Physics, Chinese Academy of Sciences, Beijing 100029, China
[3] School of Atmospheric Sciences, Sun Yat-sen University, Zhuhai 519082, China
[4] Guangdong Province Key Laboratory for Climate Change and Natural Disaster Studies, Sun Yat-sen University, Zhuhai 519082, China
[5] Southern Marine Science and Engineering Guangdong Laboratory (Zhuhai), Zhuhai 519082, China
[6] Institute of Plateau Meteorology, China Meteorological Administration, Chengdu 610072, China
* Correspondence: wg800110@aliyun.com

Abstract: Using observational data of soil moisture from the third Tibetan Plateau Experiment for atmospheric science (TIPEX III), the seasonal and diurnal variations characteristics of soil moisture at different depths of 5–160 cm from seven stations were analyzed, with emphasis on the comparative analysis of the differences of soil moisture between different sites and the differences of the synergistic relationship between soil moisture and temperature. The soil moisture was wet in the southeast and dry in the northwest. The studied sites were Lhari, Biru, Nyainrong, Amdo, Nagqu, Baingoin and Seng-ge Kambab in descending order, according to the soil moisture. The seasonal variation of soil moisture at the different sites showed a significant three-peak structure, which was more obvious in the shallow layer than in the deep layer. The first peak occurred from March to May, which was mainly due to the soil thawing in spring. The other two peaks corresponded to the two rainy seasons in the plateau. Soil moisture was the greatest during this rainy period. The diurnal variations of soil moisture and temperature in Amdo, Nagqu, Nyainrong and Baingoin showed a significant positive correlation in the four seasons. The soil moisture and temperature in Lhari and Biru were significantly positively correlated in winter and spring but negatively correlated in summer and autumn. The profiles of the soil moisture with depth varied greatly at different stations in different seasons. The distribution of soil water content at each observational site did not increase or decrease with depth but showed a certain high aquifer, which might be related to the types of the underlying surface and physical properties of soil. During the summer monsoon period, soil moisture in the shallow layer of 5–10 cm was higher at all observational sites. The spatial distribution of soil moisture in the plateau was more heterogeneous than that in the plain area, and only in the central part of the Tibetan Plateau, the soil moisture varied greatly from site to site. This also indicated that it was unreasonable to only use the soil moisture of several stations to represent the overall soil moisture of the region. The results provided a multi-angle observational basis for the validation of satellite data and parameterization of the numerical model of soil moisture over the Tibetan Plateau.

Keywords: soil moisture; soil temperature; seasonal and diurnal variation; vertical profile

Citation: Li, H.; Xiao, Z.; Wei, J.; Wang, G. The Seasonal and Diurnal Variation Characteristics of Soil Moisture at Different Depths from Observational Sites over the Tibetan Plateau. *Remote Sens.* **2022**, *14*, 5010. https://doi.org/10.3390/rs14195010

Academic Editor: Nicolas Baghdadi

Received: 8 August 2022
Accepted: 27 September 2022
Published: 8 October 2022

1. Introduction

The Tibetan Plateau is known as the "roof of the world" and the "third pole of the Earth" and covers about one-quarter of China's land area, with an average altitude of 4500 m. It is the largest plateau in China and the highest in the world, and the origin of many major rivers in Asia. The Tibetan Plateau has an important impact on the weather and climate of China, Asia and the world. It restricts the basic pattern of atmospheric circulation and its system in East Asia and causes abnormal weather and climate disasters [1–9]. The terrain of the Tibetan Plateau is complex, including high mountains (the altitude is above 7000 m) and deep ravines (the altitude is below 3000 m); the surface conditions are diverse, including beaches, meadows, forests, ice and snow. The process of land–air exchange over the heterogeneous underlying surface of the Tibetan Plateau is extremely complex, which brings great difficulties to the correct understanding of atmospheric processes and the accurate prediction of weather and climate processes over the Tibetan Plateau. From May to August 1978, China conducted the first Tibetan Plateau atmospheric scientific experiment. The second Tibetan Plateau atmospheric scientific experiment was conducted to deeply study the process of land–air exchange during the period from May to August 1998. In these experiments, soil temperature and moisture were targeted as important basic observational items, which are essential for the process of land–atmosphere exchange.

Soil moisture is a physical quantity that indicates the degree of soil wetness. It is an important variable in the parameterization scheme of the land surface process. Its variation changes the physical properties of the surface and then affects the energy and water exchange between the earth and the atmosphere. It gradually affects the troposphere through the near-surface and boundary layers and is an important influencing factor on atmospheric circulation and climate change [10–14]. The importance of the role of soil moisture in the climate system is second only to sea surface temperature. For the climate system over land, it even exceeds the role of sea surface temperature [15]. Previous studies have shown that 65% of land precipitation comes from land–surface evaporation, which largely depends on soil moisture [16,17]. Numaguti [18] pointed out through numerical simulations that about 71% of precipitation from June to August near Nagqu came from land evaporation, and the soil water supply for evaporation mainly came from atmospheric precipitation recharge.

The soil in the Qinghai-Tibet Plateau mainly consists of clay loam and loam, and the soil sandiness is enhanced with depth. Due to unique soil characteristics and geographical environment, the variation of soil temperature and humidity in this region is very large [19]. The temporal and spatial variation of soil moisture plays an important role in the water cycle of the Tibetan Plateau, so it is of great significance to study the distribution characteristics of soil temperature and moisture in the Tibetan Plateau. The seasonal variation of soil moisture is mainly affected by soil physical properties and soil water budget. In the seasonal frozen soil environment of the Tibetan Plateau, soil temperature and precipitation have significant effects on soil moisture. In general, in the wet season, soil moisture is significantly affected by local precipitation. If water income is higher than consumption, soil moisture will increase, and vice versa. In the dry season, soil moisture is greatly affected by the intensity of water evaporation caused by soil temperature. Yang et al. [20–24] studied the characteristics of diurnal, annual and spatial changes in soil temperature and moisture in the northern Tibet Plateau by using the data obtained from GAME-Tibet and analyzed the role of the freezing and thawing processes in the dry-wet season transition and the changes in heat distribution. Gao et al. [25] used a SiB2 (Simple Biosphere Mode2) model to simulate surface energy distribution, soil temperature and moisture conditions on the underlying surface of low grassland in northern Tibet and obtained reasonable results. Wang et al. [26], based on the observation results of Tuotuohe station on the northern Tibetan Plateau, showed that changes in soil moisture associated with the freezing and thawing processes were closely related to the transformation of dry and wet seasons and the amount of precipitation in wet seasons on the Tibetan Plateau. Wan et al. [27] analyzed in detail the changes in soil moisture at different time scales at BJ station near Nagqu in the central Qinghai-Tibet Plateau, and the results showed that the changes in soil temperature and moisture were closely related at different time scales. Limited by the lack of observation data, the above studies mainly focused on the analysis of a single site or typical underlying surface, while the comparative study of soil temperature and humidity observation data in different regions of the plateau is very scarce.

Due to the difficult conditions and the scarcity of observation stations in the plateau area, the difficulty of soil temperature and moisture observation and research under the complex terrain of the plateau is much greater than that in other areas. The observation time, space and physical quantities are very limited, and the data are very scarce. Due to the complex topography and underlying surface characteristics of the plateau, the representativeness of the observation stations is limited, as well as the uncertainties of satellite inversion products on the plateau, which restrict our correct understanding of the various characteristics of soil moisture at different time scales on the Tibetan Plateau. In order to compensate for the lack of observational data on the Tibetan Plateau, in 2014, the China Meteorological Administration (CMA), in collaboration with many domestic institutions, launched the third Tibetan Plateau atmospheric scientific experiment. The experimental sites were more widely distributed, and the data were the latest and most comprehensive, which provides us with an important database for the study of land–air energy exchange over the Qinghai-Tibet Plateau. Many significant results have been obtained in the study of land–air energy transport by using this boundary layer observation data [28–32]. Li et al. [33] analyzed the seasonal and diurnal variation characteristics of soil moisture at different depths using 28 stations in Nagqu. However, their analysis was mainly based on the regional average in the Nagqu region, while there was a lack of comparative studies on the differences among stations in different regions of the plateau. Based on the observational data for soil temperature and moisture at different depths from 5–160 cm from the third Tibetan Plateau atmospheric scientific experiment from December 2014 to December 2015, this study analyzed the seasonal variation characteristics of soil moisture at different depths at seven stations on the plateau, focusing on the comparative analysis of the differences in soil moisture at different stations and the differences between the synergistic changes in soil temperature and moisture.

2. Materials and Methods

The data used in this study were soil temperature and moisture data from TIPEX III obtained from December 2014 to December 2015. The boundary layer observatories were obtained in Amdo, Seng-ge Kambab, Baingoin, Biru, Lhari, Nyingchi, Namco, Nagqu and Nyainrong over the Tibetan Plateau. It should be noted that the Nyingchi and Namco stations were excluded from our study after the data quality control steps were completed. Therefore, Amdo, Nagqu, Nyainrong, Baingoin, Biru, Lhari and Seng-ge Kambab were finally selected for analysis in this study. These seven observational sites are mainly distributed in the western and central part of the Tibetan Plateau, with the latest data and relatively complete observations of meteorological elements, which provides an important database for studying the characteristics of soil temperature and moisture in the Tibetan Plateau. The vertical depths of soil temperature and moisture in Amdo and Nagqu stations were 5, 10, 20, 40, 80 and 160 cm. The available vertical depths of Biru, Lhari and Baingoin stations were 5, 10, 20, 40 and 100 cm. Those of Nyainrong station were 5, 10, 20, 50 and 100 cm, with those of Seng-ge Kambab station at 5, 10, 20, 40 and 80 cm. The soil moisture sensors were CAMPBELL CS616 (Campbell Scientific, Inc., Logan, UT, USA), and the soil temperature sensors were CAMPBELL 109 (Campbell Scientific, Inc., Logan, UT, USA). Soil temperature and moisture detectors were placed at 5 or 6 different depths at each observation site. The probes were horizontally inserted at different depths to obtain soil moisture and temperature data. Data were collected every 10 min. For analysis, data were processed as 30 min averages. Soil moisture is the volumetric water content, with the unit cm^3/cm^3, and the unit of soil temperature is °C. The distribution of observation stations is shown in Figure 1, and the detailed geographic information of observation stations is shown in Table 1.

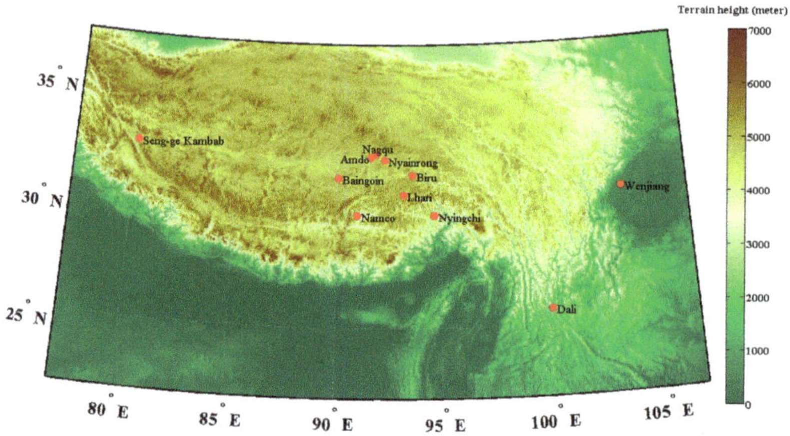

Figure 1. Topography of the Tibetan Plateau, with 9 plateau stations denoted by red dots. It should be noted that the Nyingchi and Namco stations were excluded from our study after the data quality control steps were completed.

Table 1. Geographical location information of the seven observation sites on the Qinghai-Tibet Plateau.

Stations	Location	Elevation (m)	Plateau Regions
Amdo	91.6°E, 32.2°N	4695	central plateau
Seng-ge Kambab	80.1°E, 32.5°N	4350	northwest plateau
Nyainrong	92.3°E, 32.1°N	4730	central plateau
Baingoin	90.1°E, 31.4°N	4700	central plateau
Biru	93.7°E, 31.5°N	4408	central plateau
Lhari	93.2°E,30.7°N	4500	central plateau
Nagqu	91.9°E, 32.4°N	4509	central plateau

3. Results

3.1. *Seasonal Variation of Soil Moisture*

3.1.1. Seasonal Variation of Soil Moisture at Different Observational Sites

Figure 2 shows the time series of daily accumulated precipitation at different stations over the Tibetan Plateau from December 2014 to December 2015. Amdo, Nagqu, Nyainrong, Baingoin, Biru, and Lhari are all located in the central region of the Tibetan Plateau, influenced by the South Asian summer monsoon; summer and autumn are the main flood seasons on the Tibetan Plateau, and the annual precipitation is mostly concentrated from June to September. Seng-ge Kambab is located in the alpine desert area in the western part of the plateau, with very little rainfall, only a few days in the summer. In general, the variation of precipitation in the central plateau was relatively consistent, and there were two rainy seasons. Biru and Lhari are located in the southeast and had the most precipitation; Nyainrong station was the next, then followed by Amdo and Nagqu stations, and Baingoin station had less precipitation. The rainy season started in mid-May in Biru and Lhari and in mid-June in other central stations. The rainy season started one month earlier in Biru and Lhari than in other central stations.

The geographical location of the Biru and Lhari stations is very close, and the precipitation and its variation trends were relatively consistent. The precipitation in Biru and Lhari began to increase in early April, and there was a rainy period from early April to early May, but the rainfall was not large. Biru and Lhari officially entered the rainy season in mid-May. The first rainy season of Biru station lasted from mid-May to mid-July, and the second rainy season lasted from early August to mid-September with abundant rainfall. The rainy seasons in Nyainrong, Amdo, Nagqu and Baingoin were rather consistent, and they officially entered the rainy season in the middle of June. The first rainy season is from mid-June to mid-July, and the second rainy season is from early August to mid-September.

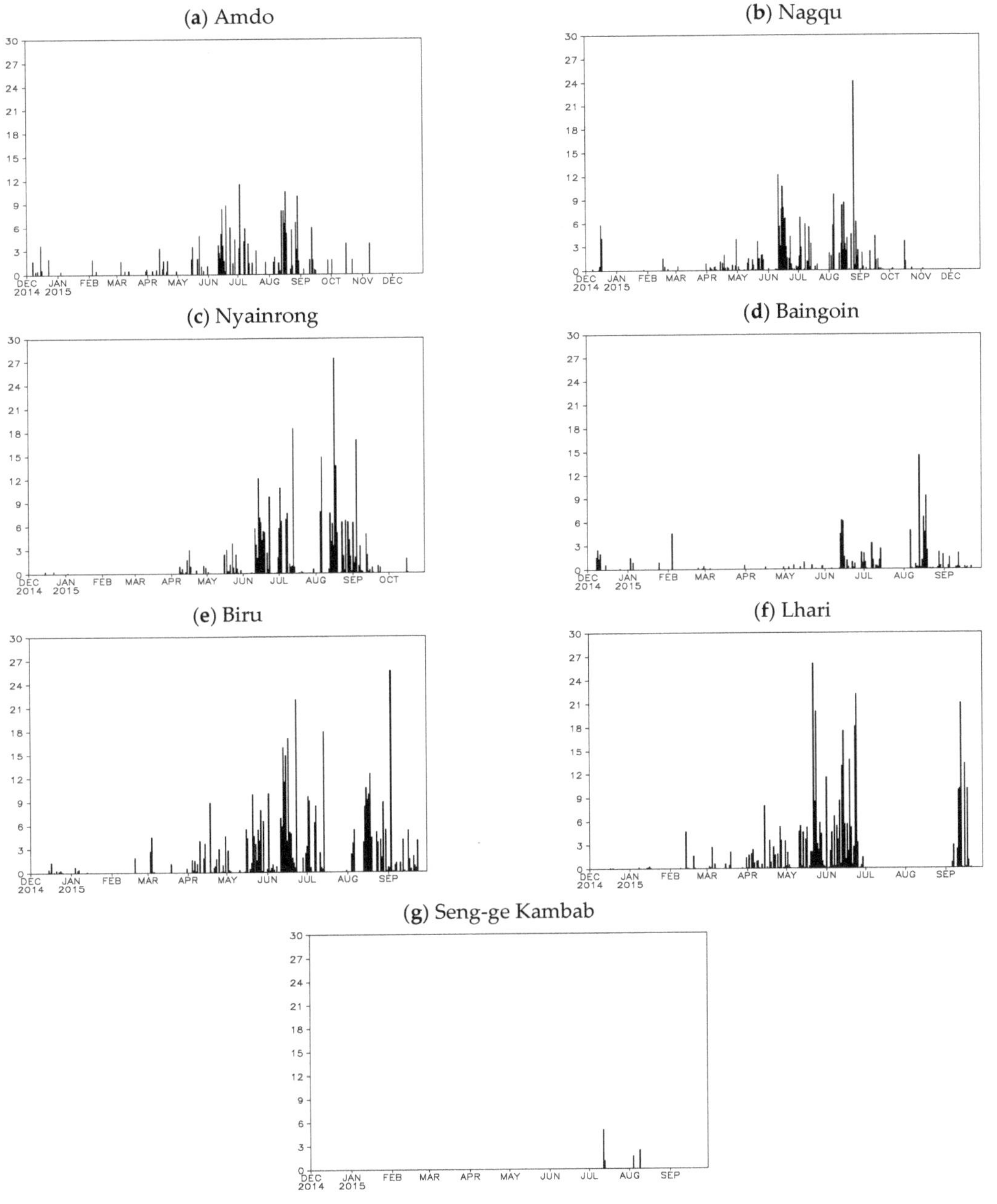

Figure 2. Daily accumulated precipitation of seven stations over Tibetan Plateau from December 2014 to December 2015 (unit: mm).

Figure 3 shows the variations in the daily mean soil moisture at different depths and from different observational sites on the plateau from December 2014 to December 2015. The spatial distribution of the soil moisture showed the characteristics of wet in the southeast and dry in the northwest. The soil moisture over the Tibetan Plateau had remarkable seasonal variation. In general, in winter and spring, the shallow soil moisture was small, while the deep soil moisture was large. In summer and autumn, due to the increase in precipitation, soil moisture was greater in the shallow layer and lesser in the deep layer.

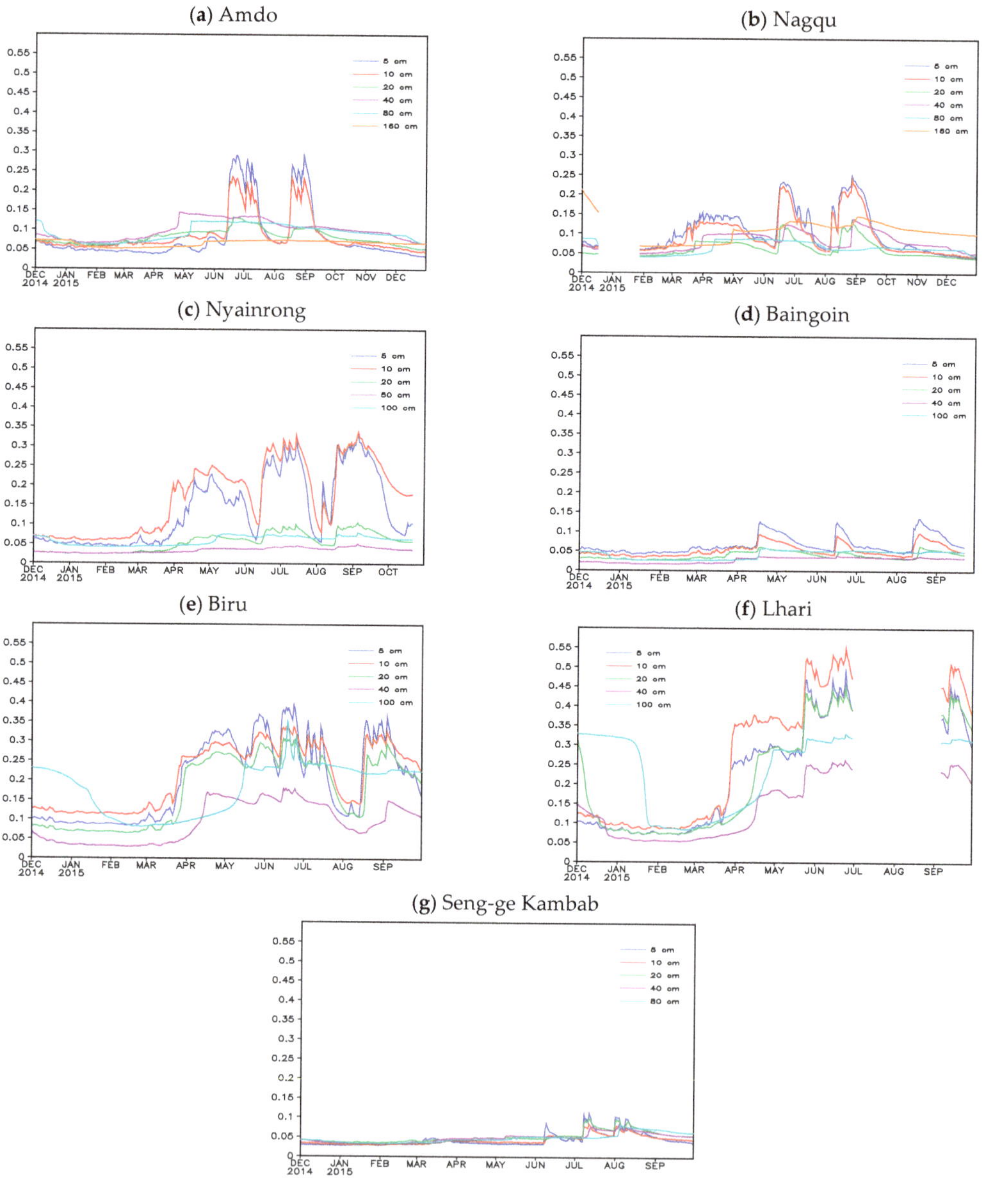

Figure 3. Time series of daily mean soil moisture at different depths (unit: cm^3/cm^3). The lines break in the graphs represents the missing data, the same below.

In winter and early spring, the soil moisture at different depths was all in the low stage, with a value of less than 0.15 cm^3/cm^3 most of the time. In winter and early spring, there was mainly solid precipitation in the Qinghai-Tibet Plateau, which had little impact on soil moisture, and the soil moisture was relatively stable.

In April, the soil moisture in all layers increased for the first time, and it reached the first high-value period from early April to the end of May, which was mainly related to the rise of spring temperature and the beginning of melting. Since the soil retained a large amount of water in its frozen state in winter, the soil showed a high water content when it completed ablation. In May, with the rise in temperature, the ice in the soil completely melted, and the precipitation began to rapidly increase. At this time, the soil moisture in all layers rapidly increased. It is worth noting that before the rainy season on the plateau, the soil moisture at each site showed a certain decline in early June (Biru and Lhari in mid-May). This was because the rainy season had not yet begun, and the rainfall was still relatively low, while the incident radiation continued to strengthen and the soil heated faster. Therefore, evaporation in the shallow layer was larger, and the soil water consumption was greater than the water income, which led to the rapid decline of the soil moisture in the shallow layer. This result was consistent with Li et al. [33].

Except for the Seng-ge Kambab station, all the other stations are located in the central region of the Qinghai-Tibet Plateau. They are all affected by the South Asian summer monsoon; three-quarters of the annual average precipitation was concentrated from June to August. Summer and autumn are the main flood seasons on the Qinghai-Tibet Plateau, and with the arrival of the plateau's rainy season, the soil moisture in the surface layer (5–10 cm) was affected by liquid precipitation, and the soil moisture content rapidly increased, resulting in high soil moisture. On average, there were two peaks of soil moisture in each layer in the flood season, corresponding to the two rainy seasons. Between the two rainy seasons, soil moisture in all layers significantly decreased from the end of July to the beginning of August. This was because, with the end of the first rainy season, all stations were in an intermittent period of precipitation. The precipitation decreased rapidly, and the daily precipitation of each station was less than 5 mm. At this time, the Tibetan Plateau was in the high-temperature period of the year, and the soil evaporation was large, resulting in the rapid decline of soil moisture during this period and the decrease of 5–10 cm depth was the most significant, which was almost equal to the soil moisture in winter and spring. After early October, with the retreat of the summer monsoons over the Tibetan Plateau, the precipitation over the Tibetan Plateau sharply decreased, and the soil moisture significantly decreased correspondingly, entering the attenuation period of soil moisture. By comparing Figures 2 and 3, it can be seen that as the shallow soil melted and stabilized in early April, the plateau entered the wet season, and the seasonal increase of soil moisture caused by soil freezing and thawing in the plateau area was earlier than the start of the plateau rainy season, and the earlier time varied at different stations.

The seasonal variation of soil moisture at different sites showed a significant three-peak structure, which was more obvious in the shallow layer than in the deep layer, and the peak time at different sites was slightly different. The soil moisture at Amdo station began to increase in mid-April, and it reached the first peak period from mid-April to early June, which was mainly related to soil thawing in spring. After entering the rainy season in June, the shallow soil moisture (5–10 cm) showed two significant peak periods from mid-June to mid-July and from early August to mid-September, which was completely consistent with the two rainy seasons of Amdo station. The maximum soil moisture of 5 cm depth at Amdo station in the rainy season was about 0.3 cm^3/cm^3. The soil moisture at 160 cm depth was the lowest in the rainy season and changed little throughout the year, with a value of about 0.07 cm^3/cm^3.

The seasonal variation of soil moisture at the Nagqu station was basically the same as that at the Amdo station, but the first peak period of soil moisture occurred from early March to mid-May, which was earlier than that at the Amdo station. This meant that the seasonal increase of soil moisture caused by soil freezing and thawing was earlier than that of the Amdo station. The two high-value periods of soil moisture at Nagqu station in the flood season had good correspondence with the two rainy seasons. The soil moisture at Nagqu station was slightly lower than that at Amdo station, and the maximum soil moisture at 5 cm depth was about 0.25 cm^3/cm^3. The soil moisture at 80 cm depth fluctuated gently throughout the year, and the value was the smallest, which was about 0.08 cm^3/cm^3 in flood season.

The high-value period of soil moisture at Nyainrong station was due to soil thawing in spring and occurred from the end of March to the beginning of June. The two high-value periods corresponding to the two rainy seasons were from mid-June to the end of July and from mid-August to the end of September. The soil moisture of Nyainrong was higher than that of the Amdo and Nagqu stations, and the highest soil moisture at 10 cm depth at the Nyainrong station was about 0.34 cm^3/cm^3 in the rainy season.

The three high-value periods of soil moisture at Baingoin station were from mid-April to mid-May, from mid-June to early July, and from mid-August to mid-September, with little difference in the three peak values. The soil moisture of Baingoin station was lower than that of the Amdo, Nagqu and Nyainrong stations, with the highest value of 0.14 cm^3/cm^3 at 5 cm deep.

The soil moisture at Biru station was higher than that at Nyainrong, Amdo and Nagqu. The seasonal increase of soil moisture associated with soil freezing and thawing occurred from the end of March to the beginning of May. The two high-value periods associated with the rainy season were from mid-May to mid-July and from mid-August to the end of September, respectively. The maximum value of soil moisture of 5 cm depth can reach 0.4 cm^3/cm^3 in the rainy seasons.

The soil moisture at Lhari station was higher than that at Biru station, and the variation trend of soil moisture at each layer was consistent. The seasonal increase of soil moisture caused by soil freezing and thawing occurred from the end of March to the middle of May. From the middle of May to the end of June, the soil moisture was in a high-value period, the data from early July to early September were missing, and the soil moisture was still in the high-value period from early to mid-September. The soil moisture at 10 cm deep at the Lhari station was the highest, and the highest value could reach 0.55 cm^3/cm^3 in the rainy season.

The soil moisture at Seng-ge Kambab was very small, with few fluctuations throughout the year, and the maximum value in the rainy season was only about 0.1 cm^3/cm^3. The soil moisture increased slightly in mid-June, which was related to the melting of frozen soil. It increased again in mid-July and early August, which was related to a small amount of precipitation in Seng-ge Kambab during these days.

Through the comparative analysis of soil moisture changes at different depths at seven stations (Figure 4), the spatial distribution of soil moisture showed the characteristics of dry in the northwest and wet in the southeast. To be more specific, the stations with soil moisture at 5 and 10 cm deep from large to small were Lhari, Biru, Nyainrong, Amdo, Nagqu, Baingoin and Seng-ge Kambab. Lhari and Biru are located in the southeast, with abundant precipitation, so the soil moisture was the largest, followed by Nyainrong, Amdo and Nagqu; Baingoin is located in the west, with less rainfall and lower soil moisture. Seng-ge Kambab station is located in the alpine desert area in the northwest of the plateau, with very little precipitation, so the soil moisture was very low and stable throughout the year with little fluctuation. Soil moisture at 10 cm depth in Lhari and Nyainrong was higher than that at 5 cm depth but decreased at other sites.

The sites with soil moisture at 20 cm deep from large to small were Lhari, Biru, Amdo, Nagqu, Nyainrong, Seng-ge Kambab and Baingoin. Lhari and Biru were much higher than the other sites. The soil moisture in Nyainrong, Amdo and Nagqu decreased very fast. The soil moisture at Nyainrong dropped to less than that at Amdo and Nagqu, and the soil moisture at Baingoin dropped to less than that at Amdo station. The sequence of sites with soil moisture from large to small at 40 cm deep was the same as that at 20 cm deep; Seng-ge

Kambab and Baingoin were the smallest and remained stable throughout the year with almost no change. Nyainrong station had no observation data at a depth of 40 cm.

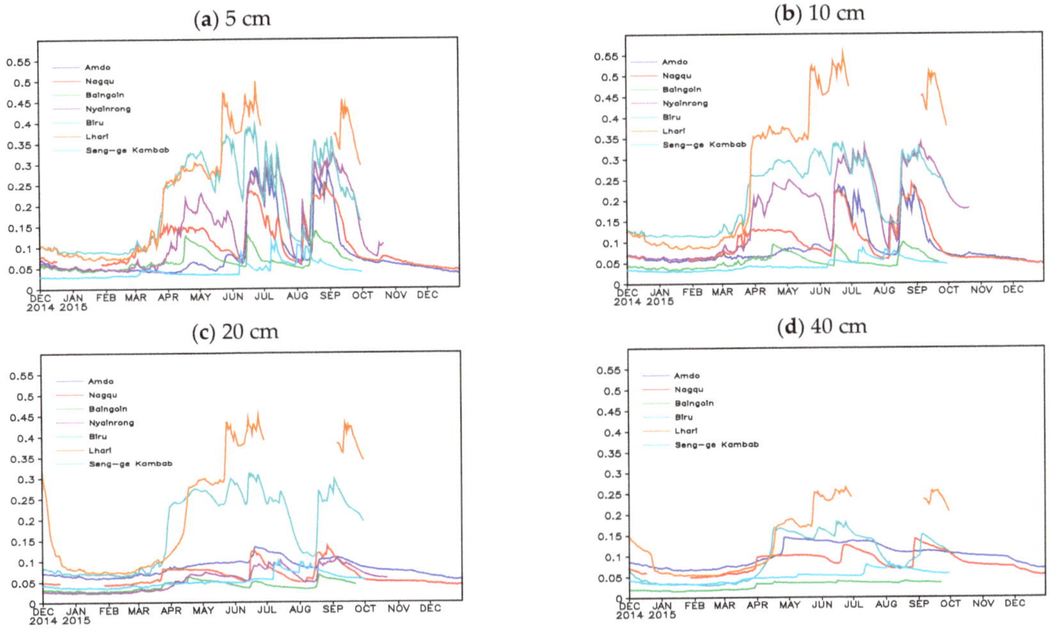

Figure 4. Time series of daily mean soil moisture at different depths for the seven observational sites (unit: cm^3/cm^3).

In terms of spatial distribution, soil moisture decreased from the southeast to the northwest. More specifically, the sites with soil moisture of 5 and 10 cm deep from large to small were Lhari, Biru, Nyainrong, Amdo, Nagqu, Baingoin and Seng-ge Kambab. Except for the Seng-ge Kambab station, the seasonal variation of soil moisture at other sites showed the characteristics of three peaks, which were more obvious in the shallow layer than in the deep layer. The first high-value period occurred from March to May, which was mainly related to the seasonal thawing of soil caused by the rising temperatures in spring. Summer and autumn are the main rainy seasons on the plateau, and there were two obvious high-value periods of soil moisture which corresponded to the two rainy seasons. During this period, the soil moisture was the greatest, reaching the highest value of the whole year. It should be noted that due to the different geographical locations, the period of concentrated precipitation at different sites was slightly different, so the period of a high value of soil moisture was also different. In particular, the rainy season started in mid-May in Lhari and Biru and in mid-June in other central stations. The rainy season started one month earlier in Lhari and Biru than in other central stations. Therefore, the first peak period of soil moisture in flood season at Lhari and Biru stations was one month earlier than that at other central stations.

In addition, Lhari and Biru stations are located to the southeast and are the wettest, so the frozen soil held much more water in winter than the other stations, and the water content of the soil rose very high when the soil was completely melted. Therefore, the seasonal increase of soil moisture caused by soil freezing and thawing in the Lhari and Biru stations was much larger than that in other stations. The soil moisture in Biru station after soil ablation was almost the same as that in flood season, while that in Lhari station was slightly lower than that in flood season.

The seasonal increase of soil moisture caused by soil freezing and thawing was 1–2 months earlier than the beginning of summer precipitation over the plateau, indicating that in spring, with the increase in temperature, the soil melted, the soil water content increased, and the soil moisture was transmitted to the atmosphere through surface evapotranspiration. Therefore, the atmospheric humidity increased, providing favorable water vapor conditions for the occurrence of precipitation on the plateau, and the plateau entered the wet season. This also indicated that during the transition from the dry season to the wet season, the contribution of soil moisture to precipitation could not be ignored. From April to May, soil moisture transferred water to the atmosphere through evaporation, which provided important conditions for the outbreak of the summer monsoon, thus affecting the time of seasonal transition over the Qinghai-Tibet Plateau.

3.1.2. Seasonal Variation of Soil Temperature at Different Observational Sites

Since the soil on the Tibetan Plateau is frozen for a long time in a year, soil moisture has significant seasonal variation. Soil freezing is the response of soil moisture to soil temperature, and soil water will have phase transformation within a certain range of soil temperature, causing changes in soil moisture. In addition, the surface will evaporate under the heating of solar radiation, which will affect the change in soil temperature.

Figure 5 shows the variation of daily mean soil temperature of different sites at different depths on the plateau. The seasonal variation of soil temperature in each layer was significant. It was frozen from early November to the end of March of the next year, and the temperature was the lowest in mid-January. It is worth noting that the soil freezing time obtained in this study is about 5 months, which is 1 month less than the freezing time obtained in previous studies based on the northern Tibetan Plateau, which may be related to the significant warming of northern China in recent years [24,26]. Taking Amdo station as an example, the soil temperature at 5 cm deep reached the lowest value in January, which was −14 °C, and the temperature in the freezing period increased with depth. Thawing began at the end of March, and the soil temperature reached the maximum temperature from July to August, and the maximum temperature of the soil at 5 cm deep reached 15 °C. In the non-freezing period, the soil temperature decreased with depth. The changing trend of soil temperature above 40 cm deep was basically the same, and the fluctuation was obvious. The seasonal variation amplitude of soil temperature at 5 cm deep in the surface layer was the largest, and with the deepening of the depth, the seasonal variation amplitude of soil temperature gradually decreased. The seasonal fluctuation amplitude of soil temperature at 160 cm deep was the smallest, and the freezing time was shorter. The effects of solar radiation and water evaporation on soil temperature gradually decreased with the depth.

The seasonal variation trend of the soil temperature at different depths at the seven stations was quite consistent. Seng-ge Kambab station had the largest seasonal variation, and Biru station had the smallest. The soil temperature reached the lowest in mid-January, and the temperature from low to high was Baingoin, Amdo, Seng-ge Kambab, Nyainrong, Lhari and Biru. In summer, the soil temperature from high to low was Seng-ge Kambab, Nagqu, Baingoin, Biru, Lhari, Amdo and Nyainrong. During the freezing period, the soil temperature increased with depth, while the soil temperature in summer decreased with depth. As the soil temperature was affected by solar radiation and water evaporation, this effect gradually decreased with depth. Therefore, the fluctuation of soil temperature was the largest at 5 cm depth, and with the deepening of the depth, the fluctuation amplitude gradually decreased (Figure 6).

3.2. Relationship between the Diurnal Variation of Soil Moisture and Soil Temperature

Amdo, Nagqu, Nyainrong and Baingoin are located in the northern part of the central plateau, and their geographical positions are very close to each other. Therefore, the relationship between the diurnal variation of soil temperature and moisture at these four stations was quite consistent, and there was a significant positive correlation between the four seasons (see Figure 7, Amdo station). The maximum correlation coefficient between soil moisture and temperature in Amdo and Nagqu occurred in the summer but was relatively small in autumn. The correlation coefficient between soil moisture and temperature of Nyainrong station was the largest in winter and weaker in autumn. The correlation coefficient of the Nyainrong site was at the maximum in winter and weak in autumn. The soil temperature and moisture at Baingoin station were significantly positively correlated in four seasons (figure omitted).

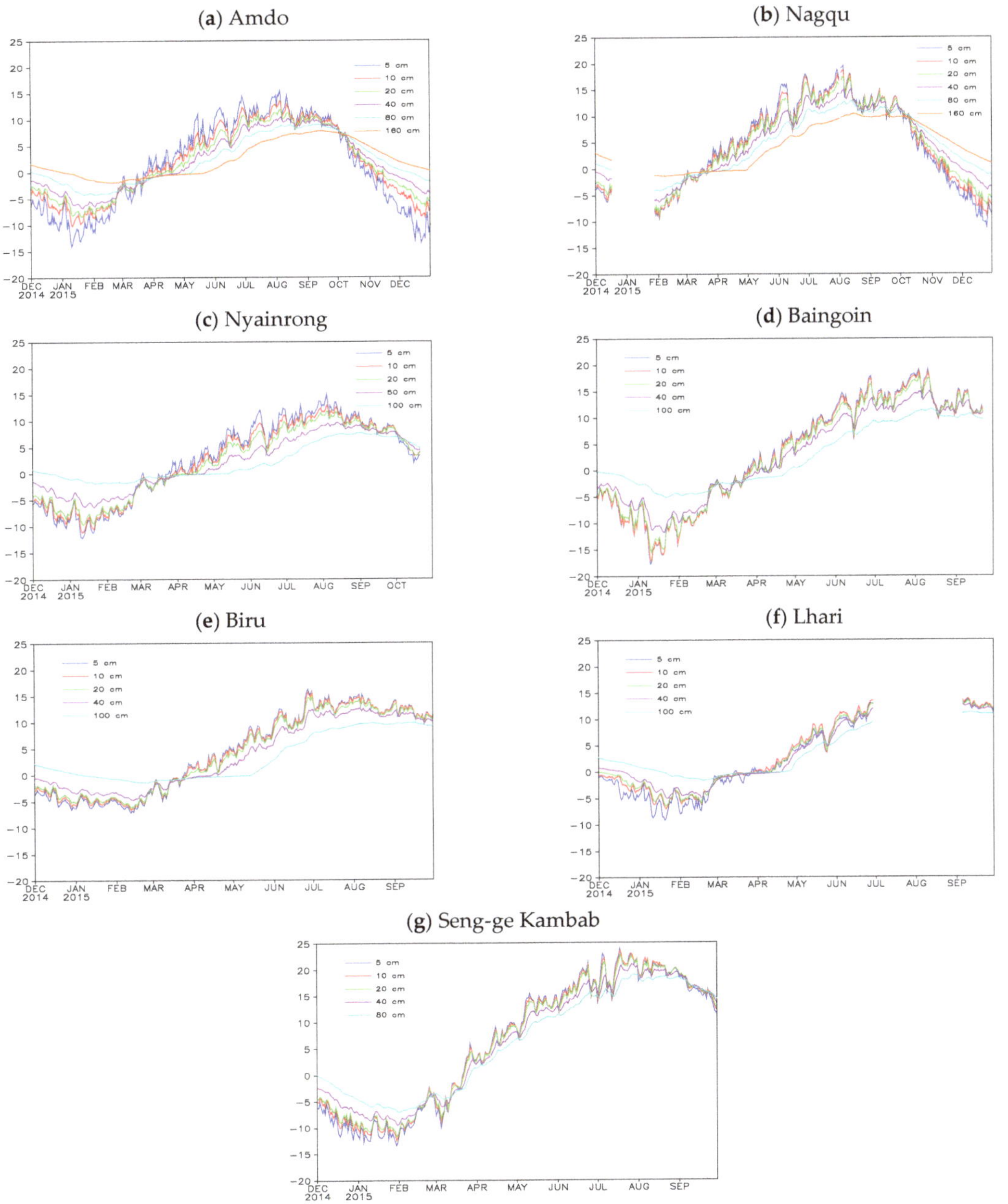

Figure 5. Time series of daily mean soil temperature at different depths (unit: °C).

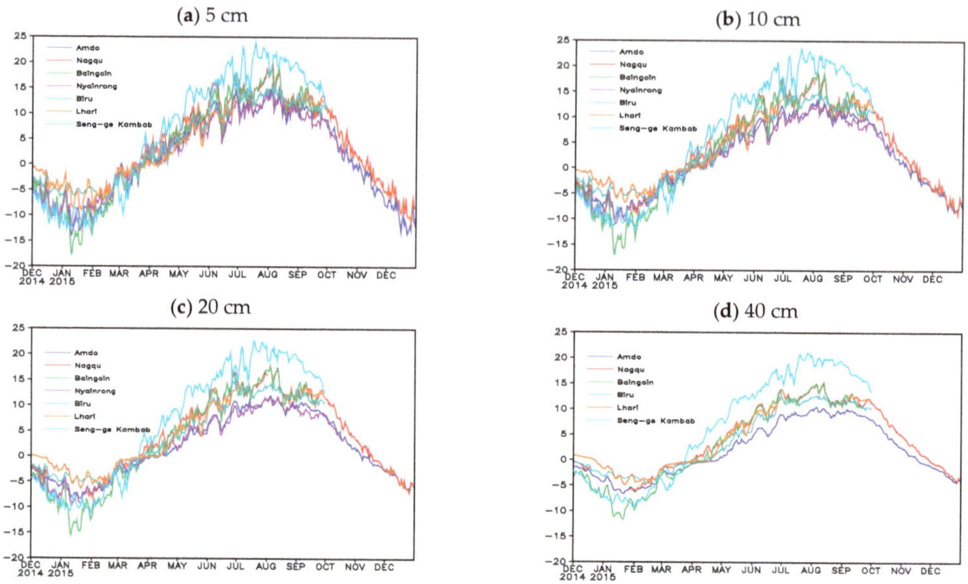

Figure 6. Time series of daily mean soil temperature at different depths for the seven observational sites (unit: °C).

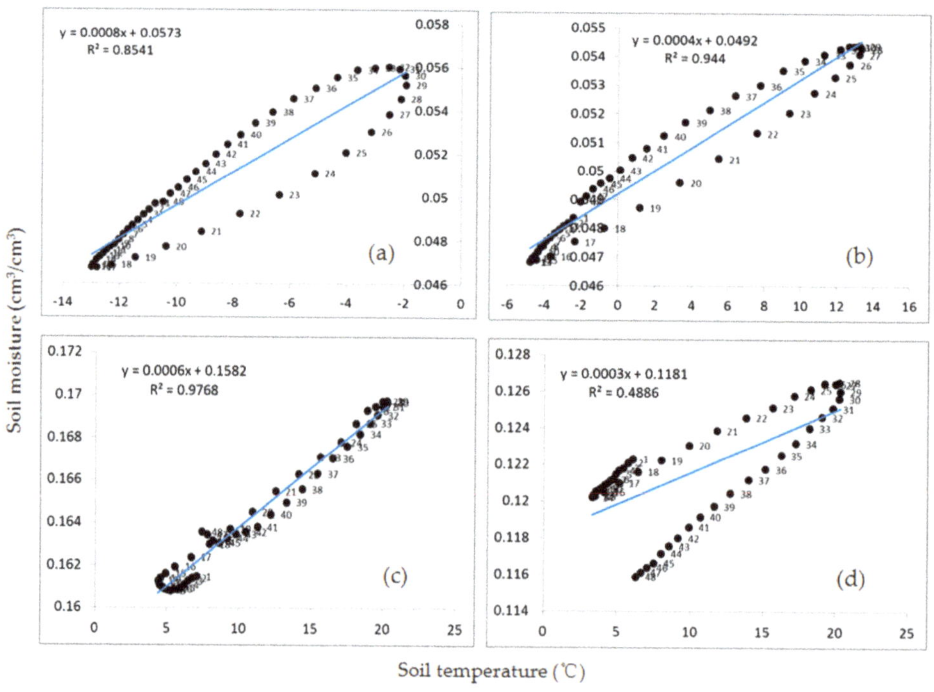

Figure 7. Scatter diagrams and linear fitting lines of diurnal variation of soil temperature and moisture at 5 cm deep at Amdo station. (The solid blue lines are the linear fitting line and the numbers next to the black dots represent the time sequence, the data interval is 30 min, from 00:00 to 23:30; there are 48 numbers in total.) (**a**) Winter (December–February), (**b**) Spring (March–May), (**c**) Summer (June–August) and (**d**) Autumn (September).

Figure 7 shows the fitting relationship between the mean diurnal variation of soil temperature and moisture at 5 cm deep at the Amdo station in winter (December–February), spring (March–May), summer (June–August) and autumn (September). As can be seen from the figure, there was a significant positive correlation between soil moisture and temperature in the four seasons at Amdo station, with the highest correlation coefficient of 0.988 in summer, and a relatively small correlation coefficient of 0.699 in autumn, both of which have passed a 99% reliability test. In the winter and spring, there were frequent melting and freezing processes on the soil surface, which affected the diurnal variation of soil temperature and moisture, and the soil moisture w sensitive to the change in soil temperature. In winter, spring and autumn, it was mainly affected by incident radiation. During the day, the temperature of the soil surface was heated by radiation and began to rise in the morning and reached a peak in the afternoon. With the rise in soil temperature, the frozen water partially melted during the day, and the upper soil moisture gradually increased in the morning and reached the maximum in the afternoon. Then, as the incident radiation decreased, the soil temperature dropped, the surface layer cooled at night, and the melted water froze; the soil moisture also slowly decreased.

The correlation between the diurnal variation of soil temperature and moisture in the Lhari and Biru stations was basically the same. In winter and spring, soil moisture and soil temperature were positively correlated. However, there was a significant negative correlation between summer and autumn (autumn here is only in September) (see Figure 8 for Lhari station). The negative correlation of the Lhari station in winter and autumn was much more significant than that of the Biru station, and the correlation coefficients were as high as −0.943 and −0.967, respectively, both of which have passed a 99% reliability test. The relationship between the diurnal changes in soil temperature and moisture at the Lhari and Biru stations in summer and autumn was opposite to that at the above four stations. This was because Lhari and Biru stations are located in the south of the central plateau, and the precipitation in summer and autumn is more abundant than that in other stations. The surface soil water is liquid. The soil moisture is sensitive to precipitation and is relatively less affected by incident radiation. As the surface soil is relatively wet, evaporation increases with the increase of soil temperature, resulting in a significant decrease in surface soil moisture.

At the Seng-ge Kambab station, there was a significant positive correlation between soil moisture and soil temperature in winter and spring, a significant negative correlation in summer, and little correlation in autumn. Seng-ge Kambab is located in the western part of the plateau, with little precipitation. Therefore, in summer, the soil temperature was higher than that of the other stations. With the increase in soil temperature, evaporation was very great, and the surface soil moisture significantly decreased (figure omitted).

3.3. Vertical Distribution of Soil Moisture and Soil Temperature

Figure 9 shows the vertical profiles of averaged soil temperature and soil moisture in four seasons at Amdo station. In winter, the variation of soil moisture with depth at Amdo station presents a step-like change, with the largest soil moisture at a depth of 80 cm, and soil moisture is similar at a depth of 20 cm and 160 cm (Figure 9a). Soil moisture increased with depth from the surface to 80 cm and then began to decrease. At the same time, the soil temperature increased with depth in winter. In spring (Figure 9b), soil moisture also increased first and then decreased with depth, and soil moisture was the highest at 40 cm. While the soil temperature decreased with depth, above 40 cm, soil temperature and moisture inversely changed with depth, while below 40 cm, both had the same change trend with depth. The variation trend of soil moisture in summer and autumn was very consistent (Figure 9c,d), which showed a step-like decline with the deepening of the depth. The soil moisture was the highest at the shallow layer of 5 cm and the lowest at 160 cm. In summer and autumn, the changes in soil temperature and moisture with depth were quite consistent; with the deepening of the depth, the soil temperature and moisture both decreased.

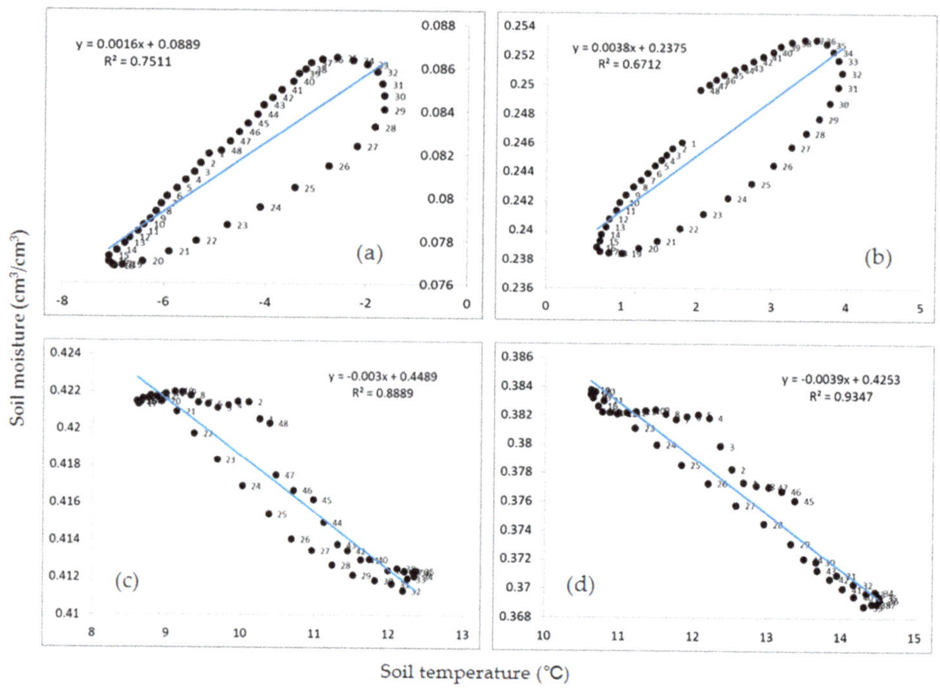

Figure 8. Same as Figure 7, but for the Lhari station: (**a**) winter, (**b**) spring, (**c**) summer and (**d**) autumn.

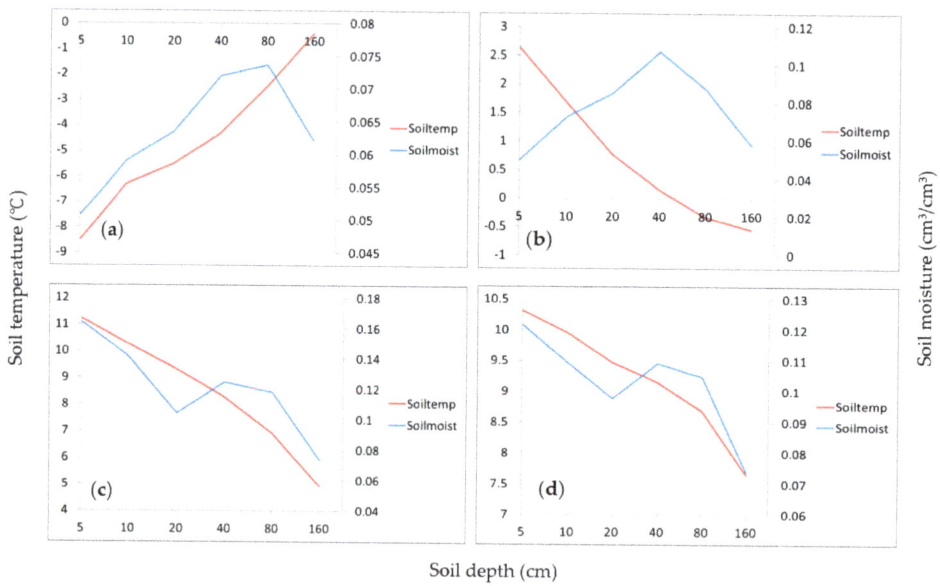

Figure 9. Profiles of soil temperature (red line) and soil moisture (blue line) at Amdo Station, (**a**) winter (December–February), (**b**) spring (March–May), (**c**) summer (June–August) and (**d**) autumn (September).

The soil moisture at Nagqu station in the winter decreased first and then increased with the depth, and the soil moisture at 160 cm depth was the greatest (Figure 10). In spring, summer and autumn, the soil moisture showed a step-like change, with the soil moisture decreasing first and then increasing. The soil moisture was the greatest at 5 cm depth and the smallest at 80 cm depth. The variation of soil temperature with depth at Nagqu station was consistent with that at Amdo Station.

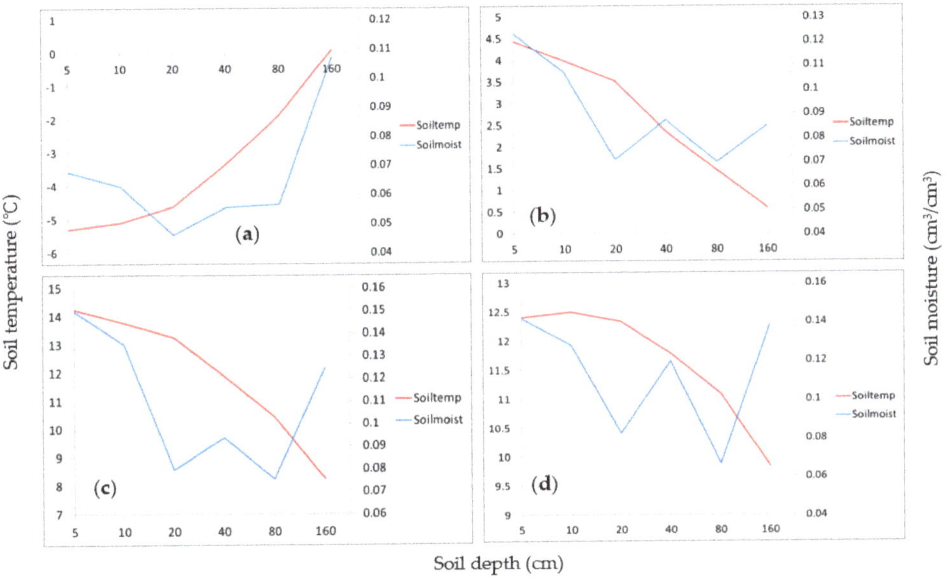

Figure 10. Same as Figure 9, but for the Nagqu station: (**a**) winter, (**b**) spring, (**c**) summer and (**d**) Autumn.

The vertical changes in soil moisture at the four sites of Nyainrong, Baingoin, Biru, and Lhari were generally consistent. In the four seasons, the soil moisture decreased first and then increased with soil depth. The soil moisture in the four seasons at Nyainrong station was the largest at 10 cm depth, followed by 5 cm, and the smallest at 50 cm (figure omitted). The soil moisture in the four seasons at Baingoin station had the maximum value at 5 cm deep and the minimum value at 40 cm deep (no observations at 50 cm) (figure omitted). In Biru and Lhari, the soil moisture was the greatest at 100 cm in the winter. In spring, summer and autumn, soil moisture was the greatest at 10 cm, followed by 5 cm, and the smallest at 40 cm (see Figure 11, no observations at 50 cm). Seng-ge Kambab station had very little precipitation, so the soil moisture gently changed with depth (figure omitted).

It was concluded that the profiles of soil moisture with depth greatly vary at different stations and in different seasons. Zhang et al. [34] found that the vertical distribution of soil moisture was very complex, and there were both regional and soil texture differences in the vertical distribution. Under the condition of clay loam and loam in the Tibetan Plateau region, the porosity of the soil in different regions greatly varies. Yang and Ma [35] showed that at a depth of 40 cm at the Namco station, 20 cm at the Everest station and 60 cm at the Southeast Tibet station, there was a relatively high aquifer, which was related to the type of underlying surface and physical properties of soil, such as vegetation coverage, soil texture and porosity. In the analysis of this observation data, the difference was also evident. The distribution of soil water content at each observation site did not increase or decrease with depth but showed a certain high aquifer. This distribution had a great influence on the freezing and thawing processes of the soil and the spatial and temporal distribution of soil temperature. During the influential period of the summer monsoon, the soil's water content at each observation site was higher in the shallow layer of 5–10 cm. This was also consistent with the research results of Yang et al. [23,24].

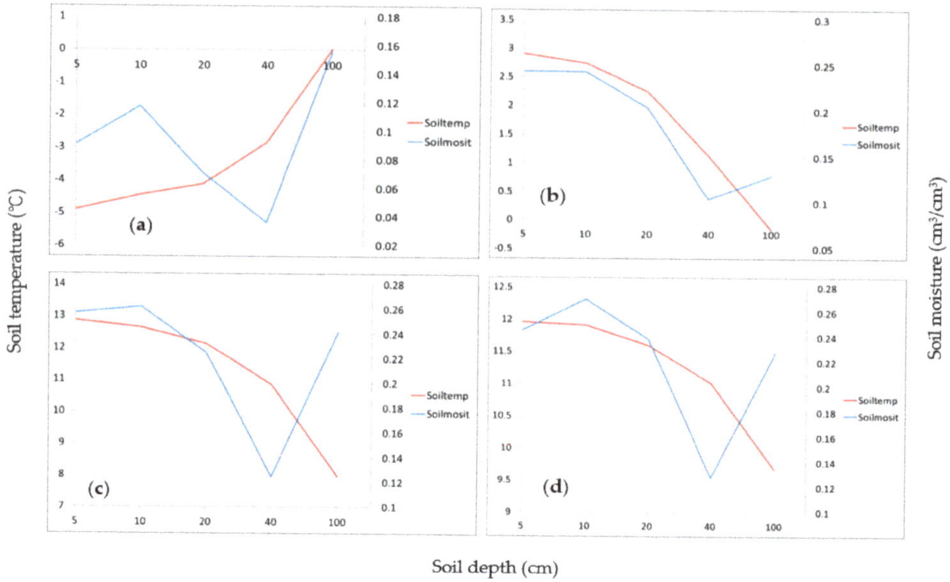

Figure 11. Same as Figure 9, but for the Biru station. (**a**) Winter. (**b**) Spring. (**c**) Summer. (**d**) Autumn.

The vertical changes in soil temperature at the seven stations were basically the same, and they all increased with depth in winter, which meant that the temperature in the deep layer was higher than that in the shallow layer, and the temperature difference between the deep and shallow layers was the largest in winter. In spring, summer and autumn, the soil temperature significantly decreased with soil depth, and the soil temperature of the shallow soil was higher than that of deep soil. The soil temperature in Seng-ge Kambab increased with depth in autumn.

4. Discussion

The heterogeneity of soil moisture on the plateau was greater than that in the plain area. In the center of the plateau, the soil moisture and temperature at different sites show different variation characteristics. Only in the central part of the Qinghai-Tibet Plateau, the soil moisture was wet in the southeast and dry in the northwest, and the spatial heterogeneity was very strong. To be more specific, Lhari and Biru are located in the southeast and had more rainfall, so the soil moisture was the highest. Among them, Lhari station is the most southeast, and the soil moisture was higher than that of Biru station. Then followed by Nyainrong, which is located in the northeast compared to the other stations, and the soil moisture in Amdo and Nagqu stations was relatively lower, and their locations are in the north. Baingoin is located in the west, with less rainfall, and the soil moisture was much smaller than that of other central stations and slightly greater than that of the Seng-ge Kambab station in the northwest. The sites in descending order of soil moisture were Lhari, Biru, Nyainrong, Amdo, Nagqu, Baingoin and Seng-ge Kambab. For the soil moisture of each layer, there were great differences between the different sites, which also showed that it was unreasonable to only use the soil moisture of several stations to represent the entire region. In the climate system model, the freezing and thawing processes related to soil moisture and the calculation of sensible heat play an important role in the simulation results. The differences in soil moisture among different stations also illustrated the necessity of a high-resolution grid for accurately simulating the climate of the Tibetan Plateau.

The vertical distribution of soil moisture in the Qinghai-Tibet Plateau is very complex. The vertical profile of soil moisture is affected not only by regional differences but also by soil properties. Also, the soil porosity greatly varies in different areas. In the analysis of this observation data, the variation profile of soil moisture with depth at different stations in different seasons also showed great differences. For example, the soil moisture at Amdo station reached a maximum of 80 cm deep in winter, 40 cm deep in spring, and 5 cm deep in summer and autumn. The soil moisture at Nagqu station was the highest at 160 cm deep in winter and 5 cm in spring, summer and autumn. The soil moisture at Biru station reached the maximum at 10 cm depth in spring, summer and autumn. The distribution of soil water content at each observation site did not increase nor decrease with depth but showed a certain high aquifer, which was related to the underlying surface and the physical properties of the soil, such as vegetation cover, soil texture and porosity. The differences among different stations are remarkable, which has great guiding significance for the construction and improvement of the parameterization process related to soil properties in the land-surface model. In this study, the seasonal changes and vertical distribution characteristics of soil moisture at different depths from seven stations on the plateau were analyzed and focused on a comparative analysis of the soil moisture differences at different stations, as well as the differences in the relationship between soil temperature and moisture, providing a multi-angle observational basis for satellite data verification and model parameterization.

Also, it was found that there was a significant positive correlation between soil temperature and moisture at 5 cm in the north of the central plateau (Amdo, Nagqu, Nyainrong, Baingoin) in summer, while there was a significant negative correlation between soil temperature and moisture at the southeast of the central plateau (Lhari, Biru). In other words, the relationship between soil temperature and moisture at 5 cm in the "south-north" sites of the central plateau was the opposite in summer. The reason was not clear, which may be related to the amount of precipitation caused by geographical location (less precipitation at the northern site, more precipitation at the southern site), and may also be related to the type of underlying surface and soil texture, which are issues that need further study.

The observational data used in this study were only one-year data, and the relevant conclusions must be verified with longer observation data in the future. Also, due to the limitation of observational data, only September was used to represent autumn in this paper. Due to the large spatio-temporal differences in soil moisture and temperature on the plateau, continuous and in-depth observation and data analysis on a larger spatial and temporal scale is needed in the future to comprehensively reveal the distribution and variation of soil moisture on the plateau.

In recent years, many studies have used the analysis method of coefficient of variation (CV) to describe some significant patterns of mean domain soil moisture content, and in terms of spatial variation, the relation between CV and soil moisture often shows a hysteresis pattern [36–38]. In the next step, we will use this method of CV to study whether hysteresis is also observed in this observational data and how that might differ between the different rainfall events. Also, the frequency distribution of CV and temporal mean soil moisture will be analyzed to display the seasonal variations for each soil depth at different observation sites over the Tibetan Plateau.

5. Conclusions

Using the observational data of soil moisture and temperature from the third Tibetan Plateau Experiment for atmospheric science (TIPEX III), the seasonal and diurnal variation characteristics of soil moisture at different depths of 5–160 cm at seven stations in the Qinghai-Tibet Plateau from December 2014 to December 2015 were analyzed. This study focused on the comparative analysis of the variations of soil moisture among different sites, as well as the differences in the synergistic changes between soil moisture and soil temperature at different sites. The conclusions are as follows:

(1) The spatial distribution of soil moisture showed the characteristics of wet in the southeast and dry in the northwest. More specifically, according to the soil moisture at 5 and 10 cm deep, the studied sites were Lhari, Biru, Nyainrong, Amdo, Nagqu,

Baingoin and Seng-ge Kambab in descending order. Lhari and Biru are located in the southeast, and the precipitation was relatively abundant, so the soil moisture was relatively high, followed by Nyainrong, Amdo and Nagqu. Baingoin is located to the west of other central stations and with less rainfall, so the soil moisture was lower. Seng-ge Kambab is located in the alpine desert area in the northwest of the plateau, which was very dry, so the soil moisture was very small with little fluctuation throughout the year.

(2) The seasonal variation of soil moisture at 5–20 cm depths at all sites showed a significant three-peak structure, with the shallow layer more obvious than the deep layer. The first peak occurred from March to May, which was mainly caused by soil freezing in spring. The other two peaks were closely related to the two rainy seasons in the plateau, and the soil moisture was the highest during this period. Different stations had different rainfall periods, so the period of high soil moisture was slightly different. In particular, the first peak period of soil moisture in flood season at Lhari and Biru stations was one month earlier than that at other central stations. The seasonal increase of soil moisture caused by soil freezing and thawing in the plateau area was 1–2 months earlier than the start of the rainy season in the plateau. Soil moisture was transferred to the atmosphere through evaporation, and the atmospheric humidity increased, which provided favorable water vapor conditions for the beginning of the plateau rainy season.

(3) The seasonal variation of the soil temperature at different depths was quite consistent. The soil temperature reached the lowest in mid-January, and the stations were Baingoin, Amdo, Seng-ge Kambab, Nyainrong, Lhari and Biru in ascending order, according to soil temperature. In midsummer, they were correspondingly Seng-ge Kambab, Nagqu, Baingoin, Biru, Lhari, Amdo and Nyainrong in descending order.

(4) The diurnal variation of soil moisture and temperature at Amdo, Nagqu, Nyainrong and Baingoin had the same relationship, which showed a significant positive correlation in the four seasons. In Lhari and Biru stations, the relationship between soil moisture and temperature was basically the same. In winter and spring, soil moisture was positively correlated with soil temperature, while there was a significant negative correlation in summer and autumn, which was contrary to the above four sites. At Seng-ge Kambab station, soil moisture and temperature had a significant positive correlation in winter and spring and a significant negative correlation in summer.

(5) The profiles of soil moisture with depth varied greatly at different stations in different seasons. The distribution of soil water content at each observational site did not increase or decrease with depth but showed a certain high aquifer, which was related to the underlying surface and physical properties of soil. Under the influence of the summer monsoon, the soil's water content in the shallow layer of 5–10 cm was higher at all observation sites. The vertical profiles of soil temperature at the seven stations were basically the same, and they all increased with depth in winter. In spring, summer and autumn, soil temperature decreased with depth.

Author Contributions: Conceptualization, H.L. and Z.X.; methodology, H.L.; validation, G.W.; formal analysis, G.W.; data curation, H.L.; writing—original draft preparation, H.L.; writing—review and editing, J.W.; funding acquisition, G.W. All authors have read and agreed to the published version of the manuscript.

Funding: This research was funded by the National Natural Science Foundation of China (grant No. 42030611), the Second Tibetan Plateau Scientific Expedition and Research (STEP) program (grant No. 2019QZKK0103), the National Natural Science Foundation of China (grant No. 91937301), Guangdong Province Key Laboratory for Climate Change and Natural Disaster Studies (Grant 2020B1212060025) and the Innovation Group Project of Southern Marine Science and Engineering Guangdong Laboratory (Zhuhai) (Grant 311021001), the Heavy Rain and Drought-Flood Disasters in Plateau and Basin Key Laboratory of Sichuan Province (SCQXKJYJXMS202116), and the Opening Foundation of Plateau Atmosphere and Environment Key Laboratory of Sichuan Province (grant No. PAEKL-2020-C7).

Data Availability Statement: The data that support the findings of this study are available from the first author upon request (Hongyi Li, lihongyi@cma.gov.cn).

Acknowledgments: The authors appreciate all of the hard work done by researchers attending the third Tibetan Plateau Experiment for atmospheric science (TIPEX III).

Conflicts of Interest: The authors declare that they have no known competing financial interests or personal relationships that could have appeared to influence the work reported in this paper.

References

1. Ye, D.; Zhang, J. A preliminary experimental simulation on the heating effect of the Tibetan Plateau on the general circulation over Eastern Asia in summer. *Sci. Sin.* **1974**, *17*, 397–420.
2. Ye, D.; Gao, Y. *Meteorology of the Qinghai-Xizang Plateau;* Science Press: Beijing, China, 1979; 278p.
3. Kuo, H.; Qian, Y. Influence of the Tibetan Plateau on cumulative and diurnal changes of weather and climate in summer. *Mon. Weather Rev.* **1981**, *109*, 2337–2356. [CrossRef]
4. Zhang, J.; Zhu, B.; Zhu, F.; Weng, D.; Sun, G.; Lü, J.; Peng, Y.; Wang, Y. *The Progress Introduction on the Study of Tibetan Plateau Meteorology;* Science Press: Beijing, China, 1988; 268p.
5. Qian, Y. Numerical simulation of the effects of underlying surfaces on the climate change. *Chin. J. Atmos. Sci.* **1993**, *17*, 283–293.
6. Ji, G.; Gu, B.; Lü, L. Characteristics of atmospheric heating field over northern Qinghai-Xizang Plateau. *Plateau Meteorol.* **2002**, *21*, 238–242.
7. Ji, G.; Shi, X.; Gao, W. The variation of surface heating field over Northern Qinghai-Tibet Plateau and its effect on climate. *Plateau Meteorol.* **2001**, *20*, 239–244.
8. Wang, C.; Dong, W.; Wei, Z. Study on relationship between the frozen-thaw process in Qinghai-Xizang Plateau and circulation in East-Asia. *Chin. J. Geophys.* **2003**, *46*, 309–316. [CrossRef]
9. Wu, G.; Sun, S.; Chen, W. *The Impact of Qinghai-Tibet Plateau and Northwest Arid Areas on Climate Disasters;* China Meteorological Press: Beijing, China, 2003; 207p.
10. Yeh, T.; Wetherald, R.; Manabe, S. The effect of soil moisture on the short-term climate and hydrology change-a numerical experiment. *Mon. Weather Rev.* **1984**, *112*, 474–490. [CrossRef]
11. Liu, Y.; Ye, D.; Ji, J. Effect of soil moisture and vegetation on climate (I): Theoretic analysis on short-term climate anomaly persistence. *Sci. China (Ser. B)* **1992**, *22*, 441–448.
12. Liu, Y.; Ye, D.; Ji, J. Effect of soil moisture and vegetation on climate II): Numerical experiment on short-term climatic anomaly persistence. *Sci. China (Ser. B)* **1992**, *22*, 554–560.
13. Ma, Z.; Fu, C.; Xie, L.; Chen, W.; Tao, S. Some problems in the study on the relationship between soil moisture and climatic change. *Adv. Earth Sci.* **2001**, *16*, 563–566.
14. Gao, Y.; Peng, W.; Wang, W. Influence of spring soil and condition on later regional precipitation simulation. *J. Glaciol. Geocryol.* **2011**, *33*, 1055–1063.
15. Guo, W.; Ma, Z.; Wang, H. Soil moisture-An important factor of seasonal precipitation prediction and its application. *Clim. Environ. Res.* **2007**, *12*, 20–27.
16. Chahine, M. The hydrological cycle and its influence on climate. *Nature* **1992**, *359*, 373–380. [CrossRef]
17. Trenberth, K. Atmospheric Moisture Recycling: Role of Advection and Local Evaporation. *J. Clim.* **1999**, *12*, 1368–1381. [CrossRef]
18. Numaguti, A. Origin and recycling processes of precipitating water over the Eurasian continent: Experiments using an atmospheric general circulation model. *J. Geophys. Res.* **1999**, *104*, 1957–1972. [CrossRef]
19. Zhang, L.; Lü, H.; Wang, L.; Yang, B. Spatial-temporal characteristics of soil moisture in China. *Acta Geogr. Sin.* **2016**, *71*, 1494–1508.
20. Yang, M.; Yao, T.; Ding, Y.; Wang, S.; Chen, X.; Toshio, K. The daily variation of the soil temperature in different seasons at site D110 in the Northern Part of Xizang Plateau. *Sci. Geogr. Sin.* **1999**, *19*, 570–574.
21. Yang, M.; Yao, T.; Wang, S.; Ding, Y.; Chen, X.; Shen, Y.; Toshio, K. The features of soil temperature and moisture on Northern Tibetan Plateau. *Geogr. Res.* **1999**, *18*, 312–317.
22. Yang, M.; Yao, T.; Toshio, K. Variation features of soil temperature in Northern Tibetan Plateau. *J. Mt. Sci.* **2000**, *18*, 13–17.
23. Yang, M.; Yao, T.; Gou, X. Distribution features of soil moisture-energy and melting-freezing processes along the Qiinghai-Xizang Road. *Prog. Nat. Sci.* **2000**, *10*, 443–450.
24. Yang, M.; Yao, T.; He, Y. The role of soil moisture-energy distribution and melting-freezing processes on seasonal shift in Tibetan Plateau. *J. Mt. Sci.* **2002**, *20*, 536–558.
25. Gao, Z.; Chae, N.; Kim, J.; Hong, J.; Choi, T.; Lee, H. Modeling of surface energy partitioning, surface temperature, and soil wetness in the Tibetan prairie using the Simple Biosphere Model2 (SiB2). *J. Geophys. Res.* **2004**, *109*, D06102. [CrossRef]
26. Wang, C.; Shang, D. Effect of the variation of the soil temperature and moisture in the transition from dry-season to wet-season over northern Tibet Plateau. *Plateau Meteorol.* **2007**, *26*, 677–685.
27. Wan, G.; Yang, M.; Wang, X.; Chen, X. Variations in soil moisture at different time scales of BJ site on the central Tibetan Plateau. *Chin. J. Soil Sci.* **2012**, *43*, 286–293.

28. Wang, Y.; Xu, X.; Liu, H.; Li, Y.; Li, Y.; Hu, Z.; Gao, X.; Ma, Y.; Sun, J.; Lenschow, D.H.; et al. Analysis of land surface parameters and turbulence characteristics over the Tibetan Plateau and surrounding region. *J. Geophys. Res. Atmos.* **2016**, *121*, 9540–9560. [CrossRef]
29. Zhao, P.; Xu, X.; Chen, F.; Guo, X.; Zheng, X.; Liu, L.; Hong, Y.; Li, Y.; La, Z.; Peng, H.; et al. The third atmospheric scientific experiment for understanding the Earth-Atmosphere coupled system over the Tibetan Plateau and its effects. *Bull. Am. Meteor. Soc.* **2018**. [CrossRef]
30. Zhao, P.; Li, Y.; Guo, X.; Xu, X.; Liu, Y.; Tang, S.; Xiao, W.; Shi, C.; Ma, Y.; Yu, X.; et al. The Tibetan Plateau surface-atmosphere coupling system and its weather and climate effects: The Third Tibetan Plateau Atmospheric Scientific Experiment. *Acta Meteorol. Sin.* **2018**, *76*, 833–860. [CrossRef]
31. Li, N.; Zhao, P.; Wang, J.; Deng, Y. Estimation of surface heat fluxes over the central Tibetan Plateau using the maximum entropy production model. *J. Geophys. Res. Atmos.* **2019**, *124*, 6827–6840. [CrossRef]
32. Li, N.; Zhao, P.; Wang, J.; Deng, Y. The Long-Term Change of Latent Heat Flux over the Western Tibetan Plateau. *Atmosphere* **2020**, *11*, 262. [CrossRef]
33. Li, B.; Zhang, M.; Tang, S.; Dong, L. Variations of Soil moisture over Nagqu at different time scales based on network observations. *Acta Meteorol. Sin. Acta* **2018**, *76*, 1040–1052.
34. Zhang, X.; Wu, X. Vertical character of soil moisture in China. *Acta Meteorol. Sin.* **2004**, *62*, 51–61.
35. Yang, J.; Ma, Y. Soil Temperature and Moisture Features of Typical Underlying Surface in the Tibetan Plateau. *J. Glaciol. Geocryol.* **2012**, *34*, 813–820.
36. Ivanov, V.Y.; Fatichi, S.; Jenerette, G.D.; Espeleta, J.F.; Troch, P.A.; Huxman, T.E. Hysteresis of soil moisture spatial heterogeneity and the "homogenizing" effect of vegetation. *Water Resour. Res.* **2010**, *46*, W09521. [CrossRef]
37. Srivastava, A.; Saco, P.M.; Rodriguez, J.F.; Kumari, N.; Chun, K.P.; Yetemen, O. The role of landscape morphology on soil moisture variability in semi-arid ecosystems. *Hydrol. Process.* **2021**, *35*, e13990. [CrossRef]
38. Fatichi, S.; Katul, G.G.; Ivanov, V.Y.; Pappas, C.; Paschalis, A.; Consolo, A.; Kim, J.; Burlando, P. Abiotic and biotic controls of soil moisture spatiotemporal variability and the occurrence of hysteresis. *Water Resour. Res.* **2015**, *51*, 3505–3524. [CrossRef]

remote sensing

MDPI

Technical Note

Glacier Mass Loss Simulation Based on Remote Sensing Data: A Case Study of the Yala Glacier and the Qiyi Glacier in the Third Pole

Ruzhen Yao [1,2,*] and Jiancheng Shi [2]

1 Department of Earth Sciences, University of Gothernburg, 41320 Gothenburg, Sweden
2 Aerospace Information Research Institute, Chinese Academy of Sciences, Beijing 100053, China
* Correspondence: ruzhen.yao@gu.se

Abstract: The climate warming over the Third Pole is twice as large as that in other regions and glacier mass loss is considered to be more intensive in the region. However, due to the vast geographical differences, the characteristics of glacier mass loss might be very different between different parts of the Third Pole, such as between the southern and northern Third Pole. It is, therefore, very important to clarify the characteristics of glacier mass loss between different parts of the Third Pole, particularly between the southern and northern Third Pole. We selected the Yala Glacier in the Central Himalayas and the Qiyi Glacier in the Qilian Mountains to study the different characteristics of glacier mass loss between the southern and northern Third Pole using remote sensing data and in situ data. Based on the results, we found that the Yala Glacier has not only been in a status of mass loss but also in a status of intensive and accelerating mass loss. Our analysis showed that the average multi-year mass loss of the Yala Glacier is −736 mm w.e.a^{-1}, with a maximum of −1815 mm w.e.a^{-1}. At the same time, the Qiyi Glacier has experienced a mild glacier mass loss process compared with the Yala Glacier. The Qiyi Glacier's mass loss is −567 mm w.e.a^{-1} with a maximum of −1516 mm w.e.a^{-1}. Our results indicate that the mass loss of the Yala Glacier is much stronger than that of the Qiyi Glacier. The major cause of the stronger mass loss of the Yala Glacier is from the decrease of glacier accumulation associated with precipitation decrease under the weakening Indian monsoon. Other factors have also contributed to the more intensive mass loss of the Yala Glacier.

Keywords: Third Pole; in-situ and remote sensing data; modeling; glacier mass

Citation: Yao, R.; Shi, J. Glacier Mass Loss Simulation Based on Remote Sensing Data: A Case Study of the Yala Glacier and the Qiyi Glacier in the Third Pole. *Remote Sens.* **2022**, *14*, 5190. https://doi.org/10.3390/rs14205190

Academic Editors: Massimo Menenti, Yaoming Ma, Li Jia and Lei Zhong

Received: 4 September 2022
Accepted: 13 October 2022
Published: 17 October 2022

1. Introduction

Glaciers are important water resources for the whole world [1,2]. The Third Pole (TP) stores the world's largest glacier mass outside Antarctica and the Arctic [2–4]. Abnormal climate warming over the TP is twice as large as that in other regions [5–7]. Under the background of the intensive glacier melt at global scale caused by global warming [8,9], the glacier melt in the TP is more intensive because of abnormal warming [10–15]. The intensive glacier melt in the TP causes runoff increase in the short term, benefiting water resources supply in the downstream areas [16–19]. However, on a longer time scale, glacier melt ultimately causes substantial runoff decrease and impacts the water supply, and more than 2 billion people living in the region would face water shortage [16]. The direct impact of glacier melt on water resource is its close relationship to the rivers of the Asian Water Tower such as the Yangtze, Yellow, Mekong, Ganges and Indus Rivers, as well as large inland rivers such as the Tarim, Amu Darya, and SYR Darya. Glacier melt also causes global sea level rising [4,20] through river discharge contribution. Intensive glacier melt is also directly related to emerging disasters such as ice collapse, which is a new serious phenomenon of intensive glacier mass loss [20–22]. It is, therefore, very important to know the status of glacier mass loss in the TP and to clarify the mass loss difference between the southern and northern TP for a comprehensive understanding of its water resource supplying capability and for the hazard risk management of the societies.

A previous study pointed out spatial heterogeneity of glacier mass loss in the TP [23]. Remote-sensing monitoring showed that glaciers experienced intense mass loss and area shrinkage in the southern TP, particularly in the Himalayas, with glacier mass loss from 0.21 to 0.38 m w.e. a^{-1} in different periods [24–27], with glacier area reduction of $-0.18 \pm 0.07\%$ a^{-1} [28] during the 1970s–2010s. In the Central Himalayas, the mass loss was -0.40 ± 0.09 m w.e. a^{-1} in 2011–2013 for the Yala Glacier [29], and -0.20 ± 0.30 m w.e. a^{-1} for the other four glaciers in 2014 [30]. In the Qilian Mountains located in the northern TP, the average mass loss was -0.24 ± 0.03 m w.e. a^{-1} during 1961–2010 [31]. The accelerated melt in the Qilian Mountains began in the early 1990s, as was confirmed by the observations [3,32]. The mass loss in the Qiyi Glacier reached -0.29 ± 0.22 m w.e. a^{-1} during 2000–2009 [33]. In the Qilian Mountains, there was rapid glacier mass loss in the eastern but slow glacier mass loss in the central and in the western [31,34]. Recently, more intensive mass loss of $-0.43\pm$ 0.03 m w.e. a^{-1} was estimated for glaciers in the Lenglongling Mountains in the very eastern of the Qilian Mountains during 1972–2016, whereas less mass loss of -0.27 ± 0.07 and -0.28 ± 0.03 m w.e. a^{-1} was estimated for glaciers in the Beida River basin in the central and Shule River basin in the western [35–38].

The differences in glacier regimes are very important for understanding the differences of mass loss between the southern TP. In the southern TP, most glaciers are maritime temperate glaciers. However, most glaciers are continental cold glaciers in the northern TP. The characteristics of the glacier regimes are determined by both the different temperature and precipitation characteristics associated with atmospheric circulation processes over the TP [3,23]. Generally speaking, glaciers in the southern TP are more sensitive to temperature variations than those in the northern TP since ice temperature is higher in the glaciers in the southern TP. Furthermore, the southern TP is dominated by the Indian monsoon, while the northern TP is dominated by the westerly. In the central Himalayas, more annual precipitation fell in summer [30] compared with that in the Qilian Mountains [39]. Due to the vast territory of the TP, different geographical zones have different climate conditions under the dominance of different atmospheric circulation processes, resulting in great differences in the characteristics of glacier mass loss. Therefore, an important scientific question is how we can clarify the major difference of the characteristics of glacier mass loss between the northern and southern TP.

In order to quantitatively study the differences in the characteristics of glacier mass loss between the northern and southern TP, we selected two representative glaciers, the Yala Glacier in the southern TP and the Qiyi Glacier in the northern TP and used remote sensing data, combined with the in situ data, to model the different characteristics of glacier mass loss shown by the two different types of glaciers, and the potential mechanisms causing differences in glacier mass loss. Our study will help improve the understanding of glacier mass loss in the TP under the background of the abnormal climate warming, and deepens the understanding of the impact mechanism of climate change on glacier fluctuations over the TP.

2. Methods

2.1. Selection for the Representative Glaciers in the Southern and Northern Third Pole

The Yala Glacier (28°14′N, 85°37′E), lying on the south slope of the central Nepal Himalayas and representing maritime temperate glacier, is selected as the representative glacier for the southern TP (Figure 1). It is a summer accumulation glacier without debris, with area and length of 1.37 km^2 and 1.5 km, respectively. The total area of all the glaciers in the region is 87.2 km^2. The glacier flows southwest from the elevation of 5681 to 5143 m a.s.l. [40]. The Yala Glacier belongs to the Indian monsoon climate, with the highest precipitation in summer months (June to September) and a rather dry winter. At the same time, glacier mass loss reaches the maximum in summer. Based on in situ measurement, the Yala Glacier has been thinning at rates of -0.69 ± 0.25 during 1982–1996 and

-0.75 ± 0.24 m a^{-1} during 1996–2009, with an accelerated mass loss since 1990s [41,42]. The glacier mass loss increased from -0.36 m w.e. in 1996 [41] to -0.81 ± 0.27 m w.e. in 2009 [40,43]. The total glacier mass loss is about 40% over the 27 years.

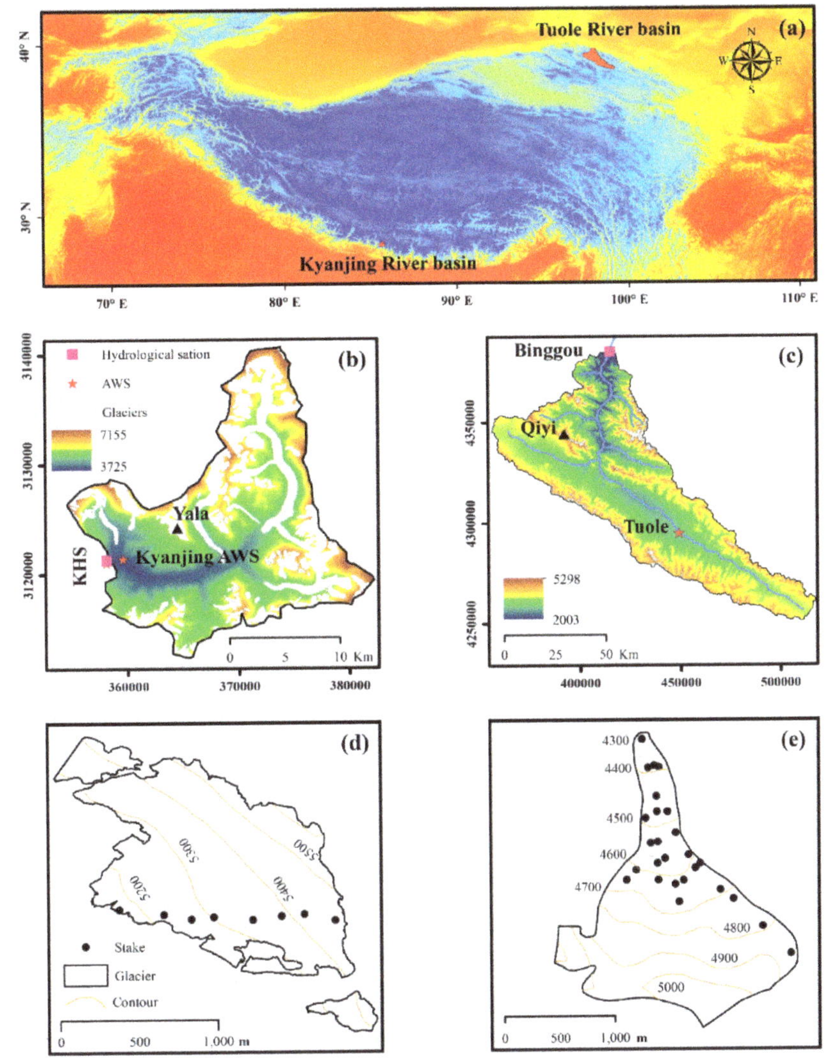

Figure 1. Locations of the Yala Glacier and Qiyi Glacier. In the figure, (**a**) is for the locations of the Yala Glacier and Qiyi Glacier, (**b**,**d**) are for the river basin where the Yalal Glacier is located and the contour map of the Yala Glacier, (**c**,**e**) for the river basin where the Qiyi Glacier is located and the contour map of the Qiyi Glacier.

The Qiyi Glacier (39°15′N, 97°45′E), lying on the north slope of the Tuolai Mountains in the Qilian Mountain, representing continental cold glacier, is selected as a representative glacier for the northern TP (Figure 1). The Qiyi Glacier is also a summer accumulation glacier and is debris free. It flows northward from an elevation of 5159 to 4304 m a.s.l., with an area of 2.76 km^2 and a length of 3.8 km [3]. The total area of all the glaciers in the region is 136 km^2. The Qiyi Glacier is located in the westerly dominated area in the northern TP and the eastern margin is affected by the East Asian monsoon. The area is mainly the continental climate and dominant by the cold air mass in winter, but occasionally affected by the East Asian monsoon in summer. Based on the in situ measurement, the equilibrium line altitude (ELA) was respectively 4600 m a.s.l. in 1970s, 4708 m a.s.l. in 1980s, 4935 m a.s.l. in 2000s [44] and 4941 m a.s.l. in 2010s [45]. In addition, the glacier mass was positive before 1980s (+0.26 m w.e. in 1970s, 0.004 m w.e. in 1980s). Glacier mass loss began after the 1990s (−0.47 m w.e. in 2000s and −0.50 m w.e. in 2010s [3,32,46]).

2.2. MODIS Temperature Data and TRMM Precipitation Data Acquisition

MODIS is the abbreviation of Moderate Resolution Imaging Spectroradiometer. There are many sensors used to acquire MODIS temperature data. The MODIS temperature data for the present study is from the Terra-MOD11A1 which provides observed results four times a day, 1:30 a.m., 10:30 a.m., 1:30 p.m., 10:30 p.m., through optical and infrared sensors. We used the Data Interpolating Empirical Orthogonal Functions (abbreviated as DINEOF method), proposed by Zhou et al. [47], to get accurate remote sensing MODIS temperature data. The MODIS temperature data we have obtained are satisfactory with RMSE of 2.91 °C and bias of −1.19 °C. Based on the interpolation in the DINEOF method, the data are still satisfactory with RMSE of 1.77 °C and bias of −0.52 °C Even under the condition of cloud day.

TRMM is the abbreviation of Tropical Rainfall Measuring Mission. TRMM sensor consists of a precipitation scanning radar (PR), a passive microwave imager (TRMM TMI) and an infrared imager (VIRS). TMI provides water vapor and the intensity of precipitation, which is therefore the core sensor for the TRMM data. The TRMM precipitation data for the present study are from TRMM 3B43. The time resolution of the TRMM 3B43 is one day and the spatial resolution is 0.25°. We used the method proposed by Zhang et al. [48] to obtain the accurate TRMM precipitation data.

2.3. Model Principle

2.3.1. Model Setup

The glacier mass loss was computed using degree day model [49].

$$M = \begin{cases} DDF_{snow/ice}T & T > T_T \\ 0 & T < T_T \end{cases} \quad (1)$$

where M is the meltwater (mm w.e.a^{-1}); $DDF_{snow/ice}$ is the degree day factor for snow and ice melting; T is the temperature; T_t is the temperature threshold, when $T > T_t$, the melting started.

Total daily precipitation at certain elevation is calculated using P_r observed at automatic weaher station (AWS) and a vertical gradient of precipitation (ΔP). Then, rainfall and snowfall were separated from total precipitation by linear interpolation using two temperature thresholds for rain (T_{rain}) and snow (T_{snow}).

In order to use the remote-sensing data to simulate glacier mass loss, we assembled four sub-models. The four sub-models driven by four different driving data sets to get different glacier mass loss: Sub-model1 was driven by the combined data set of in situ temperature and precipitation data; sub-model2 was driven by the combined data set of in situ temperature and precipitation data associated with the lapse rate calculated from the MODIS temperature data from different elevations in the Yala Glacier and Qiyi Glacier; sub-model3 was driven by the combined data set of MODIS temperature and TRMM precipitation data; sub-model4 was driven by the combined data set of in situ temperature data and TRMM precipitation data associated with the lapse rate calculated from the MODIS temperature data from different elevations in the Yala Glacier and Qiyi Glacier.

2.3.2. Model Input Data

The main inputs that the model requires are as follows: (1) The glacier outlines of the Yala and Qiyi glacier obtained from the second glacier inventory dataset of China [50] and the Randolph Glacier Inventory V6.0 (RGI6.0) [51], and the SRTM DEM (90m) is used as the glacier surface; (2) observed daily data at two AWSs are used as driving data to drive the model. In addition, MODIS temperature and MODIS temperature lapse rate were computed from MOD11A1 and MYD11A1 production, TRMM precipitation were withdraw from TRMM production; T_a, and P_{rec} are linear interpolated to each grid cell by using their vertical gradients; (3) The initial snow depth on the glacier is computed using the linear relationship between measured snow depth and elevations; (4) The density of ice and fresh snowfall are assumed to be 900 and 200 kg m^{-3}, respectively.

2.3.3. Model Calibration and Validation

The parameter ranges are summarized in Table 1. Monte Carlo simulation method is used to obtain the optimal combination of parameters for glacier mass loss simulation. The model results, with the lowest root mean square error (RMSE) between modeled glacier mass loss and observed ones and between modeled annual point mass loss and observed ones, are acceptable for selecting the optimal combination of parameters.

Table 1. List of model parameters, their initial ranges.

Symbol	Parameter	Range
ΔP	Gradient of precipitation	0–0.2 m^{-1}
TLR	Lapse rate of temperature	0.2–0.9 °C/100 m
T_{snow}	Phase threshold for snow	2–6 °C
T_{rain}	Phase threshold for rain	−4–2 °C
DDF_{snow}	Degree day factor of fresh snow	1–5 mm w.e./°C
DDF_{ice}	Degree day factor of ice	1–10 mm w.e./°C

We divided the total observations into two periods: one period (2010–2015 for the Yala Glacier and 2001–2012 for the Qiyi Glacier) was used to calibrate the model, and the other period (2016–2019 for the Yala Glacier and 2013–2018 for the Qiyi Glacier) was used to validated the models. The optimal driving data parameters of sub-models 1–4 for the Yala and Qiyi Glacier are shown in Tables 2 and 3.

Table 2. The optimal driving data parameters of the sub-models 1–4 for the Yala Glacier.

Parameters	Sub-Model 1	Sub-Model 2	Sub-Model 3	Sub-Model 4
TLR	0.46 °C	0.41 °C	0.51 °C	0.42 °C
T_{snow}	1.78 °C	2.97 °C	0.20 °C	2.87 °C
T_{rain}	3.72 °C	3.0 °C	3.49 °C	3.64 °C
DDF_{ice}	6.46 °C	5.36 °C	7.95 °C	6.45 °C
DDF_{snow}	5.08 °C	3.11 °C	5.79 °C	3.33 °C

Table 3. The optimal driving data parameters of the sub-models 1–4 for the Qiyi Glacier.

Parameters	Sub-Model 1	Sub-Model 2	Sub-Model 3	Sub-Model 4
TLR	0.78 °C		0.94 °C	
T_{snow}	1.61 °C	2.70 °C	0.55 °C	2.84 °C
T_{rain}	3.10 °C	5.62 °C	5.13 °C	5.71 °C
DDF_{ice}	4.04 °C	3.05 °C	11.10 °C	2.48 °C
DDF_{snow}	1.64 °C	0.78 °C	2.93 °C	1.25 °C

In the above two tables, temperature lapse rate (TLR) is based on MODIS temperature data, phase threshold for snow (T_{snow}), phase threshold for rain (T_{rain}), degree day factor for ice (DDF_{ice}), degree day factor for snow (DDF_{snow}) were also calculated and shown.

3. Data

In this study, we used two sets of data: remote sensing data and in situ data.

3.1. Remote Sensing Data

The remote sensing data in the study include land surface temperature based on MODIS and precipitation based on TRMM, °C.

MODIS Temperature Data

The most important parameter for glacier mass loss study is temperature. We have used the MODIS temperature data to study glacier mass loss for the Yala Glacier and Qiyi Glaicer. Since MODIS land surface temperature product has already been widely validated by scientific community, such as Wan et al. [52], Wan et al. [53] and Hu et al. [54], it can be concluded that MODIS land surface temperature can be used to study glacier mass loss in this study although some errors exist. The daily mean, minimum and maximum temperatures data from 2000 to 2008 are calculated from MODIS for the Yala Glacier and Qiyi Glacier. All the MODIS temperature data were calculated for the whole glacier rather than pixel temperature. Figure 2 shows the daily mean, minimum and maximum temperatures from 2000 to 2008 for the Yala Glacier. The MODIS temperature data indicate obvious diurnal and seasonal changes, reflecting the basic characteristics of temperature change from 2000 to 2008. The averaged minimum temperature is −4.42 °C, the averaged maximum temperature is 18.86 °C, and the mean temperature is 9.83 °C.

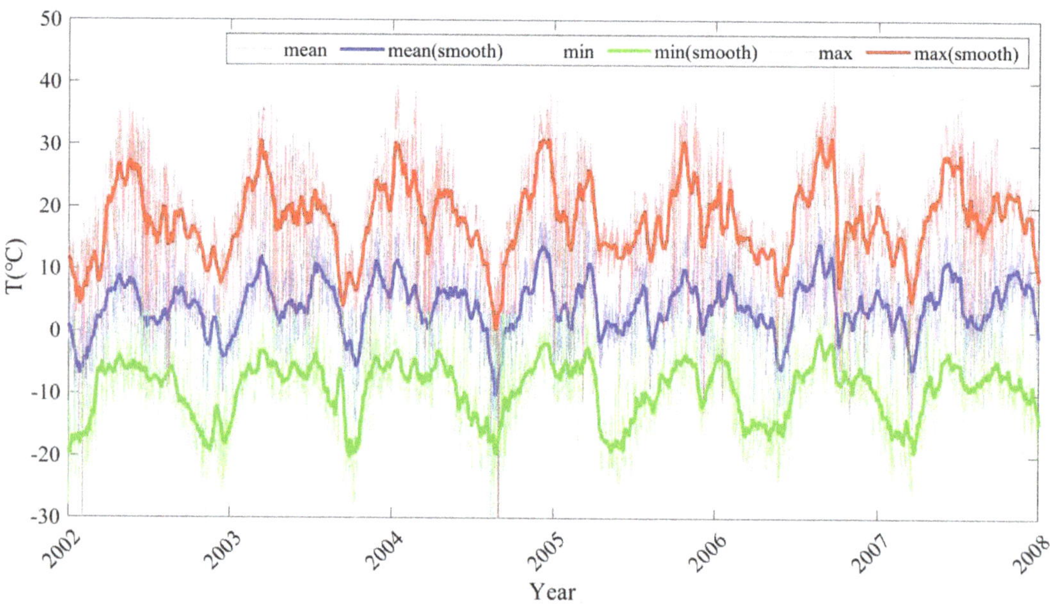

Figure 2. The glacier surface temperature data for the Yala Glacier. In the figure, we have shown the daily mean, minimum and maximum temperatures from MOD11A1 and MYD11A1 products.

Figure 3 shows the daily mean, minimum and maximum MODIS temperature data from 2000 to 2013 for the Qiyi Glacier. The MODIS temperature data indicate obvious diurnal and seasonal changes, reflecting the basic characteristics of temperature change. The MODIS temperature data clearly indicated the temperature increase from 2000 to 2013. The averaged minimum temperature is 3.11 °C, the averaged maximum temperature is 20.25 °C, and the mean temperature is 7.93 °C.

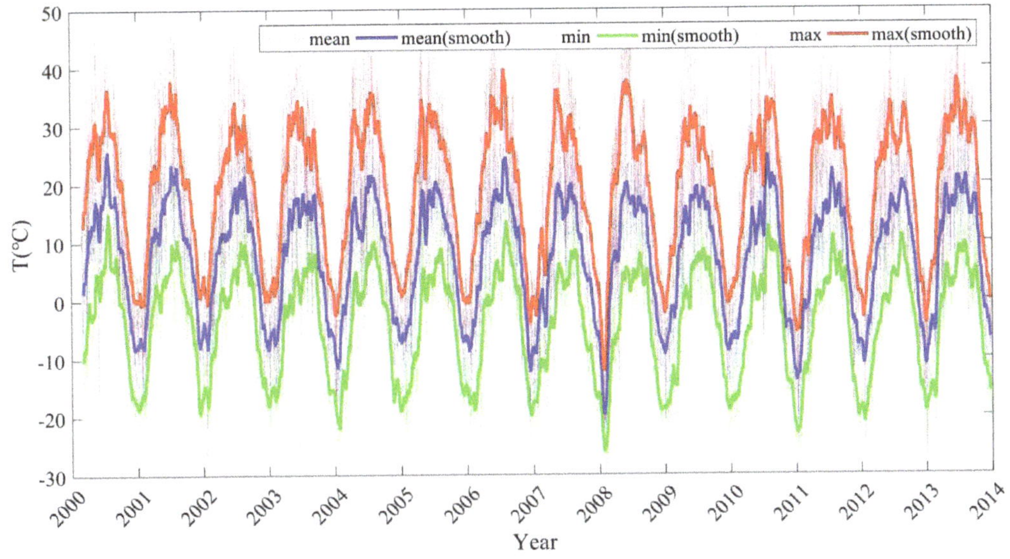

Figure 3. The glacier surface temperature data for the Qiyi Glacier. In the figure, we have shown the daily mean, minimum and maximum temperatures from MOD11A1 and MYD11A1 products.

3.2. TRMM Precipitation Data

The TRMM precipitation data for the Yala Glacier and Qiyi Glacier were obtained by: firstly, the in situ data were used to verify and correct the TRMM data from the seasonal distribution to ensure that the relative proportions of the average monthly precipitation of the two data sites in a year are stable; then, precipitation are considered with the assumption that the amount of occurrence and the amount of accumulative precipitation are proportional to the deviation correction of TRMM data.

By evaluating the daily mean, monthly mean, and annual mean precipitation and by calculating the root mean square error of the accumulated precipitation and the difference of the accumulated precipitation times, the TRMM precipitation data were calibrated. The missing data were supplemented by interpolation to ensure the completeness and consistency of the data.

In Situ Data

The in situ air temperature data were from the records of the AWS and the fixed long-term observing station. The AWS (Figure 1d) for the Yala Glacier is located at the terminus of the Yala Glacier (5058 m a.s.l.), and the fixed observing station (Kyanging, 2821 m a.s.l., Figure 1b) in the middle reaches of the Langtang River basin. The AWS (Figure 1e) for the Qiyi Glacier is located at the tongue (4408 m a.s.l.) of the Qiyi Glacier. The fixed observing station (Tuole station, 3367 m a.s.l., Figure 1c) is located in the upper reaches of the Tuole River basin. The meteorological data, including temperature, precipitation, humidity, air pressure, wind velocity and direction, were measured synchronously with glacier mass balance. In addition to the observation data by AWS, some data are from Chinese Meteorological Agency (CMA) (http://data.cma.cn/ (accessed on 20 March 2020)) and TPE data (data.tpdc.ac.cn (accessed on March 21 2020)). The discontinuous in situ measurements for the two glaciers were carried out since 1970s. The glacier mass balance was measured by measuring stakes installed in the two glaciers. The observing system included 8 stakes for the Yala Glacier (Figure 1d) and 26 stakes for the Qiyi Glacier (Figure 1e), respectively. The observations of the Yala Glacier were carried out by Fujita et al. [41,55], Racoviteanu et al. [56], ICIMOD and Acharya and Kayastha [40], and a recent observation was performed by Li (unpublished). The observations of the Qiyi Glacier were carried out by Guo et al. [57], Wang et al. [45,58], and Pu et al. [32].

4. Results and Discussions

Figure 4 shows the simulations of mass loss for the Yala Glacier by the four sub-models. It can be seen from the figure that the simulation of the sub-model3 is the best to simulate glacier mass loss and stake measured point mass loss, compared with the other sub-models in the calibration period. Although the sub-model4 is improved after adding calibrating data in the validation period, the sub-mode3 is still the best. The above analyses indicate that the sub-model using MODIS temperature data and TRMM Precipitation data at the same time has the highest simulation accuracy. Under some particular conditions, the use of MODIS temperature lapse rate might also help improve the outcomes of the sub-models.

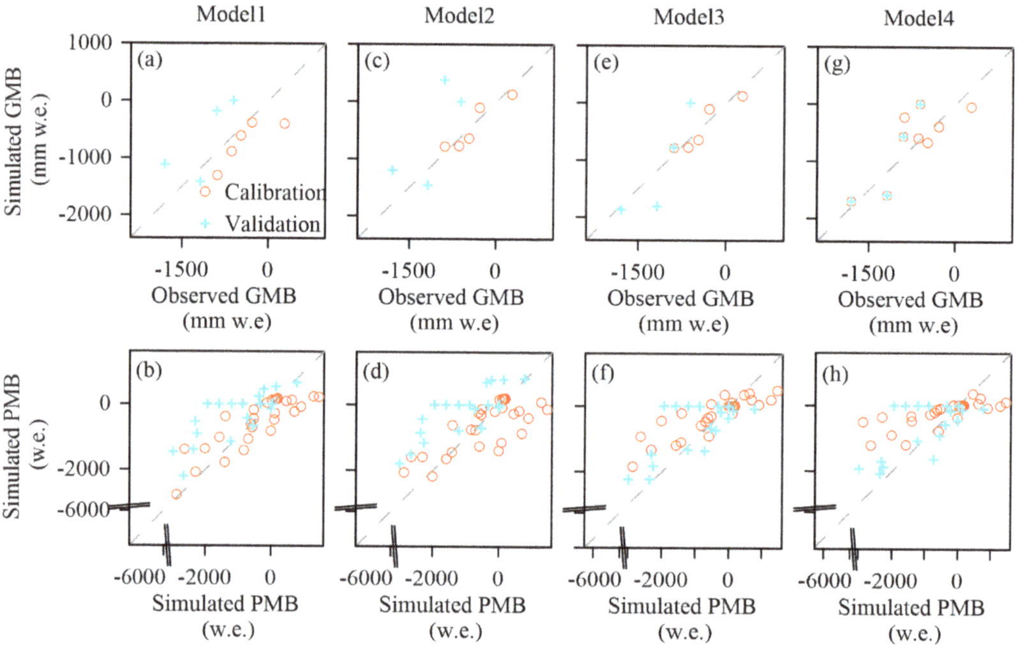

Figure 4. The modeled glacier mass loss and the stake measured mass loss as well as point mass loss of the Yala Glacier by the four sub-models. In the figure, the red circles are the data of calibration period and the blue crosses are the data of validation period; (**a**,**c**,**e**,**g**) are the simulated GMB (Glacier Mass Blalnce for the whole glacier) results of the four sub-models; (**b**,**d**,**f**,**h**) are the simulated PMB (Poin Mass Balance of the glacier).

Figure 5 shows the results of the mass loss for the Qiyi Glacier simulated by four sub-models. It can be seen from the figure that the sub-model2 is better than the other sub-models in simulating glacier mass loss and stake measured point mass loss. The sub-model3 and sub-model4 overestimated the in situ measurement of glacier mass loss during the calibration period and underestimated the glacier mass loss during the validation period, showing the inter-annual instability of TRMM precipitation data. Similar to the Yala Glacier, the Qiyi Glacier also reflects the importance of precipitation data accuracy to glacier mass loss simulation.

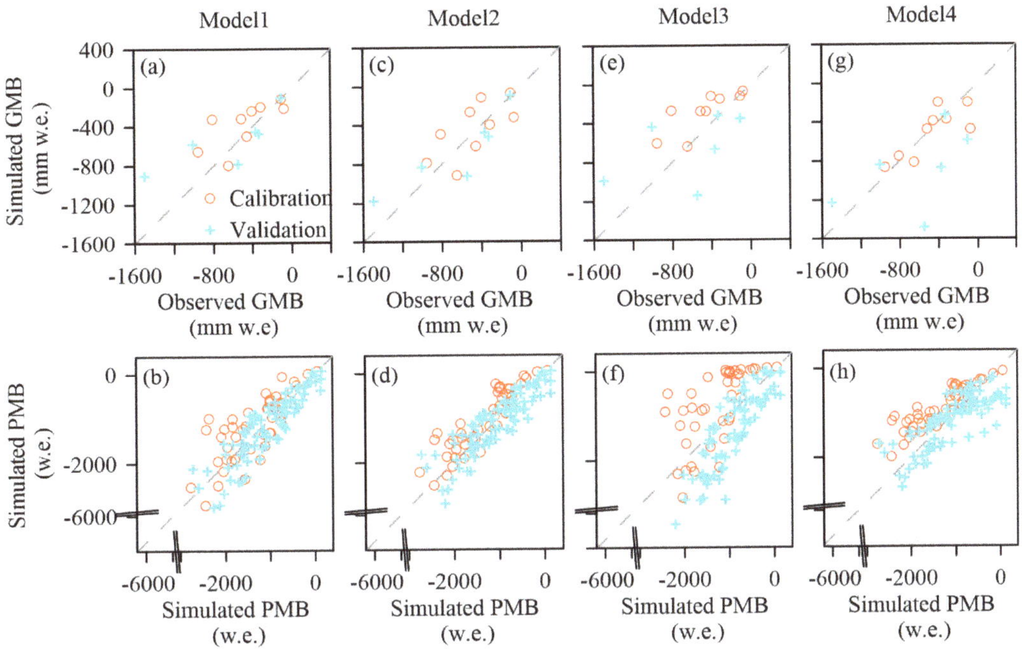

Figure 5. The modeled glacier mass loss and stake measured mass loss as well as point mass loss of the Qiyi Glacier by using the four sub-models. In the figure, the red circles are the data of calibration period and the blue crosses are the data of validation period; (**a**,**c**,**e**,**g**) are the simulated GMB (Glacier Mass Blalnce for the whole glacier) results of the four sub-models; (**b**,**d**,**f**,**h**) are the simulated PMB (Poin Mass Balance of the glacier).

Using the best sub-model for the Yala Glacier driven by the combined data set of MODIS temperature and TRMM precipitation data, we have made a regression to get the mass blance rusults for the Yala Glacier, which is shown as Figure 6a:

$$MB = -16788.91 + 3177.81 \times T_y + 1.98 \times P_y - 867.86 \times T_s - 1521.92 \times T_W - 0.92 \times P_W \quad (2)$$

In the regression, T_y is annuan mean temperature, P_y annual precipitation, T_s summer temperature, T_W winter temperature, P_W winter precipitation, R^2 0.90 and RMSE 128.8 mm w.e.

Similarly, using the best sub-model for the Qiyi Glacier driven by the combined data set of in situ temperature and precipitation data associated with the lapse rate calculated from the MODIS temperature data from different elevations, we have made a regression to get the mass blance rusults for the Qiyi Glacier, which is shown as Figure 6b:

$$MB = -124.63 - 14606.2 \times T_y + 193.62 \times P_y + 4642.50 \times T_s - 191.84 \times P_s + 9566.3 \times T_W - 189.22 \times P_W \quad (3)$$

In the regression, T_y is annuan mean temperature, P_y annual precipitation, T_s summer temperature, T_W winter temperature, P_W winter precipitation, R^2 0.82 and RMSE 226.6 mm w.e.a^{-1}.

Based on the analyses of the remote sensing data and the in situ data as well as the results through modeling using four sub-models, we have analyzed the characteristics of glacier mass loss for both the Yala Glacier located in the southern TP and the Qiyi Glacier located in the northern TP.

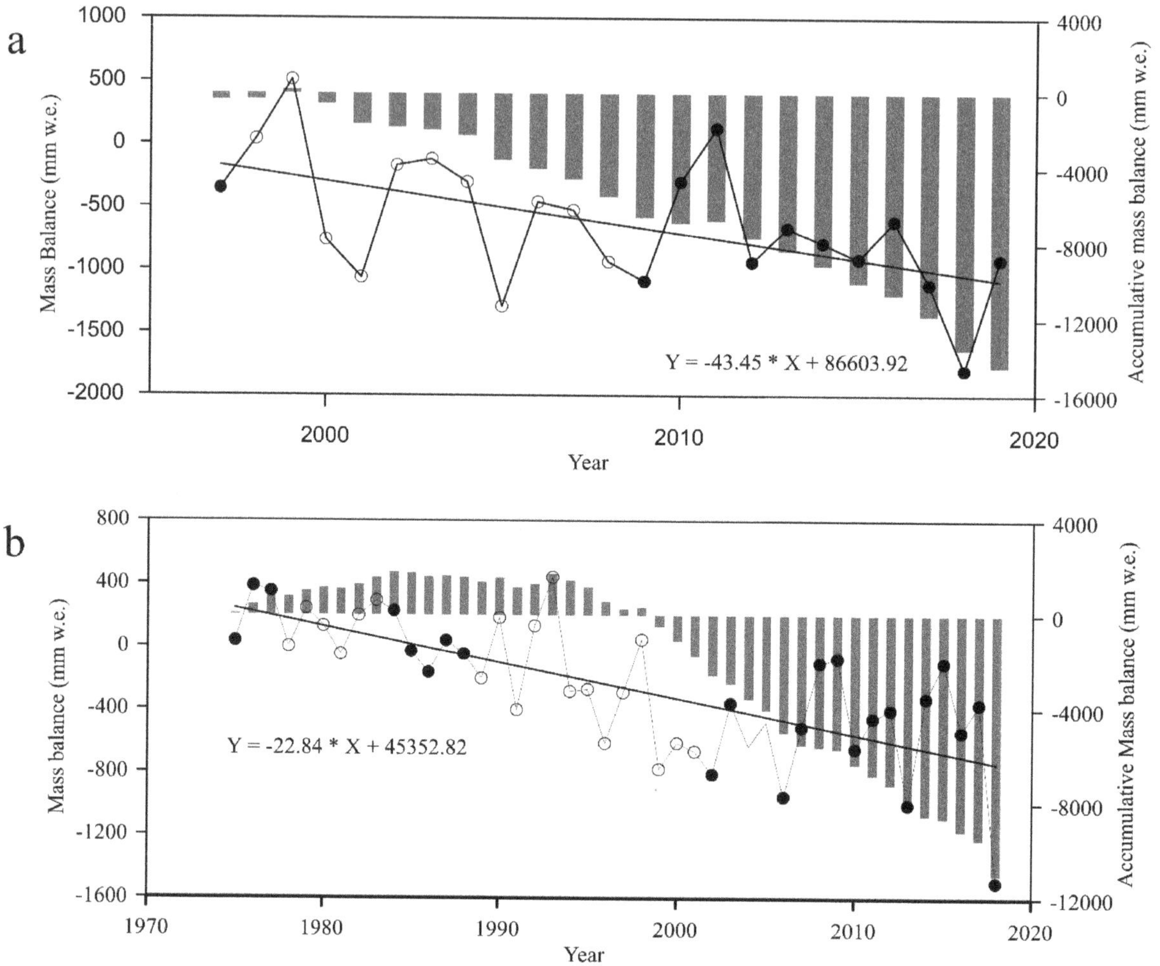

Figure 6. The modeled glacier mass balance changes of the of the Yala Glacier and Qiyi Glacier. In the figure, (**a**) is for the modeled glacier mass balance changes of the Yala Glacier and the (**b**) for the modeled glacier mass balance changes of the Qiyi Glacier.

The average multi-year mass loss of the Yala Glacier is −736 mm w.e.a^{-1}, while that of the Qiyi Glacier is −567 mm; the maximum mass loss is −1815 mm w.e.a^{-1} for the Yala Glacier and −1516 mm w.e.a^{-1} for the Qiyi Glacier. Based on a statistics of glacier mass loss comparison of the two glaciers in the same year, the mass loss of the Yala Glacier is much stronger than that of the Qiyi Glacier. From 2000 to 20018, the mass loss of the Yala Glacier is keeping at a rate −43.8 mm w.e.a^{-1}, while that of Qiyi glacier is keeping at a rate of −9.4 mm w.e.a^{-1}. The glacier mass loss of the Yala Glacier is significantly faster than that of the Qiyi Glacier on a longer time scale. The glacier mass loss of the Yala Glacier was low before 1996. It was only −357 mm w.e. in 1996, but entered into a rapid mass loss period in the 21st century, particularly after 2011, except 2012/13. At the same time, the Qiyi Glacier showed mild melt status compared with the Yala Glacier.

The major cause why the mass loss of the Yala Glacier is more intensive than that of the Qiyi Glacier is from the decrease of glacier accumulation associated with precipitation decrease under the weakening Indian monsoon and from the more intensified glacier ablation. In addition, the precipitation increases under the dominance of the westerly enhancing reduced the glacier ablation in the Qiyi Glacier. The more intensive glacier mass loss of the Yala Glacier than that of the Qiyi glacier is also related to the low latitude where glaciers receive more solar radiation, which will result in more intensive melt.

5. Conclusions

Remote sensing data and in situ data were obtained to study the glacier mass loss in the two galciers, the Yala Glacier in the Central Himalayas in the southern TP and the Qiyi Glacier in the Qilian Mountains in the northern TP. By obtaining the remote sensing data including MODIS land surface temperature and TRMM precipitation data, with calibration and validation with in situ data, the glacier mass loss were simulated. Based on the study of glacier mass loss together with the changes of temperature and precipitation, the climatological factors that resulted in the differences in the two glacier were analyzed. We concluded from the analyses that the remote sensing data are useful in the glacier mass loss study on the TP. When the remote sensing data are used to model the glacier mass loss parameters, they show that the MODIS temperature and TRMM precipitation, after a simply adjustment, can be used to drive effective models and get accurate mass loss results. The modeled results show different characteristics of glacier mass loss between the southern ansd northern parts of the TP. The Yala Glacier has not only been in a state of mass loss, but also in a status of intensive and accelerating mass loss. Our analyses have also shown that the average multi-year mass loss of the Yala Glacier is -736 mm w.e.a^{-1}, with a maximum of -1815 mm w.e.a^{-1}. At the same time, the Qiyi Glacier has experienced a mild glacier mass loss process compared with the Yala Glacier. The Qiyi Glacier's mass loss is -567 mm with a maximum of -1516 mm w.e.a^{-1}. Our results indicate that the mass loss of the Yala Glacier is much stronger and faster than that of the Qiyi Glacier.

Author Contributions: R.Y. has obtained the data, made the analyses and the figures, and written the text. J.S. has proposed the idea and designed the structure of the paper. All authors have read and agreed to the published version of the manuscript.

Funding: This research was funded by the Ministry of Science and Technology of China through the Second Tibetan Plateau Scientific Expedition and Research (STEP) program [grant no. 2019QZKK0206].

Conflicts of Interest: The authors declare no conflict of interest.

References

1. Farinotti, D.; Huss, M.; Fürst, J.J.; Landmann, J.; Machguth, H.; Maussion, F.; Pandit, A. A consensus estimate for the ice thickness distribution of all glaciers on Earth. *Nat. Geosci.* **2019**, *12*, 168–173. [CrossRef]
2. Viviroli, D.; Dürr, H.H.; Messerli, B.; Meybeck, M.; Weingartner, R. Mountains of the world water towers for humanity: Typology, mapping, and global significance. *Water Resour. Res.* **2007**, *43*, 74–79. [CrossRef]
3. Yao, T.; Thompson, L.; Yang, W.; Yu, W.; Gao, Y.; Guo, X.; Yang, X.; Duan, K.; Zhao, H.; Xu, B.; et al. Different glacier status with atmospheric circulations in Tibetan Plateau and surroundings. *Nat. Clim. Chang.* **2012**, *2*, 663–667. [CrossRef]
4. Zemp, M.; Huss, M.; Thibert, E.; Eckert, N.; McNabb, R.; Huber, J.; Barandun, M.; Machguth, H.; Nussbaumer, S.U.; Gärtner-Roer, I.; et al. Global glacier mass changes and their contributions to sea-level rise from 1961 to 2016. *Nature* **2019**, *568*, 382–386. [CrossRef] [PubMed]
5. Liu, X.; Cheng, Z.; Yan, L.; Yin, Z. Elevation dependency of recent and future minimum surface air temperature trends in the Tibetan Plateau and its surroundings. *Glob. Planet. Chang.* **2009**, *68*, 164–174. [CrossRef]
6. Liu, X.; Yin, Z.; Shao, X.; Qin, N. Temporal trends and variability of daily maximum and minimum, extreme temperature events, and growing season length over the eastern and central Tibetan Plateau during 1961–2003. *J. Geophys. Res. Atmos.* **2006**, *111*, D19109. [CrossRef]
7. Yao, T.; Xue, Y.; Chen, D.; Chen, F.; Thompson, L.; Cui, P.; Koike, T.; Lau, W.K.M.; Lettenmaier, D.; Mosbrugger, V.; et al. Recent Third Pole's rapid warming accompanies cryospheric melt and water cycle intensification and interactions between monsoon and environment: Multi-disciplinary approach with observation, modeling and analysis. *Bull. Am. Meteorol. Soc.* **2018**, *100*, 423–444. [CrossRef]
8. Hugonnet, R.; McNabb, R.; Berthier, E.; Menounos, B.; Nuth, C.; Girod, L.; Farinotti, D.; Huss, M.; Dussaillant, I.; Brun, F.; et al. Accelerated global glacier mass loss in the early twenty-first century. *Nature* **2021**, *592*, 726–731. [CrossRef]
9. Huss, M.; Hock, R. Global-scale hydrological response to future glacier mass loss. *Nat. Clim. Chang.* **2018**, *8*, 135–140. [CrossRef]
10. Kraaijenbrink, P.D.A.; Bierkens, M.F.P.; Lutz, A.F.; Immerzeel, W.W. Impact of a global temperature rise of 1.5 degrees Celsius on Asia's glaciers. *Nature* **2017**, *549*, 257–260. [CrossRef]

11. Rounce, D.R.; Hock, R.; Shean, D.E. Glacier mass change in High Mountain Asia through 2100 using the open-source Python Glacier Evolution Model (PyGEM). *Front. Earth Sci.* **2020**, *7*, 331. [CrossRef]
12. Kraaijenbrink, P.D.A.; Stigter, E.E.; Yao, T.; Immerzeel, W.W. Climate change decisive for Asia's snow meltwater supply. *Nat. Clim. Chang.* **2021**, *11*, 591–597. [CrossRef]
13. Zhang, Z.; Jiang, L.; Liu, L.; Sun, Y.; Wang, H. Annual glacier-wide mass balance (2000–2016) of the interior Tibetan Plateau reconstructed from MODIS albedo products. *Remote Sens.* **2018**, *10*, 1031–1036. [CrossRef]
14. Potocki, M.; Mayewsky, P.A.; Mathews, T.; Perry, L.B.; Schwikowski, M.; Tait, A.M.; Korotkikh, E.; Clifford, H.M.; Kang, S.; Sherpa, T.C.; et al. Mt. Everest's highest glacier is a sentinel for accelerating ice loss. *npj Clim. Atmos. Sci.* **2022**, *5*, 7. [CrossRef]
15. Maurer, J.M.; Schaefer, J.M.; Rupper, S.; Corley, A. Acceleration of ice loss across the Himalayas over the past 40 years. *Sci. Adv.* **2019**, *5*, eaav7266. [CrossRef]
16. Immerzeel, W.W.; Lutz, A.F.; Andrade, M.; Bahl, A.; Biemans, H.; Bolch, T.; Hyde, S.; Brumby, S.; Davies, B.J.; Elmore, A.C.; et al. Importance and vulnerability of the world's water towers. *Nature* **2020**, *577*, 364–369. [CrossRef]
17. Biemans, H.; Siderius, C.; Lutz, A.F.; Nepal, S.; Ahmad, B.; Hassan, T.; von Bloh, W.; Wijngaard, R.R.; Wester, P.; Shrestha, A.B.; et al. Importance of snow and glacier mass loss water for agriculture on the Indo-Gangetic Plain. *Nat. Sustain.* **2019**, *2*, 594–601. [CrossRef]
18. Pritchard, H.D. Asia's shrinking glaciers protect large populations from drought stress. *Nature* **2019**, *569*, 649–654. [CrossRef]
19. Yao, T.; Bolch, T.; Chen, D.; Gao, J.; Immerzeel, W.W.; Piao, S.; Su, F.; Thompson, L.; Wada, Y.; Wang, L.; et al. The imbalance of the Asian Water Tower. *Nat. Rev. Earth Environ.* **2022**, *3*, 618–632. [CrossRef]
20. Jacob, T.; Wahr, J.; Pfeffer, W.T.; Swenson, S. Recent contributions of glaciers and ice caps to sea level rise. *Nature* **2012**, *482*, 514–518. [CrossRef]
21. Veh, G.; Korup, O.; Walz, A. Hazard from Himalayan glacier lake outburst floods. *Proc. Natl. Acad. Sci. USA* **2020**, *117*, 907–912. [CrossRef] [PubMed]
22. Kääb, A.; Leinss, S.; Gilbert, A.; Bühler, Y.; Gascoin, S.; Evans, S.G.; Bartelt, P.; Berthier, E.; Brun, F.; Chao, W.; et al. Massive collapse of two glaciers in western Tibet in 2016 after surge-like instability. *Nat. Geosci.* **2018**, *11*, 114–120. [CrossRef]
23. Yao, T.; Yu, W.; Wu, G.; Xu, B.; Yang, W.; Zhao, H.; Wang, W.; Li, S.; Wang, N.; Li, Z.; et al. Glacier anomalies and relevant disaster risks on the Tibetan Plateau and surroundings. *Chin. Sci. Bull.* **2019**, *64*, 2770–2782. (In Chinese)
24. Ragettli, S.; Bolch, T.; Pellicciotti, F. Heterogeneous glacier thinning patterns over the last 40 years in Langtang Himal, Nepal. *Cryosphere* **2016**, *10*, 2075–2097. [CrossRef]
25. Zhou, Y.; Li, Z.; Li, J.; Zhao, R.; Ding, X. Glacier mass balance in the Qinghai-Tibet Plateau and its surroundings from the mid-1970s to 2000 based on Hexagon KH-9 and SRTM DEMs. *Remote Sens. Environ.* **2018**, *210*, 96–112. [CrossRef]
26. King, O.; Bhattacharya, A.; Bhambri, R.; Bolch, T. Glacial lakes exacerbate Himalayan glacier mass loss. *Sci. Rep.* **2019**, *9*, 18145. [CrossRef]
27. Lovell, A.M.; Carr, J.R.; Stokes, C.R. Spatially variable glacier changes in the Annapurna Conservation area, Nepal, 2000 to 2016. *Remote Sens.* **2019**, *11*, 1452. [CrossRef]
28. Zhao, X.; Wang, X.; Wei, J.; Jiang, Z.; Zhang, Y.; Liu, S. Spatiotemporal variability of glacier changes and their controlling factors in the Kanchenjunga region, Himalaya based on multi-source remote sensing data from 1975 to 2015. *Sci. Total Environ.* **2022**, *745*, 140995. [CrossRef]
29. Miles, E.S.; Willis, I.; Buri, P.; Steiner, J.F.; Arnold, N.S.; Pellicciotti, F. Surface pond energy absorption across four Himalayan glaciers accounts for 1/8 of total catchment ice loss. *Geophys. Res. Lett.* **2018**, *45*, 10464–10473. [CrossRef]
30. Bonekamp, P.N.J.; de Kok, R.J.; Collier, E.; Immerzeel, W.W. Contrasting meteorological drivers of the glacier mass balance between the Karakoram and Central Himalaya. *Front. Earth Sci.* **2019**, *7*, 107. [CrossRef]
31. Sun, M.; Liu, S.; Yao, X.; Guo, W.; Xu, J. Glacier changes in the Qilian Mountains in the past half-century: Based on the revised First and Second Chinese Glacier Inventory. *J. Geograph. Sci.* **2018**, *28*, 206–220. [CrossRef]
32. Pu, J.; Yao, T.; Duan, K.; Sakai, A.; Fujita, K.; Matsuda, Y. Mass balance of the Qiyi glacier in the Qilian mountains: A new observation. *J. Glaciol. Geocryol.* **2005**, *27*, 199–206. (In Chinese)
33. Wang, Y.; Ren, J.; Qin, D.; Qin, X. Regional glacier volume changes derived from satellite data: A case study in the Qilian Mountains. *J. Glaciol. Geocryol.* **2013**, *35*, 583–592. (In Chinese)
34. He, J.; Wang, N.; Chen, A.; Yang, X.; Hua, T. Glacier changes in the Qilian Mountains, Northwest China, between the 1960s and 2015. *Water* **2019**, *11*, 623. [CrossRef]
35. Cao, B.; Pan, B.; Wen, Z.; Guan, W.; Li, K. Changes in glacier mass in the Lenglongling Mountains from 1972 to 2016 based on remote sensing data and modeling. *J. Hydrol.* **2019**, *578*, 124010. [CrossRef]
36. Wang, S.; Yao, T.; Tian, L.; Pu, J. Glacier mass variation and its effect on surface runoff in the Beida River catchment during 1957–2013. *J. Glaciol.* **2017**, *63*, 523–534. [CrossRef]
37. Zhang, X.; Li, H.; Zhang, Z.; Wu, Q.; Zhang, S. Recent glacier mass balance and area changes from DEMs and Landsat Images in Upper Reach of Shule River Basin, northeastern edge of Tibetan Plateau during 2000 to 2015. *Water* **2018**, *10*, 796. [CrossRef]
38. Wang, X.; Tang, Z.G.; Wang, J.; Wang, X.; Wei, J. Monitoring of snowline altitude at the end of melting season in High Mountain Asia based on MODIS snow cover products. *Acta Geogr. Sin.* **2020**, *75*, 470–484. (In Chinese)

39. Li, Z.; Yuan, R.; Feng, Q.; Zhang, B.; Lv, Y.; Li, Y.; Wei, W.; Chen, W.; Ning, T.; Gui, J.; et al. Climate background, relative rate, and runoff effect of multiphase water transformation in Qilian Mountains, the third pole region. *Sci. Total Environ.* **2019**, *663*, 315–328. [CrossRef]
40. Acharya, A.; Kayastha, R.B. Mass and energy balance estimation of Yala Glacier (2011–2017), Langtang Valley, Nepal. *Water* **2019**, *11*, 6. [CrossRef]
41. Fujita, K.; Takeuchi, N.; Seko, K. Glaciological observations of Yala Glacier in Langtang Valley, Nepal Himalayas, 1994 and 1996. *Bull. Glacier Res.* **1998**, *16*, 75–78.
42. Sugiyama, S.; Fukui, K.; Fujita, K.; Tone, K.; Yamaguchi, S. Changes in ice thickness and flow velocity of Yala Glacier, Langtang Himal, Nepal, from 1982 to 2009. *Ann. Glaciol.* **2013**, *54*, 157–162. [CrossRef]
43. Baral, P.; Kayastha, R.B.; Immerzeel, W.W.; Pradhananga, N.S.; Bhattarai, B.C.; Shahi, S.; Galos, S.; Springer, C.; Joshi, S.P.; Mool, P.K. Preliminary results of mass-balance observations of Yala Glacier and analysis of temperature and precipitation gradients in Langtang Valley, Nepal. *Ann. Glaciol.* **2014**, *55*, 9–14. [CrossRef]
44. Wang, N.; He, J.; Pu, J.; Jiang, X.; Jing, Z. Variations in equilibrium line altitude of the Qiyi Glacier, Qilian Mountains, over the past 50 years. *Chin. Sci. Bull.* **2010**, *55*, 3810–3817. [CrossRef]
45. Wang, S.; Yao, T.; Pu, J. Spatial and temporal variations in mass balance of Qiyi Glacier in Qilian Mountains. *J. Nat. Resour.* **2020**, *35*, 399–412. (In Chinese)
46. Fujita, K.; Ageta, Y. Effect of summer accumulation on glacier mass balance on the Tibetan Plateau revealed by mass-balance model. *J. Glaciol.* **2000**, *46*, 244–252. [CrossRef]
47. Zhou, W.; Peng, B.; Shi, J.; Wang, T.; Dhital, Y.P.; Yao, R.; Yu, Y.; Lei, Z.; Zhao, R. Estimating High Resolution Daily Air Temperature Based on Remote Sensing Products and Climate Reanalysis Datasets over Glacierized Basins: A Case Study in the Langtang Valley, Nepal. *Remote Sens.* **2017**, *9*, 959. [CrossRef]
48. Zhang, S.; Wang, D.; Qin, Z.; Zheng, Y.; Guo, J. Assessment of the GPM and TRMM precipitation products using the Rain Gauge Network over the Tibetan Plateau. *J. Meteorol. Res.* **2018**, *32*, 324–336. [CrossRef]
49. Hock, R. Glacier melt: A review of processes and their modelling. *Progr. Phys. Geogr.* **2005**, *29*, 362–391. [CrossRef]
50. Guo, W.; Xu, J.; Liu, S.; Shangguan, D.; Wu, L.; Yao, X.; Zhao, J.; Liu, Q.; Jiang, Z.; Li, P. *The Second Glacier Inventory Dataset of China (Version 1.0)*; Cold and Arid Regions Science Data Center: Lanzhou, China, 2014.
51. RGI Consortium. *Randolph Glacier Inventory—A Dataset of Global Glacier Outlines, Version 6*; National Snow and Ice Data Center: Boulder, CO, USA, 2017; p. 576. [CrossRef]
52. Wan, Z.; Zhang, Y.; Zhang, Q.; Li, Z.L. Quality assessment and validation of the MODIS global land surface temperature. *Int. J. Remote Sens.* **2004**, *25*, 261–274. [CrossRef]
53. Wan, Z.M. New refinements and validation of the MODIS Land-Surface Temperature/Emissivity products. *Remote Sens. Environ.* **2008**, *112*, 59–74. [CrossRef]
54. Hu, Y.; Zhong, L.; Ma, Y.; Zou, M.; Xu, K.; Huang, Z.; Feng, L. Estimation of the land surface temperature over the Tibetan Plateau by using Chinese FY-2C Geostationary Satellite Data. *Sensors* **2018**, *18*, 376. [CrossRef] [PubMed]
55. Fujita, K. Effect of precipitation seasonality on climatic sensitivity of glacier mass balance. *Earth Planet. Sci. Lett.* **2008**, *276*, 14–19. [CrossRef]
56. Racoviteanu, E.; Rittger, K.; Armstrong, R. An automated approach for estimating snowline altitudes in the Karakoram and Eastern Himalaya from remote sensing. *Front. Earth Sci.* **2019**, *7*, 220. [CrossRef]
57. Guo, Z.; Wang, N.; Shen, B.; Gu, Z.; Wu, Y.; Chen, A. Spatiotemporal trends in glacier snowline altitude at the end of the melt season in the Qilian Mountains, China. *Remote Sens.* **2021**, *13*, 4935. [CrossRef]
58. Wang, S.; Pu, J.; Wang, N. Study on mass balance and sensitivity to climate change in summer on the Qiyi Glacier, Qilian Mountains. *Sci. Cold Arid. Reg.* **2012**, *4*, 281–287.

MDPI
St. Alban-Anlage 66
4052 Basel
Switzerland
www.mdpi.com

Remote Sensing Editorial Office
E-mail: remotesensing@mdpi.com
www.mdpi.com/journal/remotesensing

www.ingramcontent.com/pod-product-compliance
Lightning Source LLC
Chambersburg PA
CBHW061606220926
51850CB00005B/439
* 9 7 8 3 0 3 6 5 9 1 3 6 0 *